Ubiquitin and Autophagy

Ubiquitin and Autophagy

Editor

Taras Y. Nazarko

MDPI • Basel • Beijing • Wuhan • Barcelona • Belgrade • Manchester • Tokyo • Cluj • Tianjin

Editor
Taras Y. Nazarko
Department of Biology
Georgia State University
Atlanta
United States

Editorial Office
MDPI
St. Alban-Anlage 66
4052 Basel, Switzerland

This is a reprint of articles from the Special Issue published online in the open access journal *Cells* (ISSN 2073-4409) (available at: www.mdpi.com/journal/cells/special_issues/ubiquitin_autophagy).

For citation purposes, cite each article independently as indicated on the article page online and as indicated below:

LastName, A.A.; LastName, B.B.; LastName, C.C. Article Title. *Journal Name* **Year**, *Volume Number*, Page Range.

ISBN 978-3-0365-1342-3 (Hbk)
ISBN 978-3-0365-1341-6 (PDF)

Contents

About the Editor

Taras Y. Nazarko

Taras Y. Nazarko (Associate Professor) runs an active autophagy lab in the Department of Biology at Georgia State University (Atlanta, GA). Most of his academic career to date has been devoted to the mechanistic understanding of autophagy, an important membrane trafficking pathway that delivers cytoplasm to a vacuole/lysosome for degradation and recycling. His training and expertise are in the selective autophagy pathways, especially the selective autophagy of peroxisomes (pexophagy) and lipid droplets (lipophagy). Together with his former mentors, students, national and international collaborators, he has discovered several new autophagy-related genes, such as *ATG26*, *ATG35*, *ATG37* and *TRS85*, and authored 24 peer-reviewed publications in the autophagy field.

cells

MDPI

Editorial

Special Issue on "Ubiquitin and Autophagy"

Taras Y. Nazarko

Department of Biology, Georgia State University, Atlanta, GA 30303, USA; tnazarko@gsu.edu;
Tel.: +1-404-413-5349

The Special Issue of *Cells* on "Ubiquitin and Autophagy" is a tribute to the multifaceted role of ubiquitin and autophagic ubiquitin-like (UBL) proteins in the autophagy-related (ATG) pathways. Ubiquitin is a small regulatory protein that is used to modify other proteins in the process called ubiquitination. The specificity of ubiquitination depends on ubiquitin ligases, the enzymes that place ubiquitin on specific substrates. They are counteracted by ubiquitin proteases that perform deubiquitination. As a result of ubiquitination of a substrate and ubiquitination of ubiquitin itself, proteins become polyubiquitinated with various ubiquitin chains and degraded via the ubiquitin-proteasome system (UPS), autophagy-lysosomal pathway or endo-lysosomal pathway. The polyubiquitination of proteins in protein aggregates and at the surface of organelles or intracellular pathogens often tags these subcellular structures for sequestration (by the double-membrane vesicles, autophagosomes) and delivery to the lysosomes for degradation and recycling by the diverse selective autophagy pathways.

The "Ubiquitin and Autophagy" Special Issue features 13 papers: seven research articles [1–7] and six reviews [8–13]. All of them are at the intersection of ubiquitin-related processes and autophagy, including the roles of: (1) ATG and UBL proteins in the UPS and autophagy [2,3], (2) ubiquitin-binding autophagic receptor, p62, in autophagy signaling [4], (3) LC3-interacting region (LIR) of connexins in binding to UBL proteins [6], (4) proteasomal deubiquitinating enzyme, PSMD14, in autophagy [7], (5) sorting nexins in the UPS, autophagy and endocytosis [8], (6) ubiquitin and UBL proteins in selective autophagy [9], as well as (7) structures and interactions of the UBL proteins of Atg8-family [13]. In addition to the studies on ubiquitin and autophagy in cell culture, several groups used model organisms, such as laboratory mice [1,5], nematode *Caenorhabditis elegans* [2], and social amoeba *Dictyostelium discoideum* [3]. Besides, the review article of Ma and colleagues discusses the interplay between the UPS and autophagy in plants [10]. Although most of the studies explored the fundamental molecular and cellular mechanisms, some of them have interesting implications for human diseases, such as cancer [4] and leukodystrophy [5]. On top of that, the review article of Watanabe et al. is dedicated to the role of ubiquitin and autophagy in neurodegenerative diseases [12].

Given a complex relationship between the UPS and autophagy that started to be appreciated recently (and was one of the motivations for this Special Issue), it is not surprising that many research articles have a common sub-theme of "UPS-autophagy crosstalk". For example, Kang et al. reported that decreased proteasomal degradation of TonEBP protein under the ER stress conditions in β-cells is responsible for increased autophagosome formation, decreased accumulation of protein aggregates and better cell survival [1]. However, the decreased cleavage of K63-ubiquitin chains by proteasomal PSMD14 promotes retention of ATG9A and RAB1A proteins in Golgi apparatus and blocks autophagy via the reduced Golgi-to-ER retrograde transport [7]. Furthermore, deficient autophagy due to knockdown or knockout of the genes encoding ATG and UBL proteins in *C. elegans* and *D. discoideum* negatively affects the UPS suggesting that UPS is not always compensating for the lack of autophagy in vivo. Moreover, a fully functional UPS might depend on autophagy in the tissue- and organism-specific manner [2,3]. Interestingly,

Citation: Nazarko, T.Y. Special Issue on "Ubiquitin and Autophagy". *Cells* **2021**, *10*, 116. https://doi.org/10.3390/cells10010116

Received: 2 January 2021
Accepted: 5 January 2021
Published: 10 January 2021

Lin et al. described the impairment of both the UPS and autophagy in the cellular and murine (twitcher mice) models of globoid cell leukodystrophy [5].

Another underlying sub-theme of the issue revolves around p62 and its functions. While p62 serves as a ubiquitin-binding receptor for many selective autophagy pathways and accumulates on the ubiquitin-positive aggregates in the brains of twitcher mice [5], it also negatively regulates TLR4 signaling and autophagy by disrupting the intermediate TRAF6-BECN1 signaling complex and inhibiting BECN1 ubiquitination in cancer cells what reduces their migration and invasion after TLR4 stimulation [4]. Finally, Catarino et al. reported an alternative, ubiquitin- and p62-independent, mechanism for bridging connexins with the UBL proteins of Atg8-family, LC3B and GABARAP. The direct binding of connexin, Cx43, to LC3/GABARAP proteins via its LIR motif ensures efficient degradation of Cx43 by the p62-mediated selective autophagy [6].

Many researchers from the ubiquitin and autophagy fields will find this issue interesting either due to the thought-provoking original findings or thoughtful summaries of the literature. If you are new to the field or look for a general overview on the theme of "Ubiquitin and Autophagy" before diving into the more specific area of research, the review article of Klionsky and colleagues discusses a wide range of topics, including the autophagic UBL conjugation systems, UPS-autophagy interplay, ubiquitin signaling in selective autophagy, and autophagy regulation by ubiquitination/deubiquitination of the main players [11]. Therefore, this can serve as a starting point. I hope you enjoy reading our collection of papers on "Ubiquitin and Autophagy", and share my excitement about new developments at the intersection of ubiquitin and autophagy fields.

Acknowledgments: I am grateful to all the colleagues who contributed their manuscripts to this Special Issue. Many thanks to the Section Managing Editor, Eric Wang, for his invaluable help.

Conflicts of Interest: The author declares no conflict of interest.

References

1. Kang, H.J.; Yoo, E.J.; Lee, H.H.; An, S.M.; Park, H.; Lee-Kwon, W.; Choi, S.Y.; Kwon, H.M. TonEBP Promotes β-Cell Survival under ER Stress by Enhancing Autophagy. *Cells* **2020**, *9*, 1928. [CrossRef] [PubMed]
2. Jha, S.; Holmberg, C.I. Tissue-Specific Impact of Autophagy Genes on the Ubiquitin-Proteasome System in *C. elegans*. *Cells* **2020**, *9*, 1858. [CrossRef] [PubMed]
3. Karow, M.; Fischer, S.; Messling, S.; Konertz, R.; Riehl, J.; Xiong, Q.; Rijal, R.; Wagle, P.; Clemen, C.S.; Eichinger, L. Functional Characterisation of the Autophagy ATG12~5/16 Complex in *Dictyostelium discoideum*. *Cells* **2020**, *9*, 1179. [CrossRef] [PubMed]
4. Kim, M.J.; Min, Y.; Im, J.S.; Son, J.; Lee, J.S.; Lee, K.Y. p62 is Negatively Implicated in the TRAF6-BECN1 Signaling Axis for Autophagy Activation and Cancer Progression by Toll-Like Receptor 4 (TLR4). *Cells* **2020**, *9*, 1142. [CrossRef] [PubMed]
5. Lin, D.S.; Ho, C.S.; Huang, Y.W.; Wu, T.Y.; Lee, T.H.; Huang, Z.D.; Wang, T.J.; Yang, S.J.; Chiang, M.F. Impairment of Proteasome and Autophagy Underlying the Pathogenesis of Leukodystrophy. *Cells* **2020**, *9*, 1124. [CrossRef] [PubMed]
6. Catarino, S.; Ribeiro-Rodrigues, T.M.; Sa Ferreira, R.; Ramalho, J.; Abert, C.; Martens, S.; Girao, H. A Conserved LIR Motif in Connexins Mediates Ubiquitin-Independent Binding to LC3/GABARAP Proteins. *Cells* **2020**, *9*, 902. [CrossRef] [PubMed]
7. Bustamante, H.A.; Cereceda, K.; Gonzalez, A.E.; Valenzuela, G.E.; Cheuquemilla, Y.; Hernandez, S.; Arias-Munoz, E.; Cerda-Troncoso, C.; Bandau, S.; Soza, A.; et al. The Proteasomal Deubiquitinating Enzyme PSMD14 Regulates Macroautophagy by Controlling Golgi-to-ER Retrograde Transport. *Cells* **2020**, *9*, 777. [CrossRef] [PubMed]
8. Hanley, S.E.; Cooper, K.F. Sorting Nexins in Protein Homeostasis. *Cells* **2021**, *10*, 17. [CrossRef] [PubMed]
9. Vainshtein, A.; Grumati, P. Selective Autophagy by Close Encounters of the Ubiquitin Kind. *Cells* **2020**, *9*, 2349. [CrossRef] [PubMed]
10. Su, T.; Yang, M.; Wang, P.; Zhao, Y.; Ma, C. Interplay between the Ubiquitin Proteasome System and Ubiquitin-Mediated Autophagy in Plants. *Cells* **2020**, *9*, 2219. [CrossRef] [PubMed]
11. Yin, Z.; Popelka, H.; Lei, Y.; Yang, Y.; Klionsky, D.J. The Roles of Ubiquitin in Mediating Autophagy. *Cells* **2020**, *9*, 2025. [CrossRef] [PubMed]
12. Watanabe, Y.; Taguchi, K.; Tanaka, M. Ubiquitin, Autophagy and Neurodegenerative Diseases. *Cells* **2020**, *9*, 2022. [CrossRef] [PubMed]
13. Wesch, N.; Kirkin, V.; Rogov, V.V. Atg8-Family Proteins-Structural Features and Molecular Interactions in Autophagy and beyond. *Cells* **2020**, *9*, 2008. [CrossRef] [PubMed]

cells

MDPI

Review

The Roles of Ubiquitin in Mediating Autophagy

Zhangyuan Yin [1,2,†], Hana Popelka [1,†], Yuchen Lei [1,2,†], Ying Yang [1,2,†] and Daniel J. Klionsky [1,2,*]

1 Life Sciences Institute, University of Michigan, Ann Arbor, MI 48109, USA; zyyin@umich.edu (Z.Y.); popelka@umich.edu (H.P.); yclei@umich.edu (Y.L.); yingyan@umich.edu (Y.Y.)
2 Department of Molecular, Cellular and Developmental Biology, University of Michigan, Ann Arbor, MI 48109, USA
* Correspondence: klionsky@umich.edu
† These authors contributed equally to this work.

Received: 3 August 2020; Accepted: 28 August 2020; Published: 2 September 2020

check for updates

Abstract: Ubiquitination, the post-translational modification essential for various intracellular processes, is implicated in multiple aspects of autophagy, the major lysosome/vacuole-dependent degradation pathway. The autophagy machinery adopted the structural architecture of ubiquitin and employs two ubiquitin-like protein conjugation systems for autophagosome biogenesis. Ubiquitin chains that are attached as labels to protein aggregates or subcellular organelles confer selectivity, allowing autophagy receptors to simultaneously bind ubiquitinated cargos and autophagy-specific ubiquitin-like modifiers (Atg8-family proteins). Moreover, there is tremendous crosstalk between autophagy and the ubiquitin-proteasome system. Ubiquitination of autophagy-related proteins or regulatory components plays significant roles in the precise control of the autophagy pathway. In this review, we summarize and discuss the molecular mechanisms and functions of ubiquitin and ubiquitination, in the process and regulation of autophagy.

Keywords: autophagy; lysosome; selective autophagy; ubiquitin; ubiquitination

1. Introduction

The proper balance between synthesis and degradation that maintains cellular homeostasis is essential for all eukaryotic cells. Proteins, metabolites and organelles are continuously generated and degraded in a delicate equilibrium to support cellular growth, function, development and survival. Disruption of the degradation/recycling process could cause the accumulation of damaged, or superfluous proteins and organelles, which can in turn perturb related cellular processes, harm the cells and even induce dysfunction at the organ level. Therefore, it is of great importance to understand the mechanisms and regulation of the two major intracellular degradation pathways: autophagy and the ubiquitin-proteasome system (UPS).

Autophagy is a dynamic recycling process that involves degradation of cytoplasmic components in the vacuole/lysosome. There are three primary types of autophagy: microautophagy, chaperone-mediated autophagy and macroautophagy, which mainly differ in the types of cargo they degrade and how these cargos are delivered [1]. This review focuses on macroautophagy (hereafter referred to as autophagy), which is characterized by the de novo formation of a transient double-membraned compartment, termed a phagophore, that sequesters intracellular cargo including cytosol, protein aggregates and organelles. The phagophore matures into a closed autophagosome that subsequently fuses with the vacuole/lysosome, allowing degradation of the enclosed cargo, and release of the breakdown products for reuse by the cell. Genetic screens in yeast identified a set of autophagy-related (*ATG*) genes and pioneered the understanding of the autophagy machinery [2]; not surprisingly, the genes encoding the core machinery are highly conserved among more complex eukaryotes. Through reverse genetic approaches in various cellular and animal models, more physiological and pathophysiological roles of autophagy have been uncovered [3]. Although autophagy primarily acts as an inducible rapid adaptation mechanism to cope with environmental adversity, such as starvation or hypoxia, the constitutive turnover of cytoplasmic contents through basal autophagy in nutrient-rich conditions is also crucial for cellular homeostasis.

Because any portion of the cytoplasm can be randomly engulfed by phagophores, autophagy has long been considered nonselective. However, many studies have revealed that aggregates, organelles and invading pathogens can be targeted in a highly selective manner. Selective autophagy occurs constitutively but can also be induced in response to damaged or dysfunctional organelles and changing nutrient conditions. The selectivity of cargo recognition relies on the unique cargo receptors, which tether cargos to the phagophore by the simultaneous binding of cargo and ubiquitin-like (UBL) modifiers, comprised of the Atg8-family proteins [4]. Of note, ubiquitination of cargos/substrates is often employed as a label to confer selectivity through the interaction of this moiety with receptors.

Ubiquitination is the process of attaching ubiquitin, a highly conserved 76 amino acid globular protein, to protein substrates via covalent conjugation [5]. This modification involves a multistep reaction that requires the sequential action of three types of enzymes: E1 (ubiquitin-activating enzyme), E2 (ubiquitin-conjugating enzyme) and E3 (ubiquitin ligase). The E1 activates ubiquitin and forms a ubiquitin adenylate. Next, activated ubiquitin is transferred to a cysteinyl residue in the E2. Finally, the E3 transfers ubiquitin to the Lys residues on substrates, although sometimes Cys, Ser and Thr residues are the sites of attachment [6–8]. For many organisms, there is only one E1, but a variety of E2 and especially E3 enzymes [5]. For example, in *S. cerevisiae*, there are approximately 80 genes found or predicted to encode E3 ligases; in humans, this number was estimated to be approximately 600–700 [9]. Consequently, substrate recognition and much of the selectivity are dependent on E3. Ubiquitin that has already been conjugated to a substrate can be further modified by post-translational modifications, such as additional ubiquitination, phosphorylation, acetylation and SUMOylation (covalent attachment of small UBL SUMO proteins), resulting in increased complexity of the ubiquitin signaling [10].

Ubiquitin signals can be recognized, processed and bound by specialized ubiquitin-binding domains (UBDs), which will direct the substrates to downstream processes [11]. For example, the most common fate of poly-ubiquitinated proteins is to be recognized and delivered by receptors to the 26S proteasome for degradation, which enables the rapid selective recycling of thousands of different proteins, and thus in turn affects many aspects of cellular physiology. Specific interaction between ubiquitinated substrates and selective autophagy receptors facilitates autophagic degradation. Finally, ubiquitin–UBD interactions can also modulate non-degradative processes, including DNA repair, regulation of protein activity/location and immune-signaling pathways [12]. Ubiquitination is a dynamic and reversible process; the attached ubiquitin can be removed by de-ubiquitinating enzymes, thus allowing for spatiotemporal regulation of proteolysis and cell signaling [13].

In this review, we describe the two ubiquitin-like conjugation systems in autophagy and further discuss the potential ubiquitin-like folds in Atg5. We also consider the molecular and functional crosstalk between autophagy and the UPS. Last, we review and discuss available evidence for the roles of ubiquitination in autophagy, in particular, ubiquitin-dependent signals in selective autophagy, and how ubiquitination and de-ubiquitination regulate the levels of Atg proteins and autophagy regulators.

2. Ubiquitin-like Conjugation Systems in Autophagy

The autophagy pathway adopted the structural architecture of ubiquitin in a few proteins of the core machinery (i.e., those proteins that are required for autophagosome formation). Two of these ubiquitin-like proteins, Atg8 and Atg12 in yeast and MAP1LC3 (microtubule-associated protein 1 light chain 3—abbreviated LC3 hereafter)/GABARAP (GABA type A receptor-associated protein) and ATG12 in more complex eukaryotes, are substrates of two parallel, highly conserved ubiquitin-like conjugation reactions that are mediated by E1-E2-E3-like enzymes (Figure 1) [14–18]. Before Atg8/LC3/GABARAP enters this enzymatic cascade, the protein is primed by Atg4 (yeast)/ATG4 (more complex eukaryotes), which removes a C-terminal extension to expose a glycine residue. Subsequently, Atg7/ATG7 acts as an E1-like enzyme activating the inert C terminus of both UBL proteins, which enables covalent linkage of their C terminus to the catalytic cysteine of Atg7/ATG7. Afterwards, the E1-like protein delivers UBLs to the active site of E2-like (conjugating) enzymes, which differ depending on the ubiquitin-like conjugation system. The Atg12/ATG12 conjugation system utilizes Atg10/ATG10, whereas the conjugation system of Atg8-family proteins relies on Atg3/ATG3. The E2 Atg10/ATG10 directly interacts with Atg5/ATG5, allowing downstream ligation of a C-terminal glycine residue on Atg12/ATG12 to a single conserved lysine on Atg5/ATG5 in an E3-independent manner. The resulting Atg12–Atg5/ATG12–ATG5 conjugate binds noncovalently to Atg16/ATG16L1 and together, the complex acts as an E3-like (ligase) enzyme in the Atg8-family protein conjugation reaction [19]. The E3-like Atg12–Atg5-Atg16 in yeast, and analogously ATG12–ATG5-ATG16L1 in more complex eukaryotes, promotes transfer of Atg8-family proteins from the E2-like enzyme to the primary amino group of phosphatidylethanolamine (PE) on the phagophore membrane [17,20]. Structures and molecular mechanisms of interactions within the Atg7-Atg8-Atg3 and Atg7-Atg12-Atg10 complexes were revealed in detail by several studies [21–26], and are comprehensively summarized in a structure-focused review [17].

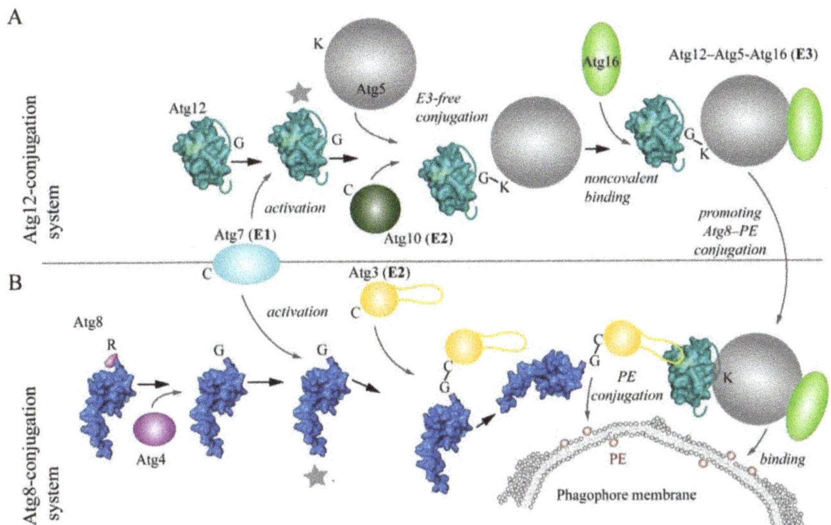

Figure 1. Schematic representation of the Atg8- and Atg12-conjugation systems in yeast. (**A**) Atg12 is activated by Atg7, the E1-like (activating) enzyme, and then transferred to Atg10, the E2-like (conjugating) enzyme. The Atg12–Atg10 intermediate interacts with Atg5, where the conserved lysine residue is covalently conjugated to the Atg12 C terminus in the E3-free (ligase) reaction. The Atg12–Atg5 binds noncovalently to Atg16. The resulting Atg12–Atg5-Atg16 complex acts as the E3-like enzyme in the Atg8–PE conjugation reaction. (**B**) Atg8 enters the conjugation reaction unprimed, due to the presence of a C-terminal arginine. Atg4 primes Atg8 by removing this last residue, leaving a C-terminal glycine exposed. As in the case of Atg12, Atg7 activates the Atg8 ubiquitin-like (UBL) domain by C-terminal adenylation, and then transfers Atg8 to the catalytic cysteine in the active site of Atg3. The Atg12–Atg5-Atg16 complex (E3-like) interacts with the Atg8~Atg3 intermediate, where a long flexible loop of Atg3 binds to a hydrophobic cavity on the surface of Atg12. This interaction enhances ligation of the C-terminal Gly in Atg8 to PE on the phagophore membrane. Atg8 and Atg12 are visualized using the crystal structures PDB ID: 2KQ7 and PDB ID: 3W1S, respectively. Gray stars denote activated states of molecules.

A very recent study elaborated on one of these mechanisms, specifically, on structural flexibility of Atg3 homologs [27]. Zheng et al. showed that a peptide within a long intrinsically disordered loop of Atg3 functions as an allosteric switch that restrains the catalytic loop of the molecule until binding of E1- or E3-like enzymes to this peptide releases, via intramolecular interactions, the inactive conformation [27]. Progress has also been made using biophysical and biochemical approaches that revealed a new molecular mechanism promoting lipidation of Atg8 homologs. These studies showed that human ATG3 and ATG16L1 utilize an amphipathic helix at the N terminus for insertion into the lipid bilayer [28,29]. New insights into the E3-like complex recently unmasked its novel functions, showing a more significant role than was previously ascribed to this complex. One study revealed that the E3-like complex binds autophagy cargo receptors. Specifically, Atg19 or Atg34 in yeast, and SQSTM1/p62 (sequestosome 1), CALCOCO2/NDP52 (calcium binding and coiled-coil domain 2) or OPTN (optineurin) in human cells utilize the Atg8-interacting motif (AIM)/LC3-interacting region (LIR) motif to interact with Atg5/ATG5. In vitro assessments with Atg19 showed that the Atg5-Atg19 AIM interaction competes with the Atg8-Atg19 AIM interaction [30]. Another study proposed that the Atg12–Atg5-Atg16 complex from yeast is involved in autophagy induction, because the Atg12 N terminus binds Atg17, a scaffolding subunit of the Atg1 autophagy induction complex [31].

3. Unexplored Ubiquitin-Like Folds in Atg5

Numerous crystal structures show that all LIR peptides, as well as the AIM sequences, follow the same binding principle when they interact with Atg8/LC3/GABARAP. The UBL fold of Atg8-family proteins creates two hydrophobic pockets (called W and L sites) on the flanking surface area of the β2 strand, and each AIM/LIR tetrapeptide, in an intrinsically disordered domain, adopts the β strand to form an intermolecular β-sheet with the β2 of Atg8 homologs (Figure 2A) [4,32]. This binding mechanism is not entirely unique in autophagy. A recent crystallographic study of the human E3-like enzyme in a complex with ATG3 revealed that ATG12 shares with LC3 not only the UBL fold, but also a mechanism for the binding of a rope-like structure of its binding partner [26]. Specifically, a tetrapeptide ($AADM_{157}$) in an intrinsically disordered loop of ATG3 adopts a β strand and forms an intermolecular β sheet with the β2 strand of ATG12 (Figure 2B). The UBL fold of human ATG12 creates a hydrophobic cavity on the surface of the β2 strand (PDB ID: 4NAW) that corresponds to the L site in LC3 and that fits M157 of the ATG3 tetrapeptide. There is no cavity on ATG12 corresponding to the LC3 W site. Similarity of the hydrophobic pockets of LC3 and ATG12 was interpreted to indicate a common ancestor protein from which both UBL folds evolved [26].

It is often overlooked that Atg5/ATG5 is another autophagy structure that adopts, in part, a ubiquitin-like fold. The protein is composed of two UBL domains connected by a single helical-rich domain (PDB ID: 2DYM or 2DYO and 4NAW) [26,33]. As mentioned in the previous section, Atg5/ATG5 was recently found in a study by Fracchiolla et al. to bind the AIM/LIR motifs of autophagy receptors [30]. The similarity of LC3 and ATG12 in binding of disordered regions opens a question of whether the UBL domain(s) in Atg5/ATG5 may also create hydrophobic pockets on the flanking surface area of the β2 strand for binding of the LIR/AIM tetrapeptide via an intermolecular β sheet, in analogy to Atg8/LC3/GABARAP. Fracchiolla et al. proposed one AIM-binding site away from the β2 strand in Atg5 and the other AIM-binding area on the Atg5 surface that requires movement of the Atg12 molecule to expose the hydrophobic surface for AIM binding [30]. However, such a movement of Atg12 is unlikely because of an extended interface between Atg5 and Atg12 that is important for the E3-like activity [23–25], and that was suggested to maintain Atg12 in a fixed position [34]. Thus, it remains unclear what type of structure the AIM/LIR tetrapeptide adopts when it interacts with Atg5/ATG5.

The crystal structure of the human ATG3-ATG12–ATG5-ATG16N (PDB ID: 4NAW) complex (Figure 2B), as well as the Atg12–Atg5-Atg16N complex from yeast (PDB ID: 3W1S, not depicted in Figure 2) shows a long β2 strand in the UBL-A, but not UBL-B domain of Atg5/ATG5. Does the AIM/LIR motif of autophagy receptors form an intermolecular β-sheet with the β2 strand of Atg5/ATG5, as it does with the β2 strand of Atg8/LC3/GGABARAP? Apparently, this question remains unresolved, and represents a research potential for future crystallographic studies. Along these lines, there are other questions in regard to the UBL folds in Atg5/ATG5. For example, why does Atg5/ATG5 possess two UBL domains? Another question is whether the presence of the well-formed β2 strand in Atg5/ATG5 UBL-A predisposes this domain for binding to the AIM/LIR. If so, what is the purpose of UBL-B? Apparently, autophagy-specific UBL folds have functional potentials, which, once elucidated, can reveal unknown mechanisms in autophagy.

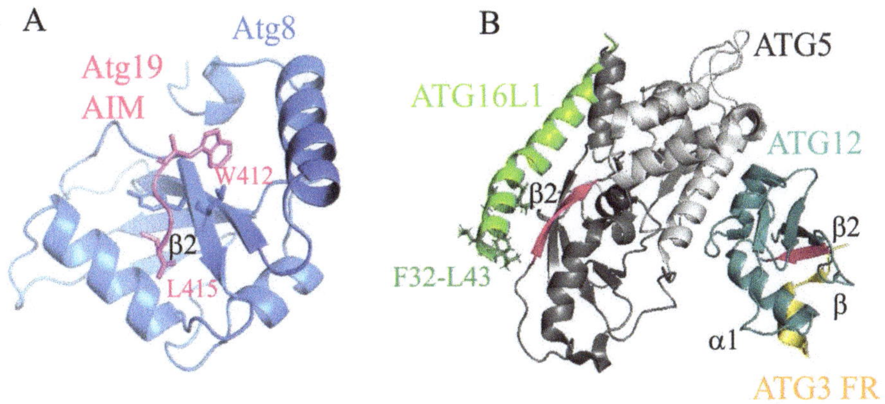

Figure 2. Binding of the autophagy related 8 (Atg8)-interacting motif (AIM)/microtubule associated protein 1 light chain 3 (LC3)-interacting region (LIR) sequences to the ubiquitin-like folds in the autophagy machinery. (**A**) The Atg8 ubiquitin-like (UBL) folds in a complex with the Atg19 AIM peptide (PDB ID: 2ZPN). Hydrophobic amino acid residues in the AIM motif (WEEL) are inserted into the two hydrophobic cavities (W and L site) on the flanking surface areas of the β2 strand. The Atg19 AIM tetrapeptide and the Atg8 β2 strand form an intermolecular β-sheet, a secondary structure found in all canonical AIM/LIR-Atg8/LC3/GABA type A receptor-associated protein (GABARAP) interactions. (**B**) The ATG3-ATG12–ATG5-ATG16N complex (PDB ID: 4NAW) in ribbon representation. When ATG3 binds to ATG12, the β2 strand (pink) of ATG12 forms an intermolecular β sheet with the β strand (orange) of the $AADM_{157}$ tetrapeptide in the disordered region of ATG3 (ATG3 FR). The UBL-A domain (dark gray) on ATG5 possesses the β2 strand (pink). It remains to be elucidated if the ATG5 β2 forms an intermolecular β sheet with the β strand of the LIR motifs in autophagy receptors. Note that the ATG16L1 helical region spanning amino acid residues F32-L43 forms an amphipathic helix that inserts into a lipid bilayer [28,29], instead of binding to ATG5 that is seen in the crystal structure in the absence of membranes. FR, flexible region.

4. Complex Interplay between the Ubiquitin-Proteasome System and Autophagy

Eukaryotic cells maintain protein homoeostasis by utilizing two pathways for protein degradation, the UPS and autophagy. The UPS targets individual short-lived proteins, whereas autophagy clears cells of long-lived proteins, protein complexes and protein aggregates. The UPS and autophagy were initially considered as two independent degradative pathways. However, for more than a decade, various studies have reported discoveries demonstrating a significant crosstalk between the ubiquitin-proteasome system and autophagy [35–42]. An interplay between these two degradative pathways occurs at multiple levels. One mechanism of communication is that changes in one proteolytic pathway induce changes in the activity of the other pathway. One of the first proteins discovered to be associated with this mechanism is HDAC6 (histone deacetylase 6), a microtubule-associated enzyme mediating autophagy induction in response to impaired UPS function [43]. Cancer cells treated with proteasome inhibitors also upregulate autophagy [44] and knockdown of two proteasomal ubiquitin receptors, PSMD4/S5a and ADRM1, in HeLa cells promotes compensatory SQSTM1/p62-mediated autophagy that clears accumulated polyubiquitinated substrates [45].

Further mechanisms activating autophagy due to inhibition of the UPS in more complex eukaryotes are N-terminal arginylation of endoplasmic reticulum chaperones, TP53/p53 (tumor protein p53) accumulation in the nucleus [37] and the unfolded-protein response [46]. Specifically, arginylation ensures recognition of ER-residing chaperones by the ZZ domain of SQSTM1/p62, leading to its conformational change, oligomerization and targeting to autophagy. In the case of TP53, a subpopulation of this protein accumulating in the nucleus acts as a transcription factor for autophagy-related genes [37]. Evidence for an opposite shift, that is autophagy-to-proteasome, is rather weak [35,38]. Some studies from yeast and human cells suggest that the proteasome is activated in response to pharmacological or genetic disruption of autophagy [47,48]. However, this view is contradicted by the finding that inhibition of autophagy in HeLa cells causes impaired UPS function due to accumulation of SQSTM1/p62, which would normally be degraded by autophagic clearance. Ubiquitinated substrates are sequestered by accumulated SQSTM1/p62, causing delayed delivery to the proteasome with otherwise unaffected activity [49]. A suggestion that some autophagic substrates are too large to be degraded by the proteasome is another plausible argument against bi-directional compensation in this direction [38].

A second aspect of communication between the UPS and autophagy is when a component of one degradative pathway regulates a subunit of, or becomes a substrate for, the other pathway. For instance, Atg16 in the slime mold *D. discoideum* can directly bind 19S proteasomal subunits Psmd1 and Psmd2 for lysosomal degradation [50]. Conversely, the 20S core proteasomal cylinder regulates autophagosome-lysosome fusion by degrading two autophagic soluble NSF attachment protein receptor (SNARE) proteins in human cells, SNAP29 (synaptosome-associated protein 29) and STX17 (syntaxin 17), involved in the fusion process [51]. Another example of this type of regulation is seen with the involvement of a deubiquitinating enzyme, USP14 (ubiquitin-specific peptidase 14). This protein directly interacts with the autophagy regulator UVRAG (UV radiation resistance associated), prolongs its half-life and upregulates the autophagy flux [52]. In contrast, USP14 activity that removes activating ubiquitins from BECN1 (beclin 1) downregulates autophagy [53]. Thus, the USP14–UVRAG axis has an opposite effect to the USP14–BECN1 axis. Proteaphagy, autophagic degradation of dysfunctional proteasomes [54,55], is also an important example of cross-regulation (described in detail below).

Other means of crosstalk of the two degradative pathways are chaperones and co-chaperones. For instance, decreased expression of the co-chaperone BAG1 (BAG cochaperone 1) at the expense of the expression of the co-chaperone BAG3 was proposed to promote a switch from proteasomal to autophagic degradation [56]. Another example is Cdc48/p97, a ubiquitin-binding chaperone that is regulated by many co-factors and that is an essential component of the UPS [57]. This chaperone is also important for the function of the autophagy pathway in yeast and mammals, and is indispensable for several types of selective autophagy in yeast [58].

The UPS and autophagy also communicate via sharing cellular components. For example, the transcription factor FOXO3 (forkhead box O3) participates in transcriptional co-regulation of both proteasomal and autophagic degradation [59]. The E3 ubiquitinase PRKN (parkin RBR E3 ubiquitin protein ligase) marks substrates for the proteasome but plays also an important role in mitophagy [60]. A major point where components are shared, but then substrates are specifically directed to a particular pathway, is the ubiquitination machinery: modifying various substrates with different types of ubiquitin moieties can determine which degradative pathway is utilized [61].

One type of substrate that illustrates this ubiquitin-dependent divergence is misfolded proteins, which can be degraded by either autophagy or the UPS. What are the molecular mechanisms and determinants that decide which pathway will be used? The type of linkage between ubiquitin moieties in a chain, known as the ubiquitin code [61], used to be considered a single key factor. The lysine (K) 48 linkage was proposed to be utilized by the UPS, whereas K27- or K63-linked ubiquitin chains were thought to mark protein aggregates or mitochondria for degradation by autophagy [62,63]. In agreement with this hypothesis, the K63 linkage is preferentially recognized by SQSTM1/p62 or

NBR1, receptors acting in autophagy on ubiquitinated substrates [64]. However, increasing evidence suggests that the ubiquitin code itself is much more complex [65], and may be involved only in pre-disposition to one or the other pathway. The ubiquitin code alone is insufficient as a single factor for the choice of degradative pathway. A later study of NBR1 showed that this receptor has no preference for K48-linked or K63-linked ubiquitin chains [66]. Furthermore, experiments using the baker's yeast *S. cerevisiae* demonstrated that Dsk2, a UPS receptor for soluble ubiquitinated proteins, and Cue5, an autophagic receptor for insoluble protein aggregates, have no preference for K48-linked or K63-linked ubiquitin chains [67]. Instead, Dsk2 exhibits much higher affinity for ubiquitin than Cue5, because Cue5 needs to self-interact and assemble into higher-order oligomers to stably bind to ubiquitin. A similar phenomenon was observed for other autophagy receptors, such as SQSTM1/p62 and NBR1 [68].

Thus, ubiquitin-binding receptors are additional players in the navigation of ubiquitinated substrates to the UPS or autophagy. These receptors are regulated via post-translational modifications (PTMs) or by intra- and/or inter-molecular interactions of receptor domains. Both types of regulation affect oligomerizations status, and, thereby, affinity of receptors for ubiquitin. The most common PTMs are ubiquitination and phosphorylation. These modifications, summarized in a recent review [41], on ubiquitin-binding receptors positively or negatively regulate their function in the UPS and autophagy, and thereby, affect the choice of degradative pathway. Intra- and inter-molecular interactions of receptors involve their major functional domains. With the exception of Cue5-TOLLIP (CUET) proteins, which bind ubiquitin via a coupling of ubiquitin conjugation to ER degradation (CUE) domain, most ubiquitin-binding receptors acting in the UPS or autophagy possess a ubiquitin-associated domain (UBA), a tightly-packed fold of approximately 40 amino acids comprised of three α-helices separated by two flexible regions. The first and third helix make hydrophobic contacts with a ubiquitin molecule. Studies show that UBA domains of human or yeast ubiquitin-binding receptors are engaged in intra- or inter-molecular interactions with their UBL or PB1 domains, and thereby, modulate an oligomeric state [41]. For example, Dsk1 from *S. cerevisiae* dimerizes through the UBA–UBL domain interaction [69,70]. In principle, ubiquitin and UBL or PB1 domains competing for the interaction with UBA can open or lock a receptor in a conformation that affects an oligomerization state, which in turn influences its involvement in the proteasomal or autophagic pathway. The consequence of this mechanism for misfolded proteins is that single polypeptides are degraded by the proteasome, whereas misfolded protein aggregates require oligomerized receptors for autophagic clearance by the vacuole/lysosome.

In conclusion, there is no single, specific signal targeting substrates to the UPS or autophagy. The two pathways are significantly intertwined on multiple levels (Figure 3), making the network of communication very complex. Despite all recent discoveries, we are only at the beginning of disclosing the UPS-autophagy interplay.

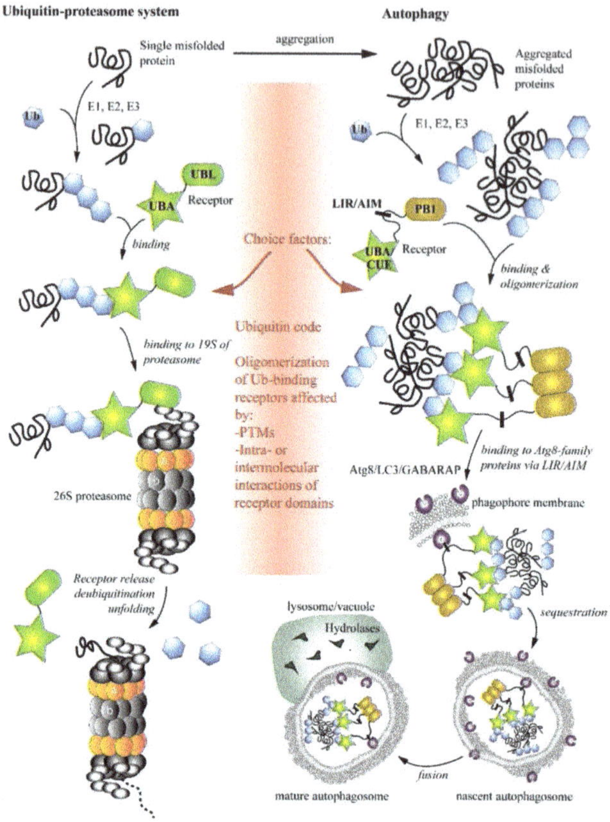

Figure 3. Degradation of misfolded proteins by the UPS and autophagy. Misfolded proteins are modified by the ubiquitination machinery involving the E1-E2-E3 enzymatic cascade. In the UPS (*left*), a mono- or poly-ubiquitin moiety on a single misfolded protein is recognized by a ubiquitin-associated (UBA) domain of a ubiquitin-binding receptor acting in the UPS, for example, Dsk2 in yeast. This complex is targeted to the 26S proteasome, where 19S subunits recognize a ubiquitin-like (UBL) domain of a receptor. After receptor release, deubiquitination, and unfolding of a misfolded protein, the 20S core cylinder composed of α and β subunits loads the protein for degradation. In autophagy (*right*), ubiquitin chains of misfolded protein aggregates are recognized by UBA domains of receptors acting in autophagy, for example, Cue5 in yeast. Oligomerization of these receptors via their UBA and Phox and Bem1 (PB1) domains is essential in high-affinity binding of receptors to ubiquitin. At the same time, the LIR/AIM motif of each receptor must be accessible for binding to Atg8-family proteins, and, thereby, targeting the receptor-substrate complex to the autophagy machinery. Specifically, the LIR/AIM motif binds to two hydrophobic pockets on the surface of Atg8/LC3/GABARAP that decorate the phagophore membrane. Expansion of the phagophore ultimately leads to sequestration of the cargo (that is, substrates with their corresponding receptors) by the nascent autophagosome that is decorated on the inner and outer membrane by Atg8-family proteins. After release of various proteins from the outer membrane, the mature autophagosome fuses with a degradative organelle, the vacuole in yeast and plants, and lysosomes in more complex eukaryotes, where hydrolases break down the cargo. The major factors (*red*) affecting which pathway will be used for misfolded-protein degradation are the ubiquitin code and ubiquitin-binding receptors. Binding affinity of each receptor to ubiquitin is determined by its oligomerization status that is affected by post-translational modifications (PTMs) and/or interactions between UBA, and UBL and PB1 domains of receptors.

5. Ubiquitin and Selective Autophagy in Yeast

As mentioned above, the ubiquitin code and ubiquitin-binding receptors are essential factors to determine which pathway will be used for degradation. Besides the UPS, there are multiple types of selective autophagy exploiting ubiquitin-binding receptors to target cargos for degradation. Depending on the specific cargo that is targeted and digested, selective autophagy can be distinguished as mitophagy, pexophagy, aggrephagy, ribophagy, proteaphagy, nucleophagy and the cytoplasm-to-vacuole targeting (Cvt) pathway, etc. (see below for detailed discussions) [71].

Of note, the molecular mechanisms of selective autophagy in yeast and mammals are conserved to some extent. Mounting evidence suggests that the ubiquitination of cargos is a critical mechanism for selective autophagy [72–74]. In ubiquitination-dependent selective autophagy, the ubiquitin on the cargo is usually exposed to the cytosol and is captured by ubiquitin-binding receptors, through a ubiquitin-binding domain. The receptors can interact with the ubiquitin-like protein Atg8, or the homologous LC3/GABARAP proteins, through an AIM or LIR, and thus recruit cargo to the phagophore membrane [74]. The main difference between yeast and mammals in this process is whether the action of a scaffold protein is required. In yeast, the interaction between specific cargo receptors for selective autophagy and Atg8 often needs the assistance of the scaffold protein Atg11, whereas no scaffold protein has been reported in more complex eukaryotes [75]. It is noteworthy that in addition to the critical interaction with receptors and Atg8 in selective autophagy, Atg11 can also associate with Atg9, Ypt1, Atg20, Atg1, the Atg17-Atg31-Atg29 complex and the Atg12–Atg5 conjugation system [76], demonstrating its multiple roles in mediating cargo selection, membrane trafficking, autophagosome biogenesis and phagophore expansion. To cover the current knowledge regarding various types of selective autophagy, we summarize the E3 ligase, scaffold and receptor proteins for each type of selective autophagy in yeast (Table 1) and mammals (Table 2).

Table 1. Cargo, receptor and E3 ligase of selective autophagy in yeast.

Pathway	Cargo	Receptors	E3 Ligase
Cytoplasm-to-vacuole targeting (Cvt) pathway	prApe1, Ams1 and Ape4	Atg19 and Atg34	-
Mitophagy	Mitochondria	Atg32	-
Pexophagy	Peroxisomes (*S. cerevisiae*)	Atg36	-
	Peroxisomes (*K. phaffii/P. pastoris*)	Atg30	-
Aggrephagy	Protein aggregates	Cue5	Rsp5
Ribophagy	Ribosome	-	-
Proteaphagy	Proteasome	Cue5	-

Table 2. Cargo, receptor and E3 ligase of selective autophagy in mammals.

Pathway	Cargo	Receptors	E3 Ligase
Mitophagy	Mitochondria	SQSTM1/p62, BNIP3L/Nix, OPTN FUNDC1, PHB2, CALCOCO2/NDP52	PRKN/PARK2/parkin
Pexophagy	Peroxisomes	SQSTM1/p62, NBR1	PEX2
Aggrephagy	Protein aggregates	SQSTM1/p62, NBR1, TOLLIP	-
Ribophagy	Ribosome	NUFIP1	-
Proteaphagy	Proteasome	SQSTM1/p62	-

BNIP3L, BCL2 interacting protein 3 like; FUNDC1, FUN14 domain containing 1; NUFIP1, nuclear FMR1 interacting protein 1; PHB2, prohibitin 2; TOLLIP, toll interacting protein.

In the following sections, we mainly focus on the roles of ubiquitin and receptors in selective autophagy in yeast.

5.1. Mitophagy

Mitophagy describes the specific autophagy process that engulfs and degrades damaged or superfluous mitochondria. Atg32 is the key receptor protein identified in yeast in the process of cargo recognition in mitophagy [77]. Atg32 is a mitochondrial outer membrane protein, and it connects mitochondria with the phagophore assembly site (PAS), the location of phagophore nucleation, through its association with Atg11 and Atg8 during mitophagy. With regard to mitophagy, the role of ubiquitination in cargo selectivity has been more clearly revealed in mammalian cells by the elucidation of the PINK1-PRKN/PARK2/parkin-mediated mitophagy mechanism [78]. Although no direct evidence has shown that Atg32 is a ubiquitin-binding receptor, a recent study reported that Atg32 is ubiquitinated at K282 [79], allowing a precise control of mitophagy.

In addition, a ubiquitin-dependent regulatory mechanism of mitophagy is also suggested by the suppression effect of the Ubp3-Bre5 de-ubiquitination complex on mitophagy [80]. Interestingly, although mitophagy is upregulated when genes encoding components of the Ubp3-Bre5 complex are deleted, other types of selective autophagy, including ribophagy and the Cvt pathway, are impaired in the absence of Ubp3 or Bre5, implying an intriguing role of Upb3-Bre5 in regulating distinct types of selective autophagy.

5.2. Cvt Pathway

The Cvt pathway is a yeast-specific biosynthetic pathway that utilizes autophagic machinery to deliver resident hydrolases, such as Ape1, Ams1 and Ape4, to the vacuole. Precursor Ape1 (prApe1) and Ams1 oligomerize after synthesis, and subsequently bind as large complexes to the receptor Atg19. prApe1 binding occurs in a propeptide-dependent manner [81]. In contrast, Ape4 does not self-assemble into a complex, but still interacts with Atg19 at a site different from prApe1 and Ams1 [82]. The Atg19 receptor interacts with the scaffold protein Atg11 through Hrr25-dependent phosphorylation, to recruit the Cvt complex (i.e., the cargo bound to Atg19) to the PAS. Similarly, the paralog of Atg19, Atg34, also acts as a receptor for Ams1, but not prApe1 or Ape4, under starvation conditions [83].

Interestingly, although no evidence has demonstrated that Atg19 and Atg34 are ubiquitin-binding receptors, Atg19 is ubiquitinated in vivo, and the attached ubiquitin is constitutively removed by the Ubp3-Bre5 complex [84].

5.3. Pexophagy

Pexophagy selectively degrades peroxisomes, organelles that are responsible for multiple biological functions, including lipid metabolism, purine catabolism, bile acid synthesis, etc. Atg30 and Atg36 are the receptors for pexophagy in the methylotrophic yeast *Komagataella phaffii*/Pichia pastoris and S. cerevisiae, respectively. Atg30 recognizes and binds to the peroxisomal membrane proteins Pex3 and Pex14 in *K. phaffii*/P. pastoris, and Atg36 binds to Pex3 in S. cerevisiae [85]. Thus, Pex3 and Pex14 are considered to function in part as docking factors to localize Atg30 and Atg36 on peroxisomes. Notably, however, instead of affecting the docking of Atg30 on peroxisomes, disrupting the Atg30-binding domain of Pex3 impairs the interaction between Atg30 and Atg11 [86], suggesting an additional regulatory role of Pex3 in pexophagy.

In mammalian cells, induction of pexophagy leads to the accumulation of ubiquitinated PEX5 at the peroxisomal membrane, following by binding of the ubiquitin-binding receptor NBR1 [87]; however, there is no equivalent mechanism in yeast [88,89], suggesting that an alternative signal instead of ubiquitination exists to target peroxisomes in this organism.

5.4. Aggrephagy

Aggrephagy selectively clears protein aggregates containing poly-glutamatine (poly-Q). Unlike pexophagy, there is evidence for ubiquitin-dependent aggrephagy in yeast, and Rsp5 is the E3 ligase that is responsible for the ubiquitination of protein aggregates in this process. The ubiquitin-binding receptor for aggrephagy in yeast is Cue5, which possesses a ubiquitin-binding CUE domain and an AIM domain [90]. Interestingly, Rsp5 can be immunoprecipitated with Cue5, and Cue5 itself is ubiquitin-regulated by Rsp5 as well.

As ubiquitinated protein aggregates are common substrates for both aggrephagy and the UPS, the precise mechanism by which cells orchestrate the pathways is under debate and investigation [41]. As described previously, many factors, including the position of Lys residues, the length of ubiquitin chains, the specificity of the receptors and receptor oligomerization, play highly complicated roles in the interplay between the UPS and aggrephagy.

5.5. Ribophagy

Unlike other forms of selective autophagy, where cargo ubiquitination typically marks substrates for removal, ubiquitination of the 60S ribosome protein Rpl25 by the ribosome-associated E3 ligase Rkr1/Ltn1 in yeast cells protects ribosomes from being targeted by selective autophagy. Instead, de-ubiquitination of Rpl25 on the same site by the Ubp3-Bre5 de-ubiquitinase triggers selective engulfment of the large ribosomal subunit by ribophagy during nutrient starvation conditions [91]. Further studies show that the chaperone-like protein Cdc48 and its ubiquitin-binding adaptor Doa1/Ufd3, both of which interact with Ubp3-Bre5, are also required for efficient ribophagy [92]. *PRO1*, the gene encoding gamma-glutamyl kinase, which genetically interacts with *UBP3*, is also involved in ribophagy [93]. However, how de-ubiquitinated Rpl25 is recognized by the phagophore, and whether there is another conventional ubiquitinated-cargo-receptor pair existing, remain unknown. Intriguingly, the Ubp3-Bre5 complex is only required for 60S ribosome but not 40S ribosome degradation, suggesting there is another independent ribophagy pathway and posing the question of how subunits of ribosomes are recycled separately.

In mammalian cells, NUFIP1 (nuclear FMR1-interacting protein 1) was identified as a selective ribophagy receptor [94]. In conditions of MTOR (mechanistic target of rapamycin kinase) complex 1 (MTORC1) inhibition, NUFIP1 binds to and co-migrates with the 60S ribosomal subunit; however, the ligand, and the nature of any MTORC1-mediated modification of the ligand, remain unknown.

5.6. Proteaphagy

Proteaphagy is the autophagic degradation of proteasomes, which represents another major point of crosstalk between autophagy and the UPS. Proteaphagy in yeast has been observed under basal conditions, nitrogen starvation and chemically or genetically induced proteasome inactivation. Proteaphagy during both nitrogen starvation and proteasome inhibition depends on Snx4/Atg24, which cooperates with sorting nexins Snx41 and Snx42 to mediate the nuclear export and agglomeration of proteasomes into cytoplasmic puncta before targeting to the vacuole [95]; however, the mechanisms in these two conditions are quite different. Following proteasome inactivation, proteasomes undergo extensive ubiquitination by an unknown E3 ligase, and accumulate in cytoplasmic insoluble protein deposits (IPODs) through chaperone Hsp42 delivery [55]. The receptor Cue5 binds to ubiquitinated proteasomes via its CUE domain, and to Atg8 via its AIM domain, thereby tethering aggregated proteasomes to the expanding phagophore [96]. In contrast, nitrogen deprivation-induced proteaphagy depends on the core autophagy machinery but not on Cue5 or Hsp42. In addition, the 20S proteolytic core particles (CPs) and 19S regulatory particles (RPs) of the proteasome are dissociated and separately targeted in nitrogen starvation-induced proteaphagy. The de-ubiquitinating enzyme complex Ubp3-Bre5 promotes autophagy of CPs but not RPs, and whether other receptors are involved is as yet unknown [97]. Intriguingly, during carbon starvation, the proteasome holo-complexes dissociate to CPs and RPs, which subsequently are exported to the cytoplasm with the aid of Blm10 and Spg5 respectively, and sequestered in proteasome storage granules to protect them from autophagy [98].

In contrast to yeast, amino acid starvation-induced polyubiquitination of proteasomes in more complex eukaryotes mostly involves RP subunits RPN1, RPN2, PSMD4/RPN10 and ADRM1/RPN13 [99].

This ubiquitination is essential for sequestration by the phagophore through the receptor SQSTM1/p62 via its ubiquitin-associated domain.

6. Ubiquitination in the Regulation of Autophagy

Protein degradation by autophagy and the UPS is crucial for proteostasis in the cell. Because the proteins carrying out these two pathways also need to be maintained/induced to a proper level in response to different conditions, it is not surprising that components or regulators of one system are degraded/recycled by the other. We have discussed the example of proteaphagy, in which entire proteasomes are the substrates of selective autophagy. Similarly, Atg proteins and their regulators can be ubiquitinated and subjected to proteasomal degradation by the UPS. Below, we review how ubiquitination serves as a means to control autophagic activity. Because there are only a few studies done in yeast, we will mainly focus on mammalian cells in this section (Figure 4).

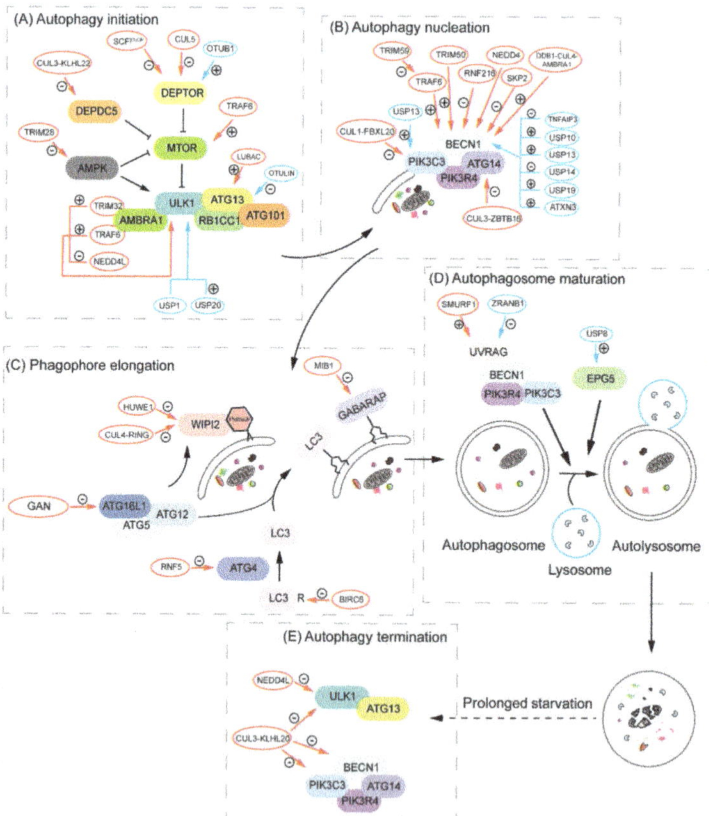

Figure 4. Ubiquitination in regulating autophagy-related proteins. All of the processes and proteins illustrated are shown for mammalian cells. (**A**) Autophagy initiation: autophagy is initiated by the inhibition of MTOR (mechanistic target of rapamycin kinase). Another energy-sensitive kinase, AMP-activated protein kinase (AMPK), is also involved in autophagy initiation through inhibiting MTOR and phosphorylating ULK1 (unc-51 like autophagy activating kinase 1) followed by the formation of a ULK1-ATG13-RB1CC1 (RB1 inducible coiled-coil 1)-ATG101 complex. During this process, MTOR, MTOR regulators, including DEPTOR (DEP domain containing MTOR interacting protein) and DEPDC5 (DEP domain containing 5, GATOR1 subcomplex subunit), AMPK, ULK1 and

ATG13 can be regulated by ubiquitination and de-ubiquitination, which will further affect autophagy initiation. (**B**) Autophagy nucleation: following autophagy initiation, PtdIns3P, which is critical for the localization of phosphatidylinositol-3-phosphate (PtdIns3P)-binding proteins and the further recruitment of other ATG proteins, is generated by a phosphatidylinositol 3-kinase (PtdIns3K) complex at the phagophore. Some of the components of this PtdIns3K complex, including PI3KC3/VPS34 (phosphatidylinositol 3-kinase catalytic subunit type 3), BECN1 (beclin 1) and ATG14 can be ubiquitinated, and the modification on these proteins will affect their stability or function and further regulate autophagy activity. (**C**) Phagophore expansion: WIPI2 (WD repeat domain, phosphoinositide interacting 2), a PtdIns3P-binding protein, is recruited to the phagophore, which is followed by LC3 lipidation. The latter requires the function of two ubiquitin-like conjugation systems, which have been discussed in detail in the previous sections. For the purpose of better indicating how the ubiquitination system regulates phagophore expansion, we only show part of the process, including LC3 C-terminal processing by ATG4 and its conjugation to PE by the ATG12–ATG5-ATG16L1 complex. Ubiquitination on WIPI2, ATG16L1, ATG4, LC3 and GABARAP promotes their degradation and negatively regulates autophagy. (**D**) Autophagosome maturation: the phagophore expands and matures into a double-membrane structure termed an autophagosome, which will ultimately fuse with a lysosome. This process is reported to be facilitated by a UVRAG (UV radiation resistance associated)-containing complex and EPG5. These proteins can be regulated by ubiquitination and de-ubiquitination, thereby regulating autophagosome maturation. (**E**) Autophagy termination: after prolonged starvation, autophagy is terminated by the downregulation of the ULK1 and PtdIns3K complexes, which is mediated by the ubiquitination and degradation of some of the components in these two complexes, including ULK1, BECN1 and PI3KC3/VPS34. E3 ubiquitin ligases are shown in red circles and deubiquitinating enzymes are in blue. "+" indicates that the ubiquitination or de-ubiquitination enhances the stability or promotes the activity of the target protein; "-" refers to the degradation or functional inhibition of the substrate. AMBRA1, autophagy and beclin 1 regulator 1; ATXN3, ataxin 3; CUL3, cullin 3; DDB1, damage specific DNA binding protein 1; EPG5, ectopic P-granules autophagy protein 5 homolog; FBXL20, F-box and leucine rich repeat protein 20; GAN, gigaxonin; HUWE1, HECT, UBA and WWE domain containing E3 ubiquitin protein ligase 1; KLHL22, kelch like family member 22; LUBAC, linear ubiquitin chain assembly complex; MIB1, mindbomb E3 ubiquitin protein ligase 1; NEDD4L, NEDD4 like E3 ubiquitin protein ligase; OTUB1, OUT deubiquitinase, ubiquitin aldehyde binding 1; PIK3R4, phosphoinositide-3-kinase regulatory subunit 4; RNF216, ring finger protein 216; SKP2, S-phase kinase associated protein 2; SMURF1, SMAD specific E3 ubiquitin protein ligase 1; TNFAIP3, TNF alpha induced protein 3; TRAF6, TNF receptor associated factor 6; TRIM32, tripartite motif containing 32; USP1, ubiquitin specific peptidase 1; ZBTB16, zinc finger and BTB domain containing 16; ZRANB1, zinc finger RANBP2-type containing 1.

6.1. The Ubiquitination System and Autophagy Initiation

In yeast, the target of rapamycin complex 1 (TORC1), which includes either Tor1 or Tor2, Kog1, Lst8 and Tco89, works as the master regulator of cell growth and metabolism [100]. Under nutrient-rich conditions, TORC1 inhibits autophagy through phosphorylating Atg13, which regulates its interaction with Atg1. The inhibition of TORC1 activity, for instance, by starvation or rapamycin, leads to the dephosphorylation of Atg13 and the assembly of the Atg1-Atg13-Atg17 complex, which initiates autophagy [101]. The corresponding mammalian rapamycin-sensitive MTORC1 also works as the master regulator of autophagy. MTORC1 consists of three major parts: MTOR, RPTOR/RAPTOR (regulatory-associated protein of MTOR complex 1) and MLST8 (MTOR-associated protein, LST8 homolog) [102–104]. The activity of MTORC1 can be regulated by the nutrient status of the cell. Nutrient sufficiency activates lysosome-localized RRAG (Ras related GTP binding) GTPases, which will bind to RPTOR and recruit the complex to the lysosome for its activation by RHEB (RAS homolog, mTORC1 binding) [105]. Activated MTORC1 phosphorylates both ATG13 and ULK1/2 (homolog of yeast Atg1), inhibiting the kinase activity. In response to starvation, MTORC1 is inactivated, leading to dephosphorylation of ULK1. At this point, ULK1 will phosphorylate ATG13, RB1CC1/FIP200 and ATG101, and initiate autophagy [106–110].

AMP-activated protein kinase (AMPK), another important energy-sensing kinase, is activated by metabolic stress or ATP consumption and promotes the catabolic pathway. In accordance with this, the yeast AMPK homolog, Snf1, is a positive regulator of nitrogen-starvation-induced autophagy, possibly through its phosphorylation of Atg1, and is essential for glucose-starvation-induced autophagy [111,112]. In mammalian cells, AMPK also works as an autophagy regulator through two mechanisms. When the energy level is low, AMPK is activated and phosphorylates TSC2 (TSC complex subunit 2) and promotes its activity as a GTPase-activating protein, which will inactivate RHEB, thereby inhibiting MTOR activation [113]. Meanwhile, AMPK phosphorylates RPTOR, which is another way to inhibit MTORC1 activity [114]. Besides the inhibition of MTORC1, AMPK also directly phosphorylates ULK1, which induces the formation of the ULK1 complex and autophagy initiation [115]. Overall, the initiation of autophagy requires the formation of the ULK1 complex, which is regulated by the energy-sensing kinases, MTORC1 and AMPK. Both of these kinases, as well as some of the components of the ULK1 complex, can be regulated by ubiquitination, and we will discuss how ubiquitination regulates autophagy initiation in the following subsections.

6.1.1. Ubiquitination, Tor and MTOR Complex 1

Using a temperature-sensitive mutant yeast, Tor2, which is deficient in binding to Kog1, resulting in its degradation and the inactivation of TORC1, Hu et al. found that overexpression of ubiquitin prevents the degradation of Kog1, attenuates TORC1 inactivation and suppresses the growth defect seen at the non-permissive temperature [116,117]. Interestingly, the regulation of TORC1 activity by ubiquitin is not mediated by the normal polyubiquitination, but possibly through the non-covalent binding between ubiquitin and Kog1. However, how ubiquitin protects Kog1 from degradation still remains unknown and needs further investigation [116,117].

In mammalian cells, MTORC1 can also be regulated by ubiquitination, but compared with yeast, the mechanism is more complex. As mentioned above, MTOR activation requires its recruitment to the lysosome by RRAG GTPases [105]. In response to amino acids, the E3 ubiquitin ligase TRAF6 (TNF receptor-associated factor 6) is recruited to MTORC1 by SQSTM1, and this interaction is necessary for the translocation of MTORC1 to the lysosome and its subsequent activation. TRAF6 catalyzes K63-linked polyubiquitination of MTOR and regulates MTORC1 activation. Along these lines, the deletion of *TRAF6* leads to enhanced autophagic flux [118]. Recently, it was reported that the CUL3 (cullin 3)-KLHL22 (kelch-like family member 22) E3 ligase promotes K48-linked polyubiquitination and degradation of DEPDC5 (DEP domain-containing 5, GATOR1 subcomplex subunit), which inhibits RRAG GTPase activity under nutrient-deprivation conditions. Therefore, KLHL22 is critical for the activation of MTOR and works as a negative regulator of autophagy [119].

In addition to the major components of MTORC1 mentioned above, DEPTOR (DEP domain containing MTOR-interacting protein) is also an important protein in the MTOR complex, and is able to inhibit MTOR kinase activity [120]. Therefore, MTOR and autophagy activity can also be regulated by the ubiquitination of DEPTOR. DEPTOR interacts with BTRC/βTrCP (beta-transducin repeat containing E3 ubiquitin protein ligase), an F-box protein responsible for the specificity of the Skp1, Cullin, F-box (SCF)-containing complex E3 ligase. In response to a growth signal, DEPTOR is phosphorylated by different kinases, including RPS6KB1/S6K1, RPS6KA1/RSK1 and CSNK1A1/CK1, resulting in ubiquitination by the SCF E3 ligase and degradation. The accumulation of DEPTOR upon BTRC knockdown promotes autophagy through inhibiting MTOR activity [121–123]. Additionally, DEPTOR is ubiquitinated by CUL5 and degraded under normal growth conditions, which will lead to the activation of MTOR and inhibition of autophagy. However, when autophagy is induced, AMBRA1 (autophagy and beclin 1 regulator 1) stabilizes DEPTOR by inhibiting CUL5 activity so that the inhibition of MTOR activity is reinforced [124,125]. Furthermore, OTUB1 (OTU deubiquitinase, ubiquitin aldehyde binding 1) is reported to interact with, remove the polyubiquitin chain from, and stabilize DEPTOR, which suppresses MTOR activity and ultimately regulates autophagy [126].

6.1.2. Ubiquitination of ULK1

ULK1 ubiquitination is mediated by various E3 ubiquitin ligases. For example, ULK1 is regulated by the TRAF6 E3 ligase. Upon autophagy induction, MTOR inactivation leads to the dephosphorylation of AMBRA1 and its interaction with TRAF6, which catalyzes K63-linked ubiquitination and promotes ULK1 stabilization, self-association and function [127]. AMBRA1 also interacts with another E3 ligase, TRIM32 (tripartite motif-containing 32), which in turn binds ULK1 and promotes K63-linked polyubiquitination of ULK1 in an AMBRA1-dependent manner. TRIM32 is necessary for the activation of ULK1 and autophagy induction during atrophic stimuli, such as dexamethasone treatment [128]. Apart from the ubiquitin ligase itself, C1QBP/p32 (complement C1q binding protein), a chaperone-like protein, impairs K48-linked, but promotes K63-linked, polyubiquitination of ULK1, and is thereby important for ULK1 stabilization and the initiation of both autophagy and mitophagy [129].

Some types of ubiquitination result in ULK1 degradation. NEDD4L (NEDD4-like E3 ubiquitin protein ligase) ubiquitinates ULK1, promoting its degradation, whereas the deletion of *NEDD4L* stimulates autophagy activity [130]. A mitochondrially localized E3 ligase, MUL1 (mitochondrial E3 ubiquitin protein ligase 1), promotes selenite-induced mitophagy but, paradoxically, MUL1 can also lead to the polyubiquitination and degradation of ULK1 [48]. In *Caenorhabditis elegans*, UNC-51/ULK1 is ubiquitinated by RPM-1, an atypical RING E3 ubiquitin ligase, which drives the degradation of UNC-51 and restriction of autophagy in the nervous system [131]. Besides the E3 ligases mentioned above, ULK1 is also ubiquitinated by other enzymes, but this modification usually occurs during prolonged starvation and leads to the termination of autophagy (discussed in detail below).

De-ubiquitination also plays an important role in regulating ULK1 activity. USP20 can bind, de-ubiquitinate and stabilize ULK1 at a basal level and plays an important role in autophagy initiation. However, USP20 will release from ULK1 at a later stage after autophagy initiation, and this dissociation leads to enhanced ULK1 degradation [132]. ULK1 de-ubiquitination is also controlled by USP1, a de-ubiquitinating enzyme that removes the K63-linked ubiquitin chain. The deletion of *USP1* results in the formation of ULK1 insoluble aggregates, and impaired canonical autophagic flux [133].

As mentioned above, AMPK is also a key factor to maintain energy homeostasis by upregulating autophagy, and it can also be regulated by ubiquitination. In yeast, the SAGA (Spt-Ada-Gcn5-acetyltransferase) complex, a histone modifier, can de-ubiquitinate Snf1 in a Ubp8-dependent manner. The deletion of *UBP8* leads to the hyperubiquitination of Snf1, which corresponds to a decrease in Snf1 phosphorylation, and results in the degradation of Snf1 [134]. However, the physiological role of this regulation, for instance, whether the regulation of Snf1 through ubiquitination affects autophagy activity, is unclear. Similarly, in mammalian cells, AMPK can be regulated by the ubiquitination system, and this modification can further regulate autophagy activity. MAGEA3 (MAGE family member A3)/MAGEA6 are frequently reactivated in cancer cells and bind to the TRIM28 E3 ligase, which ubiquitinates PRKAA1/AMPKα1 (protein kinase AMP-activated catalytic subunit alpha 1), promoting the degradation of AMPK and the inhibition of autophagy [135]. One of the components of the ULK1 complex, ATG13, can also be regulated by ubiquitination. Recently, Chu et al. [136] reported that the linear ubiquitin chain assembly complex (LUBAC) and the de-ubiquitinating enzyme OTULIN (OTU deubiquitinase with linear linkage specificity) promote and remove the linear ubiquitin on ATG13, respectively. The linear ubiquitin on ATG13 is important for protecting it from being degraded, and therefore plays a role in autophagy initiation. Accordingly, in OTULIN knockdown cells, autophagy initiation is promoted; however, the prolonged existence of ubiquitinated ATG13 inhibits autophagosome maturation. These findings together demonstrate that the ubiquitination system plays a critical role in regulating autophagy initiation.

6.2. The Ubiquitination System and Autophagy Nucleation

Autophagy nucleation refers to the mobilization of groups of molecules to the phagophore assembly site, which will further expand to form the phagophore and sequester autophagic cargos. In yeast, the class III phosphatidylinositol 3-kinase (PtdIns3K) complex is one of the key factors that are recruited to the PAS upon initiation. This complex includes five distinct proteins: the lipid kinase Vps34, the regulatory kinase Vps15, Vps30/Atg6, Atg14 and Atg38, which together generate phosphatidylinositol-3-phosphate (PtdIns3P) to direct the localization of some PtdIns3P-binding proteins and the further recruitment of other autophagy-related proteins [137]. In mammalian cells, the mechanism is similar: autophagy nucleation also depends on a PIK3C3/VPS34 (phosphatidylinositol 3-kinase catalytic subunit type 3)-containing PtdIns3K complex, including PIK3R4/VPS15 (phosphoinositide-3-kinase regulatory subunit 4), BECN1, ATG14 and NRBF2 [138]. Several studies show that some of these proteins can be regulated by post-translational modification, including ubiquitination, which further controls autophagy activity [139–145].

6.2.1. Ubiquitination of PIK3C3/VPS34

The CUL1-SKP1 (S-phase kinase-associated protein 1) complex, with its adaptor FBXL20 (F-box and leucine-rich repeat protein 20), which interacts with PIK3C3/VPS34, mediates the ubiquitination and degradation of PIK3C3/VPS34 during DNA-damage-induced mitotic arrest, thus suppressing autophagy [146]. Moreover, Liu et al. reported that in *C. elegans*, CHN-1, an E3 ubiquitin ligase, interacts with VPS-34 and together with UBC-13 (an E2 ubiquitin-conjugating enzyme) and UEV-1 (a non-catalytic E2 variant), catalyzes the K63-linked ubiquitination of PIK3C3/VPS34 at K348 and K352, which is important for the stabilization of the protein: the loss of UBC-13 or CHN-1 partially impairs autophagy [147]. De-ubiquitination also plays an important role in regulating the activity of PIK3C3/VPS34. Recently, it was demonstrated that NEDD4/NEDD4-1 (NEDD4 E3 ubiquitin protein ligase) undergoes K29-linked auto-ubiquitination at K1279, recruiting USP13 to PIK3C3/VPS34, which reduces K48-linked ubiquitination of PIK3C3/VPS34 at K419 and promotes its stabilization. Therefore, NEDD4 functions as a positive regulator of autophagy by targeting PIK3C3/VPS34 [148].

6.2.2. Ubiquitination of BECN1

TRAF6 can catalyze K63-linked ubiquitination of BECN1 at K117, which inhibits its interaction with BCL2 and facilitates lipopolysaccharide-induced autophagy [149]. Furthermore, the K63-linked ubiquitination mediated by TRAF6 can be inhibited by TRIM59 in non-small cell lung cancer cells. TRIM59 promotes the K48-linked ubiquitination and degradation of TRAF6, thereby inhibiting the ubiquitination of BECN1 and complex formation with PIK3C3/VPS34 [150]. Another protein from the TRIM family, TRIM50, also catalyzes the ubiquitination of BECN1 in a K63-dependent manner and induces its interaction with ULK1, thus functioning as a positive regulator of autophagy [151].

AMBRA1 interacts with BECN1 to regulate autophagy [152]. DDB1 (damage specific DNA binding protein 1)-CUL4 E3 ubiquitin ligase complex, together with the receptor AMBRA1, contributes to the K63-liked ubiquitination of BECN1. The K63-linked ubiquitination on K437 of BECN1 enhances its interaction with PIK3C3/VPS34 to promote the latter's lipid kinase activity [153]. Moreover, RNF216 (ring finger protein 216), an E3 ubiquitin ligase, interacts with BECN1 and drives its K48-linked ubiquitination, which promotes the degradation of BECN1 and inhibits autophagy [154]. Additionally, NEDD4 also mediates the ubiquitination of BECN1, but the role of ubiquitination mediated by NEDD4 is controversial. One study showed that NEDD4 controls K11- and K63-linked polyubiquitination, with the former enhancing the degradation of BECN1 [155]. However, another study demonstrated that NEDD4 interacts with BECN1 and mediates its K6- and K27-linked ubiquitination, which stabilizes BECN1 and promotes autophagy. Meanwhile, the knockdown of NEDD4 results in enhanced K48-linked ubiquitination of BECN1 and impaired autophagy [156]. Therefore, the role of NEDD4-mediated ubiquitination of BECN1, whether it depends on different conditions, and how it affects autophagy, still need further investigation. Recently, another E3 ligase, SKP2, was found to execute K48-linked polyubiquitination at K402 of BECN1, which induces its degradation and inhibits autophagy [157].

De-ubiquitinating enzymes are also involved in regulating BECN1 activity. Contrary to TRAF6, the de-ubiquitinating enzyme TNFAIP3/A20 (TNF alpha induced protein 3) reduces K63-linked ubiquitination of BECN1, inhibiting the autophagy induced by lipopolysaccharide [149]. Ubiquitin on BECN1 can also be removed by USP10 and USP13, which maintains the stability of the BECN1-PIK3C3/VPS34 complex, and BECN1 in turn can regulate the stability of USP10 and USP13 [158]. Jin et al. reported that USP19 stabilizes BECN1 by removing the K11-linked ubiquitin chains of BECN1 at K437, thus positively regulating autophagy [159]. In contrast, USP14 is responsible for moving the K63-linked ubiquitination on BECN1, resulting in autophagy inhibition [160]. Furthermore, another de-ubiquitinating enzyme, ATXN3 (ataxin 3), interacts with BECN1 and removes the K48-linked polyubiquitin chain from K402, which protects BECN1 from proteasome degradation, and enables autophagy [161].

Besides PIK3C3/VPS34 and BECN1, some other proteins in this complex can be regulated by ubiquitination. ATG14 is ubiquitinated by the ZBTB16 (zinc finger and BTB domain containing 16)-CUL3-RBX1/Roc1 (ring-box 1) E3 ubiquitin ligase complex in response to G-protein-coupled signaling, which leads to the degradation of ATG14 and the inhibition of autophagy [162]. In addition, AMBRA1 can be ubiquitinated in a K48-linked manner by RNF2 at K45, leading to the degradation of AMBRA1 and reduced PIK3C3/VPS34 activity [163]. Apart from the core components of the PtdIns3K complex, some of its regulators are also found to be regulated by ubiquitination. PRKN mono-ubiquitinates and stabilizes BCL2, which results in more interaction between BECN1 and BCL2, thus suppressing autophagy [164]. Moreover, it has been reported that the RAS-like GTPase RALB (RAS like proto-oncogene B) and its effector EXOC8/Exo84 (exocyst complex component 8) drive the assembly of the BECN1-PIK3C3/VPS34 complex [165], and USP33 can de-ubiquitinate RALB under starvation conditions to induce the formation of the RALB-EXOC8-BECN1 complex, which is crucial for autophagosome formation [166]. Therefore, the ubiquitination system provides diverse ways to regulate autophagy by stabilizing or inducing degradation of the PIK3C3/VPS34-BECN1 complex and its regulators.

6.3. The Ubiquitination System and Phagophore Expansion

After nucleation, one of the essential processes in autophagosome formation is the recruitment of the PtdIns3P-binding protein WIPI2 (WD repeat domain, phosphoinositide-interacting 2; a homolog of yeast Atg18), followed by ATG12–ATG5-ATG16L1 complex-mediated LC3 lipidation and phagophore expansion. In this section, we mainly focus on how these proteins that are involved in phagophore expansion are regulated by ubiquitination.

WIPI2 can be ubiquitinated by HUWE1 (HECT, UBA and WWE domain containing E3 ubiquitin protein ligase), which leads to the degradation of WIPI2 and autophagy inhibition. The HUWE1-mediated ubiquitination depends on WIPI2 phosphorylation by MTOR at S395, and this provides another explanation for how MTOR inhibits autophagy [167]. A more recent study demonstrated that WIPI2 is a substrate of CUL4-RING ubiquitin ligase, which is activated upon mitosis induction and promotes the K48-linked polyubiquitination of WIPI2 and its degradation, again inhibiting autophagy [168].

In the ubiquitin-like conjugation system, the protease Atg4 is necessary for priming Atg8 for lipidation by cleaving the last arginine residue to expose the penultimate glycine. In mammalian cells, there are four ATG4 proteins: RNF5 ubiquitinates ATG4B and induces its proteasome-mediated degradation, which further limits LC3 lipidation and basal autophagy activity [169]. ATG16L1, which interacts with ATG12–ATG5, is also essential for phagophore expansion [24], and triggers the binding of the complex to the membrane, specifying the location of LC3 lipidation [170]. Ubiquitination of ATG16L1 is mediated by the GAN (gigaxonin) E3 ligase, and the ubiquitination results in the degradation of ATG16L1. The deletion of *GAN* leads to the aggregation of ATG16L1, which impairs LC3 lipidation and autophagosome biogenesis [171]. LC3B can be monoubiquitinated and targeted for degradation by the ubiquitin-activating enzyme UBA6 (ubiquitin-like modifier-activating enzyme 6) and the hybrid ubiquitin-conjugating enzyme/ubiquitin ligase BIRC6 (baculoviral IAP repeat containing 6), which negatively regulates autophagy through reducing the amount of LC3B [172]. One of the Atg8-family proteins in mammalian cells, GABARAP, can be ubiquitinated by MIB1 (mindbomb E3 ubiquitin protein ligase 1), but the centriolar satellite protein PCM1 (pericentriolar material 1) binds GABARAP and protects it from MIB1-mediated ubiquitination and degradation, promoting GABARAP-positive autophagosome formation [173].

6.4. The Ubiquitination System and Autophagosome Maturation

As noted above, the expanding phagophore ultimately matures into a double-membrane autophagosome, which then fuses with a vacuole in yeast or a lysosome in mammalian cells for the degradation of the cargo. UVRAG is important in regulating autophagosome maturation by interacting with BECN1, PIK3C3 and RUBCN (rubicon autophagy regulator) [174–176]. A recent study showed that an E3 ubiquitin ligase, SMURF1 (SMAD specific E3 ubiquitin protein ligase 1), induces K29- and K33-linked polyubiquitin of UVRAG at K517 and K559, which inhibits its association with the inhibitor RUBCN, and thereby promotes autophagosome maturation. In contrast, the de-ubiquitinating enzyme, ZRANB1 (zinc finger RANBP2-type-containing 1) antagonizes SMURF1-mediated ubiquitination and functions as a negative regulator of autophagy [177]. In addition, EPG5 (ectopic P-granules autophagy protein 5 homolog) functions as a RAB7 (RAB7, member of the RAS oncogene family) effector responsible for the specificity of autophagosome and lysosome or late endosome fusion [178]. The de-ubiquitinating enzyme USP8 removes non-classical K63-linked ubiquitin chains from EPG5 at K252, which leads to a strengthened interaction between EPG5 and LC3. The overexpression of USP8 results in enhanced autophagy flux in both normal and starvation conditions in embryonic stem cells [179].

6.5. The Ubiquitination System and Autophagy Termination

Even though autophagy is essential in keeping the homeostasis of cells under stress conditions such as starvation, the failure of autophagy termination will also lead to cellular damage or death [180]. In addition to the MTOR reactivation that occurs in response to the autophagy-dependent release of intracellular nutrients from the lysosome, ubiquitination-mediated degradation of key proteins in autophagy also plays an important role in terminating this process. AMBRA1 is ubiquitinated by DDB1-CUL4 under nutrient-rich conditions, and this ubiquitination ultimately lowers the amount of AMBRA1. Upon starvation, DDB1-CUL4 transiently dissociates from AMBRA1, which then stabilizes DEPTOR and promotes autophagy through inhibiting CUL5 activity, as mentioned above. However, during prolonged starvation, DDB1-CUL4 re-associates with AMBRA1, driving the ubiquitination and degradation of this protein and leading to autophagy termination [124,125]. Furthermore, CUL3, together with its adaptor KLHL20, ubiquitinates ULK1 and controls the degradation of ATG13, PIK3C3/VPS34, BECN1 and ATG14 in a direct or indirect mechanism under conditions of prolonged starvation, therefore regulating the amplitude and termination of autophagy [181]. The NEDD4L E3 ubiquitin ligase also participates in autophagy termination. During prolonged starvation, NEDD4L mediates the ubiquitination of ULK1 at K925 and K933, which leads to the degradation of ULK1 [182]. These facts indicate that many ubiquitin enzymes cooperate to regulate the protein levels of several essential genes in autophagy during prolonged stress conditions, which contributes to autophagy termination.

6.6. Ubiquitination of Autophagy Regulators

Apart from the master regulators such as MTOR and AMPK, some downstream regulators, including transcription factors, control the expression of essential autophagy genes and keep autophagy activity at a proper level. Several transcription factors are regulated by ubiquitination; therefore, autophagy can be regulated by ubiquitination of these transcription factors.

In yeast, Ume6, together with the corepressor Sin3 and the histone deacetylase Rpd3, works as a negative regulator of *ATG8* transcription, and the deletion of *UME6* increases the level of Atg8 and autophagy activity [183]. Ume6 is reported to be a substrate of the E3 ubiquitin ligase APC/C^{Cdc20} (anaphase promoting complex-cell division cycle 20), which is required for the degradation of this protein. Even though in this study, the authors only showed the role of this ubiquitination in meiosis, the fact that Ume6-Cdc20 association is regulated by PKA, another kinase involved in autophagy control [184], implies that the ubiquitination of Ume6 may also function in autophagy regulation [185].

In mammalian cells, TP53/p53 (tumor protein p53) is a transcription factor that plays a dual role in regulating autophagy [186]. Nuclear-localized TP53 stimulates the transcription of several genes, such as *ATG2*, *ATG4*, *ATG7* and *ATG10*. Conversely, cytosolic TP53 suppresses autophagy through inhibiting AMPK and activating MTOR [187]. However, the stimulation of autophagy, such as occurs through rapamycin treatment and starvation, can induce the degradation of TP53, which depends on the E3 ubiquitin ligase MDM2/HDM2 (MDM2 proto-oncogene), and the inhibition of TP53 degradation inhibits autophagy activation [188].

Apart from TP53, the stability and activity of some other transcription factors that regulate autophagy are also controlled by ubiquitination in mammalian cells [189]. For instance, FOXO family proteins control the transcription of some *ATG* genes, and some of the FOXO proteins can be regulated by ubiquitination. SKP2, COP1 (COP1 E3 ubiquitin ligase) and MDM2 mediate the polyubiquitination of FOXO1 (forkhead box O1) and thus promote its degradation [190–192]. A de-ubiquitinating enzyme, USP7, removes the mono-ubiquitination on FOXO1 and negatively regulates its transcriptional activity [193]. In addition, FOXO3 can be ubiquitinated by MDM2 and SKP2, both of which mediate its degradation via the proteasome [194]. E2F1 (E2F transcription factor 1) can induce autophagy by stimulating the transcription of several *ATG* genes, including *ATG1* and *ATG5*, and the knockdown of E2F1 will impair DNA-damage-induced autophagy [193]. Two de-ubiquitinating enzymes, PSMD14/POH1 (proteasome 26S subunit, non-ATPase 14) and UCHL5/UCH37 (ubiquitin C-terminal hydrolase L5), remove K63-linked poly-ubiquitin chains on E2F1, leading to the stabilization and enhanced activity of E2F1, respectively [195,196]. However, it was reported by a later study that BIRC2/cIAP1 E3-ligase promotes the K63-linked poly-ubiquitination at K161/164 on E2F1, which is associated with DNA-damage-induced stabilization of this protein [197]. Therefore, the role of DNA-damage-associated K63-linked ubiquitination on E2F1 and the subsequent outcome need further study. Even though there is no report showing whether or how these ubiquitination modifications directly regulate autophagy, we cannot exclude the possibility that the ubiquitination of these transcription factors also plays a role in controlling autophagy activity. The ubiquitination status of these transcription factors under different conditions, and whether and how they affect the transcription of genes involved in autophagy, still need further investigation.

7. Conclusions and Perspectives

Autophagy and the UPS are the essential degradation programs for cellular homeostasis in all eukaryotes. These two pathways communicate with, compensate and regulate each other in a coordinated manner to efficiently and effectively maintain the intracellular catabolism at a proper level. Interestingly, ubiquitination, which has been initially considered to be a specialized tag for the UPS, is involved in regulating multiple aspects of autophagy, from the nucleation of the phagophore, to autophagosome fusion with the lysosome. From the structural point of view, adopting the ubiquitin architecture into the ubiquitin-like folds of three proteins (namely Atg8, Atg12 and Atg5 in *S. cerevisiae* and their homologs in more complex eukaryotes) playing a central role in the core autophagy machinery, illustrates the significance of this architecture for the autophagy pathway. The UBL fold of Atg12/ATG12 provides a base for binding of Atg3/ATG3, an interaction that is indispensable for efficient lipidation of Atg8-family proteins. The UBL folds of Atg8/LC3/GABARAP and Atg5/ATG5 create binding platforms for a vast number of receptors and adaptors that connect the autophagy machinery to other components in yeast and mammalian cells. Furthermore, two different subsets of receptors utilize two different motifs, AIM/LIR or ubiquitin-interacting motif/UIM, that can bind simultaneously to two different docking sites (LDS or UDS) on the UBL fold of Atg8-family proteins [96]. Such a complexity is further amplified by an intrinsically disordered nature of the Atg3/ATG3 or receptor and adaptor sequences, which bind to the autophagic UBLs. These intrinsically disordered regions offer tremendous potential for PTMs and structural plasticity, and, together with the UBLs, yield remarkable versatility that is available to the autophagy pathway, and that is worth exploring through further research.

Although autophagy was initially thought to be non-selective, tremendous studies in recent years have identified and uncovered the mechanisms of multiple selective autophagy pathways [71]. Ubiquitination is an indispensable signal to initiate some types of selective autophagy. In fact, the specificity of many selective autophagy pathways is determined by ubiquitination and the binding of receptor proteins. One intriguing question is what determines which pathway recycles the ubiquitinated substrate. Some studies suggest that the ubiquitin chain type is the deciding factor, but other studies identified the same ubiquitin chain type in both pathways [37,40,198]. The receptor recognition and binding partly accounts for the choice of autophagy, however, this still does not explain why the binding to receptors occurs before diverting the substrate into the UPS. Another possibility is that further post-translational modifications of the ubiquitin chains is part of the signal and is involved in the process of recognition and preferential binding. Further studies are needed to understand what is fully encoded by ubiquitin signaling. It will also be interesting to understand how the ubiquitinated substrates are selected to represent the entire organelle; that is, does part of the specificity come from the ubiquitinated cargo?

Numerous studies in the past have shown the involvement and importance of the UPS in the regulation of many cellular processes [199], including autophagy. This can occur through either direct turnover of Atg proteins or modulating upstream signals that control autophagy. Because autophagy often works as an adaptive response to environmental stimuli such as starvation, it is of great importance to understand how the UPS is integrated to regulate autophagy in a spatio-temporal correct manner. In some of the studies mentioned above, autophagy regulation through ubiquitination or de-ubiquitination of critical proteins is related to certain types of pathophysiologies, such as cancer and infectious diseases, some of which are valid drug targets. For instance, it was found that in breast tumors of patients, there is an increasing level of KLHL22 and a corresponding decreasing level of DEPDC5. More interestingly, the deletion of *KLHL22* suppresses the growth rate of breast cancer cells, and *KLHL22* knockout also prevents tumor growth in mice. All these findings indicate that KLHL22 may work as an oncogene, and the small molecules inhibiting its activity may work as drugs to treat breast cancer [119]. Additionally, K29- and K33-linked ubiquitination on UVRAG, which induces autophagosome maturation, promotes the degradation of EGFR and inhibits the growth of hepatocellular carcinoma. This study reveals the role of an autophagy-related ubiquitination on UVRAG in tumor growth, indicating another therapeutic potential [177]. Besides cancer, autophagy-related ubiquitination is also involved in other diseases. For example, TRIM32 positively regulates autophagy through ubiquitinating ULK1, whereas disease-associated mutant forms of the protein are deficient in promoting autophagy. This matches the defective autophagy seen in myoblasts derived from limb-girdle muscular dystrophy type 2H/LGMD2H patients, implying that TRIM32-dependent autophagy may have an important correlation with this disease [128]. Another study also shows that SKP2i, an SKP2 inhibitor, inhibits the replication of MERS-CoV (Middle East respiratory syndrome coronavirus) through enhancing the BECN1 level and autophagy activity, and these findings suggest the potential of some SKP2 inhibitors to be used in the clinical treatment of viral infection [157]. Still another example is that an E3 ligase, TNFAIP3, and ATG16L1, which participates in autophagy, have been found to be associated with inflammatory bowel disease in a genome-wide association study [200,201]; a recent analysis shows that these two proteins interact and regulate the level of each other in a compensatory way to maintain the integrity of the intestinal epithelial barrier, and the loss of both proteins leads to chronic intestinal inflammation [202]. Therefore, whether and how these different types of regulation are implicated in human diseases, and whether there exists the potential to target them in therapeutic strategies, are worth further study.

Although tremendous progress has been made in understanding the involvement of ubiquitination in autophagy, there are still many more questions that need to be addressed. One common question, as mentioned above, is the specific recognition of the substrates, which is mostly determined by E3 ligases. Therefore, in the journey of completing our understanding of the degradation network, it is of great significance to identify the E3 ligases involved in selective autophagy and the associated regulatory steps.

Funding: This work was funded by the National Institutes of Health grant number GM131919, and the APC was paid using discretionary funds.

Conflicts of Interest: The authors declare no conflict of interest.

Abbreviations

AIM	Atg8-interacting motif
Atg	autophagy-related
CPs	core particles
CUE	coupling of ubiquitin conjugation to ER degradation
Cvt	cytoplasm-to-vacuole targeting
LIR	LC3-interacting region
MTORC1	mechanistic target of rapamycin kinase complex 1
PAS	phagophore assembly site
prApe1	precursor aminopeptidase I
PtdIns3K	phosphatidylinositol 3-kinase
PtdIns3P	phosphatidylinositol-3-phosphate
PTMs	post-translational modifications
RPs	regulatory particles
UBA	ubiquitin-associated domain
UBD	ubiquitin-binding domain
UBL	ubiquitin-like
UPS	ubiquitin-proteasome system

References

1. Mizushima, N.; Komatsu, M. Autophagy: Renovation of Cells and Tissues. *Cell* **2011**, *147*, 728–741. [CrossRef] [PubMed]
2. Klionsky, D.J.; Cregg, J.M.; A Dunn, W.; Emr, S.D.; Sakai, Y.; Sandoval, I.V.; Sibirny, A.; Subramani, S.; Thumm, M.; Veenhuis, M.; et al. A Unified Nomenclature for Yeast Autophagy-Related Genes. *Dev. Cell* **2003**, *5*, 539–545. [CrossRef]
3. Yin, Z.; Pascual, C.; Klionsky, D.J. Autophagy: Machinery and regulation. *Microb. Cell* **2016**, *3*, 588–596. [CrossRef] [PubMed]
4. Rogov, V.; Dötsch, V.; Johansen, T.; Kirkin, V. Interactions between Autophagy Receptors and Ubiquitin-like Proteins Form the Molecular Basis for Selective Autophagy. *Mol. Cell* **2014**, *53*, 167–178. [CrossRef] [PubMed]
5. Pickart, C.M.; Eddins, M.J. Ubiquitin: Structures, functions, mechanisms. *Biochim. Biophys. Acta* **2004**, *1695*, 55–72. [CrossRef] [PubMed]
6. Callis, J. The Ubiquitination Machinery of the Ubiquitin System. *Arab. Book* **2014**, *12*, e0174. [CrossRef]
7. Wang, X.; Herr, R.A.; Chua, W.-J.; Lybarger, L.; Wiertz, E.J.; Hansen, T.H. Ubiquitination of serine, threonine, or lysine residues on the cytoplasmic tail can induce ERAD of MHC-I by viral E3 ligase mK3. *J. Cell Biol.* **2007**, *177*, 613–624. [CrossRef]
8. Cadwell, K.; Coscoy, L. Ubiquitination on Nonlysine Residues by a Viral E3 Ubiquitin Ligase. *Science* **2005**, *309*, 127–130. [CrossRef]
9. Li, W.; Bengtson, M.H.; Ulbrich, A.; Matsuda, A.; Reddy, V.A.; Orth, A.; Chanda, S.K.; Batalov, S.; Joazeiro, C.A. Genome-Wide and Functional Annotation of Human E3 Ubiquitin Ligases Identifies MULAN, a Mitochondrial E3 that Regulates the Organelle's Dynamics and Signaling. *PLoS ONE* **2008**, *3*, e1487. [CrossRef]
10. Swatek, K.N.; Komander, D. Ubiquitin modifications. *Cell Res.* **2016**, *26*, 399–422. [CrossRef]
11. Husnjak, K.; Dikic, I. Ubiquitin-Binding Proteins: Decoders of Ubiquitin-Mediated Cellular Functions. *Annu. Rev. Biochem.* **2012**, *81*, 291–322. [CrossRef] [PubMed]
12. Grabbe, C.; Dikic, I. Functional Roles of Ubiquitin-Like Domain (ULD) and Ubiquitin-Binding Domain (UBD) Containing Proteins. *Chem. Rev.* **2009**, *109*, 1481–1494. [CrossRef] [PubMed]

13. Komander, D.; Clague, M.J.; Urbé, S. Breaking the chains: Structure and function of the deubiquitinases. *Nat. Rev. Mol. Cell Biol.* **2009**, *10*, 550–563. [CrossRef] [PubMed]
14. Mizushima, N.; Noda, T.; Yoshimori, T.; Tanaka, Y.; Ishii, T.; George, M.D.; Klionsky, D.J.; Ohsumi, M.; Ohsumi, Y. A protein conjugation system essential for autophagy. *Nature* **1998**, *395*, 395–398. [CrossRef]
15. Ichimura, Y.; Kirisako, T.; Takao, T.; Satomi, Y.; Shimonishi, Y.; Ishihara, N.; Mizushima, N.; Tanida, I.; Kominami, E.; Ohsumi, M.; et al. A ubiquitin-like system mediates protein lipidation. *Nature* **2000**, *408*, 488–492. [CrossRef]
16. Tanida, I.; Ueno, T.; Kominami, E. LC3 conjugation system in mammalian autophagy. *Int. J. Biochem. Cell Biol.* **2004**, *36*, 2503–2518. [CrossRef]
17. Klionsky, D.J.; Schulman, B.A. Dynamic regulation of macroautophagy by distinctive ubiquitin-like proteins. *Nat. Struct. Mol. Biol.* **2014**, *21*, 336–345. [CrossRef]
18. Mizushima, N. The ATG conjugation systems in autophagy. *Curr. Opin. Cell Biol.* **2020**, *63*, 1–10. [CrossRef]
19. Hanada, T.; Noda, N.N.; Satomi, Y.; Ichimura, Y.; Fujioka, Y.; Takao, T.; Inagaki, F.; Ohsumi, Y. The Atg12-Atg5 Conjugate Has a Novel E3-like Activity for Protein Lipidation in Autophagy. *J. Biol. Chem.* **2007**, *282*, 37298–37302. [CrossRef]
20. Sakoh-Nakatogawa, M.; Matoba, K.; Asai, E.; Kirisako, H.; Ishii, J.; Noda, N.N.; Inagaki, F.; Nakatogawa, H.; Ohsumi, Y. Atg12–Atg5 conjugate enhances E2 activity of Atg3 by rearranging its catalytic site. *Nat. Struct. Mol. Biol.* **2013**, *20*, 433–439. [CrossRef]
21. Yamada, Y.; Suzuki, N.N.; Hanada, T.; Ichimura, Y.; Kumeta, H.; Fujioka, Y.; Ohsumi, Y.; Inagaki, F. The Crystal Structure of Atg3, an Autophagy-related Ubiquitin Carrier Protein (E2) Enzyme that Mediates Atg8 Lipidation. *J. Biol. Chem.* **2007**, *282*, 8036–8043. [CrossRef] [PubMed]
22. Taherbhoy, A.M.; Tait, S.W.; Kaiser, S.E.; Williams, A.H.; Deng, A.; Nourse, A.; Hammel, M.; Kurinov, I.; Rock, C.O.; Green, D.R.; et al. Atg8 Transfer from Atg7 to Atg3: A Distinctive E1-E2 Architecture and Mechanism in the Autophagy Pathway. *Mol. Cell* **2011**, *44*, 451–461. [CrossRef] [PubMed]
23. Kaiser, S.E.; Mao, K.; Taherbhoy, A.M.; Yu, S.; Olszewski, J.L.; Duda, D.M.; Kurinov, I.; Deng, A.; Fenn, T.D.; Klionsky, D.J.; et al. Noncanonical E2 recruitment by the autophagy E1 revealed by Atg7–Atg3 and Atg7–Atg10 structures. *Nat. Struct. Mol. Biol.* **2012**, *19*, 1242–1249. [CrossRef] [PubMed]
24. Otomo, C.; Metlagel, Z.; Takaesu, G.; Otomo, T. Structure of the human ATG12~ATG5 conjugate required for LC3 lipidation in autophagy. *Nat. Struct. Mol. Biol.* **2013**, *20*, 59–66. [CrossRef] [PubMed]
25. Noda, N.N.; Fujioka, Y.; Hanada, T.; Ohsumi, Y.; Inagaki, F. Structure of the Atg12–Atg5 conjugate reveals a platform for stimulating Atg8–PE conjugation. *EMBO Rep.* **2012**, *14*, 206–211. [CrossRef]
26. Metlagel, Z.; Otomo, C.; Takaesu, G.; Otomo, T. Structural basis of ATG3 recognition by the autophagic ubiquitin-like protein ATG12. *Proc. Natl. Acad. Sci. USA* **2013**, *110*, 18844–18849. [CrossRef]
27. Zheng, Y.; Qiu, Y.; Grace, C.R.R.; Liu, X.; Klionsky, D.J.; Schulman, B.A. A switch element in the autophagy E2 Atg3 mediates allosteric regulation across the lipidation cascade. *Nat. Commun.* **2019**, *10*, 3600. [CrossRef]
28. Hervás, J.H.; Landajuela, A.; Antón, Z.; Shnyrova, A.V.; Goñi, F.M.; Alonso, A. Human ATG3 binding to lipid bilayers: Role of lipid geometry, and electric charge. *Sci. Rep.* **2017**, *7*, 15614. [CrossRef]
29. Lystad, A.H.; Carlsson, S.R.; De La Ballina, L.R.; Kauffman, K.J.; Nag, S.; Yoshimori, T.; Melia, T.J.; Simonsen, A. Distinct functions of ATG16L1 isoforms in membrane binding and LC3B lipidation in autophagy-related processes. *Nat. Cell. Biol.* **2019**, *21*, 372–383. [CrossRef]
30. Fracchiolla, D.; Sawa-Makarska, J.; Zens, B.; De Ruiter, A.; Zaffagnini, G.; Brezovich, A.; Romanov, J.; Runggatscher, K.; Kraft, C.; Zagrovic, B.; et al. Mechanism of cargo-directed Atg8 conjugation during selective autophagy. *eLife* **2016**, *5*, e18544. [CrossRef]
31. Harada, K.; Kotani, T.; Kirisako, H.; Sakoh-Nakatogawa, M.; Oikawa, Y.; Kimura, Y.; Hirano, H.; Yamamoto, H.; Ohsumi, Y.; Nakatogawa, H. Two distinct mechanisms target the autophagy-related E3 complex to the pre-autophagosomal structure. *eLife* **2019**, *8*, 8. [CrossRef] [PubMed]
32. Popelka, H.; Klionsky, D.J. Analysis of the native conformation of the LIR/AIM motif in the Atg8/LC3/GABARAP-binding proteins. *Autophagy* **2015**, *11*, 2153–2159. [CrossRef] [PubMed]
33. Matsushita, M.; Suzuki, N.N.; Obara, K.; Fujioka, Y.; Ohsumi, Y.; Inagaki, F. Structure of Atg5.Atg16, a Complex Essential for Autophagy. *J. Biol. Chem.* **2007**, *282*, 6763–6772. [CrossRef] [PubMed]
34. Cappadocia, L.; Lima, C.D. Ubiquitin-like Protein Conjugation: Structures, Chemistry, and Mechanism. *Chem. Rev.* **2018**, *118*, 889–918. [CrossRef]
35. Park, C.; Cuervo, A.M. Selective autophagy: Talking with the UPS. *Cell Biophys.* **2013**, *67*, 3–13. [CrossRef]

36. Schreiber, A.; Peter, M. Substrate recognition in selective autophagy and the ubiquitin–proteasome system. *Biochim. Biophys. Acta* **2014**, *1843*, 163–181. [CrossRef]
37. Ji, C.H.; Kwon, Y.T. Crosstalk and Interplay between the Ubiquitin-Proteasome System and Autophagy. *Mol. Cells* **2017**, *40*, 441–449. [CrossRef]
38. Dikic, I. Proteasomal and Autophagic Degradation Systems. *Annu. Rev. Biochem.* **2017**, *86*, 193–224. [CrossRef]
39. Nam, T.; Han, J.H.; Devkota, S.; Lee, H.-W. Emerging Paradigm of Crosstalk between Autophagy and the Ubiquitin-Proteasome System. *Mol. Cells* **2017**, *40*, 897–905.
40. Kocaturk, N.M.; Gozuacik, D. Crosstalk Between Mammalian Autophagy and the Ubiquitin-Proteasome System. *Front. Cell Dev. Biol.* **2018**, *6*, 128. [CrossRef]
41. Zientara-Rytter, K.; Subramani, S. The Roles of Ubiquitin-Binding Protein Shuttles in the Degradative Fate of Ubiquitinated Proteins in the Ubiquitin-Proteasome System and Autophagy. *Cells* **2019**, *8*, 40. [CrossRef] [PubMed]
42. Shin, W.H.; Park, J.H.; Chung, K.C. The central regulator p62 between ubiquitin proteasome system and autophagy and its role in the mitophagy and Parkinson's disease. *BMB Rep.* **2020**, *53*, 56–63. [CrossRef] [PubMed]
43. Pandey, U.B.; Nie, Z.; Batlevi, Y.; McCray, B.A.; Ritson, G.P.; Nedelsky, N.B.; Schwartz, S.L.; DiProspero, N.A.; Knight, M.A.; Schuldiner, O.; et al. HDAC6 rescues neurodegeneration and provides an essential link between autophagy and the UPS. *Nature* **2007**, *447*, 859–863. [CrossRef] [PubMed]
44. Wu, W.K.K.; Cho, C.H.; Lee, C.W.; Wu, Y.C.; Yu, L.; Li, Z.J.; Wong, C.C.M.; Li, H.T.; Zhang, L.; Ren, S.X.; et al. Macroautophagy and ERK phosphorylation counteract the antiproliferative effect of proteasome inhibitor in gastric cancer cells. *Autophagy* **2010**, *6*, 228–238. [CrossRef]
45. Demishtein, A.; Fraiberg, M.; Berko, D.; Tirosh, B.; Elazar, Z.; Navon, A. SQSTM1/p62-mediated autophagy compensates for loss of proteasome polyubiquitin recruiting capacity. *Autophagy* **2017**, *13*, 1697–1708. [CrossRef]
46. Hetz, C.; Chevet, E.; Oakes, S.A. Proteostasis control by the unfolded protein response. *Nat. Cell Biol.* **2015**, *17*, 829–838. [CrossRef]
47. Wang, X.J.; Yu, J.; Wong, S.; Cheng, A.S.; Chan, F.K.; Ng, S.S.-M.; Cho, C.H.; Sung, J.J.; Wu, W.K.K. A novel crosstalk between two major protein degradation systems: Regulation of proteasomal activity by autophagy. *Autophagy* **2013**, *9*, 1500–1508. [CrossRef]
48. Athané, A.; Buisson, A.; Challier, M.; Beaumatin, F.; Manon, S.; Bhatia-Kiššová, I.; Camougrand, N. Insights into the relationship between the proteasome and autophagy in human and yeast cells. *Int. J. Biochem. Cell Biol.* **2015**, *64*, 167–173. [CrossRef]
49. Korolchuk, V.I.; Mansilla, A.; Menzies, F.M.; Rubinsztein, D.C. Autophagy Inhibition Compromises Degradation of Ubiquitin-Proteasome Pathway Substrates. *Mol. Cell* **2009**, *33*, 517–527. [CrossRef]
50. Xiong, Q.; Fischer, S.; Karow, M.; Müller, R.; Meßling, S.; Eichinger, L. ATG16 mediates the autophagic degradation of the 19S proteasomal subunits PSMD1 and PSMD2. *Eur. J. Cell Biol.* **2018**, *97*, 523–532. [CrossRef]
51. Njomen, E.; Tepe, J. Regulation of Autophagic Flux by the 20S Proteasome. *Cell Chem. Biol.* **2019**, *26*, 1283–1294.e5. [CrossRef] [PubMed]
52. Kim, E.; Park, S.; Lee, J.H.; Mun, J.Y.; Choi, W.H.; Yun, Y.; Lee, J.; Kim, J.H.; Kang, M.-J.; Lee, M.J. Dual Function of USP14 Deubiquitinase in Cellular Proteasomal Activity and Autophagic Flux. *Cell Rep.* **2018**, *24*, 732–743. [CrossRef] [PubMed]
53. Min, Y.; Lee, S.; Kim, M.-J.; Chun, E.; Lee, K.-Y. Ubiquitin-Specific Protease 14 Negatively Regulates Toll-Like Receptor 4-Mediated Signaling and Autophagy Induction by Inhibiting Ubiquitination of TAK1-Binding Protein 2 and Beclin 1. *Front. Immunol.* **2017**, *8*, 1827. [CrossRef] [PubMed]
54. Marshall, R.S.; Li, F.; Gemperline, D.C.; Book, A.J.; Vierstra, R.D. Autophagic Degradation of the 26S Proteasome Is Mediated by the Dual ATG8/Ubiquitin Receptor RPN10 in Arabidopsis. *Mol. Cell* **2015**, *58*, 1053–1066. [CrossRef] [PubMed]
55. Marshall, R.S.; McLoughlin, F.; Vierstra, R.D. Autophagic Turnover of Inactive 26S Proteasomes in Yeast Is Directed by the Ubiquitin Receptor Cue5 and the Hsp42 Chaperone. *Cell Rep.* **2016**, *16*, 1717–1732. [CrossRef]

56. Minoia, M.; Boncoraglio, A.; Vinet, J.; Morelli, F.F.; Brunsting, J.F.; Poletti, A.; Krom, S.; Reits, E.; Kampinga, H.H.; Carra, S. BAG3 induces the sequestration of proteasomal clients into cytoplasmic puncta: Implications for a proteasome-to-autophagy switch. *Autophagy* **2014**, *10*, 1603–1621. [CrossRef]
57. Finley, D.; Ulrich, H.D.; Sommer, T.; Kaiser, P. The Ubiquitin–Proteasome System of Saccharomyces cerevisiae. *Genetics* **2012**, *192*, 319–360. [CrossRef]
58. Dargemont, C.; Ossareh-Nazari, B. Cdc48/p97, a key actor in the interplay between autophagy and ubiquitin/proteasome catabolic pathways. *Biochim. Biophys. Acta* **2012**, *1823*, 138–144. [CrossRef]
59. Zhao, J.; Brault, J.J.; Schild, A.; Cao, P.; Sandri, M.; Schiaffino, S.; Lecker, S.H.; Goldberg, A.L. FoxO3 Coordinately Activates Protein Degradation by the Autophagic/Lysosomal and Proteasomal Pathways in Atrophying Muscle Cells. *Cell Metab.* **2007**, *6*, 472–483. [CrossRef]
60. Lazarou, M.; Sliter, D.A.; Kane, L.A.; Sarraf, S.; Wang, C.; Burman, J.L.; Sideris, D.P.; Fogel, A.I.; Youle, R.J. The ubiquitin kinase PINK1 recruits autophagy receptors to induce mitophagy. *Nature* **2015**, *524*, 309–314. [CrossRef]
61. Komander, D.; Rape, M. The Ubiquitin Code. *Annu. Rev. Biochem.* **2012**, *81*, 203–229. [CrossRef] [PubMed]
62. Tan, J.M.; Wong, E.; Kirkpatrick, D.S.; Pletnikova, O.; Ko, H.S.; Tay, S.-P.; Ho, M.W.; Troncoso, J.; Gygi, S.P.; Lee, M.; et al. Lysine 63-linked ubiquitination promotes the formation and autophagic clearance of protein inclusions associated with neurodegenerative diseases. *Hum. Mol. Genet.* **2008**, *17*, 431–439. [CrossRef] [PubMed]
63. Geisler, S.; Holmstrom, K.M.; Skujat, D.; Fiesel, F.C.; Rothfuss, O.C.; Kahle, P.J.; Springer, W. PINK1/Parkin-mediated mitophagy is dependent on VDAC1 and p62/SQSTM1. *Nat. Cell Biol.* **2010**, *12*, 119–131. [CrossRef] [PubMed]
64. Kirkin, V.; Lamark, T.; Johansen, T.; Dikic, I. NBR1 cooperates with p62 in selective autophagy of ubiquitinated targets. *Autophagy* **2009**, *5*, 732–733. [CrossRef]
65. Kravtsova-Ivantsiv, Y.; Sommer, T.; Ciechanover, A. The Lysine48-Based Polyubiquitin Chain Proteasomal Signal: Not a Single Child Anymore. *Angew. Chem. Int. Ed. Engl.* **2013**, *44*, 192–198. [CrossRef]
66. Walinda, E.; Morimoto, D.; Sugase, K.; Konuma, T.; Tochio, H.; Shirakawa, M. Solution Structure of the Ubiquitin-associated (UBA) Domain of Human Autophagy Receptor NBR1 and Its Interaction with Ubiquitin and Polyubiquitin. *J. Biol. Chem.* **2014**, *289*, 13890–13902. [CrossRef]
67. Lu, K.; Brave, F.D.; Jentsch, S. Receptor oligomerization guides pathway choice between proteasomal and autophagic degradation. *Nat. Cell Biol.* **2017**, *19*, 732–739. [CrossRef]
68. Wurzer, B.; Zaffagnini, G.; Fracchiolla, D.; Turco, E.; Abert, C.; Romanov, J.; Martens, S. Oligomerization of p62 allows for selection of ubiquitinated cargo and isolation membrane during selective autophagy. *eLife* **2015**, *4*, 08941. [CrossRef]
69. Sasaki, T.; Funakoshi, M.; Endicott, J.A.; Kobayashi, H. Budding yeast Dsk2 protein forms a homodimer via its C-terminal UBA domain. *Biochem. Biophys. Res. Commun.* **2005**, *336*, 530–535. [CrossRef]
70. Lowe, E.D.; Hasan, N.; Trempe, J.-F.; Fonso, L.; Noble, M.E.; Endicott, J.A.; Johnson, L.N.; Brown, N.R. Structures of the Dsk2 UBL and UBA domains and their complex. *Acta Crystallogr. Sect. D Biol. Crystallogr.* **2006**, *62*, 177–188. [CrossRef]
71. Gatica, D.; Lahiri, V.; Klionsky, D.J. Cargo recognition and degradation by selective autophagy. *Nat. Cell Biol.* **2018**, *20*, 233–242. [CrossRef] [PubMed]
72. Yang, Y.; Klionsky, D.J. Autophagy and disease: Unanswered questions. *Cell Death Differ.* **2020**, *27*, 858–871. [CrossRef] [PubMed]
73. Feng, L.; Zhang, J.; Zhu, N.; Ding, Q.; Zhang, X.; Yu, J.; Qiang, W.; Zhang, Z.; Ma, Y.; Huang, D.; et al. Ubiquitin ligase SYVN1/HRD1 facilitates degradation of the SERPINA1 Z variant/α-1-antitrypsin Z variant via SQSTM1/p62-dependent selective autophagy. *Autophagy* **2017**, *13*, 686–702. [CrossRef] [PubMed]
74. Wauer, T.; Simicek, M.; Schubert, A.; Komander, D. Mechanism of phospho-ubiquitin-induced PARKIN activation. *Nature* **2015**, *524*, 370–374. [CrossRef]
75. Matscheko, N.; Mayrhofer, P.; Rao, Y.; Beier, V.; Wollert, T. Atg11 tethers Atg9 vesicles to initiate selective autophagy. *PLoS Biol.* **2019**, *17*, e3000377. [CrossRef]
76. Zientara-Rytter, K.; Subramani, S. Mechanistic Insights into the Role of Atg11 in Selective Autophagy. *J. Mol. Biol.* **2020**, *432*, 104–122. [CrossRef]
77. Kanki, T.; Wang, K.; Cao, Y.; Baba, M.; Klionsky, D.J. Atg32 Is a Mitochondrial Protein that Confers Selectivity during Mitophagy. *Dev. Cell* **2009**, *17*, 98–109. [CrossRef]

78. Gladkova, C.; Maslen, S.L.; Skehel, J.M.; Komander, D. Mechanism of parkin activation by PINK1. *Nature* **2018**, *559*, 410–414. [CrossRef]
79. Vigie, P.; Gonzalez, C.; Manon, S.; Bhatia-Kissova, I.; Camougrand, N. Ubiquitination of the yeast receptor Atg32 modulates mitophagy. *bioRxiv* **2019**, 652933. [CrossRef]
80. Muller, M.; Kötter, P.; Behrendt, C.; Walter, E.; Scheckhuber, C.Q.; Entian, K.-D.; Reichert, A.S. Synthetic Quantitative Array Technology Identifies the Ubp3-Bre5 Deubiquitinase Complex as a Negative Regulator of Mitophagy. *Cell Rep.* **2015**, *10*, 1215–1225. [CrossRef]
81. Scott, S.V.; Guan, J.; Hutchins, M.U.; Kim, J.; Klionsky, D.J. Cvt19 Is a Receptor for the Cytoplasm-to-Vacuole Targeting Pathway. *Mol. Cell* **2001**, *7*, 1131–1141. [CrossRef]
82. Yuga, M.; Gomi, K.; Klionsky, D.J.; Shintani, T. Aspartyl Aminopeptidase Is Imported from the Cytoplasm to the Vacuole by Selective Autophagy inSaccharomyces cerevisiae. *J. Biol. Chem.* **2011**, *286*, 13704–13713. [CrossRef] [PubMed]
83. Watanabe, Y.; Noda, N.N.; Kumeta, H.; Suzuki, K.; Ohsumi, Y.; Inagaki, F. Selective Transport of α-Mannosidase by Autophagic Pathways STRUCTURAL BASIS FOR CARGO RECOGNITION BY Atg19 AND Atg34. *J. Biol. Chem.* **2010**, *285*, 30026–30033. [CrossRef] [PubMed]
84. Baxter, B.K.; Abeliovich, H.; Zhang, X.; Stirling, A.G.; Burlingame, A.L.; Goldfarb, D.S. Atg19p Ubiquitination and the Cytoplasm to Vacuole Trafficking Pathway in Yeast. *J. Biol. Chem.* **2005**, *280*, 39067–39076. [CrossRef]
85. Motley, A.M.; Nuttall, A.J.M.; Hettema, E.H. Pex3-anchored Atg36 tags peroxisomes for degradation inSaccharomyces cerevisiae. *EMBO J.* **2012**, *31*, 2852–2868. [CrossRef]
86. Burnett, S.F.; Farré, J.-C.; Nazarko, T.Y.; Subramani, S. Peroxisomal Pex3 Activates Selective Autophagy of Peroxisomes via Interaction with the Pexophagy Receptor Atg30. *J. Biol. Chem.* **2015**, *290*, 8623–8631. [CrossRef]
87. Deosaran, E.; Larsen, K.B.; Hua, R.; Sargent, G.; Wang, Y.; Kim, S.; Lamark, T.; Jauregui, M.; Law, K.; Lippincott-Schwartz, J.; et al. NBR1 acts as an autophagy receptor for peroxisomes. *J. Cell Sci.* **2013**, *126*, 939–952. [CrossRef]
88. Nuttall, J.M.; Motley, A.M.; Hettema, E.H. Deficiency of the exportomer components Pex1, Pex6, and Pex15 causes enhanced pexophagy in Saccharomyces cerevisiae. *Autophagy* **2014**, *10*, 835–845. [CrossRef]
89. Germain, K.; Kim, P.K. Pexophagy: A Model for Selective Autophagy. *Int. J. Mol. Sci.* **2020**, *21*, 578. [CrossRef]
90. Lu, K.; Psakhye, I.; Jentsch, S. Autophagic Clearance of PolyQ Proteins Mediated by Ubiquitin-Atg8 Adaptors of the Conserved CUET Protein Family. *Cell* **2014**, *158*, 549–563. [CrossRef]
91. Kraft, C.; Deplazes, A.; Sohrmann, M.; Peter, M. Mature ribosomes are selectively degraded upon starvation by an autophagy pathway requiring the Ubp3p/Bre5p ubiquitin protease. *Nat. Cell Biol.* **2008**, *10*, 602–610. [CrossRef] [PubMed]
92. Ossareh-Nazari, B.; Bonizec, M.; Cohen, M.M.; Dokudovskaya, S.; Delalande, F.; Schaeffer, C.; Van Dorsselaer, A.; Dargemont, C. Cdc48 and Ufd3, new partners of the ubiquitin protease Ubp3, are required for ribophagy. *EMBO Rep.* **2010**, *11*, 548–554. [CrossRef] [PubMed]
93. Tatehashi, Y.; Watanabe, D.; Takagi, H. gamma-Glutamyl kinase is involved in selective autophagy of ribosomes in Saccharomyces cerevisiae. *FEBS Lett.* **2016**, *590*, 2906–2914. [CrossRef] [PubMed]
94. Wyant, G.A.; Abu-Remaileh, M.; Frenkel, E.M.; Laqtom, N.N.; Dharamdasani, V.; Lewis, C.A.; Chan, S.H.; Heinze, I.; Ori, A.; Sabatini, D.M. NUFIP1 is a ribosome receptor for starvation-induced ribophagy. *Science* **2018**, *360*, 751–758. [CrossRef]
95. Nemec, A.A.; Howell, L.A.; Peterson, A.K.; Murray, M.A.; Tomko, R.J. Autophagic clearance of proteasomes in yeast requires the conserved sorting nexin Snx4. *J. Biol. Chem.* **2017**, *292*, 21466–21480. [CrossRef]
96. Marshall, R.S.; Hua, Z.; Mali, S.; McLoughlin, F.; Vierstra, R.D. ATG8-Binding UIM Proteins Define a New Class of Autophagy Adaptors and Receptors. *Cell* **2019**, *177*, 766–781.e24. [CrossRef]
97. Waite, K.A.; De La Mota-Peynado, A.; Vontz, G.; Roelofs, J. Starvation Induces Proteasome Autophagy with Different Pathways for Core and Regulatory Particles. *J. Biol. Chem.* **2016**, *291*, 3239–3253. [CrossRef]
98. Marshall, R.S.; Vierstra, R.D. Proteasome storage granules protect proteasomes from autophagic degradation upon carbon starvation. *eLife* **2018**, *7*, e34532. [CrossRef]
99. Cohen-Kaplan, V.; Livneh, I.; Avni, N.; Fabre, B.; Ziv, T.; Kwon, Y.T.; Ciechanover, A. p62- and ubiquitin-dependent stress-induced autophagy of the mammalian 26S proteasome. *Proc. Natl. Acad. Sci. USA* **2016**, *113*, E7490–E7499. [CrossRef]

100. Loewith, R.; Hall, M.N. Target of rapamycin (TOR) in nutrient signaling and growth control. *Genetics* **2011**, *189*, 1177–1201. [CrossRef]
101. Kamada, Y.; Funakoshi, T.; Shintani, T.; Nagano, K.; Ohsumi, M.; Ohsumi, Y. Tor-Mediated Induction of Autophagy via an Apg1 Protein Kinase Complex. *J. Cell Biol.* **2000**, *150*, 1507–1513. [CrossRef] [PubMed]
102. Kim, D.-H.; Sarbassov, D.D.; Ali, S.M.; King, J.E.; Latek, R.R.; Erdjument-Bromage, H.; Tempst, P.; Sabatini, D.M. mTOR Interacts with Raptor to Form a Nutrient-Sensitive Complex that Signals to the Cell Growth Machinery. *Cell* **2002**, *110*, 163–175. [CrossRef]
103. Hara, K.; Maruki, Y.; Long, X.; Yoshino, K.-I.; Oshiro, N.; Hidayat, S.; Tokunaga, C.; Avruch, J.; Yonezawa, K. Raptor, a Binding Partner of Target of Rapamycin (TOR), Mediates TOR Action. *Cell* **2002**, *110*, 177–189. [CrossRef]
104. Kim, D.-H.; Sarbassov, D.D.; Ali, S.M.; Latek, R.R.; Guntur, K.V.P.; Erdjument-Bromage, H.; Tempst, P.; Sabatini, D.M. GbetaL, a positive regulator of the rapamycin-sensitive pathway required for the nutrient-sensitive interaction between raptor and mTOR. *Mol. Cell* **2003**, *11*, 895–904. [CrossRef]
105. Sancak, Y.; Peterson, T.R.; Shaul, Y.D.; Lindquist, R.A.; Thoreen, C.C.; Bar-Peled, L.; Sabatini, D.M. The Rag GTPases Bind Raptor and Mediate Amino Acid Signaling to mTORC1. *Science* **2008**, *320*, 1496–1501. [CrossRef]
106. Ganley, I.G.; Lam, D.H.; Wang, J.; Ding, X.; Chen, S.; Jiang, X. ULK1·ATG13·FIP200 Complex Mediates mTOR Signaling and Is Essential for Autophagy. *J. Biol. Chem.* **2009**, *284*, 12297–12305. [CrossRef]
107. Jung, C.H.; Jun, C.B.; Ro, S.-H.; Kim, Y.-M.; Otto, N.M.; Cao, J.; Kundu, M.; Kim, D.-H. ULK-Atg13-FIP200 Complexes Mediate mTOR Signaling to the Autophagy Machinery. *Mol. Biol. Cell* **2009**, *20*, 1992–2003. [CrossRef]
108. Hosokawa, N.; Hara, T.; Kaizuka, T.; Kishi, C.; Takamura, A.; Miura, Y.; Iemura, S.-I.; Natsume, T.; Takehana, K.; Yamada, N.; et al. Nutrient-dependent mTORC1 Association with the ULK1–Atg13–FIP200 Complex Required for Autophagy. *Mol. Biol. Cell* **2009**, *20*, 1981–1991. [CrossRef]
109. Mercer, C.A.; Kaliappan, A.; Dennis, P.B. A novel, human Atg13 binding protein, Atg101, interacts with ULK1 and is essential for macroautophagy. *Autophagy* **2009**, *5*, 649–662. [CrossRef]
110. Zachari, M.; Ganley, I.G. The mammalian ULK1 complex and autophagy initiation. *Essays Biochem.* **2017**, *61*, 585–596. [CrossRef]
111. Wang, Z.; Wilson, W.A.; Fujino, M.A.; Roach, P.J. Antagonistic Controls of Autophagy and Glycogen Accumulation by Snf1p, the Yeast Homolog of AMP-Activated Protein Kinase, and the Cyclin-Dependent Kinase Pho85p. *Mol. Cell. Biol.* **2001**, *21*, 5742–5752. [CrossRef] [PubMed]
112. Yi, C.; Tong, J.; Lu, P.; Wang, Y.; Zhang, J.; Sun, C.; Yuan, K.; Xue, R.; Zou, B.; Li, N.; et al. Formation of a Snf1-Mec1-Atg1 Module on Mitochondria Governs Energy Deprivation-Induced Autophagy by Regulating Mitochondrial Respiration. *Dev. Cell* **2017**, *41*, 59–71.e4. [CrossRef] [PubMed]
113. Inoki, K.; Zhu, T.; Guan, K.-L. TSC2 Mediates Cellular Energy Response to Control Cell Growth and Survival. *Cell* **2003**, *115*, 577–590. [CrossRef]
114. Gwinn, D.M.; Shackelford, D.B.; Egan, D.F.; Mihaylova, M.M.; Méry, A.; Vasquez, D.S.; Turk, B.E.; Shaw, R.J. AMPK Phosphorylation of Raptor Mediates a Metabolic Checkpoint. *Mol. Cell* **2008**, *30*, 214–226. [CrossRef] [PubMed]
115. Kim, J.; Kundu, M.; Viollet, B.; Guan, K.-L. AMPK and mTOR regulate autophagy through direct phosphorylation of Ulk1. *Nat. Cell Biol.* **2011**, *13*, 132–141. [CrossRef]
116. Hu, K.; Guo, S.; Yan, G.; Yuan, W.; Zheng, Y.; Jiang, Y. Ubiquitin regulates TORC1 in yeast Saccharomyces cerevisiae. *Mol. Microbiol.* **2016**, *100*, 303–314. [CrossRef]
117. Jiang, Y. Regulation of TORC1 by ubiquitin through non-covalent binding. *Curr. Genet.* **2016**, *62*, 553–555. [CrossRef]
118. Linares, J.F.; Duran, A.; Yajima, T.; Pasparakis, M.; Moscat, J.; Diaz-Meco, M.T. K63 Polyubiquitination and Activation of mTOR by the p62-TRAF6 Complex in Nutrient-Activated Cells. *Mol. Cell* **2013**, *51*, 283–296. [CrossRef]
119. Chen, J.; Ou, Y.; Yang, Y.; Li, W.; Xu, Y.; Xie, Y.; Liu, Y. KLHL22 activates amino-acid-dependent mTORC1 signalling to promote tumorigenesis and ageing. *Nature* **2018**, *557*, 585–589. [CrossRef]
120. Peterson, T.R.; Laplante, M.; Thoreen, C.C.; Sancak, Y.; Kang, S.A.; Kuehl, W.M.; Gray, N.S.; Sabatini, D.M. DEPTOR Is an mTOR Inhibitor Frequently Overexpressed in Multiple Myeloma Cells and Required for Their Survival. *Cell* **2009**, *137*, 873–886. [CrossRef]

121. Zhao, Y.; Xiong, X.; Sun, Y. DEPTOR, an mTOR inhibitor, is a physiological substrate of SCF(betaTrCP) E3 ubiquitin ligase and regulates survival and autophagy. *Mol. Cell* **2011**, *44*, 304–316. [CrossRef] [PubMed]
122. Gao, D.; Inuzuka, H.; Tan, M.-K.M.; Fukushima, H.; Locasale, J.W.; Liu, P.; Wan, L.; Zhai, B.; Chin, Y.R.; Shaik, S.; et al. mTOR drives its own activation via SCF(betaTrCP)-dependent degradation of the mTOR inhibitor DEPTOR. *Mol. Cell* **2011**, *44*, 290–303. [CrossRef] [PubMed]
123. Duan, S.; Skaar, J.R.; Kuchay, S.; Toschi, A.; Kanarek, N.; Ben-Neriah, Y.; Pagano, M. mTOR generates an auto-amplification loop by triggering the betaTrCP- and CK1alpha-dependent degradation of DEPTOR. *Mol. Cell* **2011**, *44*, 317–324. [CrossRef] [PubMed]
124. Antonioli, M.; Albiero, F.; Nazio, F.; Vescovo, T.; Perdomo, A.B.; Corazzari, M.; Marsella, C.; Piselli, P.; Gretzmeier, C.; Dengjel, J.; et al. AMBRA1 Interplay with Cullin E3 Ubiquitin Ligases Regulates Autophagy Dynamics. *Dev. Cell* **2014**, *31*, 734–746. [CrossRef]
125. McEwan, D.G.; Dikic, I. Cullins keep autophagy under control. *Dev. Cell* **2014**, *31*, 675–676. [CrossRef]
126. Zhao, L.; Wang, X.; Yu, Y.; Deng, L.; Chen, L.; Peng, X.; Jiao, C.; Gao, G.; Tan, X.; Pan, W.; et al. OTUB1 protein suppresses mTOR complex 1 (mTORC1) activity by deubiquitinating the mTORC1 inhibitor DEPTOR. *J. Biol. Chem.* **2018**, *293*, 4883–4892. [CrossRef]
127. Nazio, F.; Strappazzon, F.; Antonioli, M.; Bielli, P.; Cianfanelli, V.; Bordi, M.; Gretzmeier, C.; Dengjel, J.; Piacentini, M.; Fimia, G.M.; et al. mTOR inhibits autophagy by controlling ULK1 ubiquitylation, self-association and function through AMBRA1 and TRAF6. *Nat. Cell Biol.* **2013**, *15*, 406–416. [CrossRef]
128. Di Rienzo, M.; Antonioli, M.; Fusco, C.; Liu, Y.; Mari, M.; Orhon, I.; Refolo, G.; Germani, F.; Corazzari, M.; Romagnoli, A.; et al. Autophagy induction in atrophic muscle cells requires ULK1 activation by TRIM32 through unanchored K63-linked polyubiquitin chains. *Sci. Adv.* **2019**, *5*, eaau8857. [CrossRef]
129. Jiao, H.; Su, G.-Q.; Dong, W.; Zhang, L.; Xie, W.; Yao, L.-M.; Chen, P.; Wang, Z.-X.; Liou, Y.-C.; You, H. Chaperone-like protein p32 regulates ULK1 stability and autophagy. *Cell Death Differ.* **2015**, *22*, 1812–1823. [CrossRef]
130. Lee, D.-E.; Yoo, J.E.; Kim, J.; Kim, S.; Kim, S.; Lee, H.; Cheong, H. NEDD4L downregulates autophagy and cell growth by modulating ULK1 and a glutamine transporter. *Cell Death Dis.* **2020**, *11*, 1–17. [CrossRef]
131. Crawley, O.; Opperman, K.J.; Desbois, M.; Adrados, I.; Borgen, M.A.; Giles, A.C.; Duckett, D.R.; Grill, B. Autophagy is inhibited by ubiquitin ligase activity in the nervous system. *Nat. Commun.* **2019**, *10*, 1–17. [CrossRef]
132. Kim, J.H.; Seo, D.; Kim, S.; Choi, D.W.; Park, J.S.; Ha, J.; Choi, J.; Lee, J.; Jung, S.M.; Seo, K.; et al. The deubiquitinating enzyme USP20 stabilizes ULK1 and promotes autophagy initiation. *EMBO Rep.* **2018**, *19*, e44378. [CrossRef] [PubMed]
133. Raimondi, M.; Cesselli, D.; Di Loreto, C.; La Marra, F.; Schneider, C.; Demarchi, F. USP1 (ubiquitin specific peptidase 1) targets ULK1 and regulates its cellular compartmentalization and autophagy. *Autophagy* **2018**, *15*, 613–630. [CrossRef] [PubMed]
134. Wilson, M.A.; Koutelou, E.; Hirsch, C.; Akdemir, K.; Schibler, A.; Barton, M.; Dent, S.R. Ubp8 and SAGA Regulate Snf1 AMP Kinase Activity. *Mol. Cell. Biol.* **2011**, *31*, 3126–3135. [CrossRef] [PubMed]
135. Pineda, C.T.; Ramanathan, S.; Tacer, K.F.; Weon, J.L.; Potts, M.B.; Ou, Y.-H.; White, M.A.; Potts, P.R. Degradation of AMPK by a Cancer-Specific Ubiquitin Ligase. *Cell* **2015**, *160*, 715–728. [CrossRef]
136. Chu, Y.; Kang, Y.; Yan, C.; Yang, C.; Zhang, T.; Huo, H.; Liu, Y. LUBAC and OTULIN regulate autophagy initiation and maturation by mediating the linear ubiquitination and the stabilization of ATG13. *Autophagy* **2020**, 1–16. [CrossRef]
137. Kihara, A.; Noda, T.; Ishihara, N.; Ohsumi, Y. Two Distinct Vps34 Phosphatidylinositol 3–Kinase Complexes Function in Autophagy and Carboxypeptidase Y Sorting inSaccharomyces cerevisiae. *J. Cell Biol.* **2001**, *152*, 519–530. [CrossRef]
138. Funderburk, S.F.; Wang, Q.J.; Yue, Z. The Beclin 1–VPS34 complex—At the crossroads of autophagy and beyond. *Trends Cell Biol.* **2010**, *20*, 355–362. [CrossRef]
139. Furuya, T.; Kim, M.; Lipinski, M.M.; Li, J.; Kim, D.; Lu, T.; Shen, Y.; Rameh, L.; Yankner, B.; Tsai, L.-H.; et al. Negative Regulation of Vps34 by Cdk Mediated Phosphorylation. *Mol. Cell* **2010**, *38*, 500–511. [CrossRef]
140. Eisenberg-Lerner, A.; Kimchi, A. PKD is a kinase of Vps34 that mediates ROS-induced autophagy downstream of DAPk. *Cell Death Differ.* **2012**, *19*, 788–797. [CrossRef]

141. Yang, Y.; Fiskus, W.; Yong, B.; Atadja, P.; Takahashi, Y.; Pandita, T.K.; Wang, H.-G.; Bhalla, K.N. Acetylated hsp70 and KAP1-mediated Vps34 SUMOylation is required for autophagosome creation in autophagy. *Proc. Natl. Acad. Sci. USA* **2013**, *110*, 6841–6846. [CrossRef] [PubMed]
142. Zalckvar, E.; Berissi, H.; Mizrachy, L.; Idelchuk, Y.; Koren, I.; Eisenstein, M.; Sabanay, H.; Pinkas-Kramarski, R.; Kimchi, A. DAP-kinase-mediated phosphorylation on the BH3 domain of beclin 1 promotes dissociation of beclin 1 from Bcl-XL and induction of autophagy. *EMBO Rep.* **2009**, *10*, 285–292. [CrossRef] [PubMed]
143. Wang, R.C.; Wei, Y.; An, Z.; Zou, Z.; Xiao, G.; Bhagat, G.; White, M.; Reichelt, J.; Levine, B. Akt-Mediated Regulation of Autophagy and Tumorigenesis Through Beclin 1 Phosphorylation. *Science* **2012**, *338*, 956–959. [CrossRef] [PubMed]
144. Russell, R.C.; Tian, Y.; Yuan, H.; Park, H.W.; Chang, Y.-Y.; Kim, J.; Kim, H.; Neufeld, T.P.; Dillin, A.; Guan, K.-L. ULK1 induces autophagy by phosphorylating Beclin-1 and activating VPS34 lipid kinase. *Nat. Cell Biol.* **2013**, *15*, 741–750. [CrossRef] [PubMed]
145. Wei, Y.; Zou, Z.; Becker, N.; Anderson, M.; Sumpter, R.; Xiao, G.; Kinch, L.; Koduru, P.; Christudass, C.S.; Veltri, R.W.; et al. EGFR-mediated Beclin 1 phosphorylation in autophagy suppression, tumor progression, and tumor chemoresistance. *Cell* **2013**, *154*, 1269–1284. [CrossRef] [PubMed]
146. Xiao, J.; Zhang, T.; Xu, D.; Wang, H.; Cai, Y.; Jin, T.; Liu, M.; Jin, M.; Wu, K.; Yuan, J. FBXL20-mediated Vps34 ubiquitination as a p53 controlled checkpoint in regulating autophagy and receptor degradation. *Genes Dev.* **2015**, *29*, 184–196. [CrossRef]
147. Liu, J.; Li, M.; Li, L.; Chen, S.; Wang, X. Ubiquitination of the PI3-kinase VPS-34 promotes VPS-34 stability and phagosome maturation. *J. Cell Biol.* **2018**, *217*, 347–360. [CrossRef]
148. Xie, W.; Jin, S.; Wu, Y.; Xian, H.; Tian, S.; Liu, D.-A.; Guo, Z.; Cui, J. Auto-ubiquitination of NEDD4-1 Recruits USP13 to Facilitate Autophagy through Deubiquitinating VPS34. *Cell Rep.* **2020**, *30*, 2807–2819.e4. [CrossRef]
149. Shi, C.-S.; Kehrl, J.H. TRAF6 and A20 regulate lysine 63-linked ubiquitination of Beclin-1 to control TLR4-induced autophagy. *Sci. Signal.* **2010**, *3*, ra42. [CrossRef]
150. Han, T.; Guo, M.; Gan, M.; Yu, B.; Tian, X.; Wang, J.-B. TRIM59 regulates autophagy through modulating both the transcription and the ubiquitination of BECN1. *Autophagy* **2018**, *14*, 2035–2048. [CrossRef]
151. Fusco, C.; Mandriani, B.; Di Rienzo, M.; Micale, L.; Malerba, N.; Cocciadiferro, D.; Sjøttem, E.; Augello, B.; Squeo, G.M.; Pellico, M.T.; et al. TRIM50 regulates Beclin 1 proautophagic activity. *Biochim. Biophys. Acta Mol. Cell Res.* **2018**, *1865*, 908–919. [CrossRef] [PubMed]
152. Fimia, G.M.; Stoykova, A.; Romagnoli, A.; Giunta, L.; Di Bartolomeo, S.; Nardacci, R.; Corazzari, M.; Fuoco, C.; Ucar, A.; Schwartz, P.; et al. Ambra1 regulates autophagy and development of the nervous system. *Nature* **2007**, *447*, 1121–1125. [CrossRef] [PubMed]
153. Xia, P.; Wang, S.; Du, Y.; Zhao, Z.; Shi, L.; Sun, L.; Huang, G.; Ye, B.; Li, C.; Dai, Z.; et al. WASH inhibits autophagy through suppression of Beclin 1 ubiquitination. *EMBO J.* **2013**, *32*, 2685–2696. [CrossRef] [PubMed]
154. Xu, C.; Feng, K.; Zhao, X.; Huang, S.; Cheng, Y.; Qian, L.; Wang, Y.; Sun, H.; Jin, M.; Chuang, T.-H.; et al. Regulation of autophagy by E3 ubiquitin ligase RNF216 through BECN1 ubiquitination. *Autophagy* **2014**, *10*, 2239–2250. [CrossRef]
155. Platta, H.W.; Abrahamsen, H.; Thoresen, S.B.; Stenmark, H. Nedd4-dependent lysine-11-linked polyubiquitination of the tumour suppressor Beclin 1. *Biochem. J.* **2012**, *441*, 399–406. [CrossRef] [PubMed]
156. Pei, G.; Buijze, H.; Liu, H.; Moura-Alves, P.; Goosmann, C.; Brinkmann, V.; Kawabe, H.; Dorhoi, A.; Kaufmann, S.H. The E3 ubiquitin ligase NEDD4 enhances killing of membrane-perturbing intracellular bacteria by promoting autophagy. *Autophagy* **2017**, *13*, 2041–2055. [CrossRef]
157. Gassen, N.C.; Niemeyer, D.; Muth, D.; Corman, V.M.; Martinelli, S.; Gassen, A.; Hafner, K.; Papies, J.; Mösbauer, K.; Zellner, A.; et al. SKP2 attenuates autophagy through Beclin1-ubiquitination and its inhibition reduces MERS-Coronavirus infection. *Nat. Commun.* **2019**, *10*, 5770. [CrossRef]
158. Liu, J.; Xia, H.; Kim, M.; Xu, L.; Li, Y.; Zhang, L.; Cai, Y.; Norberg, H.V.; Zhang, T.; Furuya, T.; et al. Beclin1 Controls the Levels of p53 by Regulating the Deubiquitination Activity of USP10 and USP13. *Cell* **2011**, *147*, 223–234. [CrossRef]
159. Jin, S.; Tian, S.; Chen, Y.; Zhang, C.; Xie, W.; Xia, X.; Cui, J.; Wang, R.-F. USP 19 modulates autophagy and antiviral immune responses by deubiquitinating Beclin-1. *EMBO J.* **2016**, *35*, 866–880. [CrossRef]
160. Xu, D.; Shan, B.; Sun, H.; Xiao, J.; Zhu, K.; Xie, X.; Li, X.; Liang, W.; Lu, X.; Qian, L.; et al. USP14 regulates autophagy by suppressing K63 ubiquitination of Beclin 1. *Genes Dev.* **2016**, *30*, 1718–1730. [CrossRef]

161. Ashkenazi, A.; Bento, C.F.; Ricketts, T.; Vicinanza, M.; Siddiqi, F.; Pavel, M.; Squitieri, F.; Hardenberg, M.C.; Imarisio, S.; Menzies, F.M.; et al. Polyglutamine tracts regulate beclin 1-dependent autophagy. *Nature* **2017**, *545*, 108–111. [CrossRef] [PubMed]
162. Zhang, T.; Dong, K.; Liang, W.; Xu, D.; Xia, H.; Geng, J.; Najafov, A.; Liu, M.; Li, Y.; Han, X.; et al. G-protein-coupled receptors regulate autophagy by ZBTB16-mediated ubiquitination and proteasomal degradation of Atg14L. *eLife* **2015**, *4*, 06734. [CrossRef] [PubMed]
163. Xia, P.; Wang, S.; Huang, G.; Du, Y.; Zhu, P.; Li, M.; Fan, Z. RNF2 is recruited by WASH to ubiquitinate AMBRA1 leading to downregulation of autophagy. *Cell Res.* **2014**, *24*, 943–958. [CrossRef] [PubMed]
164. Chen, D.; Gao, F.; Li, B.; Wang, H.; Xu, Y.; Zhu, C.; Wang, G. Parkin Mono-ubiquitinates Bcl-2 and Regulates Autophagy. *J. Biol. Chem.* **2010**, *285*, 38214–38223. [CrossRef] [PubMed]
165. Bodemann, B.O.; Orvedahl, A.; Cheng, T.; Ram, R.R.; Ou, Y.-H.; Formstecher, E.; Maiti, M.; Hazelett, C.C.; Wauson, E.M.; Balakireva, M.; et al. RalB and the Exocyst Mediate the Cellular Starvation Response by Direct Activation of Autophagosome Assembly. *Cell* **2011**, *144*, 253–267. [CrossRef] [PubMed]
166. Simicek, M.; Lievens, S.; Laga, M.; Guzenko, D.; Aushev, V.N.; Kalev, P.; Baietti, M.F.; Strelkov, S.V.; Gevaert, K.; Tavernier, J.; et al. The deubiquitylase USP33 discriminates between RALB functions in autophagy and innate immune response. *Nat. Cell Biol.* **2013**, *15*, 1220–1230. [CrossRef]
167. Wan, W.; You, Z.; Zhou, L.; Xu, Y.; Peng, C.; Zhou, T.; Yi, C.; Shi, Y.; Liu, W. mTORC1-Regulated and HUWE1-Mediated WIPI2 Degradation Controls Autophagy Flux. *Mol. Cell* **2018**, *72*, 303–315.e6. [CrossRef]
168. Lu, G.; Yi, J.; Gubas, A.; Wang, Y.-T.; Wu, Y.; Ren, Y.; Wu, M.; Shi, Y.; Ouyang, C.; Tan, H.W.S.; et al. Suppression of autophagy during mitosis via CUL4-RING ubiquitin ligases-mediated WIPI2 polyubiquitination and proteasomal degradation. *Autophagy* **2019**, *15*, 1917–1934. [CrossRef]
169. Kuang, E.; Okumura, C.Y.M.; Sheffy-Levin, S.; Varsano, T.; Shu, V.C.-W.; Qi, J.; Niesman, I.R.; Yang, H.-J.; López-Otín, C.; Yang, W.Y.; et al. Regulation of ATG4B Stability by RNF5 Limits Basal Levels of Autophagy and Influences Susceptibility to Bacterial Infection. *PLoS Genet.* **2012**, *8*, e1003007. [CrossRef]
170. Fujita, N.; Itoh, T.; Omori, H.; Fukuda, M.; Noda, T.; Yoshimori, T. The Atg16L Complex Specifies the Site of LC3 Lipidation for Membrane Biogenesis in Autophagy. *Mol. Biol. Cell* **2008**, *19*, 2092–2100. [CrossRef]
171. Scrivo, A.; Codogno, P.; Bomont, P. Gigaxonin E3 ligase governs ATG16L1 turnover to control autophagosome production. *Nat. Commun.* **2019**, *10*, 780. [CrossRef] [PubMed]
172. Jia, R.; Bonifacino, J.S. Negative regulation of autophagy by UBA6-BIRC6-mediated ubiquitination of LC3. *eLife* **2019**, *8*, 8. [CrossRef]
173. Joachim, J.; Razi, M.; Judith, D.; Wirth, M.; Calamita, E.; Encheva, V.; Dynlacht, B.D.; Snijders, A.P.; O'Reilly, N.; Jefferies, H.B.; et al. Centriolar Satellites Control GABARAP Ubiquitination and GABARAP-Mediated Autophagy. *Curr. Biol.* **2017**, *27*, 2123–2136.e7. [CrossRef] [PubMed]
174. Liang, C.; Lee, J.-S.; Inn, K.-S.; Gack, M.U.; Li, Q.; Roberts, E.A.; Vergne, I.; Deretic, V.; Feng, P.; Akazawa, C.; et al. Beclin1-binding UVRAG targets the class C Vps complex to coordinate autophagosome maturation and endocytic trafficking. *Nat. Cell Biol.* **2008**, *10*, 776–787. [CrossRef] [PubMed]
175. Zhong, Y.; Wang, Q.J.; Li, X.; Yan, Y.; Backer, J.M.; Chait, B.T.; Heintz, N.; Yue, Z. Distinct regulation of autophagic activity by Atg14L and Rubicon associated with Beclin 1–phosphatidylinositol-3-kinase complex. *Nat. Cell Biol.* **2009**, *11*, 468–476. [CrossRef] [PubMed]
176. Kim, Y.-M.; Jung, C.H.; Seo, M.; Kim, E.K.; Park, J.-M.; Bae, S.S.; Kim, D.-H. mTORC1 phosphorylates UVRAG to negatively regulate autophagosome and endosome maturation. *Mol. Cell* **2015**, *57*, 207–218. [CrossRef] [PubMed]
177. Feng, X.; Jia, Y.; Zhang, Y.; Ma, F.; Zhu, Y.; Hong, X.; Zhou, Q.; He, R.; Zhang, H.; Jin, J.; et al. Ubiquitination of UVRAG by SMURF1 promotes autophagosome maturation and inhibits hepatocellular carcinoma growth. *Autophagy* **2019**, *15*, 1130–1149. [CrossRef]
178. Wang, Z.; Miao, G.; Xue, X.; Guo, X.; Yuan, C.; Wang, Z.; Zhang, G.; Chen, Y.; Feng, D.; Hu, J.; et al. The Vici Syndrome Protein EPG5 Is a Rab7 Effector that Determines the Fusion Specificity of Autophagosomes with Late Endosomes/Lysosomes. *Mol. Cell* **2016**, *63*, 781–795. [CrossRef]
179. Gu, H.; Shi, X.; Liu, C.; Wang, C.; Sui, N.; Zhao, Y.; Gong, J.; Wang, F.; Zhang, H.; Li, W.; et al. USP8 maintains embryonic stem cell stemness via deubiquitination of EPG5. *Nat. Commun.* **2019**, *10*, 1465. [CrossRef]
180. Liu, Y.; Levine, B. Autosis and autophagic cell death: The dark side of autophagy. *Cell Death Differ.* **2015**, *22*, 367–376. [CrossRef]

181. Liu, C.-C.; Lin, Y.-C.; Chen, Y.-H.; Chen, C.-M.; Pang, L.-Y.; Chen, H.-A.; Wu, P.-R.; Lin, M.-Y.; Jiang, S.-T.; Tsai, T.-F.; et al. Cul3-KLHL20 Ubiquitin Ligase Governs the Turnover of ULK1 and VPS34 Complexes to Control Autophagy Termination. *Mol. Cell* **2016**, *61*, 84–97. [CrossRef] [PubMed]
182. Nazio, F.; Carinci, M.; Valacca, C.; Bielli, P.; Strappazzon, F.; Antonioli, M.; Ciccosanti, F.; Rodolfo, C.; Campello, S.; Fimia, G.M.; et al. Fine-tuning of ULK1 mRNA and protein levels is required for autophagy oscillation. *J. Cell Biol.* **2016**, *215*, 841–856. [CrossRef] [PubMed]
183. Bartholomew, C.R.; Suzuki, T.; Du, Z.; Backues, S.K.; Jin, M.; Lynch-Day, M.A.; Umekawa, M.; Kamath, A.; Zhao, M.; Xie, Z.; et al. Ume6 transcription factor is part of a signaling cascade that regulates autophagy. *Proc. Natl. Acad. Sci. USA* **2012**, *109*, 11206–11210. [CrossRef] [PubMed]
184. Stephan, J.S.; Yeh, Y.-Y.; Ramachandran, V.V.; Deminoff, S.J.; Herman, P.K. The Tor and cAMP-dependent protein kinase signaling pathways coordinately control autophagy in Saccharomyces cerevisiae. *Autophagy* **2010**, *6*, 294–295. [CrossRef] [PubMed]
185. Mallory, M.J.; Cooper, K.F.; Strich, R. Meiosis-Specific Destruction of the Ume6p Repressor by the Cdc20-Directed APC/C. *Mol. Cell* **2007**, *27*, 951–961. [CrossRef]
186. White, E. Autophagy and p53. *Cold Spring Harb. Perspect. Med.* **2016**, *6*, a026120. [CrossRef]
187. Green, D.R.; Kroemer, G. Cytoplasmic functions of the tumour suppressor p53. *Nature* **2009**, *458*, 1127–1130. [CrossRef]
188. Tasdemir, E.; Maiuri, M.C.; Galluzzi, L.; Vitale, I.; Djavaheri-Mergny, M.; D'Amelio, M.; Criollo, A.; Morselli, E.; Zhu, C.; Harper, F.; et al. Regulation of autophagy by cytoplasmic p53. *Nat. Cell Biol.* **2008**, *10*, 676–687. [CrossRef]
189. Di Malta, C.; Cinque, L.; Settembre, C. Transcriptional Regulation of Autophagy: Mechanisms and Diseases. *Front. Cell Dev. Biol.* **2019**, *7*, 114. [CrossRef]
190. Huang, H.; Regan, K.M.; Wang, F.; Wang, D.; I Smith, D.; Van Deursen, J.M.A.; Tindall, D.J. Skp2 inhibits FOXO1 in tumor suppression through ubiquitin-mediated degradation. *Proc. Natl. Acad. Sci. USA* **2005**, *102*, 1649–1654. [CrossRef]
191. Kato, S.; Ding, J.; Pisck, E.; Jhala, U.S.; Du, K. COP1 Functions as a FoxO1 Ubiquitin E3 Ligase to Regulate FoxO1-mediated Gene Expression. *J. Biol. Chem.* **2008**, *283*, 35464–35473. [CrossRef]
192. Fu, W.; Ma, Q.; Chen, L.; Li, P.; Zhang, M.; Ramamoorthy, S.; Nawaz, Z.; Shimojima, T.; Wang, H.; Yang, Y.; et al. MDM2 Acts Downstream of p53 as an E3 Ligase to Promote FOXO Ubiquitination and Degradation. *J. Biol. Chem.* **2009**, *284*, 13987–14000. [CrossRef] [PubMed]
193. Polager, S.; Ofir, M.; Ginsberg, D. E2F1 regulates autophagy and the transcription of autophagy genes. *Oncogene* **2008**, *27*, 4860–4864. [CrossRef] [PubMed]
194. Huang, H.; Tindall, D.J. Regulation of FOXO protein stability via ubiquitination and proteasome degradation. *Biochim. Biophys. Acta* **2011**, *1813*, 1961–1964. [CrossRef] [PubMed]
195. Wang, B.; Ma, A.; Zhang, L.; Jin, W.-L.; Qian, Y.; Xu, G.; Qiu, B.; Yang, Z.; Liu, Y.; Xia, Q.; et al. POH1 deubiquitylates and stabilizes E2F1 to promote tumour formation. *Nat. Commun.* **2015**, *6*, 8704. [CrossRef]
196. Mahanic, C.S.; Budhavarapu, V.; Graves, J.D.; Li, G.; Lin, W.-C. Regulation of E2 Promoter Binding Factor 1 (E2F1) Transcriptional Activity through a Deubiquitinating Enzyme, UCH37. *J. Biol. Chem.* **2015**, *290*, 26508–26522. [CrossRef]
197. Glorian, V.; Allègre, J.; Berthelet, J.; Dumetier, B.; Boutanquoi, P.-M.; Droin, N.; Kayaci, C.; Cartier, J.; Gemble, S.; Marcion, G.; et al. DNA damage and S phase-dependent E2F1 stabilization requires the cIAP1 E3-ubiquitin ligase and is associated with K63-poly-ubiquitination on lysine 161/164 residues. *Cell Death Dis.* **2017**, *8*, e2816. [CrossRef]
198. Herhaus, L.; Dikic, I. Expanding the ubiquitin code through post-translational modification. *EMBO Rep.* **2015**, *16*, 1071–1083. [CrossRef]
199. Oh, E.; Akopian, D.; Rape, M. Principles of Ubiquitin-Dependent Signaling. *Annu. Rev. Cell Dev. Biol.* **2018**, *34*, 137–162. [CrossRef]
200. Rioux, J.D.; Xavier, R.J.; Taylor, K.D.; Silverberg, M.S.; Goyette, P.; Huett, A.; Green, T.; Kuballa, P.; Barmada, M.M.; Datta, L.W.; et al. Genome-wide association study identifies new susceptibility loci for Crohn disease and implicates autophagy in disease pathogenesis. *Nat. Genet.* **2007**, *39*, 596–604. [CrossRef]
201. Musone, S.L.; Taylor, K.E.; Nititham, J.; Chu, C.; Poon, A.; Liao, W.; Lam, E.T.; Ma, A.; Kwok, P.-Y.; A Criswell, L. Sequencing of TNFAIP3 and association of variants with multiple autoimmune diseases. *Genes Immun.* **2011**, *12*, 176–182. [CrossRef] [PubMed]

202. Slowicka, K.; Serramito-Gómez, I.; Boada-Romero, E.; Martens, A.; Sze, M.; Petta, I.; Vikkula, H.K.; De Rycke, R.; Parthoens, E.; Lippens, S.; et al. Physical and functional interaction between A20 and ATG16L1-WD40 domain in the control of intestinal homeostasis. *Nat. Commun.* **2019**, *10*, 1834. [CrossRef] [PubMed]

cells

MDPI

Review

Atg8-Family Proteins—Structural Features and Molecular Interactions in Autophagy and Beyond

Nicole Wesch [1,†], Vladimir Kirkin [2,†] and Vladimir V. Rogov [1,3,4,*]

1 Institute of Biophysical Chemistry and Center for Biomolecular Magnetic Resonance, Goethe-University Frankfurt, 60438 Frankfurt am Main, Germany; Wesch@bpc.uni-frankfurt.de
2 Cancer Research UK Cancer Therapeutics Unit, The Institute of Cancer Research London, Sutton SM2 5NG, UK; vladimir.kirkin@icr.ac.uk
3 Structural Genomics Consortium, Buchmann Institute for Life Sciences, Goethe-University Frankfurt, 60438 Frankfurt am Main, Germany
4 Institute of Pharmaceutical Chemistry, Goethe-University Frankfurt, 60438 Frankfurt am Main, Germany
* Correspondence: rogov@bpc.uni-frankfurt.de; Tel.: +49-69-7982-9622
† These authors contributed equally to this work.

Received: 8 August 2020; Accepted: 27 August 2020; Published: 1 September 2020

check for updates

Abstract: Autophagy is a common name for a number of catabolic processes, which keep the cellular homeostasis by removing damaged and dysfunctional intracellular components. Impairment or misbalance of autophagy can lead to various diseases, such as neurodegeneration, infection diseases, and cancer. A central axis of autophagy is formed along the interactions of autophagy modifiers (Atg8-family proteins) with a variety of their cellular counter partners. Besides autophagy, Atg8-proteins participate in many other pathways, among which membrane trafficking and neuronal signaling are the most known. Despite the fact that autophagy modifiers are well-studied, as the small globular proteins show similarity to ubiquitin on a structural level, the mechanism of their interactions are still not completely understood. A thorough analysis and classification of all known mechanisms of Atg8-protein interactions could shed light on their functioning and connect the pathways involving Atg8-proteins. In this review, we present our views of the key features of the Atg8-proteins and describe the basic principles of their recognition and binding by interaction partners. We discuss affinity and selectivity of their interactions as well as provide perspectives for discovery of new Atg8-interacting proteins and therapeutic approaches to tackle major human diseases.

Keywords: Atg8; LC3; GABARAP; LIR motif; SAR; UBL; autophagy

1. Introduction

Autophagy is a fundamental process of degradation and recycling of cellular components to maintain homeostasis [1,2]. In concert with, but also far beyond, the ubiquitin-proteasome degradation system (UPS), autophagy serves to remove bulky cytosolic cargo, such as long-lived protein and protein complexes, lipid droplets, portions of and whole organelles, cytosolic bacteria, etc., (reviewed in [3–5]). This is achieved by enclosing the cargo by a lipid membrane and subsequent trafficking into degradative cellular compartments—vacuoles (in fungi and plants) and lysosomes (other eukaryotes). The autophagy pathway branches into three relatively independent subtypes: *microautophagy*, in which the cytoplasmic cargo is directly engulfed by invagination of the lysosomal membrane; *chaperon-mediated autophagy*, which relies on translocation of unfolded proteins across the lysosome membrane in a receptor-depended manner, [1]; and *macroautophagy* (hereafter simply *autophagy*), which makes use of a double-membrane vesicle (*autophagosome*) as a shuttle between the cytosol and the degradative compartment.

Autophagosomes are formed by multiple mechanisms (reviewed in [6]) from a primordial membrane (*phagophore* or *isolation membrane*) that elongates by adding additional lipid and engulfs cargo in either selective or non-specific manner. After expansion and closure of the membranes, in a so-called maturation step, autophagosomes fuse with the vacuoles/lysosomes, in which multiple hydrolytic enzymes digest the engulfed cargo and molecular building blocks can be released into the cytosol for recycling or downstream catabolic reactions.

This complex process is orchestrated by >36 evolutionary conserved autophagy-related (Atg) proteins. In 2016, the Nobel Prize in Physiology and Medicine was awarded to Yoshinori Ohsumi for his discovery of Atg genes in 1990s. Within Atgs, autophagy-related, ubiquitin-like modifiers (Atg8 in yeast, LC3/GABARAP proteins in mammals) play a special role. Like many other ubiquitin-like proteins (UBLs), Atg8/LC3/GABARAPs display a high structural similarity to ubiquitin and are expressed as inactive precursors that undergo proteolytic maturation (performed by Atg4 proteases) to expose their invariant *C*-terminal Gly residue participating in a ubiquitin-like substrate conjugation reaction, involving Atg7 (E1), Atg3 (E2), and Atg5~Atg12: Atg16L1 complex (E3) proteins [7–9]. Of note, Atg5 and Atg12 are themselves autophagy-related UBLs, covalently bound in a separate conjugation cascade involving Atg7 (E1) and Atg10 (E2) enzymes (reviewed in [4]). The concerted action of the UBL conjugation cascade results in the formation of a covalent bond between the *C*-terminal Gly of Atg8/LC3/GABARAPs and the amino group of the substrate lipid, phosphatidylethanolamine (PE). PE is the second most abundant phospholipid (the first one being phosphatidylcholine, PC) that participates in many cellular pathways (reviewed in [10]), populating among others endoplasmic reticulum (ER) membranes and autophagosomes. In their lipid-conjugated form, Atg8/LC3/GABARAPs become embedded in the autophagosomal inner and outer membranes [11].

Being the core Atg proteins, Atg8s and their conjugation machinery are often mutated in different experimental models to study the consequence of autophagy deficiency. *Atg8/Apg8/Aut7/Cvt5* is a non-essential gene in yeast; however, its deficiency leads to the defect in selective and non-selective forms of autophagy [12]. Prolonged stress conditions, such as nitrogen starvation, lead to decreased fitness of Atg8- (and autophagy-) deficient yeast cells [13]. Similarly, inactivation of two Atg8 homologues in *C. elegans*, *LGG-1* and *LGG-2*, leads to significant life span shortening in this organism [14]. Because of the gene duplication in Metazoans, instead of disrupting Atg8, knockout of Atg7 has been employed to study the physiological role of autophagy. Lack of Atg8 lipidation results in neurodegeneration in *Drosophila* [15] and mice [16]. Later studies, however, showed that Atg8-/Atg5-/Atg7-mediated autophagy is required for homeostasis of a majority of organ systems [17], while its disruption leads to premature aging in mice [18]. In higher plants, Atg8 and the conjugation machinery are required for resilience under adverse conditions [19]. Intriguingly, several forms of non-canonical autophagy have been described that can proceed (albeit inefficiently) without the functional Atg8 conjugation system [20–22].

Atg8/LC3/GABARAPs participate in all steps of the autophagosome biogenesis: phagophore initiation and elongation as well as autophagosome maturation and fusion with the vacuole/lysosome; they also mediate the *selectivity* of the autophagy pathway. *Selective autophagy* is an evolutionary adaptation of the bulk autophagy, aimed at sequestering and degrading specific types of cargo [23,24]. Here, the cargo is recognized by a growing number of *selective autophagy receptors* (SAR), which simultaneously bind the target and the components of the autophagosome, mainly Atg8-family proteins, triggering a chain of events resulting in cargo sequestration in the autophagosomes.

Most SARs display modular structural organization, possessing specific domains/motifs for saturated intermolecular network (reviewed in [5,25]). The size and domain content of different SARs could vary significantly; however, all known SARs seem to contain sites to mediate interactions with Atg8/LC3/GABARAPs. For instance, the SARs of the sequestosome-1-like receptor (SLR) group, such as the founding member p62/SQSTM1, interact with ubiquitin and LC3/GABARAPs, thereby mediating degradation of ubiquitin-decorated cargo, such as protein aggregates [5]. Interactions between Atg8/LC3/GABARAPs and SARs were extensively investigated in the past decade, resulting in the elegant and powerful concept of LC3-Interacting Region (LIR) motif (also known as Atg8-Interacting Motif, AIM, for the fungi and plants) [26–28].

The canonical LIR is a short (up to 10 residues) unstructured or β-stranded region within SARs (and other Atg8-interacting proteins) responsible for the efficient and selective recognition of the autophagosome-embedded Atg8/LC3/GABARAP proteins. In the case of the SARs, this interaction mediates tethering of the cargo-SAR complex to the autophagosome for the subsequent degradation. There are more than 50 structures of Atg8/LC3/GABARAPs proteins with various LIR motifs in the protein data bank (PDB), and the amount of the structural information on the Atg8/LC3/GABARAP interactions grows constantly. Based on this information, computational approaches to predict and validate LIR motifs within specific proteins of interest or within a given proteome were successfully established [29,30]. Additionally, a number of biochemical and biological methods to validate LIR motifs were developed and broadly used in research [31–34].

However, together with the obvious progress in elucidating the interactions between Atg8/LC3/GABARAPs and SARs, recent structural and functional studies revealed a number of fundamental insufficiencies in global understanding of the Atg8/LC3/GABARAPs cellular functions. For example, very little is known about the non-autophagic roles of the Atg8/LC3/GABARAPs. Also, presently it is not clear if the LIR motif is the only structural determinant mediating interaction between Atg8/LC3/GABARAPs and their interaction partners. What determines affinity and selectivity of these interactions and how can one modulate them? Is it possible to rationally modify the LIR motif in order to target a specific member of Atg8/LC3/GABARAP in vivo?

In this review, we reanalyzed available information on the Atg8-family proteins, including their specific differences in sequences, structure and functions, carefully and extensively reviewed in the past years [4,5,11,25,35–38]. However, the main goal of this work is to cast new light on the previously reported features of Atg8/LC3/GABARAP as well as to provide clarity on similarities vs. differences between the individual members of the subfamilies. We review known determinants of molecular recognition demonstrated and/or suggested for Atg8/LC3/GABARAP proteins as they perform their autophagy-related and autophagy-independent functions, but also indicate novel structural motifs that may be implicated in the Atg8/LC3/GABARAP interactions. We describe the current state with respect to the canonical and atypical LIR-motifs, their *N*- and *C*-terminal extensions and provide a structural view on the LIR-independent interactions, including helical binders and novel ubiquitin-interacting motif (UIM)-dependent (UDS)-binders. We also present a current view on the selectivity determinants (within Atg8/LC3/GABARAP proteins and LIR motifs) and expand on the potential use of highly affinitive and highly selective Atg8/LC3/GABARAP binders in research and therapy.

2. Ubiquitin, UBLs, and Atg8-Family Proteins

2.1. Structural Overview

Ubiquitin is a small globular protein of 76 residues, highly conserved from yeast to human. Its tightly folded structure, known as the "β-grasp fold," is characterized by the five-stranded β-sheet wrapped around the central α-helix [39] (Figure 1A). Another evolutionarily conserved feature of ubiquitin is its synthesis as a precursor protein that undergoes proteolytic maturation to expose a *C*-terminal Gly residue, whose carboxyl group can be conjugated to an amino group of a Lys or an *N*-terminal residue of another ubiquitin or a great variety of other proteins. A cascade of enzymatic reactions, involving activating (E1), conjugating (E2), and ligating (E3) enzymes, generates ubiquitin conjugates containing either monoubiquitin or polyubiquitin chains, whose linkage and topology are stipulated by the choice of one of the seven internal ubiquitin Lys residues (Lys6, Lys11, Lys27, Lys29, Lys33, Lys48, and Lys63). Ubiquitin-binding domain (UBD)-containing proteins interact with conjugated ubiquitin in a non-covalent fashion and act as ubiquitin receptors, which can mediate assembly of structural or signaling complexes [40]. Importantly, the ubiquitin signal can be reversed through the action of a large class of deubiquitinating enzymes (DUBs), which are proteases capable of cleaving the isopeptide bond between ubiquitin and its substrate [41].

A superfamily of ubiquitin-like (UBL) modifiers share with its founding member the characteristic fold (but not necessarily the primary sequence) as well as the ability to become covalently conjugated to a substrate, which can be either a protein (in case of e.g., SUMO, Nedd8, UFM1, and Atg12) or a lipid (in case of Atg8/LC3/GABARAP). Like with ubiquitin, activity of dedicated proteases is required to produce a mature UBL as well as to deconjugate it from the substrate for signal termination [40]. Atg12 was the first UBL found in yeast as an essential autophagy protein [42]. Atg12 becomes conjugated to Lys130 of Atg5. Atg5 itself contains two UBL domains [43], so that the Atg12~Atg5 conjugate comprises three UBL moieties, which may be important for the recruitment of other factors required for phagophore elongation. There is growing evidence that Atg5 interacts with selective autophagy receptors (SARs), such as Atg19, optineurin, p62/SQSTM1, and NDP52 [44,45]. This interaction was proposed to be mediated by AIM/LIR and stimulate Atg8 conjugation. On the other hand, clathrin-mediated vesicular trafficking, including clathrin heavy and light chains, and several clathrin adaptors, have been identified in the Atg5/Atg12 interactome [46], and earlier studies identified clathrin heavy chain as a direct interactor for GABARAP [47]. Additional work is required to clarify the exact interaction surfaces between the UBLs in the Atg12-Atg5 conjugate and partner proteins.

Intriguingly, Atg12 can also be conjugated to other proteins [48]. However, the canonical autophagic role of the Atg12~Atg5 conjugate is in the complex with Atg16, which acts as an E3 ligase for Atg8-PE conjugation, as illustrated by experiments with ectopic expression of Atg16 [49].

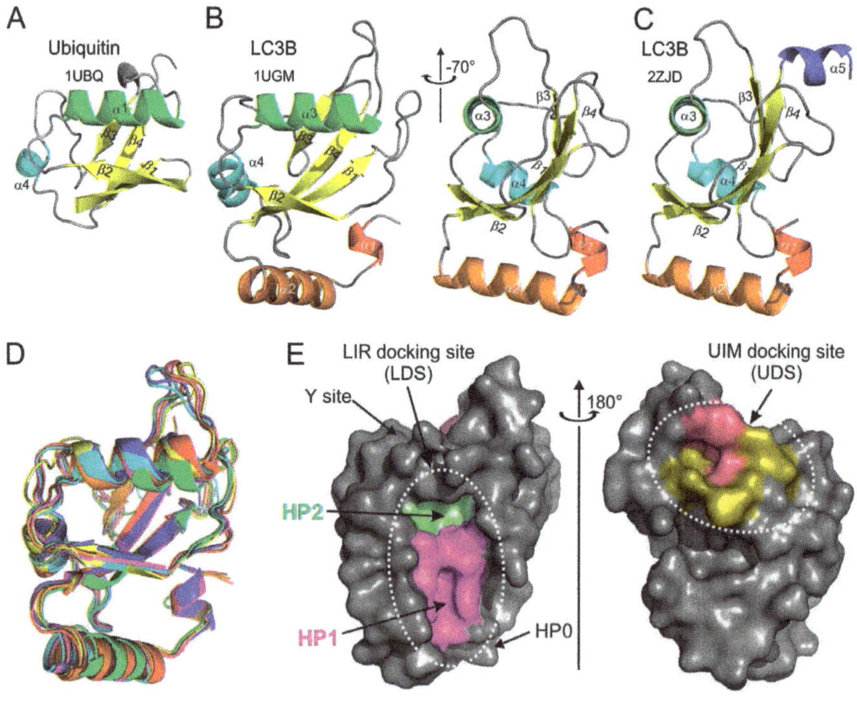

Figure 1. Structural features of ubiquitin and Atg8/LC3/GABARAP proteins. Ribbon diagrams of the (**A**) ubiquitin and (**B**) LC3B structures aligned to the common ubiquitin core (left plot in **B**). The right plot was generated by rotation of the LC3B structure by −70° around the *Y*-axis. The secondary structure

elements in both proteins are colored in rainbow color-code from α-helix α1 in LC3B (red-orange-green-cyan-blue), all β-strands are colored yellow. The PDB ID codes of each structure are presented under the protein names. (**C**) LC3B structure with *C*-terminal α-helix α5 (the same orientation as the right plot in (B)). (**D**) Structural alignment of yeast Atg8 and human LC3A, LC3B, LC3C, GABARAP, GABARAPL1, and GABARAPL2 (rainbow color-code) proteins shown as ribbon diagrams (the same orientation as the left plot in (**B**)). (**E**) Left plot: surface representation of LC3B structure (the same orientation as the left plot in (**B**)), showing the main interacting sites—HP1 (magenta) and HP2 (light green), which form the LC3 docking site (LDS). Position of additional interacting sites, like HP0 [50], Y-site [51], etc., are indicated by arrows. The alternative interacting area, the UIM docking site (UDS), is located on the opposite side of the LC3B molecule (right plot). The most relevant residues are colored dark red, additional hydrophobic residues around it are colored yellow.

The other autophagy-related UBL, Atg8, as well as its mammalian homologs, LC3 and GABARAP proteins, besides structural homology to ubiquitin reveal a significant similarity in their biogenesis. They are synthesized as precursors and undergo processing by Atg4 proteases. Mature Atg8, with the exposed *C*-terminal Gly, is activated by Atg7 (E1 enzyme), transferred to Atg3 (E2), and finally linked to the amino group of PE [52] via the Atg12~Atg5:Atg16 complex (E3). Members of the Atg8/LC3/GABARAP family are the only known UBLs that modify a lipid. Atg8-PE localizes on phagophores. Upon autophagosome maturation, Atg8 is deconjugated from the outer membrane by Atg4—a step important for the phagophore closure [53]; while Atg8 conjugated to the inner membrane of the autophagosome is delivered to the lysosome, where it is degraded together with the autophagosome and its contents. Since Atg8 (and also LC3) is associated with the autophagosome at all times throughout its biogenesis, it is used as a *bona fide* marker for autophagosome formation [1].

Structurally, the Atg8/LC3/GABARAP proteins form a family of GABARAP-like proteins within the "ubiquitin-like" superfamily (structural classification of proteins database, SCOP [54]). Similar to the ubiquitin, Atg8/LC3/GABARAP proteins possess a sheet of mixed parallel/antiparallel β-strands wrapped around a central α-helix and decorated with auxiliary α-, 3.10-helices and loops (Figure 1B).

Upon analysis of the available Atg8/LC3/GABARAP structures, we have noticed that non-processed LC3/GABARAP proteins (LC3A, LC3B, and GABRAP; no structural data is available for the non-processed LC3C; no α5 helix in Atg8) display in some structures an additional α-helix at their C-terminus (Figure 1C). This α-helix α5 is mostly associated with non-processed LC3/GABARAP proteins in complex with various LIR-motifs. It seems that the processed (and lipidated) forms of LC3/GABARAPs are not decorated with this accessorial α-helix; thus, we speculate that in the full length, non-processed forms of LC3 this α-helix could participate in some functional protein–protein interactions inside and outside of the autophagy pathway. However, existence of the helix α5 in different Atg8/LC3/GABARAP proteins, its stability in dependence of various factors, and its functional role(s) have to be thoroughly investigated. Despite low sequence correlations, Atg8, LC3, and GABARAP proteins show very high structural similarity—their structures could be overlaid upon each other with a root-mean-square deviation (RMSD) of 1.2 Å on backbone atoms (Figure 1D).

The main structural difference between Atg8/LC3/GABARAP proteins and other UBLs, which also determines the specific role of Atg8/LC3/GABARAP in autophagy, is the presence of two extra α-helices located *N*-terminally to the ubiquitin core. This *N*-terminal α-helical subdomain significantly varies in the amino-acid content among the different members of the Atg8/LC3/GABARAP family, and structural studies indicate that it displays a dynamic behavior, participating in a conformational exchange [55–57]. Consequently, this structural adaptation of Atg8/LC3/GABARAP is reflected in a set of new functions not observed for other UBLs. For instance, the *N*-terminal α-helices are essential for tubulin binding and oligomerization [55], tethering of lipid bilayers upon autophagosome maturation [58,59], and recognition of mitochondrial phospholipids [60].

Despite their flexibility, *N*-terminal α-helices are specifically aligned to the ubiquitin-like core, forming a deep hydrophobic pocket 1 (HP1, also termed W-site) together with residues of β-strand β2 (Figure 1E, left plot). This pocket binds preferentially indole-based substances, albeit with low affinity [61], and usually accommodates large sidechains of non-polar aromatic residues within the LIRs. HP1 is formed by residues D19, I23, P32, I34, K51, L53, and F108 in LC3B [27]. Another hydrophobic pocket, HP2 (L-site), is built by the solely hydrophobic residues of central α-helix α3 and β-strand β2 (F52, V54, P55, L63, I66, and I67). These two pockets form the so-called *LIR-docking site* (LDS) and mediate a vast majority of known-to-date interactions between SARs, adaptor, and scaffolding proteins with Atg8/LC3/GABARAPs. Of note, LDS occupies the Atg8/LC3/GABARAP surface on an opposite side of the well-known hydrophobic patch (L8-I44-V70) of ubiquitin [62]. On the other hand, the newly described hydrophobic patch on Agt8/LC3/GABARAP surfaces and called the *UIM-docking site* (UDS, Figure 1E, right plot) is similar to the L8-I44-V70 one on ubiquitin and is used by components of the UPS machinery (such as RPN10) and during intracellular trafficking (such as Ataxin-3 and EPS15) [63].

2.2. *Lessons from Alignment of Atg8 Family Members*

Atg8, the single autophagy modifier in yeast, gave the name to Atg8-family proteins. The number of autophagy modifiers in other organisms varies significantly, with strong expansion over the higher metazoans and plants (up to 22 Atg8 family members in some plant species). In humans, there are six Atg8 orthologs: LC3A, LC3B, LC3C (encoded by respective *MAP1LC3*, microtubule associated protein light chain 3, alpha, beta, and gamma genes); and GABARAP (γ-aminobutyric acid receptor-associated proteins), GABARAPL1 (GABARAP-like protein 1) and GABARAPL2 (also known as GATE-16). Despite the fact that LC3C and GABARAPL2 are branched into separate clades on the phylogenetic tree [64,65], human Atg8-family proteins can be broadly categorized into two groups: LC3s and GABARAPs. The sequence alignment (Figure 2) of the Atg8-family members from five different species reveals a clear similarity for the proteins within individual subfamilies. It also indicates that the proteins from the GABARAP subfamily are more evolutionary related to the Atg8 proteins than those of the LC3 subfamily. However, there are substantial differences in the sequences not only between the subfamilies but also between the individual subfamily members. This was proposed to lead to a functional segregation of the LC3 and GABARAP proteins. Indeed, the LC3 and GABARAP proteins were first identified in different compartments of human cells (microtubules for LC3 [66] and synaptic membranes for GABARAP [67]), suggesting different functions for each subfamily. It could subsequently be shown that, upon starvation-induced autophagy, LC3-subfamily proteins are responsible for the elongation of the autophagosomal membranes, while GABARAPs are acting downstream, participating in the maturation and closure steps of the autophagosome formation [68]. Recent studies showed that only GABARAP-subfamily members are important for the activation of the phagophore-priming ULK1-ATG13-FIP200 complex [69,70]. The centriolar satellites protein PCM1 binds unconjugated GABARAP and LC3C proteins to mediate their localization at the pericentriolar material and control autophagic degradation of centriolar satellites and GABARAPs [71,72]. Another example of the selective function of individual LC3/GABARAPs is the recruitment of LC3C to invading bacteria (*Salmonella typhimurium*) via the specific SAR, NDP52, important for autophagy-mediated restriction of the bacterial growth. Depletion of both (NDP52 and LC3C) proteins is followed by an inability of the cell to defend the cytosol against invasion by *S. typhimurium*, while depletion of all other LC3/GABARAPs does not affect it [73]. Investigation of the molecular mechanisms behind these selective functions (e.g., linkage between residues at specific positions within LC3/GABARAP proteins and their functions) is only the beginning.

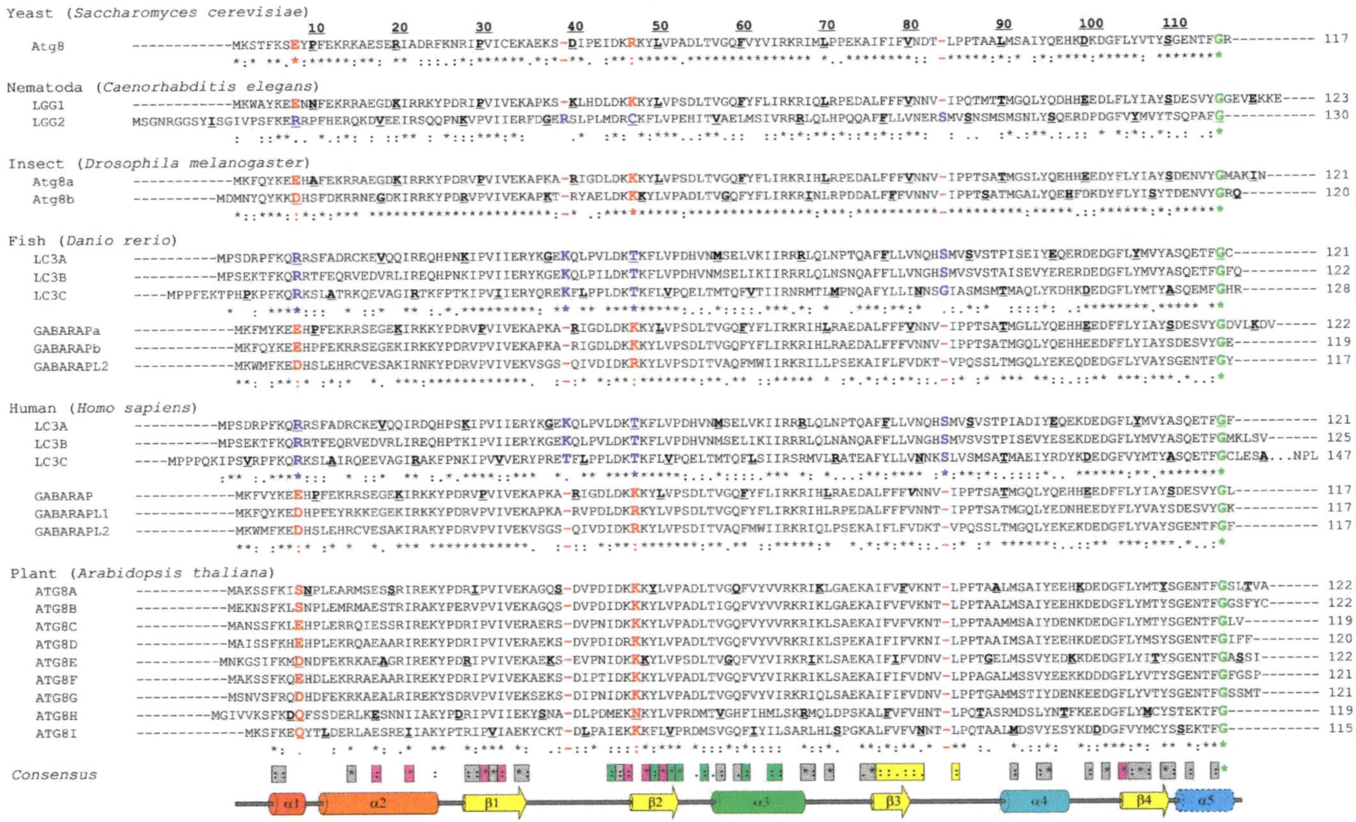

Figure 2. Sequence alignment of Atg8/LC3/GABARAP proteins. Sequence alignment of the Atg8-family members from six model species—yeast (*Saccharomyces cerevisiae*), Nematoda (*Caenorhabditis elegans*), insect (*Drosophila melanogaster*), fish (*Danio rerio*), human (*Homo sapiens*), and plant (*Arabidopsis thaliana*). Secondary structure

Figure 3. elements from the human LC3B (PDB ID 2ZJD) are shown on top (color-code as in Figure 1B,C). Every tenth residue in each sequence is marked bold/underlined, the catalytic Gly is marked green. The identity scores (asterix, * , for identical residues; colon, : , for very similar residues; dot, . , for analogous residues; space, , for residues without any similarity; dash, - , for gaps) are presented below each group of the Atg8/LC3/GABARAP. The residues (or their absence) separating GABARAP/Atg8 and LC3 protein subtypes are marked red and blue, respectively. The consensus string for all 28 proteins is presented at the bottom of alignment. The residues showed conservation are grouped within the following classes: residues participating in the protein folding (grey); residues forming HP1 (magenta); residues forming HP2 (light green); and residues forming UDS (yellow).

In plants, the number of Atg8 orthologs varies from 1 in algae to 22 in angiosperms [74]; however, the diversity of plant Atg8 proteins could be significantly higher because of multiple gene duplications in order to adapt to various adverse conditions where Atg8 proteins play a crucial role [75]. The plant Atg8 proteins also reveal significant selectivity in interaction with their interaction partners, originated from the sequence difference between Atg8 isoforms in different species [76]. Of note, all the key residues, participating in the HP1, HP2 [27], and UDS [63] are conserved, as well as the key lysine residues facilitating LIR binding: K49 and K51 in LC3B (K46 and K49 in GABARAPs). The K49 performs a gatekeeper function, regulating the entrance of the aromatic residues of canonical LIRs into the HP1 [77]. Interestingly, K49A mutation significantly enhances binding of canonical LIRs to LC3B protein [77,78], while K51A decreased or abolished it.

The *N*-terminal α-helices show significantly less conservation, which agrees with the hypothesis that these helices predetermine the selectivity of the interactions between Atg8 proteins and LIR motifs in SARs [59] and thus should be different in amino acid content for each individual family member. The few conserved residues within these α-helices participate either in folding of Atg8 proteins or in the formation of HP1. As expected, the loop regions are significantly less conserved, the relatively long loops L1, L2, and L3 show almost no identical or similar residues. Most conserved are regions of all β-strands, indicating their pivotal role in Atg8-protein folding and in the formation of HP1 and HP2.

The most significant consequence of this alignment, in our view, is the clear separation of all Atg8 proteins in LC3B and Atg8/GABARAP subtypes based on a few positions within their sequence. The first constant difference between LC3B and Atg8/GABARAP subtypes of proteins is the switch between the intramolecular electrostatic contacts for residues at positions 8 and 47 in Atg8/GABARAP (positions 10 and 50 in LC3B, respectively). For all Atg8/GABARAPs, position 8 is occupied by a negatively charged or polar residues which are able to play a role of hydrogen bond acceptors (Glu, Asp, Gln, Asn, Ser, and Thr), while position 47 is permanently used for electropositive residues (Lys and Arg), serving as the hydrogen bond donors. In contrast, in LC3 subtype, there are electropositive residues (donors) at position 10 and electronegative residues (acceptors) at position 50. These residues come closely to each other and form intramolecular hydrogen bonds or salt bridges to stabilize the Atg8/LC3/GABARAP structure and ensure a proper orientation of the *N*-terminal α-helical subdomain (Figure 4A). Importantly, these residues also form intermolecular contacts to residues in LIR motifs and, therefore, also contribute to the selectivity of Atg8 interactions with other proteins. It was shown recently that the T50 in LC3B is phosphorylated by a number of kinases [79,80]. This phosphorylation is necessary for normal autophagosome-lysosome fusion and clearance of invading bacteria [80]; inhibiting, however, the autophagic degradation of p62/SQSTM1 [79]. Both studies emphasize significant functional differences as a consequence of the different content at positions 47 (in GABARAPs) and 50/56 (in LC3A, B/LC3C). Another difference is constantly shorter long loops in Atg8 and GABARAPs. The loops between β-strands β1 and β2 (L1) and between β-strand β3 and α-helix α4 (L3) display no conservation; however, they undergo significant dynamic modulations in the free and LIR-bound forms of Atg8/LC3/GABARAPs, as was observed by NMR experiments [34,81]. Therefore, lack of one residue could in principle affect their dynamics and thus modulate selectivity to a specific LIR [50].

Besides these two constant differences, reflecting global separation of LC3 and GABARAP/Atg8 subtypes, there are a several organism-specific hotspots, regulating the difference in the recognition of LIR motifs by the subfamily members. For example, Y25 (invariant in all GABARAPs) participates frequently in the formation of intermolecular hydrogen bonds with positively charged or polar residues at position X_2 of LIR motifs. The favorable conformation of Y25 is stabilized via cation-π interactions (reviewed in [82]) with a guanidinium moiety of invariant R28 (Figure 4B). The distinct conformation of Y25 and R28 making the intermolecular hydrogen bonds more energetically favorable and thus increasing the affinity of the GABARAP: LIR binding. In LC3, there are H27/H27/F34 (LC3A/LC3B/LC3C) at position Y25, and, therefore, hydrogen bonds could not be formed in the same way, as well as the conformation of aromatic rings could not be stabilized by cation-π interactions with K30/K30/K36. Thus, this set of residues in GABARAPs predetermines the selectivity of LIRs in KBTBD6/7 [83] and PCM1 [71,84] binding to GABARAP proteins. R28 itself forms a hydrogen bond with the negatively charged or polar residues at LIR position X_3, and cation-π interactions also play a favorable role for the elevated affinity to GABARAPs [34].

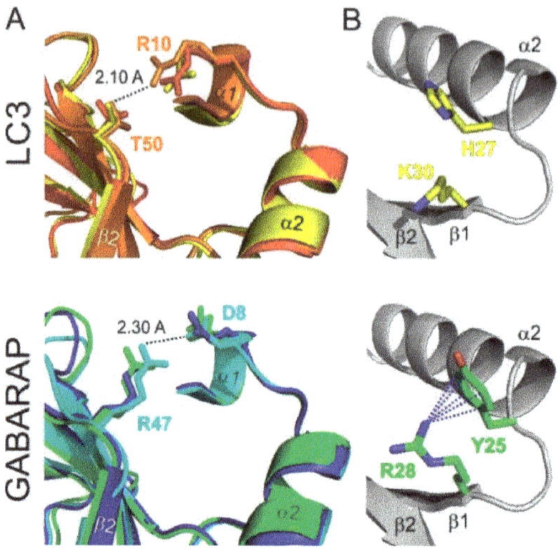

Figure 4. Structural differences between the LC3 and GABARAP proteins. (**A**) Specific structural difference between GABARAP/Atg8 and LC3 protein subtypes. Intramolecular contacts within LC3 (top) and GABARAP (bottom) proteins; color code as in Figure 1D. Involved residues are presented as sticks; the shortest distance is given for specific protein residues (indicated at the plot). (**B**) Orientation of H27 and K30 sidechains in LC3B (top) and corresponding Y25 and R28 sidechains in GABARAPs (bottom). Cation-π interactions (the non-covalent electrostatic interaction between an electron-reach face of aromatic rings and adjacent cations), stabilizing the specific orientation of Y25/R28 sidechains in GABARAPs are shown as dashed lines.

Also, post-translational modifications (PTMs) of Atg8/LC3/GABARAPs proteins may affect the affinity and selectivity of LIR:Atg8/LC3/GABARAP interactions. Phosphorylation of Atg8/LC3/GABARAPs at accessible S/T residues has been shown to regulate the selective autophagy [79,80,85–87]. Enzymatic acetylation/deacetylation of K49 and K51 regulates LC3 nuclear-cytoplasmic shuttling [88] but might equally regulate the affinity of the binding between LC3 and the LIRs within SARs, as these residues are key elements of the LIR:Atg8/LC3/GABARAP interface. Ubiquitination is another type of PTM modulating autophagy in an LC3- or GABARAP-selective manner. Monoubiquitination of LC3 (and not GABARAPs) at K51 driven by the coordinated action of ubiquitin-activating enzyme UBA6 and the hybrid ubiquitin-conjugating enzyme/ubiquitin ligase BIRC6 targets LC3 for the proteasomal degradation and negatively regulates autophagy [89]. Mib1-driven mono- and polyubiquitination of the GABARAP (not LC3) on K13 and K23 within *N*-terminal α-helical subdomain occurs through K48-chains and targets GABARAPs to proteasomal degradation [72].

3. Interactions between Atg8/LC3/GABARAP Proteins and Their Binding Partners

3.1. The LIR Concept

The LIR in mammals and the AIM in yeast were described in pioneering biochemical [28] and structural works [26,27] as short polypeptide sequences containing ~10–20 residues. Early structural studies revealed that the core LIR/AIM (LIR thereafter) sequence contains a W-X-X-L motif (where X is any residue). The LIR polypeptide of p62/SQSTM1, for instance, adopts a β-stranded conformation, forming an intermolecular parallel β-sheet with the β-strand β2 of LC3B, while the sidechains of W and L residues occupy the HP1 and HP2 of LC3B, respectively, stabilizing the complex (Figure 5A). Extensive studies in past years provided a more general core consensus, which can be described as Θ-X-X-Γ, where Θ is an aromatic (W/F/Y) and Γ is a hydrophobic (L/I/V) residue (Figure 5B). Investigations of the residues which could occupy the Θ and Γ positions (ether by analyzing the sequences of the hitherto known canonical LIR motifs [25] or by mutational 2D peptide arrays [32,84,90,91] revealed a very high conservation of the three aromatic residues in Θ. As expected from the hydrophobicity profile of HP1, a much higher abundance of solely non-polar Trp and Phe was observed in the native canonical LIRs. In contrast, partially polar Tyr residues are found in only a minority of canonical LIR motifs. It seems that Trp is the most energetically favorable residue for the Θ position. Mutation of the Tyr732 to a Trp increases the NBR1 LIR affinity to GABARAPL1 eight-folds, while the Y732F mutant showed the same affinity [92]. The Phe-containing OPTN LIR shows an eight-fold increase in affinity to LC3B when Phe178 is substituted to a Trp [81]. Of note, the lower affinity of Tyr- and Phe-containing canonical LIR motifs in both aforementioned cases might be associated with the ability of NBR1 and OPTN to regulate autophagic functions [81,92]. The Γ position is a bit less conserved and tolerates large hydrophobic residues, including canonical L/V/I and aromatic residues, except His. Apparently, smaller hydrophobic residues, such as Ala, Pro or Met, do not have enough volume to fill the HP2, while aromatic residues are too big to be docked.

A track of negatively charged residues (Glu/Asp) prior to the Θ enhances the affinity of the LIR interactions with Atg8/LC3/GABARAP. Phosphorylation of residues within the *N*-terminal flanking region of the core LIR (especially directly prior to the aromatic residue) may enhance the affinity of the SAR: Atg8/LC3/GABARAP binding [78,81,93–96] and serves as a key autophagy regulator in corresponding types of selective autophagy. Phosphorylation of the more distant residues preceding Θ also increases the affinity, however, not so strongly as at positions −3, −2, and −1. In optineurin's LIR, phosphorylation of individual Ser residues up to position −9 still increases its affinity to LC3B [81]. IKKa-mediated phosphorylation of AMBRA1 S1014 at position −6 promotes AMBRA1's binding to LC3 and GABARAP (in vitro and in vivo) and serves as a positive regulator of AMBRA1-mediated mitophagy [93]. In some cases, direct phosphorylation of the Tyr residue at the LIR Θ position leads to a weakening of the LIR:LC3/GABARAP binding affinity [94]. Additionally, phosphomimetic mutations in the LIR proximity, for a large number of investigated proteins, increase their affinity to Atg8/LC3/GABARAPs. Interestingly, phosphorylation/phosphomimicking differently affects LIR-dependent interactions with individual members of LC3 and GABARAP subfamilies, indicating that this PTM affects not only the affinity but also the selectivity of the interactions. For example, Beclin1 S93E and S96E mutations enhance binding of Beclin1 to GABARAP and GABARAPL1 three-fold, to LC3A five-fold, and to LC3C eight-fold [91]. Mutational analysis has further shown that certain LIR core sequences possess an increased affinity to the LC3 vs. GABARAP subfamilies of UBLs. This allowed to define a broad consensus for the GABARAP-interacting motif (GIM), conforming to the core sequence (W/F)-(V/I)-X-V [34].

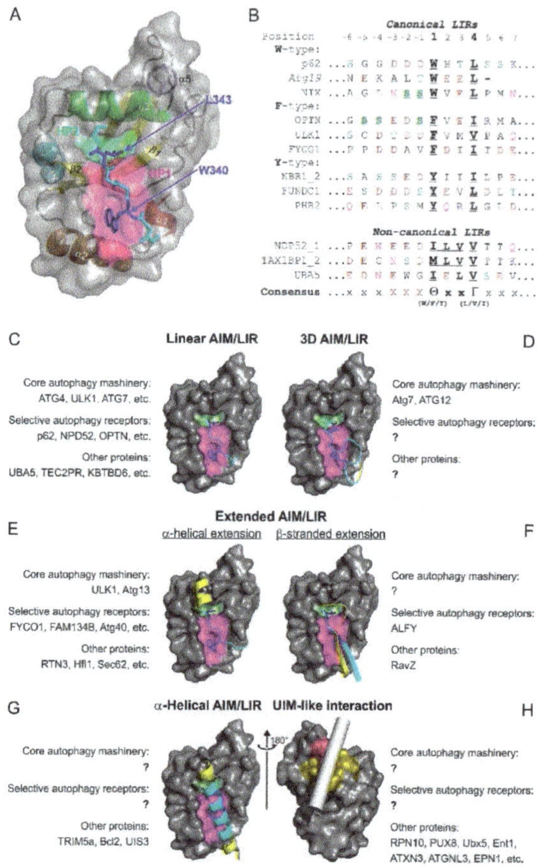

Figure 5. Atg8/LC3/GABARAP interactions with their partners. (**A**,**B**) The LIR concept. Structure of p62/SQSTM1-LIR:LC3B complex (PDB ID 2ZJD) (**A**). LC3B is shown as a semi-transparent surface with the structural elements (α-helices and β-strands) visible. p62/SQSTM1-LIR is shown as a main chain (cyan) with sidechains of core LIR residues (W340 and L343, blue) as sticks. Two hydrophobic pockets of LC3B, accommodating W340 and L343 sidechains, are shown on LC3B surface (HP1—magenta, HP2—light green). (**B**) Alignment of canonical (upper section) and non-canonical (lower section) LIR motifs with positions of residues indicated on top (from −6 to +7). Negatively charged residues (red), polar residues (orange), and phosphorylatable residues (green) are indicated over the LIR sequences. The phosphorylatable residues confirmed to be phosphorylated are marked bold/underlined. The underlined characters within core LIR sequences indicate residues whose sidechains are accommodated by HP1 and HP2. (**C**–**H**) Types of Atg8/LC3/GABARAP-interacting motifs and elements are shown as examples of known structures. For all plots, LC3B surface in orientations as in Figure 1E is shown. The interacting elements are given as ribbon diagrams for the known structures; in the case of Ubx5, the putative position of the helical ubiquitin-interacting motif (UIM) is indicated by a gray cylinder. Within the interacting elements, residues contributed with sidechains into HP1 and HP2 are shown in sticks and colored blue, residues within the close LDS contacts are colored cyan, and residues with other contacts are yellow. For each type, the names of the known interactors are indicated, as well as their functional roles sorted in groups of "core autophagy machinery," "selective autophagy receptors," and "other proteins."

3.2. Affinity and Selectivity of Interactions between Atg8/LC3/GABARAPs and LIR Motifs

Affinity (expressed in K_D values) of interactions involving Atg8/LC3/GABARAPs is not so high in comparison with affinities of the strongest biointeractions (K_D~10^{-6} nM for streptavidin:biotin binding [97], up to 10^{-3} nM for antigen:antibody [98]). The K_D values for interactions between the canonical p62/SQSTM1 LIR motif and all six human Atg8 proteins were measured by the isothermal titration calorimetry (ITC) experiments and are around 1 μM without any selectivity [92]. Still, these values are >100 times lower in comparison to the interactions between ubiquitin and UBDs (~100 μM, reviewed in [99]). Selectivity, however, cannot be expressed in terms of only one interaction. Generally, selectivity is a comparison between K_D values of the two (or more) binding processes. The higher the difference between two K_D values, the more selective is the binding. In case of interactions between Atg8/LC3/GABARAP proteins with their partners, we assume that affinity was calculated as K_D values for at least one representative member of LC3 and one representative member of GABARAP proteins under the same conditions and by the same method. This may allow us to compare the K_D values for them and state if selective interactions take place or not. Table 1 summarizes the most exemplary K_D values reported for the interactions between Atg8/LC3/GABARAP proteins and corresponding binding partners.

Table 1. K_D values measured for interactions between different LIRs and LC3/GABARAP proteins. All K_D values are given in μM.

Protein	LC3A	LC3B	LC3C	GABARAP	GABARAPL1	GABARAPL2	Method	Ref.
AnkG	0.55	0.34	2.39	2.6×10^{-3}	3.7×10^{-3}	40×10^{-3}	ITC	[100]
AnkB	3.7×10^{-3}	4.2×10^{-3}	10.5×10^{-3}	0.27×10^{-3}	0.29×10^{-3}	0.21×10^{-3}	ITC	[100]
PLEKHM1	4.22	6.33	3.45	0.55	0.77	0.93	ITC	[34]
PCM1	292	982	17.9	2.0	1.6	14.4	BLI	[84]
ULK1	5.9	48.2	2.5	50×10^{-3}	48×10^{-3}	0.53	BLI	[84]
ATG13	4.1	9.6	0.48	0.59	0.53	3.1	BLI	[84]
FIP200	281	1206	63.3	5.6	7.0	86.4	BLI	[84]
p62/SQSTM1	2.0	4.5	2.7	0.9	0.6	5.2	BLI	[84]
Ambra1	>100	>100	>100	39	>50	>100	ITC	[93]
pS1014Ambra1	(50)	(50)	(>100)	(21)	(25)	(>100)		

ITC—isothermal titration calorimetry, BLI—bio-layer interferometry.

3.3. Atg8/LC3/GABARAP Interactions: LIR and Beyond

During the last 5 years, extensive studies aimed at linking specific proteins with the autophagy pathway were undertaken. As a result of these efforts, ample knowledge on interactions between Atg8/LC3/GABARAPs and various proteins (e.g., SARs, adaptors, targets, and scaffolding factors) at thermodynamic, structural, and functional levels has accumulated and been rationally analyzed. While a majority of the interactions confirmed and refined the concept of the canonical LIR motif, a number of unusual interaction mechanisms were identified outside the LIR concept. Their existence hints at the fact that Atg8/LC3/GABARAPs could provide a significantly broader platform for binding and cellular functions.

Several major types of LIR motifs are used in functionally relevant interactions (Figure 5C–H):

3.3.1. Linear LIR-Like Motifs

The canonical LIR, as well as its variations with several types of the so-called non-canonical sequences, form the most populated group of linear LIR (Figure 5C). They are linear polypeptides from mostly unstructured protein regions with a tendency to adopt a β-stranded conformation. Discovery of a new canonical LIR is facilitated by number of programs such as iLIR [29,30] and hfAIM [101] which analyze the core 4 residue sequence Θ-X-X-Γ as well as it *N*- and *C*-terminal extensions. Unfortunately, the application of the programs is still limited to canonical LIR sequences and does not predict the non-canonical LIRs. Sequences of non-canonical LIRs are characterized by a lack of conserved aromatic residues Θ (as in the cLIR of NDP52 [73]) or an unusual spacing of the aromatic and aliphatic residues, such as that in the combined LIR/UFIM (LIR and UFM1-interacting motif) and its variations, which binds both LC3/GABARAP and UFM1 protein [50,102]. In case of cLIR from the major SAR for mitophagy and xenophagy, NDP52, with core sequence ^{133}I-L-V-V^{136} (Figure 5B), HP1 is not engaged by an aromatic sidechain but is rather conformationally adopted to form hydrophobic contacts with the I and L of the cLIR. This interaction is selective for the LC3C protein; the change of I to W increases its affinity but also results in the loss of selectivity [73]. Similar adaptation mechanism is observed for the LIR of the NDP52 homolog, TAX1BP1, which has the core sequence ^{140}M-L-V-V^{143} and mediates interactions of this SAR with both LC3 and GABARAP subfamilies [103].

The LIR/UFIM of E1 enzyme for the UBL UFM1, UBA5, with the longer core sequence ^{341}W-G-I-E-L-V^{346} exploits hydrophobic contacts of I, L, and V to occupy both HP1 and the HP2 similar to the NDP52 cLIR. However, the evolutionary conserved W engages in a new hydrophobic pocket, termed HP0, which is induced by conformational changes within GABARAPs upon the occupation of HP1 and HP2, making the affinity of the LIR/UFIM:GABARAP interactions comparable with the usual affinity for LIRs (~1 μM range). A set of artificial peptides, rationally derived from UBA5 LIR/UFIM sequence, showed a comparable or slightly higher affinity to LC3B and GABARAPL2 proteins, indicating a high potential of this non-canonical LIR sequence to mediate functionally relevant Atg8/LC3/GABARAP interactions [50].

3.3.2. Three-Dimensional (3D) Interacting Regions (3D LIRs)

Besides linear LIRs, the so-called 3D LIR motifs (some researchers suggest *through-space* or *interspaced LIR* terms) have been discovered (Figure 5D). They mimic the structural arrangement of a canonical LIR by presenting aromatic, aliphatic and negatively charged residues distant in the sequence but kept together in space by the protein's 3D structure. One such example is the 3D LIR in Atg12 [104], deduced from the structural correspondence of surface-located W185 and I111 in the yeast Atg12~Atg5 complex to W and L in the canonical Atg32 LIR. Mutational analysis confirmed that Atg12 residues W185 and I111 are indispensable for Atg8 membrane recruitment. A 3D-type interaction is also observed between the short *C*-terminal tail of Atg7 (residues 611-630) and Atg8. Spaced in sequence, F619 and I629 occupy in the 3D structure positions of Θ and Γ of the core LIR, and interact with HP1 and HP2 of Atg8, respectively. The negatively charged residues D617, D624, E625, and E628 form intermolecular ionic contacts to the usual set of R and K residues on Atg8 surface, strengthening the interaction [105].

3.3.3. Extended LIR Motifs: *C*-Terminal α-Helical Extensions

N- and *C*-terminal extensions of the canonical LIR motifs could dramatically increase affinity of their interactions with the Atg8/LC3/GABARAPs and provide a gain in selectivity. These extensions build additional intermolecular contacts outside the LDS and therefore significantly stabilize the resulting complex. A set of functionally relevant *C*-terminally extended LIRs employ an α-helical structure (Figure 5E), which stabilizes the complex via high-energy intermolecular hydrogen bonds between polar or negatively charged residues within the α-helical extension and the invariant R70 in LC3B (R67 in GABARAPs) [84,106–109]. This considerable gain in affinity is for example instrumental in selective autophagy of the ER (ERphagy) mediated in *Saccharomyces cerevisiae* by the SAR Atg40. To fragment ER membrane and load into the autophagosome, multiple membrane-resident Atg40 proteins simultaneously engage a number of Atg8 proteins, which they bind with their AIMs tightly, owing to the short helix *C*-terminal to AIMs. Importantly, this feature is conserved all the way to the mammalian ERphagy-specific SARs [110]. Increase in the affinity in some cases can reach >1000 fold. For example, α-helically extended LIR in ankyrins AnkB shows an affinity to GABARAPs in sub-nanomolar ranges (K_D values 0.21–0.29 nM) [100]. The physiological importance of the GABARAP:AnkG interaction is underscored by the study in which a LIR-deficient form of AnkG (W1989R), which fails to bind to GABARAP, leads to a massive reduction in forebrain GABAergic synapses associated with hyperexcitability of pyramidal cell and disruptions in the synchronization of neuronal networks [111].

Interestingly, LC3 proteins display a lower affinity for this LIR (K_D 3.2–10.5 nM), highlighting the functional specialization within the LC3/GABARAP family. The extended LIR in the mammalian ERphagy SAR, FAM134B, reveals an elevated affinity (K_D~25 nM), while the same LIR without a C-terminal α-helical extension shows usual affinity in sub-micromolar range (K_D~700 nM) [100].

C-terminally α-helical-extended LIRs are not the only type of the extended LIR motifs. The ATG13, PCM1, and ULK1 use linear polypeptide as the C-terminal extension of the LIR motif for selective GABARAP binding. They exploit non-polar extensions of HP2 over the central helix α3 in GABARAPs, which is more polar and even charged in LC3 as well as described above intermolecular hydrogen bonds to GABARAP R28 stabilized by cation-π interactions.

3.3.4. Extended LIR Motifs: *N*- and *C*-Terminal β-Stranded Extensions

Another type of *C*- and *N*-terminal LIR extensions, affecting affinity and selectivity of the LIR: Atg8/LC3/GABARAP interactions, are β-strands, which are separated by a loop and form additional β-strands layer in the intermolecular β-sheet (Figure 5F). The *Legionella pneumophila* effector protein RavZ, inhibiting xenophagy, has a tandem LIR motif at its *N*-terminal region (NLIR). Both parts of this tandem LIR have a canonical sequence (^{16}F-E-E-L^{19} and ^{29}F-D-L-L^{32}) and negatively charged residues prior to both F16 and F32; however, only the first LIR in the tandem provides sidechains of F16 and L19 to engage HP1 and HP2. The second part contributes in stabilization of LC3/GABARAP: NLIR complexes with intermolecular ionic interactions mediated by D30, and with a number of hydrophobic contacts mediated by L31 and L32. In this case, H27 in LC3A and LC3B and invariant Y25 in GABARAPs are the key residue in selectivity of interactions [112].

An *N*-terminal β-stranded extension was observed in the LIR of the autophagy-linked FYVE protein (ALFY), which interacts selectively with GABARAP proteins and LC3C, contributing to autophagic clearance of aggregated proteins. ALFY has the canonical LIR motif (^{3346}F-I-F-V^{3349}), whose binding to GABARAPs is stabilized by additional intermolecular hydrogen bonds between ALFY's residues E3342, D3344, and Y3351, and GABARAP residues Y25, K46, and D54, respectively. E3342 and D3344 are fixed in a favorable conformation within the additional β-strand preceding a canonical LIR motif. Invariant GABARAP residues K24/Y25/D54 are the selectivity determinants of binding. These are similar to residues K32/F33/E63 in LC3C and different to Q26/H27/H57 in LC3A/LC3B proteins, and mutations Q26K/H27Y/H57D in LC3B drastically enhance its interaction with ALFY [113].

3.3.5. α-Helical Interacting Regions

α-Helical structures could be implemented by proteins to bind Atg8/LC3/GABARAPs (Figure 5G). The α-helical coiled-coil segment of the retroviral restriction factor TRIM5α mediates a direct interaction between TRIM5α and LC3B, allowing simultaneous TRIM5α dimerization and interaction with two LC3B moieties. Besides TRIM5α's W196 and Q203, whose sidechains are located in the close proximity to the HP1 and HP2, respectively, there are additional close polar and hydrophobic contacts over the LDS stabilizing the complex. However, absence of high-energy intermolecular hydrogen bonds, which appear upon the formation of the intermolecular β-sheet make affinity of TRIM5α:LC3B interaction very low (K_D~100 μM) [114]. A similar mechanism was shown for the interaction between GABARAP and the anti-apoptotic protein Bcl-2 [115]. In the GABARAP:Bcl-2 complex structure (modelled on the NMR-titration experiment, α-helical segment of Bcl-2 (residues 10-34) covers the LDS surface of GABARAP, accommodating Bcl-2 W30 into the HP1. In addition, a recent study indicates that the *Plasmodium berghei* transmembrane protein UIS3, which blocks the processive LC3/GABARAP interactions with autophagic machinery and facilitates *Plasmodium*'s escape from the autophagic degradation cycle, also interacts with the LDS of LC3/GABARAP in an α-helical mode with K_D of 0.24 μM. However, UIS3 may use another pattern of residues for this interaction [116].

3.3.6. UIM-Like Interacting Regions

A conceptually new mechanism of protein–protein interactions for Atg8/LC3/GABARAP was discovered very recently [63]. It was shown that Atg8/LC3/GABARAPs possess the evolutionary conserved hydrophobic patch (referred to as UDS) around F79 and L81 in LC3A/B (I77 and V79 in GABARAPS), capable of accommodating amphiphilic α-helical UIM domains with an affinity in the lower micromolar range (Figure 5H). It was shown that 19 UIM-containing proteins could interact with Atg8s in *A. thaliana* in a UDS-dependent manner. In human, 6 out of 28 tested UIM-containing proteins (EPN1, EPN2, EPN3, Rabenosyn, ATXN3, and ATXN3L) interacted with both LC3 and GABARAP subfamilies via UDS binding with a different degree of selectivity [63]. The discovery of the alternative binding site on Atg8/LC3/GABARAP surface will lead to the identification of new selective autophagy receptors, adaptors and scaffolding proteins in the near future. Besides UIMs, there are more α-helical substructures which could interact with the human LC3 and GABARAP proteins in UDS-dependent way, and the future investigations should be focused on the identification of a complete set of UDS-dependent interactors and on a more detailed structural characterization of this binding type.

4. Emerging Atg8/LC3/GABARAP Interaction Motifs and Elements

It was reported that the LC3/GABARAP interactome in human contains ~400 potential candidates under basal autophagy conditions [117]. Only a small number of these proteins were validated and characterized as LC3/GABARAP binders, while validation of the rest and/or discovery of new candidates is complicated by the fact that researchers are looking for more conventional and better characterized canonical LIR motifs as the interaction determinant. This strategy, however, will not work in the light of the latest finding described above. The progress in identification of the different mechanisms for interactions of Atg8/LC3/GABARAPs allows us to suggest a number of possible, still not identified, structural motifs which could be implicated in these interactions and may serve as a starting point for new investigations.

4.1. The Anti-Parallel Intermolecular β-Sheet

For all the canonical and non-canonical LIR sequences identified to date, the orientation of the extended β-stranded conformation of the LIR peptide is parallel to the β-strand β2 in Atg8/LC3/GABARAP. The only exception was reported so far is the structure of an artificial peptide K1 in complex with GABARAP [118]. This peptide adopts a complex conformation with N- and C-termini representing an extended β-strand, connected with a 3.10 helix in the middle. The HP1 and HP2 of GABARAP are both occupied by Trp (W11 and W6, respectively), making the situation more "non-canonical." Nevertheless, one can predict that antiparallel β-stranded linear peptides (Figure 6A) with a reverse order of residues for Θ and Γ positions and with a corresponding C-terminal track of negatively charged residues after the core aromatic residue (Γ-X-X-Θ-X^--X^--X^- instead of X^--X^--X^--Θ-X-X-Γ) could efficiently bind Atg8/LC3/GABARAP proteins. Considering another non-canonical linear LIR sequences (cLIR in NDP52 and LIR/UFIM in UBA5), one can predict existence of a very high amount of LIR-like sequences representing this category. The attempt to generate (by a phage display) highly affinitive and highly selective synthetic peptides capable of binding individual members of LC3 and GABARAP subfamilies in human cells led to the generation of a number of sensor molecules; however, they all seem to contain canonical LIR motifs [119].

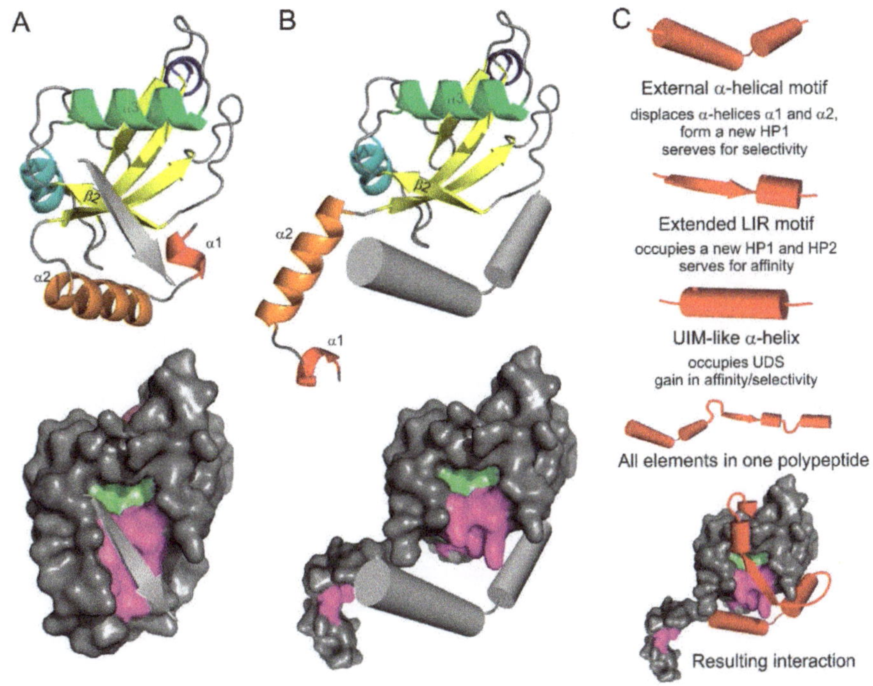

Figure 6. Emerging types of Atg8/LC3/GABARAP interacting motifs and elements: the antiparallel β-strand (**A**), and displacing α-helical structure (**B**). The interacting elements shown as grey arrows or cylinders on LC3B ribbon diagram (top) and on LC3B surface (bottom, with HP1 and HP2 indicated). (**C**) Superbinder construction. Combination of the known/hypothetical binding elements (shown as red cylinders/arrows for each type), complete polypeptide with all the elements, and the resulting position of the "superbinder" on the surface of LC3B.

4.2. *N-Terminal α-Helical Subdomain Displacement*

The *N*-terminal α-helical subdomain in Atg8/LC3/GABARAP proteins is a key evolutionary addition to the core ubiquitin-like fold to separate structurally and functionally the autophagy modifiers from any other UBLs. The α-helices show a significant conformational exchange [55,81,120] and could potentially be separated from the ubiquitin core as the truncated LC3B and GABARAPL2 proteins were still able to perform some functions, like membrane fusion [59]. The first α-helix in LC3B and GABARAPL2 was successfully swapped to emphasize their role in p62/SQSTM1 recognition [121]. Moreover, it was shown that GABARAPL1 being truncated *N*-terminally for the α-helical subdomain and the β-strand β1 could still recognize and bind a number of cognate receptors and proteins (γ2 subunit of $GABA_A$ receptor [67], human κ opioid receptor [122], gephyrin [123]).

Based on these facts one can predict that the α-helical subdomain can be displaced from the ubiquitin core of Atg8/LC3/GABARAP proteins by another α-helical structure (Figure 6B) containing a combination of residues, which are more favorable for the binding of the ubiquitin core of a particular Atg8/LC3/GABARAP protein. More aggressive conditions, which appear in close proximity to membranous structures or in cellular compartments with critical pH values, might facilitate the displacement. In this case, the amino acid content of the displacing α-helices could significantly differ from that for displaced helices α1 and α2, leading to the HP1 modulation in shape and dynamics. That could promote an effective binding of alternative LIR-like sequences from proteins which solely interact with intact Atg8/LC3/GABARAPs.

4.3. LIR-Based Atg8/LC3/GABARAP Superbinders or Combinatorial Binder

Interesting applications might be seen from the combination of all possible binding mechanisms in a single polypeptide. Taking into account that the displacing α-helical structure might serve for enhanced selectivity of the polypeptide (according to the role of displaced helices α1 and α2) and that affinity of binding is controlled by canonical or non-canonical LIR core with a C-terminal α-helical extension (like in AnkB/AnkG [100]) with a following UIM, one could design a polypeptide which would bind selectively a single member of Atg8/LC3/GABARAP family (Figure 6C). Such a polypeptide will block all the binding sites on Atg8/LC3/GABARAP surface, thus inactivating a specific autophagy modifier in cells. Coupled to a labelling group (GFP, fluorescein, Alexa, etc.), the polypeptide could reveal cellular distribution of a specific autophagy modifier.

5. Therapeutic Exploitation of the LIR: Atg8/LC3/GABARAP Interactions

With the growing appreciation of the roles that Atg8/LC3/GABARAP proteins play in the cell, most significantly in various selective autophagy pathways, therapeutic exploitation of these lipid modifiers rapidly comes into focus. It is conceivable that engineering strong and selective binders to individual members of the Atg8/LC3/GABARAP family, on the one hand, and pathological cargo proteins, on the other hand, will enable targeted degradation of complex substrates by the autophagy pathway. This especially may be applicable for multiple neurodegenerative diseases marked by accumulation of pathogenic protein aggregates, such as those containing huntingtin (characteristic of Huntington's disease), β-amyloid peptide (Alzheimer's disease), or α-synuclein (Parkinson's disease) [124]. Such SAR mimetics would have to penetrate the blood-brain barrier and access sick neurons in order to bring about their therapeutic effect. Increased affinity of a peptide or small molecule for LC3/GABARAP could be gained by applying the principles identified in the aforementioned structural studies, e.g., of very strong LIR-based binders [100] or from the emerging combined "superbinders" (Figure 6C). A preclinical proof of principle for such compounds was recently demonstrated for dual-specificity small molecules that were selected to bind mutant huntingtin and LC3B simultaneously. Excitingly, these prototypic selective autophagy inducers could lower levels of mutant huntingtin aggregates in fly and mouse neurons in vivo, showing the potential of this approach [125]. A similar approach may be used to fight infectious diseases, i.e., via enhanced targeting of intracellular pathogens to autophagosomes one could achieve both clearance of the pathogen and its improved presentation to the immune system. Thus, the linear fusion of *Mycobacterium tuberculosis* antigen LpqH in line with LC3B led to enhanced delivery of the protein to autophagosomes and lowered mycobacterial load in immunized mice challenged with a virulent form of *M. tuberculosis* [126].

Inhibition of autophagy is of interest in other disease settings, such as in cancer, where autophagy has been described to provide growth advantage to Ras-driven tumors [127]. Here, design of strong LIR-based Atg8-protein binders may compete for SARs to reduce clearance of different cargo: e.g., accumulation of damaged mitochondria by blocking NDP52- and OPTN-mediated mitophagy may lead to increased apoptosis in cancer cells, in which this suicide pathway is still active. Intriguingly, it was shown that autophagy inhibition also led to the stabilization of MHC-I antigen complexes and enhanced anti-tumor immunity in combination with immune checkpoint inhibitors. A major ubiquitin-binding SAR NBR1 was recently implicated in this process [128]. Autophagy inhibition should be done with significant cautions, as mice in which autophagy was inhibited in an inducible manner (*Atg7* knockout) developed multiple abnormalities involving brain, liver, and muscle and succumbed to infectious diseases within 2–3 months of the onset of the autophagy block [17].

Better understanding of what drives selectivity of LIR peptides to individual members of LC3/GABARAP family can help design more specific treatments in the future. Several groups performed rational design or screening to develop peptides with desirable specificities. These could be used as fluorescent sensors for detecting LC3/GABARAPs in real time, allowing live imaging of autophagosome formation [129] as well as different selective autophagy processes, such as mitophagy and xenophagy [119]. A similar approach can be adopted for screening for peptides with super high affinity toward LC3/GABARAP proteins for potential use in therapeutic applications. Structural insights into the complexes between high-affinity peptide interactors and the LC3/GABARAP core will eventually help develop peptidomimetic compounds with superior pharmacological properties and amenability to oral administration for use in patients with neurodegenerative and infectious diseases as well as cancer.

Funding: V.V.R. received funding from the SFB 1177 "Molecular and Functional Characterization of Selective Autophagy," and from the Innovative Medicines Initiative 2 Joint Undertaking (JU) under grant agreement No 875510. The JU receives support from the European Union's Horizon 2020 research and innovation programme and EFPIA and Ontario Institute for Cancer Research, Royal Institution for the Advancement of Learning McGill University, Kungliga Tekniska Hoegskolan, Diamond Light Source Limited. The SGC is a registered charity (number 1097737) that receives funds from AbbVie, Bayer Pharma AG, Boehringer Ingelheim, Canada Foundation for Innovation, Eshelman Institute for Innovation, Genentech, Janssen, Merck KGaA, Darmstadt, Germany, MSD, Ontario Ministry of Research, Innovation and Science (MRIS), Pfizer, São Paulo Research Foundation-FAPESP, Takeda, and Wellcome [106169/ZZ14/Z]. Research in the laboratory of V.K. is supported by the CRUK program grant C2739/A22897 and by a Marie Skłodowska-Curie ETN grant under the European Union's Horizon 2020 Research and Innovation Program (Grant Agreement No 765912).

Acknowledgments: The authors thank Zvulun Elazar and Terje Johansen for their critical comments and helpful discussions. The authors apologize to those researchers whose excellent work could not be cited in this manuscript.

Conflicts of Interest: The authors declare no conflict of interest.

References

1. Mizushima, N.; Yoshimori, T.; Ohsumi, Y. The role of Atg proteins in autophagosome formation. *Annu. Rev. Cell. Dev. Biol.* **2011**, *27*, 107–132. [CrossRef] [PubMed]
2. Ohsumi, Y. Historical landmarks of autophagy research. *Cell Res.* **2014**, *24*, 9–23. [CrossRef] [PubMed]
3. Dikic, I. Proteasomal and autophagic degradation systems. *Annu. Rev. Biochem.* **2017**, *86*, 193–224. [CrossRef]
4. Dikic, I.; Elazar, Z. Mechanism and medical implications of mammalian autophagy. *Nat. Rev. Mol. Cell Biol.* **2018**, *19*, 349–364. [CrossRef] [PubMed]
5. Kirkin, V.; Rogov, V.V. A diversity of selective autophagy receptors determines the specificity of the autophagy pathway. *Mol. Cell* **2019**, *76*, 268–285. [CrossRef] [PubMed]
6. Tooze, S.A.; Yoshimori, T. The origin of the autophagosomal membrane. *Nat. Cell Biol.* **2010**, *12*, 831–835. [CrossRef]
7. Kabeya, Y.; Mizushima, N.; Yamamoto, A.; Oshitani-Okamoto, S.; Ohsumi, Y.; Yoshimori, T. LC3, GABARAP and GATE16 localize to autophagosomal membrane depending on form-II formation. *J. Cell Sci.* **2004**, *117*, 2805–2812. [CrossRef]
8. Kirisako, T.; Ichimura, Y.; Okada, H.; Kabeya, Y.; Mizushima, N.; Yoshimori, T.; Ohsumi, M.; Takao, T.; Noda, T.; Ohsumi, Y. The reversible modification regulates the membrane-binding state of Apg8/Aut7 essential for autophagy and the cytoplasm to vacuole targeting pathway. *J. Cell Biol.* **2000**, *151*, 263–276. [CrossRef]
9. Ohsumi, Y. Molecular mechanism of autophagy in yeast, Saccharomyces cerevisiae. *Philos. Trans. R. Soc. Lond. B Biol. Sci.* **1999**, *354*, 1577–1581. [CrossRef]
10. Calzada, E.; Onguka, O.; Claypool, S.M. Phosphatidylethanolamine metabolism in health and disease. *Int. Rev. Cell Mol. Biol.* **2016**, *321*, 29–88. [CrossRef]
11. Birgisdottir, Å.B.; Lamark, T.; Johansen, T. The LIR motif–crucial for selective autophagy. *J. Cell Sci.* **2013**, *126*, 3237–3247. [CrossRef] [PubMed]
12. Harding, T.M.; Morano, K.A.; Scott, S.V.; Klionsky, D.J. Isolation and characterization of yeast mutants in the cytoplasm to vacuole protein targeting pathway. *J. Cell Biol.* **1995**, *131*, 591–602. [CrossRef] [PubMed]
13. Tsukada, M.; Ohsumi, Y. Isolation and characterization of autophagy-defective mutants of Saccharomyces cerevisiae. *FEBS Lett.* **1993**, *333*, 169–174. [CrossRef]
14. Alberti, A.; Michelet, X.; Djeddi, A.; Legouis, R. The autophagosomal protein LGG-2 acts synergistically with LGG-1 in dauer formation and longevity in *C. elegans*. *Autophagy* **2010**, *6*, 622–633. [CrossRef] [PubMed]
15. Juhász, G.; Érdi, B.; Sass, M.; Neufeld, T.P. Atg7-dependent autophagy promotes neuronal health, stress tolerance, and longevity but is dispensable for metamorphosis in Drosophila. *Genes Dev.* **2007**, *21*, 3061–3066. [CrossRef] [PubMed]

16. Komatsu, M.; Waguri, S.; Chiba, T.; Murata, S.; Iwata, J.-i.; Tanida, I.; Ueno, T.; Koike, M.; Uchiyama, Y.; Kominami, E.; et al. Loss of autophagy in the central nervous system causes neurodegeneration in mice. *Nature* **2006**, *441*, 880–884. [CrossRef]
17. Karsli-Uzunbas, G.; Guo, J.Y.; Price, S.; Teng, X.; Laddha, S.V.; Khor, S.; Kalaany, N.Y.; Jacks, T.; Chan, C.S.; Rabinowitz, J.D.; et al. Autophagy is required for glucose homeostasis and lung tumor maintenance. *Cancer Discov.* **2014**, *4*, 914–927. [CrossRef]
18. Cassidy, L.D.; Young, A.R.; Young, C.N.; Soilleux, E.J.; Fielder, E.; Weigand, B.M.; Lagnado, A.; Brais, R.; Ktistakis, N.T.; Wiggins, K.A.; et al. Temporal inhibition of autophagy reveals segmental reversal of ageing with increased cancer risk. *Nat. Commun.* **2020**, *11*, 1–12. [CrossRef]
19. Han, S.; Yu, B.; Wang, Y.; Liu, Y. Role of plant autophagy in stress response. *Protein Cell* **2011**, *2*, 784–791. [CrossRef]
20. Nishida, Y.; Arakawa, S.; Fujitani, K.; Yamaguchi, H.; Mizuta, T.; Kanaseki, T.; Komatsu, M.; Otsu, K.; Tsujimoto, Y.; Shimizu, S. Discovery of Atg5/Atg7-independent alternative macroautophagy. *Nature* **2009**, *461*, 654–658. [CrossRef]
21. Torii, S.; Yamaguchi, H.; Nakanishi, A.; Arakawa, S.; Honda, S.; Moriwaki, K.; Nakano, H.; Shimizu, S. Identification of a phosphorylation site on Ulk1 required for genotoxic stress-induced alternative autophagy. *Nat. Commun.* **2020**, *11*, 1–19. [CrossRef] [PubMed]
22. Tsuboyama, K.; Koyama-Honda, I.; Sakamaki, Y.; Koike, M.; Morishita, H.; Mizushima, N. The ATG conjugation systems are important for degradation of the inner autophagosomal membrane. *Science* **2016**, *354*, 1036–1041. [CrossRef] [PubMed]
23. Xie, Z.; Klionsky, D.J. Autophagosome formation: Core machinery and adaptations. *Nat. Cell Biol.* **2007**, *9*, 1102–1109. [CrossRef] [PubMed]
24. Zaffagnini, G.; Martens, S. Mechanisms of selective autophagy. *J. Mol. Biol.* **2016**, *428*, 1714–1724. [CrossRef] [PubMed]
25. Johansen, T.; Lamark, T. Selective autophagy: ATG8 family proteins, LIR motifs and cargo receptors. *J. Mol. Biol.* **2020**, *432*, 80–103. [CrossRef]
26. Ichimura, Y.; Kumanomidou, T.; Sou, Y.-s.; Mizushima, T.; Ezaki, J.; Ueno, T.; Kominami, E.; Yamane, T.; Tanaka, K.; Komatsu, M. Structural basis for sorting mechanism of p62 in selective autophagy. *J. Biol. Chem.* **2008**, *283*, 22847–22857. [CrossRef]
27. Noda, N.N.; Kumeta, H.; Nakatogawa, H.; Satoo, K.; Adachi, W.; Ishii, J.; Fujioka, Y.; Ohsumi, Y.; Inagaki, F. Structural basis of target recognition by Atg8/LC3 during selective autophagy. *Genes Cells* **2008**, *13*, 1211–1218. [CrossRef]
28. Pankiv, S.; Clausen, T.H.; Lamark, T.; Brech, A.; Bruun, J.-A.; Outzen, H.; Øvervatn, A.; Bjørkøy, G.; Johansen, T. p62/SQSTM1 binds directly to Atg8/LC3 to facilitate degradation of ubiquitinated protein aggregates by autophagy. *J. Biol. Chem.* **2007**, *282*, 24131–24145. [CrossRef]
29. Jacomin, A.-C.; Samavedam, S.; Promponas, V.; Nezis, I.P. iLIR database: A web resource for LIR motif-containing proteins in eukaryotes. *Autophagy* **2016**, *12*, 1945–1953. [CrossRef]
30. Kalvari, I.; Tsompanis, S.; Mulakkal, N.C.; Osgood, R.; Johansen, T.; Nezis, I.P.; Promponas, V.J. iLIR: A web resource for prediction of Atg8-family interacting proteins. *Autophagy* **2014**, *10*, 913–925. [CrossRef]
31. Atkinson, J.M.; Ye, Y.; Gebru, M.T.; Liu, Q.; Zhou, S.; Young, M.M.; Takahashi, Y.; Lin, Q.; Tian, F.; Wang, H.-G. Time-resolved FRET and NMR analyses reveal selective binding of peptides containing the LC3-interacting region to ATG8 family proteins. *J. Biol. Chem.* **2019**, *294*, 14033–14042. [CrossRef] [PubMed]
32. Johansen, T.; Birgisdottir, Å.; Huber, J.; Kniss, A.; Dötsch, V.; Kirkin, V.; Rogov, V. Methods for studying interactions between Atg8/LC3/GABARAP and LIR-containing proteins. *Methods Enzymol.* **2017**, *587*, 143–169. [CrossRef] [PubMed]
33. Rasmussen, M.S.; Birgisdottir, Å.B.; Johansen, T. Use of Peptide Arrays for Identification and Characterization of LIR Motifs. *Methods Mol. Biol.* **2019**, 149–161. [CrossRef]
34. Rogov, V.V.; Stolz, A.; Ravichandran, A.C.; Rios-Szwed, D.O.; Suzuki, H.; Kniss, A.; Löhr, F.; Wakatsuki, S.; Dötsch, V.; Dikic, I.; et al. Structural and functional analysis of the GABARAP interaction motif (GIM). *EMBO Rep.* **2017**, *18*, 1382–1396. [CrossRef] [PubMed]
35. Hurley, J.H.; Schulman, B.A. Atomistic autophagy: The structures of cellular self-digestion. *Cell* **2014**, *157*, 300–311. [CrossRef]

36. Noda, N.N.; Ohsumi, Y.; Inagaki, F. Atg8-family interacting motif crucial for selective autophagy. *FEBS Lett.* **2010**, *584*, 1379–1385. [CrossRef]
37. Rogov, V.; Dötsch, V.; Johansen, T.; Kirkin, V. Interactions between autophagy receptors and ubiquitin-like proteins form the molecular basis for selective autophagy. *Mol. Cell* **2014**, *53*, 167–178. [CrossRef]
38. Sora, V.; Kumar, M.; Maiani, E.; Lambrughi, M.; Tiberti, M.; Papaleo, E. Structure and dynamics in the ATG8 family from experimental to computational techniques. *Front. Cell Dev. Biol.* **2020**, *8*, 420. [CrossRef]
39. Vijay-Kumar, S.; Bugg, C.E.; Cook, W.J. Structure of ubiquitin refined at 1.8 Åresolution. *J. Mol. Biol.* **1987**, *194*, 531–544. [CrossRef]
40. Kirkin, V.; Dikic, I. Role of ubiquitin-and Ubl-binding proteins in cell signaling. *Curr. Opin. Cell Biol.* **2007**, *19*, 199–205. [CrossRef]
41. Reyes-Turcu, F.E.; Ventii, K.H.; Wilkinson, K.D. Regulation and cellular roles of ubiquitin-specific deubiquitinating enzymes. *Annu. Rev. Biochem.* **2009**, *78*, 363–397. [CrossRef] [PubMed]
42. Mizushima, N.; Noda, T.; Yoshimori, T.; Tanaka, Y.; Ishii, T.; George, M.D.; Klionsky, D.J.; Ohsumi, M.; Ohsumi, Y. A protein conjugation system essential for autophagy. *Nature* **1998**, *395*, 395–398. [CrossRef] [PubMed]
43. Matsushita, M.; Suzuki, N.N.; Obara, K.; Fujioka, Y.; Ohsumi, Y.; Inagaki, F. Structure of Atg5· Atg16, a complex essential for autophagy. *J. Biol. Chem.* **2007**, *282*, 6763–6772. [CrossRef] [PubMed]
44. Fracchiolla, D.; Sawa-Makarska, J.; Zens, B.; de Ruiter, A.; Zaffagnini, G.; Brezovich, A.; Romanov, J.; Runggatscher, K.; Kraft, C.; Zagrovic, B.; et al. Mechanism of cargo-directed Atg8 conjugation during selective autophagy. *Elife* **2016**, *5*, e18544. [CrossRef] [PubMed]
45. Bansal, M.; Moharir, S.C.; Sailasree, S.P.; Sirohi, K.; Sudhakar, C.; Sarathi, D.P.; Lakshmi, B.J.; Buono, M.; Kumar, S.; Swarup, G. Optineurin promotes autophagosome formation by recruiting the autophagy-related Atg12-5-16L1 complex to phagophores containing the Wipi2 protein. *J. Biol. Chem.* **2018**, *293*, 132–147. [CrossRef]
46. Lane, J.D.; Baines, K. The ATG5 Interactome Links Clathrin Vesicular Trafficking With The ATG8 Lipidation Machinery For Autophagosome Assembly. *bioRxiv* **2019**, 769059. [CrossRef]
47. Mohrlüder, J.; Hoffmann, Y.; Stangler, T.; Hänel, K.; Willbold, D. Identification of clathrin heavy chain as a direct interaction partner for the γ-aminobutyric acid type A receptor associated protein. *Biochemistry* **2007**, *46*, 14537–14543. [CrossRef]
48. Radoshevich, L.; Murrow, L.; Chen, N.; Fernandez, E.; Roy, S.; Fung, C.; Debnath, J. ATG12 conjugation to ATG3 regulates mitochondrial homeostasis and cell death. *Cell* **2010**, *142*, 590–600. [CrossRef]
49. Fujita, N.; Itoh, T.; Omori, H.; Fukuda, M.; Noda, T.; Yoshimori, T. The Atg16L complex specifies the site of LC3 lipidation for membrane biogenesis in autophagy. *Mol. Biol. Cell* **2008**, *19*, 2092–2100. [CrossRef]
50. Huber, J.; Obata, M.; Gruber, J.; Akutsu, M.; Löhr, F.; Rogova, N.; Güntert, P.; Dikic, I.; Kirkin, V.; Komatsu, M.; et al. An atypical LIR motif within UBA5 (ubiquitin like modifier activating enzyme 5) interacts with GABARAP proteins and mediates membrane localization of UBA5. *Autophagy* **2020**, *16*, 256–270. [CrossRef]
51. Liu, X.-M.; Yamasaki, A.; Du, X.-M.; Coffman, V.C.; Ohsumi, Y.; Nakatogawa, H.; Wu, J.-Q.; Noda, N.N.; Du, L.-L. Lipidation-independent vacuolar functions of Atg8 rely on its noncanonical interaction with a vacuole membrane protein. *Elife* **2018**, *7*, e41237. [CrossRef] [PubMed]
52. Ichimura, Y.; Kirisako, T.; Takao, T.; Satomi, Y.; Shimonishi, Y.; Ishihara, N.; Mizushima, N.; Tanida, I.; Kominami, E.; Ohsumi, M.; et al. A ubiquitin-like system mediates protein lipidation. *Nature* **2000**, *408*, 488–492. [CrossRef] [PubMed]
53. Fujita, N.; Hayashi-Nishino, M.; Fukumoto, H.; Omori, H.; Yamamoto, A.; Noda, T.; Yoshimori, T. An Atg4B mutant hampers the lipidation of LC3 paralogues and causes defects in autophagosome closure. *Mol. Biol. Cell* **2008**, *19*, 4651–4659. [CrossRef] [PubMed]
54. Andreeva, A.; Kulesha, E.; Gough, J.; Murzin, A.G. The SCOP database in 2020: Expanded classification of representative family and superfamily domains of known protein structures. *Nucleic Acids Res.* **2020**, *48*, D376–D382. [CrossRef] [PubMed]
55. Coyle, J.E.; Qamar, S.; Rajashankar, K.R.; Nikolov, D.B. Structure of GABARAP in two conformations: Implications for GABAA receptor localization and tubulin binding. *Neuron* **2002**, *33*, 63–74. [CrossRef]
56. Krichel, C.; Möckel, C.; Schillinger, O.; Huesgen, P.F.; Sticht, H.; Strodel, B.; Weiergräber, O.H.; Willbold, D.; Neudecker, P. Solution structure of the autophagy-related protein LC3C reveals a polyproline II motif on a mobile tether with phosphorylation site. *Sci. Rep.* **2019**, *9*, 1–15. [CrossRef]

57. Stangler, T.; Mayr, L.M.; Willbold, D. Solution Structure of Human GABAA Receptor-associated Protein GABARAP: Implications for Biological Function and its Regulation. *J. Biol. Chem.* **2002**, *277*, 13363–13366. [CrossRef]
58. Nakatogawa, H.; Ichimura, Y.; Ohsumi, Y. Atg8, a ubiquitin-like protein required for autophagosome formation, mediates membrane tethering and hemifusion. *Cell* **2007**, *130*, 165–178. [CrossRef]
59. Weidberg, H.; Shpilka, T.; Shvets, E.; Abada, A.; Shimron, F.; Elazar, Z. LC3 and GATE-16 N termini mediate membrane fusion processes required for autophagosome biogenesis. *Dev. Cell* **2011**, *20*, 444–454. [CrossRef]
60. Chu, C.T.; Ji, J.; Dagda, R.K.; Jiang, J.F.; Tyurina, Y.Y.; Kapralov, A.A.; Tyurin, V.A.; Yanamala, N.; Shrivastava, I.H.; Mohammadyani, D.; et al. Cardiolipin externalization to the outer mitochondrial membrane acts as an elimination signal for mitophagy in neuronal cells. *Nat. Cell Biol.* **2013**, *15*, 1197–1205. [CrossRef]
61. Thielmann, Y.; Mohrlüder, J.; Koenig, B.W.; Stangler, T.; Hartmann, R.; Becker, K.; Höltje, H.D.; Willbold, D. An indole-binding site is a major determinant of the ligand specificity of the GABA type A receptor-associated protein GABARAP. *ChemBioChem* **2008**, *9*, 1767–1775. [CrossRef] [PubMed]
62. Cook, W.J.; Jeffrey, L.C.; Kasperek, E.; Pickart, C.M. Structure of tetraubiquitin shows how multiubiquitin chains can be formed. *J. Mol. Biol.* **1994**, *236*, 601–609. [CrossRef] [PubMed]
63. Marshall, R.S.; Hua, Z.; Mali, S.; McLoughlin, F.; Vierstra, R.D. ATG8-binding UIM proteins define a new class of autophagy adaptors and receptors. *Cell* **2019**, *177*, 766–781. [CrossRef] [PubMed]
64. Jatana, N.; Ascher, D.B.; Pires, D.E.; Gokhale, R.S.; Thukral, L. Human LC3 and GABARAP subfamily members achieve functional specificity via specific structural modulations. *Autophagy* **2020**, *16*, 239–255. [CrossRef] [PubMed]
65. Shpilka, T.; Weidberg, H.; Pietrokovski, S.; Elazar, Z. Atg8: An autophagy-related ubiquitin-like protein family. *Genome Biol.* **2011**, *12*, 226. [CrossRef] [PubMed]
66. Kuznetsov, S.A.; Gelfand, V.I. 18 kDa microtubule-associated protein: Identification as a new light chain (LC-3) of microtubule-associated protein 1 (MAP-1). *FEBS Lett.* **1987**, *212*, 145–148. [CrossRef]
67. Wang, H.; Bedford, F.K.; Brandon, N.J.; Moss, S.J.; Olsen, R.W. GABA A-receptor-associated protein links GABA A receptors and the cytoskeleton. *Nature* **1999**, *397*, 69–72. [CrossRef]
68. Weidberg, H.; Shvets, E.; Shpilka, T.; Shimron, F.; Shinder, V.; Elazar, Z. LC3 and GATE-16/GABARAP subfamilies are both essential yet act differently in autophagosome biogenesis. *EMBO J.* **2010**, *29*, 1792–1802. [CrossRef]
69. Grunwald, D.S.; Otto, N.M.; Park, J.-M.; Song, D.; Kim, D.-H. GABARAPs and LC3s have opposite roles in regulating ULK1 for autophagy induction. *Autophagy* **2020**, *16*, 600–614. [CrossRef]
70. Joachim, J.; Jefferies, H.B.; Razi, M.; Frith, D.; Snijders, A.P.; Chakravarty, P.; Judith, D.; Tooze, S.A. Activation of ULK kinase and autophagy by GABARAP trafficking from the centrosome is regulated by WAC and GM130. *Mol. Cell* **2015**, *60*, 899–913. [CrossRef]
71. Holdgaard, S.G.; Cianfanelli, V.; Pupo, E.; Lambrughi, M.; Lubas, M.; Nielsen, J.C.; Eibes, S.; Maiani, E.; Harder, L.M.; Wesch, N.; et al. Selective autophagy maintains centrosome integrity and accurate mitosis by turnover of centriolar satellites. *Nat. Commun.* **2019**, *10*, 1–19. [CrossRef] [PubMed]
72. Joachim, J.; Razi, M.; Judith, D.; Wirth, M.; Calamita, E.; Encheva, V.; Dynlacht, B.D.; Snijders, A.P.; O'Reilly, N.; Jefferies, H.B.; et al. Centriolar satellites control GABARAP ubiquitination and GABARAP-mediated autophagy. *Curr. Biol.* **2017**, *27*, 2123–2136.e7. [CrossRef] [PubMed]
73. von Muhlinen, N.; Akutsu, M.; Ravenhill, B.J.; Foeglein, Á.; Bloor, S.; Rutherford, T.J.; Freund, S.M.; Komander, D.; Randow, F. LC3C, bound selectively by a noncanonical LIR motif in NDP52, is required for antibacterial autophagy. *Mol. Cell* **2012**, *48*, 329–342. [CrossRef] [PubMed]
74. Kellner, R.; De la Concepcion, J.C.; Maqbool, A.; Kamoun, S.; Dagdas, Y.F. ATG8 expansion: A driver of selective autophagy diversification? *Trends Plant Sci.* **2017**, *22*, 204–214. [CrossRef]
75. Bu, F.; Yang, M.; Guo, X.; Huang, W.; Chen, L. Multiple functions of ATG8 family proteins in plant autophagy. *Front. Cell Dev. Biol.* **2020**, *8*, 466. [CrossRef] [PubMed]
76. Zess, E.K.; Jensen, C.; Cruz-Mireles, N.; De la Concepcion, J.C.; Sklenar, J.; Stephani, M.; Imre, R.; Roitinger, E.; Hughes, R.; Belhaj, K. N-terminal β-strand underpins biochemical specialization of an ATG8 isoform. *PLoS Biol.* **2019**, *17*, e3000373. [CrossRef]

77. Suzuki, H.; Tabata, K.; Morita, E.; Kawasaki, M.; Kato, R.; Dobson, R.C.; Yoshimori, T.; Wakatsuki, S. Structural basis of the autophagy-related LC3/Atg13 LIR complex: Recognition and interaction mechanism. *Structure* **2014**, *22*, 47–58. [CrossRef]
78. Rogov, V.V.; Suzuki, H.; Marinković, M.; Lang, V.; Kato, R.; Kawasaki, M.; Buljubašić, M.; Šprung, M.; Rogova, N.; Wakatsuki, S.; et al. Phosphorylation of the mitochondrial autophagy receptor Nix enhances its interaction with LC3 proteins. *Sci. Rep.* **2017**, *7*, 1–12. [CrossRef]
79. Shrestha, B.K.; Rasmussen, M.S.; Abudu, Y.P.; Bruun, J.-A.; Larsen, K.B.; Alemu, E.A.; Sjøttem, E.; Lamark, T.; Johansen, T. NIMA-related kinase 9–mediated phosphorylation of the microtubule-associated LC3B protein at Thr-50 suppresses selective autophagy of p62/sequestosome 1. *J. Biol. Chem.* **2020**, *295*, 1240–1260. [CrossRef]
80. Wilkinson, D.S.; Jariwala, J.S.; Anderson, E.; Mitra, K.; Meisenhelder, J.; Chang, J.T.; Ideker, T.; Hunter, T.; Nizet, V.; Dillin, A.; et al. Phosphorylation of LC3 by the Hippo kinases STK3/STK4 is essential for autophagy. *Mol. Cell* **2015**, *57*, 55–68. [CrossRef]
81. Rogov, V.V.; Suzuki, H.; Fiskin, E.; Wild, P.; Kniss, A.; Rozenknop, A.; Kato, R.; Kawasaki, M.; McEwan, D.G.; Löhr, F.; et al. Structural basis for phosphorylation-triggered autophagic clearance of Salmonella. *Biochem. J.* **2013**, *454*, 459–466. [CrossRef] [PubMed]
82. Dougherty, D.A. Cation-π interactions in chemistry and biology: A new view of benzene, Phe, Tyr, and Trp. *Science* **1996**, *271*, 163–168. [CrossRef] [PubMed]
83. Genau, H.M.; Huber, J.; Baschieri, F.; Akutsu, M.; Dötsch, V.; Farhan, H.; Rogov, V.; Behrends, C. CUL3-KBTBD6/KBTBD7 ubiquitin ligase cooperates with GABARAP proteins to spatially restrict TIAM1-RAC1 signaling. *Mol. Cell* **2015**, *57*, 995–1010. [CrossRef] [PubMed]
84. Wirth, M.; Zhang, W.; Razi, M.; Nyoni, L.; Joshi, D.; O'Reilly, N.; Johansen, T.; Tooze, S.A.; Mouilleron, S. Molecular determinants regulating selective binding of autophagy adapters and receptors to ATG8 proteins. *Nat. Commun.* **2019**, *10*, 1–18. [CrossRef] [PubMed]
85. Cherra, S.J., 3rd; Kulich, S.M.; Uechi, G.; Balasubramani, M.; Mountzouris, J.; Day, B.W.; Chu, C.T. Regulation of the autophagy protein LC3 by phosphorylation. *J. Cell Biol.* **2010**, *190*, 533–539. [CrossRef] [PubMed]
86. Herhaus, L.; Bhaskara, R.M.; Lystad, A.H.; Gestal-Mato, U.; Covarrubias-Pinto, A.; Bonn, F.; Simonsen, A.; Hummer, G.; Dikic, I. TBK1-mediated phosphorylation of LC3C and GABARAP-L2 controls autophagosome shedding by ATG4 protease. *EMBO Rep.* **2020**, *21*, e48317. [CrossRef]
87. Jiang, H.; Cheng, D.; Liu, W.; Peng, J.; Feng, J. Protein kinase C inhibits autophagy and phosphorylates LC3. *Biochem. Biophys. Res. Commun.* **2010**, *395*, 471–476. [CrossRef]
88. Huang, R.; Xu, Y.; Wan, W.; Shou, X.; Qian, J.; You, Z.; Liu, B.; Chang, C.; Zhou, T.; Lippincott-Schwartz, J.; et al. Deacetylation of nuclear LC3 drives autophagy initiation under starvation. *Mol. Cell* **2015**, *57*, 456–466. [CrossRef]
89. Jia, R.; Bonifacino, J.S. Negative regulation of autophagy by UBA6-BIRC6–mediated ubiquitination of LC3. *Elife* **2019**, *8*, e50034. [CrossRef]
90. Alemu, E.A.; Lamark, T.; Torgersen, K.M.; Birgisdottir, A.B.; Larsen, K.B.; Jain, A.; Olsvik, H.; Øvervatn, A.; Kirkin, V.; Johansen, T. ATG8 family proteins act as scaffolds for assembly of the ULK complex sequence requirements for LC3-interacting region (LIR) motifs. *J. Biol. Chem.* **2012**, *287*, 39275–39290. [CrossRef]
91. Birgisdottir, Å.B.; Mouilleron, S.; Bhujabal, Z.; Wirth, M.; Sjøttem, E.; Evjen, G.; Zhang, W.; Lee, R.; O'Reilly, N.; Tooze, S.A.; et al. Members of the autophagy class III phosphatidylinositol 3-kinase complex I interact with GABARAP and GABARAPL1 via LIR motifs. *Autophagy* **2019**, *15*, 1333–1355. [CrossRef] [PubMed]
92. Rozenknop, A.; Rogov, V.V.; Rogova, N.Y.; Löhr, F.; Güntert, P.; Dikic, I.; Dötsch, V. Characterization of the interaction of GABARAPL-1 with the LIR motif of NBR1. *J. Mol. Biol.* **2011**, *410*, 477–487. [CrossRef] [PubMed]
93. Di Rita, A.; Peschiaroli, A.; Pasquale, D.; Strobbe, D.; Hu, Z.; Gruber, J.; Nygaard, M.; Lambrughi, M.; Melino, G.; Papaleo, E.; et al. HUWE1 E3 ligase promotes PINK1/PARKIN-independent mitophagy by regulating AMBRA1 activation via IKKα. *Nat. Commun.* **2018**, *9*, 1–18. [CrossRef] [PubMed]
94. Kuang, Y.; Ma, K.; Zhou, C.; Ding, P.; Zhu, Y.; Chen, Q.; Xia, B. Structural basis for the phosphorylation of FUNDC1 LIR as a molecular switch of mitophagy. *Autophagy* **2016**, *12*, 2363–2373. [CrossRef]
95. Wild, P.; Farhan, H.; McEwan, D.G.; Wagner, S.; Rogov, V.V.; Brady, N.R.; Richter, B.; Korac, J.; Waidmann, O.; Choudhary, C.; et al. Phosphorylation of the autophagy receptor optineurin restricts Salmonella growth. *Science* **2011**, *333*, 228–233. [CrossRef]

96. Zhu, Y.; Massen, S.; Terenzio, M.; Lang, V.; Chen-Lindner, S.; Eils, R.; Novak, I.; Dikic, I.; Hamacher-Brady, A.; Brady, N.R. Modulation of serines 17 and 24 in the LC3-interacting region of Bnip3 determines pro-survival mitophagy versus apoptosis. *J. Biol. Chem.* **2013**, *288*, 1099–1113. [CrossRef]
97. Sano, T.; Cantor, C.R. Cooperative biotin binding by streptavidin. Electrophoretic behavior and subunit association of streptavidin in the presence of 6 M urea. *J. Biol. Chem.* **1990**, *265*, 3369–3373.
98. Stubenrauch, K.; Wessels, U.; Essig, U.; Kowalewsky, F.; Vogel, R.; Heinrich, J. Characterization of murine anti-human Fab antibodies for use in an immunoassay for generic quantification of human Fab fragments in non-human serum samples including cynomolgus monkey samples. *J. Pharm. Biomed. Anal.* **2013**, *72*, 208–215. [CrossRef]
99. Hurley, J.H.; Lee, S.; Prag, G. Ubiquitin-binding domains. *Biochem. J.* **2006**, *399*, 361–372. [CrossRef]
100. Li, J.; Zhu, R.; Chen, K.; Zheng, H.; Zhao, H.; Yuan, C.; Zhang, H.; Wang, C.; Zhang, M. Potent and specific Atg8-targeting autophagy inhibitory peptides from giant ankyrins. *Nat. Chem. Biol.* **2018**, *14*, 778–787. [CrossRef]
101. Xie, Q.; Tzfadia, O.; Levy, M.; Weithorn, E.; Peled-Zehavi, H.; Van Parys, T.; Van de Peer, Y.; Galili, G. hfAIM: A reliable bioinformatics approach for in silico genome-wide identification of autophagy-associated Atg8-interacting motifs in various organisms. *Autophagy* **2016**, *12*, 876–887. [CrossRef] [PubMed]
102. Habisov, S.; Huber, J.; Ichimura, Y.; Akutsu, M.; Rogova, N.; Loehr, F.; McEwan, D.G.; Johansen, T.; Dikic, I.; Doetsch, V.; et al. Structural and functional analysis of a novel interaction motif within UFM1-activating enzyme 5 (UBA5) required for binding to ubiquitin-like proteins and ufmylation. *J. Biol. Chem.* **2016**, *291*, 9025–9041. [CrossRef] [PubMed]
103. Tumbarello, D.A.; Manna, P.T.; Allen, M.; Bycroft, M.; Arden, S.D.; Kendrick-Jones, J.; Buss, F. The autophagy receptor TAX1BP1 and the molecular motor myosin VI are required for clearance of salmonella typhimurium by autophagy. *PLoS Pathog* **2015**, *11*, e1005174. [CrossRef] [PubMed]
104. Kaufmann, A.; Beier, V.; Franquelim, H.G.; Wollert, T. Molecular mechanism of autophagic membrane-scaffold assembly and disassembly. *Cell* **2014**, *156*, 469–481. [CrossRef] [PubMed]
105. Noda, N.N.; Satoo, K.; Fujioka, Y.; Kumeta, H.; Ogura, K.; Nakatogawa, H.; Ohsumi, Y.; Inagaki, F. Structural basis of Atg8 activation by a homodimeric E1, Atg7. *Mol. Cell* **2011**, *44*, 462–475. [CrossRef] [PubMed]
106. Cheng, X.; Wang, Y.; Gong, Y.; Li, F.; Guo, Y.; Hu, S.; Liu, J.; Pan, L. Structural basis of FYCO1 and MAP1LC3A interaction reveals a novel binding mode for Atg8-family proteins. *Autophagy* **2016**, *12*, 1330–1339. [CrossRef]
107. Khaminets, A.; Heinrich, T.; Mari, M.; Grumati, P.; Huebner, A.K.; Akutsu, M.; Liebmann, L.; Stolz, A.; Nietzsche, S.; Koch, N.; et al. Regulation of endoplasmic reticulum turnover by selective autophagy. *Nature* **2015**, *522*, 354–358. [CrossRef]
108. Olsvik, H.L.; Lamark, T.; Takagi, K.; Larsen, K.B.; Evjen, G.; Øvervatn, A.; Mizushima, T.; Johansen, T. FYCO1 contains a C-terminally extended, LC3A/B-preferring LC3-interacting region (LIR) motif required for efficient maturation of autophagosomes during basal autophagy. *J. Biol. Chem.* **2015**, *290*, 29361–29374. [CrossRef]
109. Sakurai, S.; Tomita, T.; Shimizu, T.; Ohto, U. The crystal structure of mouse LC3B in complex with the FYCO1 LIR reveals the importance of the flanking region of the LIR motif. *Acta Cryst. F* **2017**, *73*, 130–137. [CrossRef]
110. Mochida, K.; Yamasaki, A.; Matoba, K.; Kirisako, H.; Noda, N.N.; Nakatogawa, H. Super-assembly of ER-phagy receptor Atg40 induces local ER remodeling at contacts with forming autophagosomal membranes. *Nat. Commun.* **2020**, *11*, 1–14. [CrossRef]
111. Nelson, A.D.; Caballero-Florán, R.N.; Díaz, J.R.; Hull, J.; Yuan, Y.; Li, J.; Chen, K.; Walder, K.; Lopez-Santiago, L.; Bennett, V.; et al. Ankyrin-G regulates forebrain connectivity and network synchronization via interaction with GABARAP. *Mol. Psychiatry* **2018**, 1–18. [CrossRef] [PubMed]
112. Kwon, D.H.; Kim, L.; Kim, B.-W.; Kim, J.H.; Roh, K.-H.; Choi, E.-J.; Song, H.K. A novel conformation of the LC3-interacting region motif revealed by the structure of a complex between LC3B and RavZ. *Biochem. Biophys. Res. Commun.* **2017**, *490*, 1093–1099. [CrossRef] [PubMed]
113. Lystad, A.H.; Ichimura, Y.; Takagi, K.; Yang, Y.; Pankiv, S.; Kanegae, Y.; Kageyama, S.; Suzuki, M.; Saito, I.; Mizushima, T.; et al. Structural determinants in GABARAP required for the selective binding and recruitment of ALFY to LC 3B-positive structures. *EMBO Rep.* **2014**, *15*, 557–565. [CrossRef]
114. Keown, J.R.; Black, M.M.; Ferron, A.; Yap, M.; Barnett, M.J.; Pearce, F.G.; Stoye, J.P.; Goldstone, D.C. A helical LC3-interacting region mediates the interaction between the retroviral restriction factor Trim5α and mammalian autophagy-related ATG8 proteins. *J. Biol. Chem.* **2018**, *293*, 18378–18386. [CrossRef] [PubMed]

115. Ma, P.; Schwarten, M.; Schneider, L.; Boeske, A.; Henke, N.; Lisak, D.; Weber, S.; Mohrlüder, J.; Stoldt, M.; Strodel, B.; et al. Interaction of Bcl-2 with the Autophagy-related GABAA Receptor-associated Protein (GABARAP): Biophysical characterization and functional implications. *J. Biol. Chem.* **2013**, *288*, 37204–37215. [CrossRef] [PubMed]
116. Real, E.; Rodrigues, L.; Cabal, G.G.; Enguita, F.J.; Mancio-Silva, L.; Mello-Vieira, J.; Beatty, W.; Vera, I.M.; Zuzarte-Luís, V.; Figueira, T.N.; et al. Plasmodium UIS3 sequesters host LC3 to avoid elimination by autophagy in hepatocytes. *Nat. Microbiol.* **2018**, *3*, 17–25. [CrossRef] [PubMed]
117. Behrends, C.; Sowa, M.E.; Gygi, S.P.; Harper, J.W. Network organization of the human autophagy system. *Nature* **2010**, *466*, 68–76. [CrossRef]
118. Weiergräber, O.H.; Stangler, T.; Thielmann, Y.; Mohrlüder, J.; Wiesehan, K.; Willbold, D. Ligand binding mode of GABAA receptor-associated protein. *J. Mol. Biol.* **2008**, *381*, 1320–1331. [CrossRef]
119. Stolz, A.; Putyrski, M.; Kutle, I.; Huber, J.; Wang, C.; Major, V.; Sidhu, S.S.; Youle, R.J.; Rogov, V.V.; Dötsch, V.; et al. Fluorescence-based ATG 8 sensors monitor localization and function of LC 3/GABARAP proteins. *EMBO J.* **2017**, *36*, 549–564. [CrossRef]
120. Ma, P.; Schillinger, O.; Schwarten, M.; Lecher, J.; Hartmann, R.; Stoldt, M.; Mohrlüder, J.; Olubiyi, O.; Strodel, B.; Willbold, D. Conformational polymorphism in autophagy-related protein GATE-16. *Biochemistry* **2015**, *54*, 5469–5479. [CrossRef]
121. Shvets, E.; Abada, A.; Weidberg, H.; Elazar, Z. Dissecting the involvement of LC3B and GATE-16 in p62 recruitment into autophagosomes. *Autophagy* **2011**, *7*, 683–688. [CrossRef] [PubMed]
122. Chen, C.; Li, J.-G.; Chen, Y.; Huang, P.; Wang, Y.; Liu-Chen, L.-Y. GEC1 interacts with the κ opioid receptor and enhances expression of the receptor. *J. Biol. Chem.* **2006**, *281*, 7983–7993. [CrossRef] [PubMed]
123. Kneussel, M.; Haverkamp, S.; Fuhrmann, J.C.; Wang, H.; Wässle, H.; Olsen, R.W.; Betz, H. The γ-aminobutyric acid type A receptor (GABAAR)-associated protein GABARAP interacts with gephyrin but is not involved in receptor anchoring at the synapse. *Proc. Natl. Acad. Sci. USA* **2000**, *97*, 8594–8599. [CrossRef] [PubMed]
124. Conway, O.; Akpinar, H.A.; Rogov, V.V.; Kirkin, V. Selective autophagy receptors in neuronal health and disease. *J. Mol. Biol.* **2019**. [CrossRef]
125. Li, Z.; Wang, C.; Wang, Z.; Zhu, C.; Li, J.; Sha, T.; Ma, L.; Gao, C.; Yang, Y.; Sun, Y.; et al. Allele-selective lowering of mutant HTT protein by HTT–LC3 linker compounds. *Nature* **2019**, *575*, 203–209. [CrossRef]
126. Hu, D.; Wu, J.; Zhang, R.; Chen, L.; Chen, Z.; Wang, X.; Xu, L.; Xiao, J.; Hu, F.; Wu, C. Autophagy-targeted vaccine of LC3–LpqH DNA and its protective immunity in a murine model of tuberculosis. *Vaccine* **2014**, *32*, 2308–2314. [CrossRef]
127. White, E. The role for autophagy in cancer. *J. Clin. Investig.* **2015**, *125*, 42–46. [CrossRef]
128. Yamamoto, K.; Venida, A.; Yano, J.; Biancur, D.E.; Kakiuchi, M.; Gupta, S.; Sohn, A.S.; Mukhopadhyay, S.; Lin, E.Y.; Parker, S.J.; et al. Autophagy promotes immune evasion of pancreatic cancer by degrading MHC-I. *Nature* **2020**, *581*, 100–105. [CrossRef]
129. Park, S.-W.; Jeon, P.; Jun, Y.-W.; Park, J.-H.; Lee, S.-H.; Lee, S.; Lee, J.-A.; Jang, D.-J. Monitoring LC3-or GABARAP-positive autophagic membranes using modified RavZ-based probes. *Sci. Rep.* **2019**, *9*, 1–11. [CrossRef]

cells

MDPI

Review

Selective Autophagy by Close Encounters of the Ubiquitin Kind

Anna Vainshtein [1] and Paolo Grumati [2,*]

1 Craft Science Inc., Toronto, ON L4J 7S2, Canada; anna@craftscience.ca
2 Telethon Institute of Genetics and Medicine, 80078 Pozzuoli (NA), Italy
* Correspondence: p.grumati@tigem.it

Received: 4 September 2020; Accepted: 21 October 2020; Published: 24 October 2020

check for updates

Abstract: Autophagy, a bulk degradation process within eukaryotic cells, is responsible for cellular turnover and nutrient liberation during starvation. Increasing evidence indicate that this process can be extremely discerning. Selective autophagy segregates and eliminates protein aggregates, damaged organelles, and invading organisms. The specificity of this process is largely mediated by post-translational modifications (PTMs), which are recognized by autophagy receptors. These receptors grant autophagy surgical precision in cargo selection, where only tagged substrates are engulfed within autophagosomes and delivered to the lysosome for proteolytic breakdown. A growing number of selective autophagy receptors have emerged including p62, NBR1, OPTN, NDP52, TAX1BP1, TOLLIP, and more continue to be uncovered. The most well-documented PTM is ubiquitination and selective autophagy receptors are equipped with a ubiquitin binding domain and an LC3 interacting region which allows them to physically bridge cargo to autophagosomes. Here, we review the role of ubiquitin and ubiquitin-like post-translational modifications in various types of selective autophagy.

Keywords: selective autophagy; ubiquitin; mitophagy; aggrephagy; lysophagy; xenophagy; lipophagy; nucleophagy; ER-phagy; cargo receptors

1. Introduction

The cellular life cycle is complex, having to contend with ever-changing and at times competing internal and environmental demands, can be stressful. Fail-safe degradation mechanisms are therefore required for the effective disposal of potentially toxic and harmful components and their recycling into building blocks needed for biosynthesis. These degradation systems are vital for the survival and continuity of both long-lived and dividing cells. Several cellular degradation processes have evolved to fill this need and their importance is illustrated through their conservation across evolution, and the pathology that ensues with their perturbance [1,2].

The ubiquitin-proteasome and autophagy-lysosome are the two major cellular degradation systems found in eukaryotic cells and organisms. These processes have remained conserved among species and failure of either one can result in the accumulation of toxic or damaged proteins and organelles, culminating in a number of severe pathologies including cancer, failure to thrive, degenerative diseases, and premature death [1,2]. The ubiquitin-proteasome pathway mainly relies on the 26S proteasome for the final degradation of its substrates [3]. Considered to be the more selective of the two degradation systems, proteasomal substrates are largely composed of individual proteins, requiring large complexes to be disassembled before degradation can take place [4]. The autophagy-lysosome pathway utilizes double membraned vesicles, termed autophagosomes, for the encapsulation and delivery of components to the lysosome for breakdown [5]. Autophagy substrates tend to be larger, including protein aggregates, organelles- either in their entirety or select portions, and invading pathogens. Although autophagy was historically considered to be a bulk degradation pathway, it is now universally accepted that it can be quite selective, with a wide range of substrates under its jurisdiction. Despite the two degradation mechanisms being fairly distinct, they both appear to utilize ubiquitin modification for substrate recognition [6]. It is intriguing that two seemingly independent degradation pathways, which have evolved largely different components and substrates, converge on the same PTM for cargo recognition. This suggests some crosstalk and redundancy between these pathways, with ubiquitin acting as a universal degradation signal.

2. Ubiquitin

Ubiquitin (Ub) was originally discovered for its role in tagging proteins for proteasomal degradation. Indeed, Aaron Ciechanover, Avram Hershko, and Irwin Rose received the 2004 Nobel Prize in Chemistry for their discovery and important contributions in unravelling the mechanisms of ubiquitin-mediated proteasomal degradation and the involvement of this pathway in cellular physiology. They demonstrated that ubiquitin was critical for the degradation of proteins involved in cellular regulation, notably cyclins, which play a role in determining cellular replication and mitosis.

Much like its name implies ubiquitin is ubiquitous, it is expressed in all eukaryotic cells and tissues and is evolutionary conserved across species as diverse as plants, yeast, and mammals. A small protein, composed of only 76 amino acids, it is a powerful molecular modifier that governs the fate of proteins and organelles. The ligation of ubiquitin moieties onto a substrate requires a cascade of three distinct enzymes E1, E2, and E3 [3]. E1 is a ubiquitin-activating enzyme that must first prime ubiquitin in an ATP-dependent process. The activated ubiquitin is then transferred to E2, a conjugating enzyme, and finally covalently bonded to a specific substrate with the help of E3, a ubiquitin ligating enzyme. E3 enzymes catalyze the rate-limiting step in Ub conjugation, with over 600 distinct E3 ligases identified so far; this process is extremely substrate-specific and bestows Ub with its selectivity [7]. The ubiquitin tag is then identified by downstream receptors containing a ubiquitin-binding domain (UBD) [8]. Ubiquitin can also be removed from a substrate by deubiquitinating (DUB) enzymes, which antagonize the cell's ubiquitination machinery [9]. By mediating the degradation of cellular regulators ubiquitin controls a variety of cellular processes including cell cycle, transcription, cellular signaling, metabolism, and more. This molecule is central to cellular decision making and is involved in immune response, development, cell death, and growth [3]. Deficiencies in this process can result in a number of pathologies including cancer and neurodegeneration.

In addition to its central role in proteasomal degradation, ubiquitin has been implicated in autophagy regulation and cargo selection [10]. Substrates can be either monoubiquitinated, through the addition of a single ubiquitin moiety, poly monoubiquitinated through the addition of several single ubiquitin moieties, or polyubiquitinated through the formation of polymeric ubiquitin chains [11]. Although there are multiple possible Ub linkage types, while Lysine 48 (K48) and Lysine 63 (K63)-linked chains are among the most abundant ones [11]. Where K48-linked polyubiquitination more readily targets proteins for proteasomal degradation, K63-linked polyubiquitination appears to be more strongly associated with molecular signaling including autophagy [12]. K63 ubiquitination also regulates protein function, interactions, and localization. Although ubiquitin is itself soluble and remarkably stable, in vitro studies, using uncleavable polyubiquitin moieties, showed that ubiquitin chains, the longer they are, the less stable and prone to aggregation they become [13]. This results in the formation of ubiquitin fibrils, which signal for autophagic clearance. The failure to remove these fibrils results in disease. Evidence suggests that polyubiquitination protects cells from actions by proteins that are destined for degradation, by arresting them in solid aggregates and signaling for their autophagic removal [13]. In this scenario, aberrant autophagy owing to lack of essential autophagy components such as Atg5 or Atg7 results in the accumulation of various types of ubiquitin chain topologies suggesting that autophagy is not dependent on any one specific type of ubiquitination [10].

Thus, one important function of ubiquitin is to deliver the "kiss of death" to proteins, organelles, and invading pathogens. It does so by acting as a molecular beacon for the arrest, aggregation, and removal of substrates by cellular proteolytic systems. For a long time, Ubiquitin was thought to be a lone worrier, however, several functionally and structurally related molecules, collectively termed ubiquitin-like (UBL) proteins have been identified. Among them, LC3s/GABARAPs and ATG12 are the most well-studied in the autophagy field [14,15]. Other UBLs include SUMO, FAT10, ISG15, Hub1/Ubl5, NEDD8, UFM1, and URM1. UBLs share many similarities with Ub, including the three-step conjugation process composed of E1, E2, and E3 enzymes and much like Ub, they are also involved in a growing number of cellular processes, including autophagy [16–19].

3. Autophagy

Macroautophagy, herein on termed Autophagy, is a cellular recycling mechanism responsible for cellular housekeeping and turnover during steady-state conditions and nutrient scavenging during starvation. This process involves the formation of double membraned vesicles, known as autophagosomes, around a substrate or a cytoplasmic domain, and the subsequent delivery of these vesicles to the lysosome for proteolytic degradation (Figure 1). Autophagy depends on core autophagy proteins which contribute to the initiation of autophagosome formation, the lipidation of the autophagosomal membrane protein Atg8/LC3/GABARAP, the fusion of the autophagosome to the lysosome, and finally, the degradation of the autophagosome with its cargo within the lysosomal lumen [20]. In 2016 Yoshinori Ohsumi received the Nobel prize in Physiology and Medicine for his discovery of the mechanisms of autophagy in yeast. Ohsumi systematically uncovered autophagy-related genes, classified as ATGs, and his discoveries resulted in an autophagy renaissance.

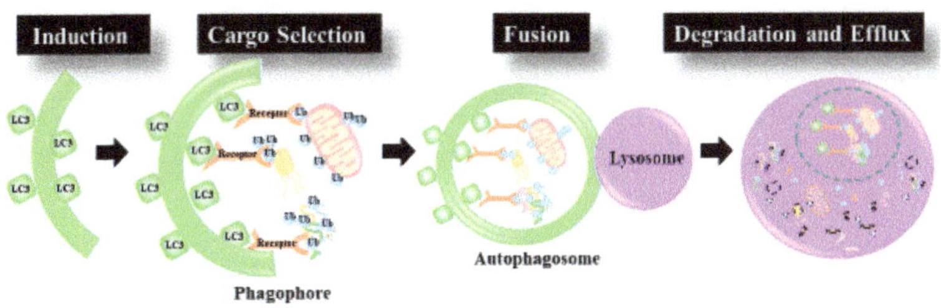

Figure 1. Autophagy Induction: first a pre-autophagosome (phagophore) is formed by the lipidation of LC3. Cargo Selection: Ubiquitinated cargo is tagged for degradation by autophagy receptors equipped with an LC3 binding motif and a ubiquitin binding domain. The Autophagosome is then elongated around the cargo to be degraded. Fusion: the mature autophagosome fuses with the lysosome. Degradation and efflux: The contents of the autophagosome are degraded by proteolytic enzymes within the lysosomal lumen and nutrients are released.

Once considered an indiscriminate bulk degradation process, autophagy is now well recognized for its selectivity. With the aid of various receptors and signaling molecules, autophagy can discern between healthy cellular components and toxic organelles, proteins, as well as invading organisms. Indeed, constitutive autophagy can be extremely selective and is responsible for the removal of cellular components, like misfolded protein aggregates or exhausted organelles, to ensure a proper physiological turnover [21]. Moreover, autophagy induced by a specific trigger can also be very selective; for instance, xenophagy induced by a pathogen infection triggers a selective autophagic response against the invading organism. Likewise, when mitochondria are chemically depolarized, mitophagy ensues and targets strictly depolarized mitochondria [22–24]. Contrarily, autophagy induced by severe nutrient depletion may not be overly selective, as its sole goal is nutrient liberation under conditions of cellular stress. With this in mind, various types of substrates are eliminated through selective autophagy and this number is continuously on the rise. These include protein aggregates (aggrephagy/proteophagy), mitochondria (mitophagy), endoplasmic reticulum (reticulophagy/ER-phagy), peroxisome (pexophagy), nucleus (nucleophagy), pathogens (xenophagy), lipids (lipophagy), and even lysosomes themselves (lysophagy) (Figure 2) [23,25]. During selective autophagy, dysfunctional or obsolete proteins and organelles are targeted for degradation by various receptors, which, in turn, entice the arrival of the autophagosome. The selectivity in this process is largely mediated by PTMs such as phosphorylation and ubiquitination, with UFMylation, ISGylation, and SUMOylation recently arising as potential PTMs, similar to Ubiquitination, that modulate autophagy [17].

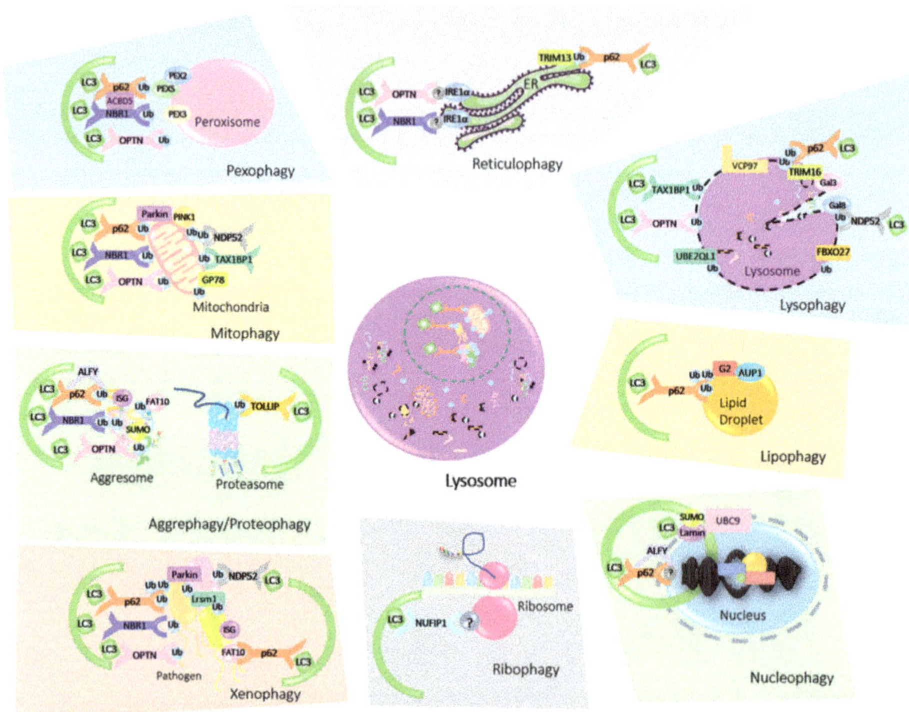

Figure 2. Mammalian Ub-mediated Organellophagy. Damaged, dysfunctional or otherwise superfluous organelles are tagged for degradation by ubiquitin (Ub) or ubiquitin-like molecular modifiers (SUMO, FAT10, ISG15). These post translational modifications (PTMs) are then recognized by autophagy cargo receptors (p62, NBR1, OPTN, NDP52, TAX1BP1, NUFIP1) which bind both the ubiquitinated cargo and the microtubule-associated protein 1B light chain 3B (LC3B). The autophagosome engulfs the receptor cargo complex and delivers it to the lysosome for breakdown. *ER-Phagy/Reticulophagy*: ER stress sensor IRE1α, is activated upon ER stress. IRE1α interacts with p62, OPTN, and NBR1 to promote reticulophagy, but whether Ub is involved in this process is unclear. Self-ubiquitination of the E3 ligase TRIM13 also recruits p62 to damaged ER. This results in the sequestration of the ER within autophagosomes and its delivery to the lysosome for elimination. *Pexophagy*: PEX2 ubiquitinates the peroxisomal import receptor PEX5 which is then recognized by autophagy receptors p62 and NBR1. PEX3 is also involved in this process and enhances peroxisomal ubiquitination, while peroxisomal acyl-CoA binding domain containing 5 (ACBD5) regulates the pexophagy receptor protein complex. OPTN was also shown to be involved in this process. *Mitophagy*: Mitochondrial damage results in the stabilization of PINK1 on the mitochondrial outer membrane, which entices the arrival of E3 ligases Parkin and GP78. These ligases, in turn, ubiquitinate a variety of substrates on the mitochondrial membrane which recruits autophagy receptors p62, NDP52, OPTN, NBR1, and TAX1BP1. *Aggrephagy/Proteophagy:* Protein aggregates are

recognized by selective autophagy receptors p62, NBR1, OPTN, and TOLLIP, with autophagy-linked FYVE protein (ALFY) acting as a scaffold protein promoting the assembly of p62 bodies. Cue5 the yeast homolog of mammalian TOLLIP binds ubiquitinated proteasomes targeting them for proteophagy. Ubiquitin and ubiquitin like PTMs (UBLs SUMO, FAT10, and ISG15) also play a role in aggrephagy. *Xenophagy:* The E3 ligases Parkin and Lrsm1 ubiquitinate invading pathogens which are then detected by p62, NBR1, NDP52, and OPTN. Other UBLs have also been demonstrated to promote xenophagy including FAT10 and ISG15. With FAT10 interacting directly with p62. Recognition by cargo receptors then promotes the incarceration of pathogens within autophagosomes and their delivery to lysosomes for elimination. *Ribophagy:* The large ribosomal subunit is targeted for autophagosomal degradation by the cargo receptor NUFIP1, however, the involvement of Ub or other UBLs in this process has not been determined. *Nucleophagy:* The selective autophagy receptor p62 and autophagic adaptor ALFY, have been documented to shuttle between the nucleus and cytoplasm, however whether Ub is involved in this process is not known. SUMOylation, was documented to drive nucleophagy where DNA damage promotes the accumulation of SUMO E2 ligase UBC9, resulting in the SUMOylation of lamin A/C. *Lipophagy:* Ancient Ubiquitous Protein 1 (AUP1) localizes to LDs and recruits the E2 ubiquitin-conjugating enzyme G2, which ubiquitinates targets on the lipid droplet, thus inviting autophagy receptor p62. *Lysophagy:* Glycans are exposed following lysosomal membrane rapture and are identified by galectins-3 and -8 (Gal3/8) or by the SCF ubiquitin ligase FBXO27. Gal3 then stimulates ubiquitination by TRIM16 and recruits the autophagy receptor p62, while Gal8 engages NDP52. The E2 ubiquitin-conjugating enzyme UBE2QL1 was also found to participate in this process, as were autophagy receptors TAX1BP1 and OPTN. The AAA-ATPase VCP97 is another essential player in the elimination of injured lysosomes.

Moreover, there are two ubiquitin-like (UBL) protein conjugation systems that act to promote the formation of the autophagosome precursor, the phagophore [26]. These are involved in the conjugation of Atg12, and ultimately, microtubule-associated protein light chain 3 (LC3) and its close relatives GABARAP and GATE16 [27–29]. One pathway involves the covalent conjugation of Atg12 to Atg5. This occurs with the help of Atg7 and Atg10, E1- and E2-like enzymes, respectively. The Atg12–Atg5 heteromer then interacts with Atg16-like 1 (Atg16L1) [30]. This complex acts as a *de facto* E3-like ligase to promote the second conjugation reaction, involving LC3. The conjugation of LC3 to phosphatidylethanolamine (PE) is essential for phagophore expansion and sealing. LC3 exists in its inactive form free in the cytosol and must first be cleaved by the protease Atg4 giving rise to LC3-I [31]. The E1-like enzyme Atg7 and the E2-like enzyme Atg3 are then tasked with priming LC3-I for its final conjugation to PE, by the E3-like complex of Atg12-Atg5. Lipid conjugation converts LC3-I, to the LC3-II form that is attached to either side of the growing phagophore membrane.

4. Selective Autophagy

The benefits of selective autophagy are vast, as it allows for the surgical removal of targeted substrates. For instance, the removal of mitochondria during hypoxia, organelle damage, or even when the organelle becomes obsolete, such as during erythrocyte maturation, all occur selectively through mitophagy [32–34]. In bulk autophagy, the ubiquitination of autophagic components often acts as a regulatory signal. However, when it comes to selective autophagy ubiquitination also acts as a signal for cargo recognition, process initiation, and rate of autophagosome formation. For instance, the ubiquitination of components of the PI3K-Beclin1 complex contributes to the regulation of autophagic induction and autophagosome formation [35,36]. Selective autophagy relies on core autophagy components but is also aided by specific selective autophagy receptors. The cargo selection step of this process is highly dependent on these receptors (Table 1).

Generally, during selective autophagy, protein aggregates, damaged organelles or portions thereof are identified and tagged for degradation by E3 Ub-ligases. Once cargos are tagged, ubiquitin acts as an "eat me" signal for autophagy receptors, which further flag targets for degradation. Autophagy receptors are also implicated in segregating and coalescing materials destined for autophagic degradation, effectively preparing them for the arrival of the phagophore. Indeed, autophagy receptors harbor a

UBL domain to sense Ub molecules, and a LIR domain to bind the LC3s/GABARAPs, which are present on autophagosomal membranes. By binding to mATG8s, cargo receptors promote the recruitment of autophagosomal membranes around the Ub-tagged materials, eventually encapsulating them completely. Mature autophagosomes are then delivered to the lysosome for degradation [37,38].

Considering that potentially all cytosolic components could be selectively degraded via lysosomes, it is plausible that selective autophagy plays a role in a number of diverse physiological processes and their associated diseases. The ability to specifically target the elimination of desired cellular constituents presents great therapeutic potential for the treatment of many diseases characterized by the accumulation of toxic materials including neurodegenerative diseases like Huntington's and Alzheimer's, lysosomal storage diseases, and mitochondrial DNA disorders [39,40]. Moreover, selective autophagy has also been implicated in chronic obstructive pulmonary disease (COPD) [41] as well as other pulmonary disorders [42]. Of note, one of the most well-studied functions of selective autophagy in human health is its role in infections [43]. Once bacteria invade the host cell, they are immediately labeled with ubiquitin chains and selectively degraded via xenophagy [44]. The removal of viral pathogens by selective autophagy is also termed virophagy [45,46]. Of note, host cells counteract some types of viral infections, like flaviviruses and Ebola virus, by directly eliminating their replicative niche in the endoplasmic reticulum, thus activating another form of selective autophagy named ER-phagy [47,48].

5. Cargo Receptors

Most autophagy cargo receptors are characterized by the presence of a ubiquitin-binding domain (UBD), enabling the detection of marked cargo, and an LC3-interacting region (LIR) allowing for interaction with LC3/GABARAP/GATE16 proteins on the autophagosomal membrane [8]. These properties permit receptors to act as middlemen, effectively bridging tagged cargos with autophagosomes. Indeed, receptors accompany their targets to the lysosome and are degraded alongside them. The UBD, composed of either a UBA, UBAN (ubiquitin-binding in ABIN and NEMO), or ubiquitin-binding zinc finger motif, can recognize various flavors of Ub chain linkages, and its affinity for ubiquitinated substrates can be further enhanced through phosphorylation [49–51]. The core of the LIR motif consists of a minimal consensus sequence W/F/Y-x-x-L/V/I, which is common to most cargo receptors [52], and the phosphorylation of this domain was also demonstrated to augment binding affinity to LC3 and related proteins [53]. Of note, some cargo receptors like NDP52 and TAX1BP1 present an unconventional LIR domain (LVV) that is missing the consensus aromatic amino acid. This non-canonical LIR is important for the function of NDP52 and TAX1BP1 in xenophagy and seems to confer the receptor preferential binding to LC3C, at least in the case of NDP52 [54,55].

Although LIR was the first domain recognized to mediate the binding of ATG8s to autophagy adaptors and receptors, another domain with a similar function has been recently characterized. Several ATG8 interactors exploit a ubiquitin-interacting motif (UIM)-like sequence that confers them with a high affinity for autophagy modifiers [56]. However, the presence of a LIR or UIM motif does not "a priori" confer cargo receptors ATG8-binding abilities. A real interaction should always be experimentally confirmed. Nevertheless, the characterization of the UIM domain greatly expands the number of cellular proteins that could serve as autophagy receptors and adaptors.

The first cargo receptor to be identified was p62 and it is undoubtedly the most versatile and well-studied receptor to date. p62 is ubiquitously expressed and mediates the removal of a wide range of ubiquitinated proteins, protein aggregates, and organelles [57]. Due to its common role in sequestering ubiquitin positive protein aggregates, it earned the name sequestosome-1 (SQSTM1) and was demonstrated to selectively bind to ubiquitinated protein aggregates such as neurofibrillary tangles, Lewy bodies, tau and α-synuclein and alike [58]. p62 contains both a UBA and an LIR domain, thus mediating the association between cargos and autophagosomes [59]. p62 itself is degraded alongside its cargo, as it accompanies the targeted components to their final destination. Indeed, conditions that are characterized by defective autophagy often display large accumulations of p62 along with ubiquitinated proteins. Interestingly, in some cases elimination of p62 in autophagy-deficient models results in protein aggregate clearance [10], suggesting a role for p62 in the coalescing of materials for autophagic degradation. In addition to p62, many other cargo receptors have been described to participate in various types of selective autophagy including NBR1 (neighbor of BRCA1 gene 1), NDP52 (nuclear dot protein 52 kDa), Optineurin (OPTN), TAX1BP1 (Tax1-binding protein 1), and TOLLIP (Toll-interacting protein) (Table 1), with the number of receptors continuing to grow as more types of selective autophagy are identified.

A screen for Atg8 interactors identified NBR1 as an autophagy receptor containing both LIR and a UBA ubiquitin-binding domain. NBR1 is recruited to Ub-positive protein aggregates and is degraded by autophagy, a process that requires a functional LIR domain [49,60–62]. NBR1 and p62 both act as selective autophagy receptors but appear to function independently as NBR1-positive aggregates are cleared by autophagosomes even in the absence of p62. The depletion of NBR1 abolishes the formation of Ub-positive p62 bodies upon the puromycin treatment of cells [61].

Similar to p62 and NBR1, Optineurin (OPTN) is a selective autophagy receptor equipped with a UBAN-type UBD and a LIR [50,53]. OPTN is present in protein inclusions observed in various neurodegenerative diseases including amyotrophic lateral sclerosis (ALS), Huntington's, Alzheimer's, Parkinson's, Creutzfeld-Jacob's, and Pick's disease. Its depletion significantly increases protein aggregation in HeLa cells and morpholino-silencing of the OPTN ortholog in zebrafish causes a motor axonopathy phenotype similar to ALS [63]. Mutations in OPTN's UBAN and LIR domains abolish its ability to bind Ub and LC3/GABARAP modifiers, respectively; these mutations do not appear to compromise its ability to colocalize with protein aggregates, suggesting that OPTN interacts with protein aggregates in a manner that is ubiquitin and LC3 -independent [53]. Indeed, OPTN recognizes various protein aggregates via its C-terminal coiled-coil domain in a ubiquitin-independent manner. Moreover, TANK1 binding kinase 1 (TBK1) phosphorylates OPTN, which regulates its ability to interact with autophagy modifiers and clear protein aggregates [53,63]. OPTN can itself be ubiquitinated by the E3 ligase HACE1 (HECT domain and ankyrin repeat-containing E3 Ub protein ligase 1) which promotes its complexing with p62 and enhances autophagy flux [64]. The HACE1-OPTN interaction synergistically suppresses the growth and tumorigenicity of lung cancer cells [64].

Thus, selective autophagy relies on receptors for the targeting and coalescing of materials into autophagosomes for degradation. Numerous such receptors have been identified, each with its own repertoire of autophagy targets.

6. Mitophagy

Mitochondria, the powerhouse of the cell, provide energy in the form of ATP to fuel most cellular activities. However, these powerful organelles also produce toxic reactive oxygen species (ROS) as a by-product of oxidative phosphorylation. This leaves them vulnerable to ROS-induced damage and dysfunction. Recycling damaged mitochondria is therefore vital for cellular health and bioenergetics. The wholesale removal of mitochondria occurs through a selective process termed mitophagy. Mitophagy is ever-ongoing at low levels and accounts for routine mitochondrial turnover which occurs approximately every 7–14 days, but may be accelerated during energetic imbalance or oxidative stress [65]. Mitophagy also participates in the pruning of healthy mitochondria during various physiological processes such as erythrocyte maturation and the elimination of paternal mitochondria during oocyte fertilization [34,66–68]. Interestingly, the stimulation of mitochondrial oxidative phosphorylation has also been demonstrated to enhance mitophagy through the small GTPase Rheb [69]. This acts as a pre-emptive mechanism to facilitate mitochondrial renewal in response to enhanced metabolic demands. Here, we provide evidence to the role of ubiquitin in targeting mitochondria for autophagic degradation, mechanisms of mitophagy have been extensively reviewed elsewhere and are beyond the scope of this review [22,24].

When mitochondria withstand irreversible damage and all other quality control mechanisms have failed to restore proper mitochondrial function, mitophagy is triggered. The loss of mitochondrial membrane potential and integrity results in the stabilization of mitochondrial health sensor PINK1 (PTEN-induced putative kinase protein 1) on the organelle's outer membrane. PINK1 serves as a docking station for the arrival of the cytosolic E3 ubiquitin ligase Parkin to damaged mitochondria. The phosphorylation of Parkin and ubiquitin chains by PINK1 ensures Parkin is retained on the surface of depolarized mitochondria effectively tagging damaged mitochondria for degradation [70–73]. To keep mitochondria stationary, PINK1 phosphorylates Miro, a component of the motor adaptor complex. Miro is responsible for anchoring the motor protein kinesin to the mitochondrial surface and facilitating mitochondrial movement along microtubule tracks. Miro phosphorylation predisposes it for degradation in a Parkin-dependent manner [74] thus, keeping mitochondria at a standstill. Moreover, Parkin also mediates the ubiquitination and degradation of mitochondrial fusion effectors, Mitofusins 1 and 2 (MFN1 and MFN2). This ensures the isolation of the damaged organelle. Parkin has also been documented to ubiquitinate a variety of other outer mitochondrial membrane proteins such as the voltage-dependent anion channels (VDACs), translocases of the outer membrane (TOMs), as well as many others [73]. The ER-associated E3 ligase Glycoprotein 78 (Gp78) was also found to ubiquitinate mitochondrial outer membrane proteins such as MFN1 and induce mitophagy of depolarized mitochondria independently of Parkin [75]. On the other hand, deubiquitinase USP30 (ubiquitin carboxyl-terminal hydrolase 30) acts in opposition to Parkin, removing Ub from depolarized mitochondria thus blocking mitophagy [76]. Defective mitophagy, resulting from deficiencies in Parkin or PINK1 can be rescued by the knockdown of USP30 [76]. The presence of ubiquitinated proteins on the mitochondrial outer membrane, act as signals for the arrival of autophagy receptors NBR1, p62, OPTN, NDP52, and TAX1BP1 [61,77–79]. However, some evidence suggests that p62 may not be required for mitochondrial degradation by autophagy [22,79,80]. Indeed, a comprehensive study demonstrated NDP52 and OPTN to be the main receptors responsible for PINK1- and Parkin-mediated mitophagy, where they also activate and recruit core autophagic machinery [79].

7. Aggrephagy/Proteophagy

Busy making proteins to sustain life, cells make mistakes in protein folding, proteins get damaged and oxidized by toxic metabolic byproducts, and certain proteins, such as polyglutamine (polyQ), are inherently prone to aggregation. Aggrephagy refers to the autophagic degradation of protein aggregates, which aberrantly accumulate due to processing mistakes, damage, or proteasomal dysfunction [81]. Protein aggregates are common in proteinopathies such as neurodegenerative diseases and aging [21]. The usual aggregate prone suspects, including amyloid-β [82], huntingtin [83,84], and alpha synuclein [85], can all be degraded through this process. Ubiquitination is vital for the autophagic removal of protein aggregates in proteinopathies [86–89], with p62, NBR1, and OPTN all participating as cargo receptors in mammals [23]. Scaffolding proteins such as ALFY also appear to play a role in this process by promoting the assembly of p62 bodies and their degradation by autophagy [90,91]. The most recently identified aggrephagy receptor, in yeast, is Cue5 (coupling of ubiquitin to ER degradation-5). Cue5 promotes aggrephagy by facilitating the interaction between Ub and Atg8 on protein aggregates. Cue5 also acts as an autophagy receptor for proteasomal subunit degradation, known as proteophagy, by binding to ubiquitinated proteasomes and promoting their removal by autophagy [92–94]. The overexpression of TOLLIP, the human homolog of Cue5 has been demonstrated to mediate the degradation of polyQ proteins in human cell lines [95].

Moreover, other UBLs including SUMOylation (the post-translational modification of proteins with small ubiquitin-like modifiers), FAT10 (the addition of the cytokine-inducible HLA-F adjacent transcript 10, also known as ubiquitin D, onto proteins) and ISGylation (the covalent attachment of Interferon-stimulated gene 15 (ISG15) to substrates) have been documented to play a role in aggrephagy. SUMOylation, like ubiquitination, involves lysine residue modifications of proteins through a three-step process involving distinct enzymes: SUMO E1 (the heterodimer SAE1 and SAE2), SUMO E2-conjugating enzyme (UBE2I/UBC9) and several E3 ligases [21,22]. SUMO-1 was found to accelerate the accumulation of autophagic vacuoles in neurons, thus increasing the production of amyloid-β (Aβ) [96]. FAT10 was found to regulate the solubility of polyQ proteins, such as huntingtin [97], and ensure the delivery of aggregated or damaged proteins to the aggresome during proteasome dysfunction by interacting with histone deacetylase 6 (HDAC6) [98]. Similarly, ISGylation marks protein aggregates for autophagic disposal and also interacts with p62 and HDAC6 to enhance aggrephagy [99].

8. ER-Phagy/Reticulophagy

The endoplasmic reticulum is a jack of many trades, involved in proper protein folding, processing and secretion, as well as calcium homeostasis, and lipid synthesis. The ER has also been documented to contribute membranes to growing autophagosomes [100–102]. The ER is a continuous membrane structure composed of an extensive network of sheets and tubules [103]. The accumulation of unfolded proteins induces an ER stress response known as the unfolded protein response (UPR). This program activates the following ER stress sensors: inositol-requiring protein 1 (IRE1), protein kinase RNA-like ER kinase (PERK) and activating transcription factor 6 (ATF6) [104], which bring protein synthesis to a halt, enhance the degradation of misfolded proteins, and increase the expression of molecular chaperones. If these actions do not resolve ER stress, the selective autophagy of the ER or ER-phagy will ensue in a final attempt to avoid the induction of apoptosis [105,106]. Through this process, the ER undergoes significant remodeling and is targeted for degradation by the following cargo receptors: FAM134B, SEC62, RTN3, CCPG1, ATL3, and TEX264 [107–113]. Thus far, the role of ubiquitin in ER-phagy, and its relationship with cargo receptors involved in this process, remains largely unknown. However, some recent data suggest that ubiquitination does indeed play a role in this process. In this scenario, the N-degron pathway mediates ER-phagy through the auto-ubiquitination of the transmembrane E3 ligase TRIM13. Once ubiquitinated, TRIM13 signals for the arrival of cargo receptor p62 thus recruiting the phagophore to the damaged organelle. This is dependent on an N-degron destabilizing residue on the N-terminal arginine (Nt-Arg) [114]. p62, in turn, facilitates the encapsulation of ER into autophagosomes and their turnover in the lysosome, by binding to ER transmembrane protein IRE1α [115,116]. IRE1α has also been demonstrated to bind other Ub binding receptors such as OPTN and NBR1 [115].

9. Ribophagy

Ribosomes are the cellular protein manufacturing facilities and can be found free-floating in the cytosol or attached to the ER. These tiny factories sustain life through continuous protein biosynthesis. During starvation, ribosomes are selectively eliminated via ribophagy to generate nucleotides/nucleosides required for cellular survival [117].

In yeast selective degradation of mature ribosomes is mediated by the Ubp3p ubiquitin protease complex composed of Ubp3p, Bre5, Cdc48, and Ufd3 [117,118]. This complex deubiquitinates Rpl25 on the large ribosomal subunit, which results in its degradation. This process is antagonized by Ltn1-mediated ubiquitination of the same residue on Rpl25, creating a dynamic and selective ribophagy signal. Therefore, ubiquitin-mediated regulation of ribophagy, in yeast, is counterintuitive to other types of selective autophagy. Where, in most other types of selective autophagy ubiquitination functions as a marker for cargo recognition, in ribophagy ubiquitination appears to be protective, and must be removed to trigger autophagic degradation, at least in yeast. Interestingly this interplay only controls the turnover of the large ribosomal subunit. This suggests ribophagy is even more selective as independent mechanisms are responsible for the degradation of the different ribosomal subunits [117].

Ribophagy has also been demonstrated to occur in human cells in response to starvation, mTOR inhibition, and treatment with cellular stressors such as sodium arsenite and reversine [119]. However, the molecular mechanisms that regulate ribophagy are not clear and some controversial theories co-exist. NUFIP1 (nuclear fragile X mental retardation–interacting protein 1) was initially characterized as a selective ribophagy receptor, responsible for mediating ribosomal degradation by directly interacting with LC3B and ribosomes during starvation [120]. Moreover, ribophagy appeared to be particularly important during nutrient depletion as it represents a source of nucleosides for starving cells [120]. Further studies on ribophagy revealed that the delivery of ribosomes to lysosomes could follow one of three paths: (i) random engulfment of ribosomes into autophagosomes during bulk autophagy; (ii) by-stander flux following ER-phagy induction and (iii) receptor-mediated selective autophagy [113,119]. Moreover, recent proteomic studies revealed that during acute nutrient stress, ribosomal protein degradation occurs mainly via a non-autophagic pathway. No obvious differences in ribosomal protein levels were detected between wild-type and *ATG7* or *RB1CC1* deficient cells following treatment with Torin 1. Moreover, similar results were obtained in NUFIP1 deficient cells, bringing the role of this protein in ribophagy into question [121].

Although the involvement of ubiquitination in ribophagy has not been determined yet, ribosomal protein ubiquitination has been demonstrated to occur under stressful conditions such as translation stalling, oxidative stress, and UPR [122–125]. Indeed, the E3 ligases HUWE1 (HECT, UBA, and WWE Domain Containing Protein 1) and UBE2O (Ubiquitin Conjugating Enzyme E2 O) have been implicated in the ubiquitination and proteasomal degradation of ribosomal proteins [126,127].

10. Pexophagy

Peroxisomes are small organelles responsible for lipid metabolism and redox regulation in most eukaryotic cells. As both producers and scavengers of reactive oxygen species (ROS), these organelles play a multipurpose role in cellular signaling and metabolism. The peroxisomal lifecycle is quite dynamic, with a half-life of only about two days. The elimination of peroxisomes takes place through a selective autophagy process termed pexophagy. Like other selective autophagy substrates, peroxisomes must be ubiquitinated and decorated by cargo receptors to be targeted for elimination.

PEX5 is an import receptor that shuttles between the peroxisomal membrane and the cytosol. Phosphorylation of PEX5 by ataxia-telangiectasia mutated (ATM) predisposes it to mono-ubiquitination, which blocks its export and signals for peroxisomal degradation by autophagy [128,129]. The E3 ubiquitin ligase Peroxin 2 (PEX2) has been implicated in the ubiquitination of PEX5, and PMP70 during starvation [130]. Moreover, PEX14, a peroxisomal membrane protein that acts as PEX5's docking partner can directly interact with LC3B-II during starvation. An additional peroxisomal import factor, PEX3, has also been documented to play a role in pexophagy, where the overexpression of PEX3 induces ubiquitination and elimination of peroxisomes by autophagy [131].

Selective autophagy receptors NBR1 [62] and p62 [132] were both shown to participate in this process. Indeed, NBR1 was demonstrated to be both sufficient and required to drive peroxisomal elimination by pexophagy, a role that requires its amphipathic α-helical J domain, ubiquitin-associated (UBA) domain, LC3-interacting region, and coiled-coil domain. p62 binding to NBR1 does not appear required for its function but does enhance pexophagy [62]. Moreover, some emerging evidence is suggesting that the peroxisomal protein ACBD5 (Acyl-CoA-binding domain containing protein 5), may be important for regulating the pexophagy receptor protein complex [133]. However, pexophagy research is still in its infancy and more studies are required to further elucidate the molecular mechanisms regulating this process.

11. Lysophagy

Lysosomes are acidic membrane-bound organelles, which contain a wide variety of hydrolytic enzymes required for the degradation of various substrates including proteins, lipids, carbohydrates, nucleic acids, and invading pathogens. Lysosomes play a key role in cellular homeostasis by controlling both cellular clearance, and energy production in response to environmental cues [134]. These organelles contain a slew of hydrolases (e.g., proteases, nucleases, esterases, polysaccharidases, and glycosidases) capable of degrading a wide spectrum of components. Therefore, upon injury the spilling of lysosomal contents poses a threat to cellular health and can culminate in inflammation and cell death [135,136].

It is not surprising then that damaged lysosomes themselves undergo a selective form of autophagy, termed lysophagy.

Several cellular events can render the lysosomal membrane prone to permeabilization including excessive cholesterol, toxic aggregates such as α-synuclein, mutant huntingtin, and Aβ or tau fibrils, as well as alterations in lipid composition and ROS, just to name a few. Physical damage to lysosomes induced by silica, lysosomotropic agents, and urate crystals have all been documented to induce lysophagy [137]. Upon lysosomal membrane rupture and failing any repair efforts, glycans that normally reside within the lysosomal lumen are exposed. These glycans are then identified by galectins-3 and -8 or by the SCF ubiquitin ligase, FBXO27, leading to the ubiquitination of lysosomal components [138,139]. Indeed, Galectin-3 (Gal3) stimulates the ubiquitination of lysosomal proteins by TRIM16 [140,141], thus recruiting the autophagy receptor p62, while Gal8 engages NDP52 [142].

Moreover, it appears that both K63- and K48-linked ubiquitination are involved in lysophagy. The E2 ubiquitin-conjugating enzyme UBE2QL1 ubiquitinates lysosomal membrane proteins with K48-linked Ub chains [4,143]. While p97 (Valosin-containing protein) translocates to the lysosome to form a complex with UBXD1, PLAA, and YOD1. This complex acts downstream of K63 ubiquitin chains and removes K48-linked ubiquitin conjugates from damaged lysosomes to promote the recruitment of autophagic membranes [4].

In this process ubiquitinated lysosomes recruit autophagy receptors which promote their engulfment within autophagosomes. They are subsequently cannibalized by healthy lysosomes [144]. Interestingly, multiple cargo receptors have been demonstrated to play a role in this process including p62 [4,145], NDP52 [142], TAX1BP1 [143], and OPTN [146]. These are essential for forming a docking platform for ULK1, which activates the core autophagic machinery and induces autophagosome formation regulated by TBK1 [146].

This fits well within the autophagy wheelhouse and can be particularly useful for host-defense. Indeed, one-way pathogens are able to establish an infection, is by breaking vacuolar membranes and escaping into the cytoplasm [147]. This has been documented during Shigella infection, where host cell-membrane remnants attached to a bacterium, activate an autophagic response resulting in pathogen removal [148].

12. Xenophagy

Xenophagy refers to the selective removal of invading pathogens and constitutes an important faction of innate and acquired immune responses. Autophagy plays a role in combating a number of pathogens including bacteria such as *Streptococcus pyogenes*, *Mycobacterium tuberculosis*, *Listeria monocytogenes*, *Shigella flexneri*, *Salmonella enterica*, and *Toxoplasma gondii*. Similarly, viruses such as Sindbis are also targeted by autophagy [149]. Like other forms of selective autophagy, pathogens are first tagged for degradation by Ub [148,150–152] and are then recognized by cargo receptors. This process also requires functional core autophagy machinery. Ubiquitin is vital for xenophagy and is responsible for linking bacteria-containing endosomes with autophagic machinery, such as Atg16L1 [145]. Some examples of E3 ubiquitin ligases that function as xenophagy regulators are: ARIH1/HHARI (ariadne homologue 1) and LRSAM1 (leucine rich repeat and sterile alpha motif containing 1) that ubiquitinate cytosolic *Salmonella Typhimurium* [153,154]. Linear ubiquitin chains have also been detected on the surface of Salmonella; therefore, the E3 complex LUBAC (linear ubiquitin chain assembly complex) and its specific DUB OTULIN (OTU Deubiquitinase With Linear Linkage Specificity) play an active role in xenophagy [155,156]. Moreover, Parkin was shown to ubiquitinate invading bacteria *Mycobacterium tuberculosis* [157]. The cargo receptors p62 [158,159], NBR1 [61], NDP52 [158,160], and OPTN [53] have all been demonstrated to facilitate the sequestration of invading bacteria in autophagic vesicles.

Interestingly, other PTMs such as interferon-α,β inducible ubiquitin-like modifiers FAT10 and ISG15 have also been implicated in xenophagy. Where FAT10 was found to decorate autophagy-targeted *Salmonella* and recruit p62, thus contributing to resistance against *Salmonella* infection in mice, through the induction of xenophagy [18]. Similarly, ISG15 was found to be essential for optimal mycobacterial immunity, by mediating the release of IFN-γ-inducing secreted molecule [161,162]. Indeed, enhanced ISGylation in cells and animals results in greater basal and infection-induced autophagy. This is largely mediated by temporary metabolic reprogramming induced by the ISGylation of mTOR, WIPI2, and AMBRA1 following infection [16].

13. Lipophagy

Autophagic degradation of lipid droplets, the organelles responsible for cellular lipid storage, is termed lipophagy. Although it was first described over a decade ago, the molecular mechanisms involved in this process remain largely elusive [163]. Lipophagy is particularly important in the liver, where it is responsible for lipid mobilization in response to a multitude of cellular stressors. Indeed, lipid droplets were found to co-localize with autophagy marker LC3B in mouse liver [163]. Later studies uncovered that Huntingtin, the protein mutated in Huntington's disease, is involved in stress-induced lipophagy and appears to act by facilitating the binding between LC3B and p62 [164]. Another protein, Ancient Ubiquitous Protein 1 (AUP1), has been shown to localize to LDs and act as a recruiter for the E2 ubiquitin conjugase G2 (Ube2g2). These findings were the first to suggest that the ubiquitination machinery localizes to LDs and promotes their degradation by autophagy [165]. Moreover, ethanol-induced lipophagy occurs through the increased presence of Ub and p62 on LDs [166]. It appears that ubiquitin may play a role in this selective autophagy process, but solid evidence to this effect, as well as the molecular mechanisms involved, remain to be illuminated.

14. Nucleophagy

The nucleus is the "brain" of the cell, entrusted with safeguarding the genome and dictating cellular activities by controlling gene expression. The health and integrity of this organelle is therefore the cell's priority. Nuclear recycling by autophagy, known as nucleophagy, is an evolutionarily conserved process among eukaryotes and is responsible for the degradation of nuclear components. Although in some multinucleated eukaryotic cells the nucleus is degraded wholesale by autophagy [167], in most eukaryotes, nucleophagy is thought to occur piecemeal, and require a series of events [168]. This process begins by first sensing and flagging damaged components which are then exported from the nucleus, encapsulated within the autophagosome, and eventually degraded by the lysosome. The precise molecular mechanisms, and whether protein ubiquitination is involved, remain largely unexplored.

In yeast nuclear recycling was documented to occur piecemeal through micronucleophagy [169,170], or by macronucleophagy with the help of the autophagy receptor Atg39 [171], under physiological and nutrient stress conditions, respectively. Nucleophagy has also been described to occur in differentiating murine and human keratinocyte, where perinuclear vesicles containing histone interacting protein and heterochromatin protein 1α, were shown to localize near Lamin A and B1 [172]. Although nucleophagy is not well understood in mammals, dysfunction in this process appears to contribute to pathologies such as cancer and neurodegeneration and is also implicated in cellular senescence and aging [168]. It is unclear, however, whether nucleophagy is an ongoing physiological process that is further triggered by pathology, as a maladaptation or protective mechanism, or rather strictly arising due to pathology. It is also unclear if Ub plays a role in this process, and thus far no autophagy receptors have been identified to participate in mammalian nucleophagy. Several autophagy-related proteins, including the selective autophagy receptor p62 and autophagic adaptor ALFY, have been documented to shuttle between the nucleus and cytoplasm [90,168,172]; however, there is no clear-cut evidence for their role in nucleophagy yet.

Interestingly, recent studies identified SUMOylation to play an important role in cellular activities including nucleophagy. DNA damage was shown to facilitate the accumulation of SUMO E2 ligase UBC9, and SUMOylation of lamin A/C [173]. This modification mediated the interaction between lamin A/C and LC3B, thus promoting nucleophagy. The knockdown of UBC9 prevented SUMOylation of lamin A/C and its interaction with LC3B, which attenuated nucleophagy and resulted in the degradation of nuclear components and leakage of nuclear DNA. SUMOylation has been previously implicated in the autophagic degradation of the polyglutamine (polyQ) protein ataxin-3, thus suggesting a role for SUMOylation in aggrephagy [19].

Table 1. Ubiquitin dependent autophagy receptors and their function.

Autophagy Receptor	Substrates	Function and Associated Pathologies	Refs
SQSTM1/p62	General autophagy Protein aggregates Mitochondria Pathogens Lipids Peroxisomes Lysosomes	The most universal autophagy receptor, p62, is involved in cellular stress response and the clearance of protein aggregates, defective organelles as well as invading pathogens. Defects in p62 are associated with Paget disease of bone, amyotrophic lateral sclerosis, and frontotemporal lobar degeneration	[58,59,80,159,174,175]
NBR1	Protein aggregates Mitochondria Peroxisomes	NBR1 is involved in aggrephagy and mitophagy, but is the main receptor for pexophagy	[60–62,130,176]
OPTN	General autophagy Protein aggregates Mitochondria Pathogens	OPTN acts as an autophagy receptor for several substrates. Its phosphorylation by TBK1 enhances its function thus facilitating the clearance of Salmonella; it acts as primary mitophagy receptor; mutations in OPTN were found to cause amyotrophic lateral sclerosis, and OPTN in present in protein inclusions found in several neurodegenerative diseases	[53,78,146]
NDP52	Mitochondria Pathogens	NDP52 interacts with LC3-C via noncanonical LIR motif and facilitates autophagosome maturation. It is the primary mitophagy receptor and collaborates with p62 for pathogen clearance	[62,158,177]
TAX1BP1	MitochondriaPathogens Lysosomes	Promotes autophagy flux in activated T cells, and is recruited to damaged lysosomes facilitating their elimination	[143,178]
TOLLIP	Protein aggregates	Facilitates degradation of protein aggregates such as huntingtin-derived polyQ proteins	[95]

15. Conclusions

Ubiquitin is a universal degradation signal used to tag proteins, organelles, and pathogens for disposal by the ubiquitin-proteasome and autophagy-lysosome systems. A large number of selective organelle-phagies are regulated, at least in part, by ubiquitination or other ubiquitin-like modifications. Ubiquitin acts as a beacon for the arrival of selective cargo receptors, and although the various selective types of autophagy rely on specific E3 ligases, of which there are upwards of 600, only a handful of Ub-dependent selective autophagy receptors have been identified thus far. Moreover, selective autophagy relies on common core autophagy machinery for the induction and formation of autophagosomes. This suggests common elements amongst the various types of selective autophagy which have likely evolved from bulk autophagy.

Despite the remarkable growth of knowledge in the field, spearheaded by talented "autophagians", there is still much work to be done. The continuous identification of new selective types of autophagy must be followed up by a detailed characterization of the molecular mechanisms governing them. Likewise, novel cargo receptors involved in the various types of selective autophagy remain to be identified and characterized. Unraveling the molecular mechanisms that regulate the spatio-temporal ubiquitin signaling that controls selective autophagy, will be of great importance, to better understand its role in physiological and pathological processes.

Author Contributions: A.V. and P.G. conceived and wrote the manuscript. All authors have read and agreed to the published version of the manuscript.

Funding: This research was funded by Telethon Foundation (TMPGCBX16TT; TMPGGLU16TT), Roche Foundation (Roche per la ricerca 2019), AFM Telethon (Trampoline Grant 2020).

Acknowledgments: We kindly acknowledge Alessio Reggio for critical comments.

Conflicts of Interest: The authors declare no conflict of interest.

References

1. Schmidt, M.; Finley, D. Regulation of proteasome activity in health and disease. *Biochim. Biophys. Acta Mol. Cell Res.* **2014**, *1843*, 13–25. [CrossRef] [PubMed]
2. Shintani, T.; Klionsky, D.J. Autophagy in health and disease: A double-edged sword. *Science* **2004**, *306*, 990–995. [CrossRef] [PubMed]
3. Kleiger, G.; Mayor, T. Perilous journey: A tour of the ubiquitin-proteasome system. *Trends Cell Biol.* **2014**, *24*, 352–359. [CrossRef] [PubMed]
4. Papadopoulos, C.; Kirchner, P.; Bug, M.; Grum, D.; Koerver, L.; Schulze, N.; Poehler, R.; Dressler, A.; Fengler, S.; Arhzaouy, K.; et al. VCP/p97 cooperates with YOD 1, UBXD 1 and PLAA to drive clearance of ruptured lysosomes by autophagy. *EMBO J.* **2017**, *36*, 135–150. [CrossRef] [PubMed]
5. Piccirillo, R.; Goldberg, A.L. The p97/VCP ATPase is critical in muscle atrophy and the accelerated degradation of muscle proteins. *EMBO J.* **2012**, *31*, 3334–3350. [PubMed]
6. Kraft, C.; Peter, M.; Hofmann, K. Selective autophagy: Ubiquitin-mediated recognition and beyond. *Nat. Cell Biol.* **2010**, *12*, 836–841. [PubMed]
7. Deshaies, R.J.; Joazeiro, C.A.P. RING domain E3 ubiquitin ligases. *Annu. Rev. Biochem.* **2009**, *78*, 399–434. [CrossRef]
8. Wild, P.; McEwan, D.G.; Dikic, I. The LC3 interactome at a glance. *J. Cell Sci.* **2014**, *127*, 3–9. [CrossRef]
9. Komander, D.; Clague, M.J.; Urbé, S. Breaking the chains: Structure and function of the deubiquitinases. *Nat. Rev. Mol. Cell Biol.* **2009**, *10*, 550–563.
10. Riley, B.E.; Kaiser, S.E.; Shaler, T.A.; Ng, A.C.Y.; Hara, T.; Hipp, M.S.; Lage, K.; Xavier, R.J.; Ryu, K.-Y.; Taguchi, K.; et al. Ubiquitin accumulation in autophagy-deficient mice is dependent on the Nrf2-mediated stress response pathway: A potential role for protein aggregation in autophagic substrate selection. *J. Cell Biol.* **2010**, *191*, 537–552.
11. Swatek, K.N.; Komander, D. Ubiquitin modifications. *Cell Res.* **2016**, *26*, 399–422. [CrossRef] [PubMed]
12. Tan, J.M.M.; Wong, E.S.P.; Kirkpatrick, D.S.; Pletnikova, O.; Ko, H.S.; Tay, S.P.; Ho, M.W.L.; Troncoso, J.; Gygi, S.P.; Lee, M.K.; et al. Lysine 63-linked ubiquitination promotes the formation and autophagic clearance of protein inclusions associated with neurodegenerative diseases. *Hum. Mol. Genet.* **2008**, *17*, 431–439. [CrossRef] [PubMed]
13. Morimoto, D.; Walinda, E.; Fukada, H.; Sou, Y.S.; Kageyama, S.; Hoshino, M.; Fujii, T.; Tsuchiya, H.; Saeki, Y.; Arita, K.; et al. The unexpected role of polyubiquitin chains in the formation of fibrillar aggregates. *Nat. Commun.* **2015**, *6*, 1–10. [CrossRef] [PubMed]
14. Paz, Y.; Elazar, Z.; Fass, D. Structure of GATE-16, Membrane Transport Modulator and Mammalian Ortholog of Autophagocytosis Factor Aut7p*. *J. Biol Chem.* **2000**, *275*, 25445–25450. [CrossRef] [PubMed]
15. Suzuki, N.N.; Yoshimoto, K.; Fujioka, Y.; Ohsumi, Y.; Inagaki, F. The crystal structure of plant ATG12 and its biological implication in autophagy. *Autophagy* **2005**, *1*, 119–126. [CrossRef]
16. Zhang, Y.; Thery, F.; Wu, N.C.; Luhmann, E.K.; Dussurget, O.; Foecke, M.; Bredow, C.; Jiménez-Fernández, D.; Leandro, K.; Beling, A.; et al. The in vivo ISGylome links ISG15 to metabolic pathways and autophagy upon Listeria monocytogenes infection. *Nat. Commun.* **2019**, *10*, 1–15. [CrossRef] [PubMed]
17. Nakka, V.P.; Mohammed, A.Q. A Critical Role for ISGylation, Ubiquitination and, SUMOylation in Brain Damage: Implications for Neuroprotection. *Neurochem. Res.* **2020**, *45*, 1975–1985. [CrossRef]
18. Spinnenhirn, V.; Farhan, H.; Basler, M.; Aichem, A.; Canaan, A.; Groettrup, M. The ubiquitin-like modifier FAT10 decorates autophagy-targeted Salmonella and contributes to Salmonella resistance in mice. *J. Cell Sci.* **2014**, *127*, 4883–4893. [CrossRef]
19. Hwang, S.P.; Lee, D.H. Autophagy mediates SUMO-induced degradation of a polyglutamine protein ataxin-3. *Animal Cells Syst.* **2017**, *21*, 169–176. [CrossRef]
20. Glick, D.; Barth, S.; Macleod, K.F. Autophagy: Cellular and molecular mechanisms. *J. Pathol.* **2010**, *221*, 3–12. [CrossRef]
21. Levine, B.; Kroemer, G. Biological Functions of Autophagy Genes: A Disease Perspective. *Cell* **2019**, *176*, 11–42. [CrossRef]
22. Youle, R.J.; Narendra, D.P. Mechanisms of mitophagy. *Nat. Rev. Mol. Cell Biol.* **2011**, *12*, 9–14. [CrossRef] [PubMed]

23. Kirkin, V.; Rogov, V.V. A Diversity of Selective Autophagy Receptors Determines the Specificity of the Autophagy Pathway. Mol. *Cell* **2019**, *76*, 268–285.
24. Georgakopoulos, N.D.; Wells, G.; Campanella, M. The pharmacological regulation of cellular mitophagy. *Nat. Chem. Biol.* **2017**, *13*, 136–146. [CrossRef]
25. Reggiori, F.; Komatsu, M.; Finley, K.; Simonsen, A. Autophagy: More than a nonselective pathway. *Int. J. Cell Biol.* **2012**, *2012*, 219625. [CrossRef] [PubMed]
26. Ohsumi, Y.; Mizushima, N. Two ubiquitin-like conjugation systems essential for autophagy. *Semin. Cell Dev. Biol.* **2004**, *15*, 231–236. [CrossRef]
27. Kabeya, Y.; Mizushima, N.; Ueno, T.; Yamamoto, A.; Kirisako, T.; Noda, T.; Kominami, E.; Ohsumi, Y.; Yoshimori, T. LC3, a mammalian homologue of yeast Apg8p, is localized in autophagosome membranes after processing. *EMBO J.* **2000**, *19*, 5720–5728. [CrossRef]
28. Ichimura, Y.; Kirisako, T.; Takao, T.; Satomi, Y.; Shimonishi, Y.; Ishihara, N.; Mizushima, N.; Tanida, I.; Kominami, E.; Ohsumi, M.; et al. A ubiquitin-like system mediates protein lipidation. *Nature* **2000**, *408*, 488–492. [CrossRef]
29. Kuma, A.; Mizushima, N.; Ishihara, N.; Ohsumi, Y. Formation of the approximately 350-kDa Apg12-Apg5.Apg16 multimeric complex, mediated by Apg16 oligomerization, is essential for autophagy in yeast. *J. Biol. Chem.* **2002**, *277*, 18619–18625. [CrossRef]
30. Ravikumar, B.; Moreau, K.; Jahreiss, L.; Puri, C.; Rubinsztein, D.C. Plasma membrane contributes to the formation of pre-autophagosomal structures. *Nat. Cell Biol.* **2010**, *12*, 747–757. [CrossRef]
31. Satoo, K.; Noda, N.N.; Kumeta, H.; Fujioka, Y.; Mizushima, N.; Ohsumi, Y.; Inagaki, F. The structure of Atg4B-LC3 complex reveals the mechanism of LC3 processing and delipidation during autophagy. *EMBO J.* **2009**, *28*, 1341–1350. [CrossRef]
32. Bellot, G.; Garcia-Medina, R.; Gounon, P.; Chiche, J.; Roux, D.; Pouysségur, J.; Mazure, N.M. Hypoxia-induced autophagy is mediated through hypoxia-inducible factor induction of BNIP3 and BNIP3L via their BH3 domains. *Mol. Cell. Biol.* **2009**, *29*, 2570–2581. [CrossRef]
33. Jin, S.M.; Youle, R.J. The accumulation of misfolded proteins in the mitochondrial matrix is sensed by PINK1 to induce PARK2/Parkin-mediated mitophagy of polarized mitochondria. *Autophagy* **2013**, *9*, 1750–1757. [CrossRef]
34. Novak, I.; Kirkin, V.; McEwan, D.G.; Zhang, J.; Wild, P.; Rozenknop, A.; Rogov, V.; Löhr, F.; Popovic, D.; Occhipinti, A.; et al. Nix is a selective autophagy receptor for mitochondrial clearance. *EMBO Rep.* **2010**, *11*, 45–51. [CrossRef]
35. Kuchay, S.; Duan, S.; Schenkein, E.; Peschiaroli, A.; Saraf, A.; Florens, L.; Washburn, M.P.; Pagano, M. FBXL2- and PTPL1-mediated degradation of p110-free p85β regulatory subunit controls the PI(3)K signalling cascade. *Nat. Cell Biol.* **2013**, *15*, 472–480. [CrossRef]
36. Abrahamsen, H.; Stenmark, H.; Platta, H.W. Ubiquitination and phosphorylation of Beclin 1 and its binding partners: Tuning class III phosphatidylinositol 3-kinase activity and tumor suppression. *FEBS Lett.* **2012**, *586*, 1584–1591. [CrossRef]
37. Stolz, A.; Ernst, A.; Dikic, I. Cargo recognition and trafficking in selective autophagy. *Nat. Cell Biol.* **2014**, *16*, 495–501. [CrossRef]
38. Shaid, S.; Brandts, C.H.; Serve, H.; Dikic, I. Ubiquitination and selective autophagy. *Cell Death Differ.* **2013**, *20*, 21–30. [CrossRef]
39. Scrivo, A.; Bourdenx, M.; Pampliega, O.; Cuervo, A.M. Selective autophagy as a potential therapeutic target for neurodegenerative disorders. *Lancet Neurol.* **2018**, *17*, 802–815. [CrossRef]
40. Dombi, E.; Mortiboys, H.; Poulton, J. Modulating Mitophagy in Mitochondrial Disease. *Curr. Med. Chem.* **2017**, *25*, 5597–5612. [CrossRef] [PubMed]
41. Mizumura, K.; Cloonan, S.M.; Nakahira, K.; Bhashyam, A.R.; Cervo, M.; Kitada, T.; Glass, K.; Owen, C.A.; Mahmood, A.; Washko, G.R.; et al. Mitophagy-dependent necroptosis contributes to the pathogenesis of COPD. *J. Clin. Investig.* **2014**, *124*, 3987–4003. [PubMed]
42. Mizumura, K.; Choi, A.M.K.; Ryter, S.W. Emerging role of selective autophagy in human diseases. *Front. Pharmacol.* **2014**, *5*, 244. [PubMed]
43. Reggio, A.; Buonomo, V.; Grumati, P. Eating the unknown: Xenophagy and ER-phagy are cytoprotective defenses against pathogens. *Exp. Cell Res.* **2020**, *396*, 112276.

44. Sharma, V.; Verma, S.; Seranova, E.; Sarkar, S.; Kumar, D. Selective autophagy and xenophagy in infection and disease. *Front. Cell Dev. Biol.* **2018**, *6*, 147. [PubMed]
45. Dong, X.; Levine, B. Autophagy and viruses: Adversaries or allies? *J. Innate Immun.* **2013**, *5*, 480–493.
46. Mijaljica, D.; Klionsky, D.J. Autophagy/virophagy: A "disposal strategy" to combat COVID-19. *Autophagy* **2020**, 1–2. [CrossRef]
47. Chiramel, A.I.; Dougherty, J.D.; Nair, V.; Robertson, S.J.; Best, S.M. FAM134B, the Selective Autophagy Receptor for Endoplasmic Reticulum Turnover, Inhibits Replication of Ebola Virus Strains Makona and Mayinga. *J. Infect. Dis.* **2016**, *214*, S319–S325.
48. Lennemann, N.J.; Coyne, C.B. Dengue and Zika viruses subvert reticulophagy by NS2B3-mediated cleavage of FAM134B. *Autophagy* **2017**, *13*, 322–332.
49. Nicot, A.S.; Lo Verso, F.; Ratti, F.; Pilot-Storck, F.; Streichenberger, N.; Sandri, M.; Schaeffer, L.; Goillot, E. Phosphorylation of NBR1 by GSK3 modulates protein aggregation. *Autophagy* **2014**, *10*, 1036–1053.
50. Richter, B.; Sliter, D.A.; Herhaus, L.; Stolz, A.; Wang, C.; Beli, P.; Zaffagnini, G.; Wild, P.; Martens, S.; Wagner, S.A.; et al. Phosphorylation of OPTN by TBK1 enhances its binding to Ub chains and promotes selective autophagy of damaged mitochondria. *Proc. Natl. Acad. Sci. USA* **2016**, *113*, 4039–4044.
51. Matsumoto, G.; Wada, K.; Okuno, M.; Kurosawa, M.; Nukina, N. Serine 403 phosphorylation of p62/SQSTM1 regulates selective autophagic clearance of ubiquitinated proteins. *Mol. Cell* **2011**, *44*, 279–289. [CrossRef] [PubMed]
52. Noda, N.N.; Kumeta, H.; Nakatogawa, H.; Satoo, K.; Adachi, W.; Ishii, J.; Fujioka, Y.; Ohsumi, Y.; Inagaki, F. Structural basis of target recognition by Atg8/LC3 during selective autophagy. *Genes Cells* **2008**, *13*, 1211–1218. [CrossRef] [PubMed]
53. Wild, P.; Farhan, H.; McEwan, D.G.; Wagner, S.; Rogov, V.V.; Brady, N.R.; Richter, B.; Korac, J.; Waidmann, O.; Choudhary, C.; et al. Phosphorylation of the autophagy receptor optineurin restricts Salmonella growth. *Science* **2011**, *333*, 228–233. [CrossRef] [PubMed]
54. Tumbarello, D.A.; Manna, P.T.; Allen, M.; Bycroft, M.; Arden, S.D.; Kendrick-Jones, J.; Buss, F. The Autophagy Receptor TAX1BP1 and the Molecular Motor Myosin VI Are Required for Clearance of Salmonella Typhimurium by Autophagy. *PLoS Pathog.* **2015**, *11*, e1005174.
55. von Muhlinen, N.; Akutsu, M.; Ravenhill, B.J.; Foeglein, Á.; Bloor, S.; Rutherford, T.J.; Freund, S.M.V.; Komander, D.; Randow, F. LC3C, bound selectively by a noncanonical LIR motif in NDP52, is required for antibacterial autophagy. *Mol. Cell* **2012**, *48*, 329–342. [CrossRef]
56. Marshall, R.S.; Hua, Z.; Mali, S.; McLoughlin, F.; Vierstra, R.D. ATG8-Binding UIM Proteins Define a New Class of Autophagy Adaptors and Receptors. *Cell* **2019**, *177*, 766–781.e24.
57. Wurzer, B.; Zaffagnini, G.; Fracchiolla, D.; Turco, E.; Abert, C.; Romanov, J.; Martens, S. Oligomerization of p62 allows for selection of ubiquitinated cargo and isolation membrane during selective autophagy. *Elife* **2015**, *4*. [CrossRef]
58. Zatloukal, K.; Stumptner, C.; Fuchsbichler, A.; Heid, H.; Schnoelzer, M.; Kenner, L.; Kleinert, R.; Prinz, M.; Aguzzi, A.; Denk, H. p62 is a common component of cytoplasmic inclusions in protein aggregation diseases. *Am. J. Pathol.* **2002**, *160*, 255–263. [CrossRef]
59. Pankiv, S.; Clausen, T.H.; Lamark, T.; Brech, A.; Bruun, J.-A.; Outzen, H.; Øvervatn, A.; Bjørkøy, G.; Johansen, T. p62/SQSTM1 binds directly to Atg8/LC3 to facilitate degradation of ubiquitinated protein aggregates by autophagy. *J. Biol. Chem.* **2007**, *282*, 24131–24145. [CrossRef]
60. Zhou, J.; Wang, J.; Cheng, Y.; Chi, Y.-J.; Fan, B.; Yu, J.-Q.; Chen, Z. NBR1-Mediated Selective Autophagy Targets Insoluble Ubiquitinated Protein Aggregates in Plant Stress Responses. *PLoS Genet.* **2013**, *9*, e1003196. [CrossRef]
61. Kirkin, V.; Lamark, T.; Sou, Y.-S.; Bjørkøy, G.; Nunn, J.L.; Bruun, J.-A.; Shvets, E.; McEwan, D.G.; Clausen, T.H.; Wild, P.; et al. A role for NBR1 in autophagosomal degradation of ubiquitinated substrates. *Mol. Cell* **2009**, *33*, 505–516. [CrossRef] [PubMed]
62. Deosaran, E.; Larsen, K.B.; Hua, R.; Sargent, G.; Wang, Y.; Kim, S.; Lamark, T.; Jauregui, M.; Law, K.; Lippincott-Schwartz, J.; et al. NBR1 acts as an autophagy receptor for peroxisomes. *J. Cell Sci.* **2013**, *126*, 939–952. [CrossRef] [PubMed]
63. Korac, J.; Schaeffer, V.; Kovacevic, I.; Clement, A.M.; Jungblut, B.; Behl, C.; Terzic, J.; Dikic, I. Ubiquitin-independent function of optineurin in autophagic clearance of protein aggregates. *J. Cell Sci.* **2013**, *126*, 580–592. [CrossRef] [PubMed]

64. Liu, Z.; Chen, P.; Gao, H.; Gu, Y.; Yang, J.; Peng, H.; Xu, X.; Wang, H.; Yang, M.; Liu, X.; et al. Ubiquitylation of Autophagy Receptor Optineurin by HACE1 Activates Selective Autophagy for Tumor Suppression. *Cancer Cell* **2014**, *26*, 106–120. [CrossRef] [PubMed]
65. Menzies, R.A.; Gold, P.H. The Turnover of Mitochondria in a Variety of Tissues of Young Adult and Aged Rats. *J. Biol. Chem.* **1971**, *246*, 2425–2429.
66. Sandoval, H.; Thiagarajan, P.; Dasgupta, S.K.; Schumacher, A.; Prchal, J.T.; Chen, M.; Wang, J. Essential role for Nix in autophagic maturation of erythroid cells. *Nature* **2008**, *454*, 232–235. [CrossRef]
67. Sato, M.; Sato, K. Degradation of paternal mitochondria by fertilization-triggered autophagy in C. elegans embryos. *Science* **2011**, *334*, 1141–1144. [CrossRef]
68. Al Rawi, S.; Louvet-Vallée, S.; Djeddi, A.; Sachse, M.; Culetto, E.; Hajjar, C.; Boyd, L.; Legouis, R.; Galy, V. Postfertilization autophagy of sperm organelles prevents paternal mitochondrial DNA transmission. *Science* **2011**, *334*, 1144–1147. [CrossRef]
69. Melser, S.; Chatelain, E.H.; Lavie, J.; Mahfouf, W.; Jose, C.; Obre, E.; Goorden, S.; Priault, M.; Elgersma, Y.; Rezvani, H.R.; et al. Rheb regulates mitophagy induced by mitochondrial energetic status. *Cell Metab.* **2013**, *17*, 719–730. [CrossRef]
70. Vives-Bauza, C.; Zhou, C.; Huang, Y.; Cui, M.; de Vries, R.L.A.; Kim, J.; May, J.; Tocilescu, M.A.; Liu, W.; Ko, H.S.; et al. PINK1-dependent recruitment of Parkin to mitochondria in mitophagy. *Proc. Natl. Acad. Sci. USA* **2010**, *107*, 378–383. [CrossRef]
71. Kondapalli, C.; Kazlauskaite, A.; Zhang, N.; Woodroof, H.I.; Campbell, D.G.; Gourlay, R.; Burchell, L.; Walden, H.; Macartney, T.J.; Deak, M.; et al. PINK1 is activated by mitochondrial membrane potential depolarization and stimulates Parkin E3 ligase activity by phosphorylating Serine 65. *Open Biol.* **2012**, *2*, 120080. [CrossRef] [PubMed]
72. Narendra, D.; Tanaka, A.; Suen, D.-F.; Youle, R.J. Parkin is recruited selectively to impaired mitochondria and promotes their autophagy. *J. Cell Biol.* **2008**, *183*, 795–803. [CrossRef] [PubMed]
73. Sarraf, S.A.; Raman, M.; Guarani-Pereira, V.; Sowa, M.E.; Huttlin, E.L.; Gygi, S.P.; Harper, J.W. Landscape of the PARKIN-dependent ubiquitylome in response to mitochondrial depolarization. *Nature* **2013**, *496*, 372–376. [CrossRef] [PubMed]
74. Wang, X.; Winter, D.; Ashrafi, G.; Schlehe, J.; Wong, Y.L.; Selkoe, D.; Rice, S.; Steen, J.; LaVoie, M.J.; Schwarz, T.L. PINK1 and Parkin target Miro for phosphorylation and degradation to arrest mitochondrial motility. *Cell* **2011**, *147*, 893–906. [CrossRef]
75. Fu, M.; St-Pierre, P.; Shankar, J.; Wang, P.T.C.C.; Joshi, B.; Nabi, I.R. Regulation of mitophagy by the Gp78 E3 ubiquitin ligase. *Mol. Biol. Cell* **2013**, *24*, 1153–1162. [CrossRef]
76. Bingol, B.; Tea, J.S.; Phu, L.; Reichelt, M.; Bakalarski, C.E.; Song, Q.; Foreman, O.; Kirkpatrick, D.S.; Sheng, M. The mitochondrial deubiquitinase USP30 opposes parkin-mediated mitophagy. *Nature* **2014**, *510*, 370–375. [CrossRef]
77. Geisler, S.; Holmström, K.M.; Skujat, D.; Fiesel, F.C.; Rothfuss, O.C.; Kahle, P.J.; Springer, W. PINK1/Parkin-mediated mitophagy is dependent on VDAC1 and p62/SQSTM1. *Nat. Cell Biol.* **2010**, *12*, 119–131. [CrossRef]
78. Wong, Y.C.; Holzbaur, E.L.F. Optineurin is an autophagy receptor for damaged mitochondria in parkin-mediated mitophagy that is disrupted by an ALS-linked mutation. *Proc. Natl. Acad. Sci. USA* **2014**, *111*, E4439–E4448. [CrossRef]
79. Lazarou, M.; Sliter, D.A.; Kane, L.A.; Sarraf, S.A.; Wang, C.; Burman, J.L.; Sideris, D.P.; Fogel, A.I.; Youle, R.J. The ubiquitin kinase PINK1 recruits autophagy receptors to induce mitophagy. *Nature* **2015**, *524*, 309–314. [CrossRef]
80. Narendra, D.; Kane, L.A.; Hauser, D.N.; Fearnley, I.M.; Youle, R.J. p62/SQSTM1 is required for Parkin-induced mitochondrial clustering but not mitophagy; VDAC1 is dispensable for both. *Autophagy* **2010**, *6*, 1090–1106. [CrossRef]
81. Lamark, T.; Johansen, T. Aggrephagy: Selective disposal of protein aggregates by macroautophagy. *Int. J. Cell Biol.* **2012**, *2012*, 736905. [CrossRef] [PubMed]
82. Pickford, F.; Masliah, E.; Britschgi, M.; Lucin, K.; Narasimhan, R.; Jaeger, P.A.; Small, S.; Spencer, B.; Rockenstein, E.; Levine, B.; et al. The autophagy-related protein beclin 1 shows reduced expression in early Alzheimer disease and regulates amyloid β accumulation in mice. *J. Clin. Investig.* **2008**, *118*, 2190–2199. [CrossRef] [PubMed]

83. Webb, J.L.; Ravikumar, B.; Atkins, J.; Skepper, J.N.; Rubinsztein, D.C. Alpha-Synuclein is degraded by both autophagy and the proteasome. *J. Biol. Chem.* **2003**, *278*, 25009–25013. [CrossRef] [PubMed]
84. Ravikumar, B.; Vacher, C.; Berger, Z.; Davies, J.E.; Luo, S.; Oroz, L.G.; Scaravilli, F.; Easton, D.F.; Duden, R.; O'Kane, C.J.; et al. Inhibition of mTOR induces autophagy and reduces toxicity of polyglutamine expansions in fly and mouse models of Huntington disease. *Nat. Genet.* **2004**, *36*, 585–595. [CrossRef]
85. Winslow, A.R.; Chen, C.W.; Corrochano, S.; Acevedo-Arozena, A.; Gordon, D.E.; Peden, A.A.; Lichtenberg, M.; Menzies, F.M.; Ravikumar, B.; Imarisio, S.; et al. α-Synuclein impairs macroautophagy: Implications for Parkinson's disease. *J. Cell Biol.* **2010**, *190*, 1023–1037. [CrossRef]
86. Ciechanover, A.; Brundin, P. The ubiquitin proteasome system in neurodegenerative diseases: Sometimes the chicken, sometimes the egg. *Neuron* **2003**, *40*, 427–446. [CrossRef]
87. Lee, Y.; Jonson, P.H.; Sarparanta, J.; Palmio, J.; Sarkar, M.; Vihola, A.; Evilä, A.; Suominen, T.; Penttilä, S.; Savarese, M.; et al. TIA1 variant drives myodegeneration in multisystem proteinopathy with SQSTM1 mutations. *J. Clin. Investig.* **2018**, *128*, 1164–1177. [CrossRef]
88. Neumann, M.; Sampathu, D.M.; Kwong, L.K.; Truax, A.C.; Micsenyi, M.C.; Chou, T.T.; Bruce, J.; Schuck, T.; Grossman, M.; Clark, C.M.; et al. Ubiquitinated TDP-43 in frontotemporal lobar degeneration and amyotrophic lateral sclerosis. *Science* **2006**, *314*, 130–133. [CrossRef]
89. Deng, H.X.; Chen, W.; Hong, S.T.; Boycott, K.M.; Gorrie, G.H.; Siddique, N.; Yang, Y.; Fecto, F.; Shi, Y.; Zhai, H.; et al. Mutations in UBQLN2 cause dominant X-linked juvenile and adult-onset ALS and ALS/dementia. *Nature* **2011**, *477*, 211–215. [CrossRef]
90. Isakson, P.; Holland, P.; Simonsen, A. The role of ALFY in selective autophagy. *Cell Death Differ.* **2013**, *20*, 12–20. [CrossRef]
91. Simonsen, A.; Birkeland, H.C.G.; Gillooly, D.J.; Mizushima, N.; Kuma, A.; Yoshimori, T.; Slagsvold, T.; Brech, A.; Stenmark, H. Alfy, a novel FYVE-domain-containing protein associated with protein granules and autophagic membranes. *J. Cell Sci.* **2004**, *117*, 4239–4251. [CrossRef] [PubMed]
92. Marshall, R.S.; Li, F.; Gemperline, D.C.; Book, A.J.; Vierstra, R.D. Autophagic Degradation of the 26S Proteasome Is Mediated by the Dual ATG8/Ubiquitin Receptor RPN10 in Arabidopsis. *Mol. Cell* **2015**, *58*, 1053–1066. [CrossRef] [PubMed]
93. Marshall, R.; McLoughlin, F.; Vierstra, R.D. Autophagic turnover of inactive 26S proteasomes in yeast is directed by the ubiquitin receptor Cue5 and the Hsp42 chaperone. *Cell Rep.* **2016**, *16*, 1717–1732. [CrossRef]
94. Cohen-Kaplan, V.; Livneh, I.; Avni, N.; Fabre, B.; Ziv, T.; Kwon, Y.T.; Ciechanover, A. p62- and ubiquitin-dependent stress-induced autophagy of the mammalian 26S proteasome. *Proc. Natl. Acad. Sci. USA* **2016**, *113*, E7490–E7499. [CrossRef]
95. Lu, K.; Psakhye, I.; Jentsch, S. Autophagic clearance of PolyQ proteins mediated by ubiquitin-Atg8 adaptors of the conserved CUET protein family. *Cell* **2014**, *158*, 549–563. [CrossRef]
96. Cho, S.J.; Yun, S.M.; Jo, C.; Lee, D.H.; Choi, K.J.; Song, J.C.; Park, S.I.; Kim, Y.J.; Koh, Y.H. SUMO1 promotes Aβ production via the modulation of autophagy. *Autophagy* **2015**, *11*, 100–112. [CrossRef]
97. Nagashima, Y.; Kowa, H.; Tsuji, S.; Iwata, A. FAT10 protein binds to polyglutamine proteins and modulates their solubility. *J. Biol. Chem.* **2011**, *286*, 29594–29600. [CrossRef]
98. Kalveram, B.; Schmidtke, G.; Groettrup, M. The ubiquitin-like modifier FAT10 interacts with HDAC6 and localizes to aggresomes under proteasome inhibition. *J. Cell Sci.* **2008**, *121*, 4079–4088. [CrossRef]
99. Nakashima, H.; Nguyen, T.; Goins, W.F.; Chiocca, E.A. Interferon-stimulated gene 15 (ISG15) and ISG15-linked proteins can associate with members of the selective autophagic process, histone deacetylase 6 (HDAC6) and SQSTM1/p62. J. *Biol. Chem.* **2015**, *290*, 1485–1495. [CrossRef]
100. Ylä-Anttila, P.; Vihinen, H.; Jokitalo, E.; Eskelinen, E.L. 3D tomography reveals connections between the phagophore and endoplasmic reticulum. *Autophagy* **2009**, *5*, 1180–1185. [CrossRef] [PubMed]
101. Axe, E.L.; Walker, S.A.; Manifava, M.; Chandra, P.; Roderick, H.L.; Habermann, A.; Griffiths, G.; Ktistakis, N.T. Autophagosome formation from membrane compartments enriched in phosphatidylinositol 3-phosphate and dynamically connected to the endoplasmic reticulum. *J. Cell Biol.* **2008**, *182*, 685–701. [CrossRef] [PubMed]
102. Matsunaga, K.; Morita, E.; Saitoh, T.; Akira, S.; Ktistakis, N.T.; Izumi, T.; Noda, T.; Yoshimori, T. Autophagy requires endoplasmic reticulum targeting of the PI3-kinase complex via Atg14L. *J. Cell Biol.* **2010**, *190*, 511–521. [CrossRef]

103. Nixon-Abell, J.; Obara, C.J.; Weigel, A.V.; Li, D.; Legant, W.R.; Xu, C.S.; Pasolli, H.A.; Harvey, K.; Hess, H.F.; Betzig, E.; et al. Increased spatiotemporal resolution reveals highly dynamic dense tubular matrices in the peripheral ER. *Science* **2016**, *354*. [CrossRef] [PubMed]
104. Hetz, C. The unfolded protein response: Controlling cell fate decisions under ER stress and beyond. *Nat. Rev. Mol. Cell Biol.* **2012**, *13*, 89–102. [CrossRef] [PubMed]
105. Grumati, P.; Dikic, I.; Stolz, A. ER-phagy at a glance. *J. Cell Sci.* **2018**, *131*. [CrossRef]
106. Stolz, A.; Grumati, P. The various shades of ER-phagy. *FEBS J.* **2019**, *286*, 4642–4649. [CrossRef]
107. Khaminets, A.; Heinrich, T.; Mari, M.; Grumati, P.; Huebner, A.K.; Akutsu, M.; Liebmann, L.; Stolz, A.; Nietzsche, S.; Koch, N.; et al. Regulation of endoplasmic reticulum turnover by selective autophagy. *Nature* **2015**, *522*, 354–358. [CrossRef] [PubMed]
108. Fumagalli, F.; Noack, J.; Bergmann, T.J.; Presmanes, E.C.; Pisoni, G.B.; Fasana, E.; Fregno, I.; Galli, C.; Loi, M.; Soldà, T.; et al. Translocon component Sec62 acts in endoplasmic reticulum turnover during stress recovery. *Nat. Cell Biol.* **2016**, *18*, 1173–1184. [CrossRef]
109. Grumati, P.; Morozzi, G.; Hölper, S.; Mari, M.; Harwardt, M.L.I.E.; Yan, R.; Müller, S.; Reggiori, F.; Heilemann, M.; Dikic, I. Full length RTN3 regulates turnover of tubular endoplasmic reticulum via selective autophagy. *Elife* **2017**, *6*. [CrossRef]
110. Smith, M.D.; Harley, M.E.; Kemp, A.J.; Wills, J.; Lee, M.; Arends, M.; von Kriegsheim, A.; Behrends, C.; Wilkinson, S. CCPG1 Is a Non-canonical Autophagy Cargo Receptor Essential for ER-Phagy and Pancreatic ER Proteostasis. *Dev. Cell* **2018**, *44*, 217–232. [CrossRef]
111. Chino, H.; Hatta, T.; Natsume, T.; Mizushima, N. Intrinsically Disordered Protein TEX264 Mediates ER-phagy Article Intrinsically Disordered Protein TEX264 Mediates ER-phagy. *Mol. Cell* **2019**, *74*, 909–921. [CrossRef]
112. Chen, Q.; Xiao, Y.; Chai, P.; Zheng, P.; Teng, J.; Chen, J. ATL3 Is a Tubular ER-Phagy Receptor for GABARAP-Mediated Selective Autophagy. *Curr. Biol.* **2019**, *29*, 846–855. [CrossRef]
113. An, H.; Ordureau, A.; Paulo, J.A.; Shoemaker, C.J.; Denic, V.; Harper, J.W. TEX264 Is an Endoplasmic Reticulum-Resident ATG8-Interacting Protein Critical for ER Remodeling during Nutrient Stress. *Mol. Cell* **2019**, *74*, 891–908. [CrossRef]
114. Ji, C.H.; Kim, H.Y.; Heo, A.J.; Lee, S.H.; Lee, M.J.; Kim, S.B.; Srinivasrao, G.; Mun, S.R.; Cha-Molstad, H.; Ciechanover, A.; et al. The N-Degron Pathway Mediates ER-phagy. *Mol. Cell* **2019**, *75*, 1058–1072. [CrossRef]
115. Tschurtschenthaler, M.; Adolph, T.E.; Ashcroft, J.W.; Niederreiter, L.; Bharti, R.; Saveljeva, S.; Bhattacharyya, J.; Flak, M.B.; Shih, D.Q.; Fuhler, G.M.; et al. Defective ATG16L1-mediated removal of IRE1α drives Crohn's disease-like ileitis. *J. Exp. Med.* **2017**, *214*, 401–422. [CrossRef]
116. Yang, H.; Ni, H.M.; Guo, F.; Ding, Y.; Shi, Y.H.; Lahiri, P.; Fröhlich, L.F.; Rülicke, T.; Smole, C.; Schmidt, V.C.; et al. Sequestosome 1/p62 protein is associated with autophagic removal of excess hepatic endoplasmic reticulum in mice. *J. Biol. Chem.* **2016**, *291*, 18663–18674. [CrossRef] [PubMed]
117. Kraft, C.; Deplazes, A.; Sohrmann, M.; Peter, M. Mature ribosomes are selectively degraded upon starvation by an autophagy pathway requiring the Ubp3p/Bre5p ubiquitin protease. *Nat. Cell Biol.* **2008**, *10*, 602–610. [CrossRef]
118. Ossareh-Nazari, B.; Bonizec, M.; Cohen, M.; Dokudovskaya, S.; Delalande, F.; Schaeffer, C.; Van Dorsselaer, A.; Dargemont, C. Cdc48 and Ufd3, new partners of the ubiquitin protease Ubp3, are required for ribophagy. *EMBO Rep.* **2010**, *11*, 548–554. [CrossRef] [PubMed]
119. An, H.; Harper, J.W. Systematic analysis of ribophagy in human cells reveals bystander flux during selective autophagy. *Nat. Cell Biol.* **2018**, *20*, 135–143. [CrossRef]
120. Wyant, G.A.; Abu-Remaileh, M.; Frenkel, E.M.; Laqtom, N.N.; Dharamdasani, V.; Lewis, C.A.; Chan, S.H.; Heinze, I.; Ori, A.; Sabatini, D.M. Nufip1 is a ribosome receptor for starvation-induced ribophagy. *Science* **2018**, *360*, 751–758. [CrossRef] [PubMed]
121. An, H.; Ordureau, A.; Körner, M.; Paulo, J.A.; Harper, J.W. Systematic quantitative analysis of ribosome inventory during nutrient stress. *Nature* **2020**, *583*, 303–309. [CrossRef]
122. Garzia, A.; Jafarnejad, S.M.; Meyer, C.; Chapat, C.; Gogakos, T.; Morozov, P.; Amiri, M.; Shapiro, M.; Molina, H.; Tuschl, T.; et al. The E3 ubiquitin ligase and RNA-binding protein ZNF598 orchestrates ribosome quality control of premature polyadenylated mRNAs. *Nat. Commun.* **2017**, *8*, 1–10. [CrossRef] [PubMed]
123. Silva, G.M.; Finley, D.; Vogel, C. K63 polyubiquitination is a new modulator of the oxidative stress response. *Nat. Struct. Mol. Biol.* **2015**, *22*, 116–123. [CrossRef]

124. Higgins, R.; Gendron, J.M.; Rising, L.; Mak, R.; Webb, K.; Kaiser, S.E.; Zuzow, N.; Riviere, P.; Yang, B.; Fenech, E.; et al. The Unfolded Protein Response Triggers Site-Specific Regulatory Ubiquitylation of 40S Ribosomal Proteins. *Mol. Cell* **2015**, *59*, 35–49. [CrossRef] [PubMed]
125. An, H.; Harper, J.W. Ribosome Abundance Control Via the Ubiquitin–Proteasome System and Autophagy. *J. Mol. Biol.* **2020**, *432*, 170–184. [CrossRef]
126. Sung, M.K.; Porras-Yakushi, T.R.; Reitsma, J.M.; Huber, F.M.; Sweredoski, M.J.; Hoelz, A.; Hess, S.; Deshaies, R.J. A conserved quality-control pathway that mediates degradation of unassembled ribosomal proteins. *Elife* **2016**, *5*. [CrossRef]
127. Yanagitani, K.; Juszkiewicz, S.; Hegde, R.S. UBE2O is a quality control factor for orphans of multiprotein complexes. *Science* **2017**, *357*, 472–475. [CrossRef]
128. Zhang, J.; Tripathi, D.N.; Jing, J.; Alexander, A.; Kim, J.; Powell, R.T.; Dere, R.; Tait-Mulder, J.; Lee, J.H.; Paull, T.T.; et al. ATM functions at the peroxisome to induce pexophagy in response to ROS. *Nat. Cell Biol.* **2015**, *17*, 1259–1269. [CrossRef]
129. Nordgren, M.; Francisco, T.; Lismont, C.; Hennebel, L.; Brees, C.; Wang, B.; van Veldhoven, P.P.; Azevedo, J.E.; Fransen, M. Export-deficient monoubiquitinated PEX5 triggers peroxisome removal in SV40 large T antigen-transformed mouse embryonic fibroblasts. *Autophagy* **2015**, *11*, 1326–1340. [CrossRef] [PubMed]
130. Sargent, G.; van Zutphen, T.; Shatseva, T.; Zhang, L.; Di Giovanni, V.; Bandsma, R.; Kim, P.K. PEX2 is the E3 ubiquitin ligase required for pexophagy during starvation. *J. Cell Biol.* **2016**, *214*. [CrossRef]
131. Yamashita, S.I.; Abe, K.; Tatemichi, Y.; Fujiki, Y. The membrane peroxin PEX3 induces peroxisome-ubiquitination-linked pexophagy. *Autophagy* **2014**, *10*, 1549–1564. [CrossRef]
132. Kim, P.K.; Hailey, D.W.; Mullen, R.T.; Lippincott-Schwartz, J. Ubiquitin signals autophagic degradation of cytosolic proteins and peroxisomes. *Proc. Natl. Acad. Sci. USA* **2008**, *105*, 20567–20574. [CrossRef]
133. Nazarko, T.Y.; Ozeki, K.; Till, A.; Ramakrishnan, G.; Lotfi, P.; Yan, M.; Subramani, S. Peroxisomal Atg37 binds Atg30 or palmitoyl-CoA to regulate phagophore formation during pexophagy. *J. Cell Biol.* **2014**, *204*, 541–557. [CrossRef]
134. Settembre, C.; Zoncu, R.; Medina, D.L.; Vetrini, F.; Erdin, S.; Erdin, S.; Huynh, T.; Ferron, M.; Karsenty, G.; Vellard, M.C.; et al. A lysosome-to-nucleus signalling mechanism senses and regulates the lysosome via mTOR and TFEB. *EMBO J.* **2012**, *31*, 1095–1108. [CrossRef]
135. de Duve, C. Lysosomes, a new group of cytoplasmic particles. *Subcell. Part.* **1959**, *60*, 128–159.
136. Boya, P.; Kroemer, G. Lysosomal membrane permeabilization in cell death. *Oncogene* **2008**, *27*, 6434–6451. [CrossRef] [PubMed]
137. Maejima, I.; Takahashi, A.; Omori, H.; Kimura, T.; Takabatake, Y.; Saitoh, T.; Yamamoto, A.; Hamasaki, M.; Noda, T.; Isaka, Y.; et al. Autophagy sequesters damaged lysosomes to control lysosomal biogenesis and kidney injury. *EMBO J.* **2013**, *32*, 2336–2347. [CrossRef]
138. Yoshida, Y.; Yasuda, S.; Fujita, T.; Hamasaki, M.; Murakami, A.; Kawawaki, J.; Iwai, K.; Saeki, Y.; Yoshimori, T.; Matsuda, N.; et al. Ubiquitination of exposed glycoproteins by SCFFBXO27 directs damaged lysosomes for autophagy. *Proc. Natl. Acad. Sci. USA.* **2017**, *114*, 8574–8579. [CrossRef] [PubMed]
139. Fiskin, E.; Bionda, T.; Dikic, I.; Behrends, C. Global Analysis of Host and Bacterial Ubiquitinome in Response to Salmonella Typhimurium Infection. *Mol. Cell* **2016**, *62*, 967–981. [CrossRef]
140. Chauhan, S.; Kumar, S.; Jain, A.; Ponpuak, M.; Mudd, M.H.; Kimura, T.; Choi, S.W.; Peters, R.; Mandell, M.; Bruun, J.A.; et al. TRIMs and Galectins Globally Cooperate and TRIM16 and Galectin-3 Co-direct Autophagy in Endomembrane Damage Homeostasis. *Dev. Cell* **2016**, *39*, 13–27. [CrossRef]
141. Aits, S.; Kricker, J.; Liu, B.; Ellegaard, A.-M.; Hämälistö, S.; Tvingsholm, S.; Corcelle-Termeau, E.; Høgh, S.; Farkas, T.; Holm Jonassen, A.; et al. Sensitive detection of lysosomal membrane permeabilization by lysosomal galectin puncta assay. *Autophagy* **2015**, *11*, 1408–1424. [CrossRef] [PubMed]
142. Thurston, T.L.M.; Wandel, M.P.; Von Muhlinen, N.; Foeglein, Á.; Randow, F. Galectin 8 targets damaged vesicles for autophagy to defend cells against bacterial invasion. *Nature* **2012**, *482*, 414–418. [CrossRef] [PubMed]
143. Koerver, L.; Papadopoulos, C.; Liu, B.; Kravic, B.; Rota, G.; Brecht, L.; Veenendaal, T.; Polajnar, M.; Bluemke, A.; Ehrmann, M.; et al. The ubiquitin-conjugating enzyme UBE 2 QL 1 coordinates lysophagy in response to endolysosomal damage. *EMBO Rep.* **2019**, *20*. [CrossRef] [PubMed]
144. Hung, Y.H.; Chen, L.M.W.; Yang, J.Y.; Yuan Yang, W. Spatiotemporally controlled induction of autophagy-mediated lysosome turnover. *Nat. Commun.* **2013**, *4*. [CrossRef]

145. Fujita, N.; Morita, E.; Itoh, T.; Tanaka, A.; Nakaoka, M.; Osada, Y.; Umemoto, T.; Saitoh, T.; Nakatogawa, H.; Kobayashi, S.; et al. Recruitment of the autophagic machinery to endosomes during infection is mediated by ubiquitin. *J. Cell Biol.* **2013**, *203*, 115–128. [CrossRef]
146. Bussi, C.; Peralta Ramos, J.M.; Arroyo, D.S.; Gallea, J.I.; Ronchi, P.; Kolovou, A.; Wang, J.M.; Florey, O.; Celej, M.S.; Schwab, Y.; et al. Alpha-synuclein fibrils recruit TBK1 and OPTN to lysosomal damage sites and induce autophagy in microglial cells. *J. Cell Sci.* **2018**, *131*. [CrossRef]
147. Goebel, W.; Kuhn, M. Bacterial replication in the host cell cytosol. *Curr. Opin. Microbiol.* **2000**, *3*, 49–53. [CrossRef]
148. Dupont, N.; Lacas-Gervais, S.; Bertout, J.; Paz, I.; Freche, B.; Van Nhieu, G.T.; van der Goot, F.G.; Sansonetti, P.J.; Lafont, F. Shigella Phagocytic Vacuolar Membrane Remnants Participate in the Cellular Response to Pathogen Invasion and Are Regulated by Autophagy. *Cell Host Microbe* **2009**, *6*, 137–149. [CrossRef]
149. Gomes, L.C.; Dikic, I. Autophagy in antimicrobial immunity. *Mol. Cell* **2014**, *54*, 224–233. [CrossRef]
150. Perrin, A.J.; Jiang, X.; Birmingham, C.L.; So, N.S.Y.; Brumell, J.H. Recognition of bacteria in the cytosol of mammalian cells by the ubiquitin system. *Curr. Biol.* **2004**, *14*, 806–811. [CrossRef]
151. van Wijk, S.J.L.; Fiskin, E.; Putyrski, M.; Pampaloni, F.; Hou, J.; Wild, P.; Kensche, T.; Grecco, H.E.; Bastiaens, P.; Dikic, I. Fluorescence-Based Sensors to Monitor Localization and Functions of Linear and K63-Linked Ubiquitin Chains in Cells. *Mol. Cell* **2012**, *47*, 797–809. [CrossRef] [PubMed]
152. Collins, C.A.; De Mazière, A.; Van Dijk, S.; Carlsson, F.; Klumperman, J.; Brown, E.J. Atg5-independent sequestration of ubiquitinated mycobacteria. *PLoS Pathog.* **2009**, *5*. [CrossRef] [PubMed]
153. Polajnar, M.; Dietz, M.S.; Heilemann, M.; Behrends, C. Expanding the host cell ubiquitylation machinery targeting cytosolic Salmonella. *EMBO Rep.* **2017**, *18*, 1572–1585. [CrossRef] [PubMed]
154. Huett, A.; Heath, R.J.; Begun, J.; Sassi, S.O.; Baxt, L.A.; Vyas, J.M.; Goldberg, M.B.; Xavier, R.J. The LRR and RING domain protein LRSAM1 is an E3 ligase crucial for ubiquitin-dependent autophagy of intracellular salmonella typhimurium. *Cell Host Microbe* **2012**, *12*, 778–790. [CrossRef] [PubMed]
155. Noad, J.; Von Der Malsburg, A.; Pathe, C.; Michel, M.A.; Komander, D.; Randow, F. LUBAC-synthesized linear ubiquitin chains restrict cytosol-invading bacteria by activating autophagy and NF-κB. *Nat. Microbiol.* **2017**, 2. [CrossRef]
156. Van Wijk, S.J.L.; Fricke, F.; Herhaus, L.; Gupta, J.; Hötte, K.; Pampaloni, F.; Grumati, P.; Kaulich, M.; Sou, Y.S.; Komatsu, M.; et al. Linear ubiquitination of cytosolic Salmonella Typhimurium activates NF-κB and restricts bacterial proliferation. *Nat. Microbiol.* **2017**, 2. [CrossRef]
157. Manzanillo, P.S.; Ayres, J.S.; Watson, R.O.; Collins, A.C.; Souza, G.; Rae, C.S.; Schneider, D.S.; Nakamura, K.; Shiloh, M.U.; Cox, J.S. The ubiquitin ligase parkin mediates resistance to intracellular pathogens. *Nature* **2013**, *501*, 512–516. [CrossRef]
158. Mostowy, S.; Sancho-Shimizu, V.; Hamon, M.A.; Simeone, R.; Brosch, R.; Johansen, T.; Cossart, P. p62 and NDP52 proteins target intracytosolic Shigella and Listeria to different autophagy pathways. *J. Biol. Chem.* **2011**, *286*, 26987–26995. [CrossRef]
159. Zheng, Y.T.; Shahnazari, S.; Brech, A.; Lamark, T.; Johansen, T.; Brumell, J.H. The Adaptor Protein p62/SQSTM1 Targets Invading Bacteria to the Autophagy Pathway. *J. Immunol.* **2009**, *183*, 5909–5916. [CrossRef]
160. Thurston, T.L.M. The tbk1 adaptor and autophagy receptor ndp52 restricts the proliferation of ubiquitin-coated bacteria. *Nat. Immunol.* **2009**, *10*, 1215–1222. [CrossRef]
161. Bogunovic, D.; Byun, M.; Durfee, L.A.; Abhyankar, A.; Sanal, O.; Mansouri, D.; Salem, S.; Radovanovic, I.; Grant, A.V.; Adimi, P.; et al. Mycobacterial disease and impaired IFN-γ immunity in humans with inherited ISG15 deficiency. *Science* **2012**, *337*, 1684–1688. [CrossRef]
162. Zhang, X.; Bogunovic, D.; Payelle-Brogard, B.; Francois-Newton, V.; Speer, S.D.; Yuan, C.; Volpi, S.; Li, Z.; Sanal, O.; Mansouri, D.; et al. Human intracellular ISG15 prevents interferon-α/β over-amplification and auto-inflammation. *Nature* **2015**, *517*, 89–93. [CrossRef]
163. Singh, R.; Kaushik, S.; Wang, Y.; Xiang, Y.; Novak, I.; Komatsu, M.; Tanaka, K.; Cuervo, A.M.; Czaja, M.J. Autophagy regulates lipid metabolism. *Nature* **2009**, *458*, 1131–1135. [CrossRef] [PubMed]
164. Rui, Y.N.; Xu, Z.; Patel, B.; Chen, Z.; Chen, D.; Tito, A.; David, G.; Sun, Y.; Stimming, E.F.; Bellen, H.J.; et al. Huntingtin functions as a scaffold for selective macroautophagy. *Nat. Cell Biol.* **2015**, *17*, 262–275. [CrossRef] [PubMed]

165. Spandl, J.; Lohmann, D.; Kuerschner, L.; Moessinger, C.; Thiele, C. Ancient ubiquitous protein 1 (AUP1) localizes to lipid droplets and binds the E2 ubiquitin conjugase G2 (Ube2g2) via its G2 binding region. *J. Biol. Chem.* **2011**, *286*, 5599–5606. [CrossRef] [PubMed]
166. Wang, L.; Zhou, J.; Yan, S.; Lei, G.; Lee, C.H.; Yin, X.M. Ethanol-triggered Lipophagy Requires SQSTM1 in AML12 Hepatic Cells. *Sci. Rep.* **2017**, *7*, 1–13. [CrossRef] [PubMed]
167. Shoji, J.Y.; Kikuma, T.; Arioka, M.; Kitamoto, K. Macroautophagy-mediated degradation of whole nuclei in the filamentous fungus Aspergillus oryzae. *PLoS ONE* **2010**, *5*. [CrossRef]
168. Papandreou, M.E.; Tavernarakis, N. Nucleophagy: From homeostasis to disease. *Cell Death Differ.* **2019**, *26*, 630–639. [CrossRef]
169. Krick, R.; Muehe, Y.; Prick, T.; Bremer, S.; Schlotterhose, P.; Eskelinen, E.L.; Millen, J.; Goldfarb, D.S.; Thumm, M. Piecemeal microautophagy of the nucleus requires the core macroautophagy genes. *Mol. Biol. Cell* **2008**, *19*, 4492–4505. [CrossRef]
170. Millen, J.I.; Krick, R.; Prick, T.; Thumm, M.; Goldfarb, D.S. Measuring piecemeal microautophagy of the nucleus in Saccharomyces cerevisiae. *Autophagy* **2009**, *5*, 75–81. [CrossRef]
171. Mochida, K.; Oikawa, Y.; Kimura, Y.; Kirisako, H.; Hirano, H.; Ohsumi, Y.; Nakatogawa, H. Receptor-mediated selective autophagy degrades the endoplasmic reticulum and the nucleus. *Nature* **2015**, *522*, 359–362. [CrossRef] [PubMed]
172. Akinduro, O.; Sully, K.; Patel, A.; Robinson, D.J.; Chikh, A.; McPhail, G.; Braun, K.M.; Philpott, M.P.; Harwood, C.A.; Byrne, C.; et al. Constitutive Autophagy and Nucleophagy during Epidermal Differentiation. *J. Investig. Dermatol.* **2016**, *136*, 1460–1470. [CrossRef]
173. Li, Y.; Jiang, X.; Zhang, Y.; Gao, Z.; Liu, Y.; Hu, J.; Hu, X.; Li, L.; Shi, J.; Gao, N. Nuclear accumulation of UBC9 contributes to SUMOylation of lamin A/C and nucleophagy in response to DNA damage. *J. Exp. Clin. Cancer Res.* **2019**, *38*, 67. [CrossRef]
174. Ji, C.H.; Kim, H.Y.; Heo, A.J.; Lee, M.J.; Park, D.Y.; Kim, D.H.; Kim, B.Y.; Kwon, Y.T. Regulation of reticulophagy by the N-degron pathway. *Autophagy* **2020**, *16*, 373–375. [CrossRef] [PubMed]
175. Rea, S.L.; Majcher, V.; Searle, M.S.; Layfield, R. SQSTM1 mutations-Bridging Paget disease of bone and ALS/FTLD. Exp. *Cell Res.* **2014**, *325*, 27–37. [CrossRef] [PubMed]
176. Fuqua, J.D.; Mere, C.P.; Kronemberger, A.; Blomme, J.; Bae, D.; Turner, K.D.; Harris, M.P.; Scudese, E.; Edwards, M.; Ebert, S.M.; et al. ULK2 is essential for degradation of ubiquitinated protein aggregates and homeostasis in skeletal muscle. *FASEB J.* **2019**, *33*, 11735–12745. [CrossRef]
177. Verlhac, P.; Grégoire, I.P.; Azocar, O.; Petkova, D.S.; Baguet, J.; Viret, C.; Faure, M. Autophagy receptor NDP52 regulates pathogen-containing autophagosome maturation. *Cell Host Microbe* **2015**, *17*, 515–525. [CrossRef]
178. Whang, M.I.; Tavares, R.M.; Benjamin, D.I.; Debnath, J.; Malynn, B.A.; Kattah, M.G.; Advincula, R.; Nomura, D.K.; Ma, A. The Ubiquitin Binding Protein TAX1BP1 Mediates Autophagosome Induction and the Metabolic Transition of Activated T Cells. *Immunity* **2017**, *46*, 405–420. [CrossRef]

cells

MDPI

Review

Ubiquitin, Autophagy and Neurodegenerative Diseases

Yoshihisa Watanabe [1,*], Katsutoshi Taguchi [2] and Masaki Tanaka [2,*]

1 Department of Basic Geriatrics, Graduate School of Medical Science, Kyoto Prefectural University of Medicine, Kyoto 602-8566, Japan
2 Department of Anatomy and Neurobiology, Graduate School of Medical Science, Kyoto Prefectural University of Medicine, Kyoto 602-8566, Japan; ktaguchi@koto.kpu-m.ac.jp
* Correspondence: y-watana@koto.kpu-m.ac.jp (Y.W.); mtanaka@koto.kpu-m.ac.jp (M.T.)

Received: 5 August 2020; Accepted: 2 September 2020; Published: 2 September 2020

check for updates

Abstract: Ubiquitin signals play various roles in proteolytic and non-proteolytic functions. Ubiquitin signals are recognized as targets of the ubiquitin–proteasome system and the autophagy–lysosome pathway. In autophagy, ubiquitin signals are required for selective incorporation of cargoes, such as proteins, organelles, and microbial invaders, into autophagosomes. Autophagy receptors possessing an LC3-binding domain and a ubiquitin binding domain are involved in this process. Autophagy activity can decline as a result of genetic variation, aging, or lifestyle, resulting in the onset of various neurodegenerative diseases. This review summarizes the selective autophagy of neurodegenerative disease-associated protein aggregates via autophagy receptors and discusses its therapeutic application for neurodegenerative diseases.

Keywords: ubiquitin; autophagy; neurodegenerative diseases; ubiquitin–proteasome system; autophagy–lysosome pathway

1. Introduction

Many neurodegenerative diseases, such as Alzheimer's disease (AD), Parkinson's disease (PD), amyotrophic lateral sclerosis (ALS), and Huntington's disease (HD), involve accumulation of harmful and aggregation-prone proteins. These aggregated proteins are known to be ubiquitinated in many neurodegenerative diseases. Although harmful proteins are immediately degraded by proteolytic systems in healthy individuals, any perturbation of these systems caused by genetic variation, aging, or lifestyle results in accumulation of harmful protein aggregates and the onset of various diseases including neurodegenerative diseases. Ubiquitination is the most important targeting signal for proteolytic systems [1]. Indeed, pathological analyses show that most of the protein inclusions and aggregates in the brains of neurodegenerative disease cases are positive for ubiquitin [2]. Recent advances in mass spectrometry technology have contributed to the characterization of ubiquitin chains and the decoding of ubiquitin signals. Ubiquitin signals are categorized as mono-ubiquitin, homotypic poly-ubiquitin, and heterotypic poly-ubiquitin [3]. Homotypic poly-ubiquitin chains are generated by conjugation of two or more ubiquitin molecules via their seven lysine residues (Lys-6, Lys-11, Lys-27, Lys-29, Lys-33, Lys-48, and Lys-63) or the initiation methionine residue (Met-1), whereas heterotypic poly-ubiquitin chains are formed by linkages of two or more different Lys residues [3]. These ubiquitin signals have roles in proteolytic functions and non-proteolytic functions, such as transcription regulation, membrane trafficking, DNA repair, and cell signaling [4]. Mutations in several autophagy related proteins, such as Parkin, PINK1, p62, and OPTN, are linked to neurodegenerative diseases. Autophagy receptors function in the selective autophagic clearance of disease-related proteins via ubiquitin signals. Thus, augmentation of autophagy is potentially a good therapeutic approach for neurodegenerative diseases. This review focuses on the role of ubiquitin signals in autophagy and their relevance to the onset of neurodegenerative diseases.

2. Neurodegenerative Diseases and Protein Aggregates

AD is a progressive neurodegenerative disorder that leads to cognitive decline [5]. The main hallmarks of AD are deposition of β-amyloid protein (Aβ) outside neurons, termed senile plaques, and in the vascular walls of the brain, and the accumulation of hyperphosphorylated Tau-protein as neurofibrillary tangles inside neurons [6]. Aβ is generated from amyloid precursor protein (APP) through sequential cleavage by β-secretase and γ-secretase complexes [7]. Studies of familial AD show that AD-causing variants in genes encoding APP and presenilins, catalytic components of the γ-secretase complex, elevate relative levels of the Aβ1–42 or Aβ1–43 isoforms of Aβ1–40 [8]. Aβ1–42 and Aβ1–43 are more aggregation-prone and cytotoxic compared with the Aβ1–40 peptide [9]. Initially, Aβ deposits are found exclusively in the neocortex and subsequently expand into the hippocampus, striatum, and brainstem [10]. Aβ is produced from APP and is mainly secreted into the extracellular space [11]. However, Aβ oligomers also accumulate intracellularly through endocytosis of secreted Aβ [12]. A triple-transgenic model of familial AD harboring transgenes expressing PS1(M146V), APP(Swe), and Tau(P301L) was defective for synaptic plasticity, including long-term potentiation, because of the accumulation of intraneuronal Aβ [13]. These findings provide evidence that extra- and intra-cellular Aβ accumulation causes cognitive impairment.

Aβ accumulation is the initial event in AD progression. Subsequently, Tau pathology develops in a delayed fashion [14]. Tau promotes the assembly of tubulin into microtubules, a component of the cytoskeleton, and lends support to neuronal morphology [15]. In normal brain, Tau contains two phosphates per molecule, while in AD, fibrillary Tau is abnormally phosphorylated (approximately eight phosphates per molecule) [16]. This hyper-phosphorylation is mediated by GSK3β, and it affects the interaction of Tau with microtubules, leading to neurodegeneration and cognitive impairment [17,18]. Accumulation of abnormally phosphorylated Tau is also causative for other neurodegenerative disorders, including frontotemporal lobar degeneration (FTD), corticobasal degeneration, and progressive supranuclear palsy [19]. Intracellular Tau aggregates have been suggested to spread though synaptic circuits in a prion-like manner [20]. Neurofibrillary tangles, a major pathological hallmark of AD, are found at early stages in the transentorhinal cortex and the entorhinal cortex, a region providing input to the hippocampal circuitry [19]. Subsequently, Tau pathology propagates to the hippocampus, the temporal cortex, and then progresses to primary motor/sensory areas [19].

Parkinson's disease and dementia with Lewy bodies are neurodegenerative diseases that are characterized by the presence of intracellular abnormal deposits called Lewy bodies and Lewy neurites. These deposits mainly consist of α-synuclein, which is a natively unfolded protein localized to the nucleus and presynaptic nerve terminals [21–23]. Several α-synuclein variants, such as missense and multiplication variants, are responsible for familial PD, suggesting that increased expression and abnormal structure of α-synuclein cause its aggregation and neurodegeneration [24,25]. Although accumulated α-synuclein is usually modified by phosphorylation, nitrosylation, glycation, and glycosylation, it remains unclear whether the aggregation is linked to its modification [26–28]. Fibrillar α-synuclein can propagate Lewy body and Lewy neurite pathology through cell-to-cell transmission leading to synaptic dysfunction and death of dopaminergic neurons in in vitro primary neurons and in the mouse brain [29–31]. Indeed, Lewy body-like inclusions were propagated in grafted embryonic nigral neurons that were transplanted into PD patients [32,33]. α-Synuclein is detected in cerebrospinal fluid of subjects with or without PD [34]. The mechanism by which α-synuclein is secreted remains unclear although the involvement of exosomal release and exocytosis via vesicles or multivesicular bodies has been reported [35–37]. Uptake of extracellular α-synuclein has been proposed to be mediated by pinocytosis and receptor-mediated endocytosis. Extracellular α-synuclein fibrils bind cell surface heparan sulfate proteoglycans and are intracellularly taken up through pinocytosis [38]. Tau fibrils are also incorporated into cells in the same manner [38]. In addition, lymphocyte-activation gene 3 (LAG3), a transmembrane protein, has high affinity for α-synuclein fibrils [39]. Interestingly, pathological Tau and Aβ species do not bind to LAG3, indicating that LAG3 is a specific receptor for α-synuclein fibrils [39]. LAG3 deficiency effectively reduces the endocytosis of α-synuclein fibrils and the propagation of PD pathology [39].

Amyotrophic lateral sclerosis (ALS) and FTD are neurodegenerative diseases characterized by motor and cognitive impairment, respectively. Both diseases have genetic and pathological overlaps. For example, variations in various genes, such as TDP-43, FUS, p62 and C9 or f72, are attributed to the etiology of both diseases, and TDP-43 pathology is often observed in both diseases (in ~97% of ALS and ~50% of FTD cases) [40,41]. In most cases of familial and sporadic ALS, immunohistochemical analysis shows that TDP-43 is included in ubiquitin-positive round and skein-like inclusions [42]. However, TDP-43 pathology is negative in other cases of familial ALS, such as in cases with SOD1 and FUS variations where inclusions are composed of SOD1 and FUS, respectively [43]. Numerous studies demonstrate that these mutant forms are aggregate-prone proteins that are cytotoxic by causing dysfunction to various cellular processes [44,45]. TDP-43, SOD1, and FUS aggregates also have prion-like seeding activity, which propagates ALS pathology [27,46–48]. TDP-43 and FUS are RNA-binding proteins that are involved in RNA and protein quality control [49]. Exposure to various stresses, such as heat shock, oxidative stress, and endoplasmic reticulum stress, induces formation of stress granules, which are dynamic assemblies of proteins and RNAs [50]. Stress granules are membrane-less organelles that contain translationally stalled mRNAs associated with translation initiation factors and multiple RNA-binding proteins, suggesting that stress granules regulate mRNA translation and stability and protect from environmental stresses [51,52]. Recent reports have indicated that variations linked to ALS-FTD in TDP-43, FUS, TIA-1, and C9 or f72 cause abnormal stress granule assembly and disassembly. For example, the TDP-43^{A382T} ALS-FTD variation causes a significant reduction in stress granule assembly in human fibroblasts [53]. In contrast, the TIA-1^{P362L} ALS-FTD variation delays stress granule disassembly and promotes the accumulation of non-dynamic stress granules [54]. These results indicate that dysfunctional stress granule dynamics might contribute to ALS pathogenesis. Moreover, mutations of p62 and OPTN were also identified in familial and sporadic ALS-FTD [55]. Both proteins are known as a ubiquitin binding protein shuttling ubiquitinated proteins for their degradation [55]. The FUS-containing inclusions are also immunoreactive with antibodies to p62 and OPTN in spinal anterior horn neurons in all sporadic ALS and in non-SOD1-familial ALS cases [55]. Recently, it was reported that ALS-FTLD-linked mutations of p62 disrupt autophagy and anti-oxidative stress pathway underlying the neurotoxicity in ALS-FTLD [56].

Other neurodegenerative diseases are also characterized by neuronal protein aggregates. Expanded polyglutamine (polyQ) tracts are aggregation-prone and expanded polyQ-containing proteins, such as huntingtin and ataxins, cause HD and spinocerebellar ataxia, respectively [57]. Huntingtin is a 348 kDa protein, and its N-terminal region contains the expandable polyQ tract [58,59]. Huntingtin undergoes post-translational modifications at multiple sites, such as phosphorylation, acetylation, sumoylation and ubiquitination, and is then cleaved by various proteases [58,60]. Cleaved N-terminal fragments with an expanded polyQ tract are released and form fibrillary aggregates or inclusion bodies [61]. Pathogenic polyQ-expanded huntingtin also has prion-like properties. Mutant huntingtin aggregates were detected in the extracellular matrix of grafted neurons in HD patient brains, indicating that pathological huntingtin can spread within the brain [62,63].

3. Ubiquitination in Protein Degradation

Accumulation of harmful proteins is a hallmark of various neurodegenerative diseases, as described above. Cells are protected from harmful proteins by protein quality control mechanisms, including molecular chaperone and protein degradation systems [64]. Eukaryotes have two major protein degradation systems, the ubiquitin–proteasome system (UPS) and the autophagy–lysosome pathway (ALP). In the UPS, ubiquitin-tagged proteins are targeted by a multi-subunit protease complex, the proteasome. The proteasome consists of a multi-catalytic proteinase complex (20S) and two regulatory complexes (19S and 11S) [65]. Misfolded proteins and short-lived proteins undergo ubiquitination by a multi-step process that requires a ubiquitin-activating (E1) enzyme, a ubiquitin-conjugating (E2) enzyme, and a ubiquitin ligase (E3) [66]. Initially, ubiquitin is activated by the E1 enzyme in an ATP-dependent manner, and activated ubiquitin is then transferred to an E2 enzyme. E3 then ligates the ubiquitin to the target protein [3]. Ubiquitinated proteins are recruited to the regulatory complex of the proteasome and are then deubiquitinated and unfolded [67]. Linearized proteins then translocate into the proteolytic chamber of the 20S proteasome and are cleaved by its six proteolytic sites [68].

The ALP is an intracellular metabolic process in which cytoplasmic proteins and organelles sequestrated by autophagosomes are degraded in lysosomes [69]. Autophagy is regulated by more than 30 autophagy regulated proteins and its core machinery is classified in four subgroups: (1) The ATG1/ULK1 complex; (2) ATG9 and its cycling system; (3) the phosphatidylinositol 3-kinase complex; (4) two ubiquitin-like conjugation systems (ATG8/LC3 and ATG12) [69]. In mammals, the ULK1 complex has an essential role in the initiation of autophagy and is directed to the endoplasmic reticulum together with ATG9 vesicles and the phosphatidylinositol 3-kinase complex [69]. DFCP1 and WIPIs are recruited to the endoplasmic reticulum membrane and promote the formation of isolation membrane [69]. Autophagosome formation is mediated by two ubiquitin-like conjugation systems, conjugation of ATG12 to ATG5 and conversion of LC3 to a phosphatidylethanolamine-conjugated membrane-bound form [69]. Finally, mature autophagosomes fuse with lysosomes, resulting in degradation of cellular components [69].

Mono-ubiquitination functions in the regulation of protein interaction, trafficking, and transcriptional activity [70] and mono-ubiquitinated proteins are degraded by the proteasome in both mammalian and yeast cells [71]. Multi-ubiquitin is generated by the sequential conjugation of ubiquitin to ubiquitin via Lys residues (Lys-6, Lys-11, Lys-27, Lys-29, Lys-33, Lys-48, and Lys-63) or Met residues (Met-1, Met-14, Met-20) [3]. Lys-11-, Lys-48-, and Lys-63-linked poly-ubiquitination act as proteolytic signals for the proteasome and autophagy (Figure 1). Lys-48-linked poly-ubiquitin is the major signal of numerous short-lived proteins and unfolded proteins for proteasomal degradation [72]. Lys-11-linked poly-ubiquitin is also involved in degradation of short-lived cell cycle proteins and in the ERAD (endoplasmic reticulum-associated degradation) pathway upon ER stress [73]. Furthermore, small aggregated proteins are selectively degraded via the autophagy–lysosome system. Lys-48- and Lys-63-linked poly-ubiquitin is required for selective sequestration of aggregated proteins into autophagosomes through autophagy receptor proteins [3]. Furthermore, accumulation of Lys-11-, Lys-48-, Lys-63-linked poly-ubiquitinated insoluble proteins was observed in the brains of *Atg5*- and *Atg7*-null mice, indicating that multiple ubiquitin signals might be involved in autophagic degradation of various cargoes (Figure 1) [74]. Autophagy receptor proteins possess both a ubiquitin-binding domain and an LC3-interacting region (LIR), and bind to various cargoes, such as protein aggregates, intracellular organelles and microbial invaders [75,76]. Ubiquitinated protein aggregates are selectively recognized by autophagy receptor p62, NBR1, OPTN, and TOLLIP [77,78]. These complexes are then associated with autophagosome protein LC3 though an LIR domain, resulting in sequestration into an isolation membrane and degradation in lysosomes [78]. Ubiquitination is also often required for selective clearance of organelles. For example, damaged mitochondria in PD are eliminated via ubiquitin-dependent PINK1-Parkin-mediated mitophagy [79]. Upon mitochondrial damage, Parkin, an E3 ubiquitin ligase, conjugates Lys-6-, Lys-11-, Lys-48-, and Lys-63-linked poly-ubiquitin to mitochondrial outer membrane proteins, and then mitochondria bind to autophagy receptors [79–81]. Furthermore, PINK1 coordinately acts upstream of Parkin in this process. Phosphorylated ubiquitin by PINK1 is required for Parkin activation [82]. These findings indicate that ubiquitin signals also have an important role in the autophagic clearance of organelles.

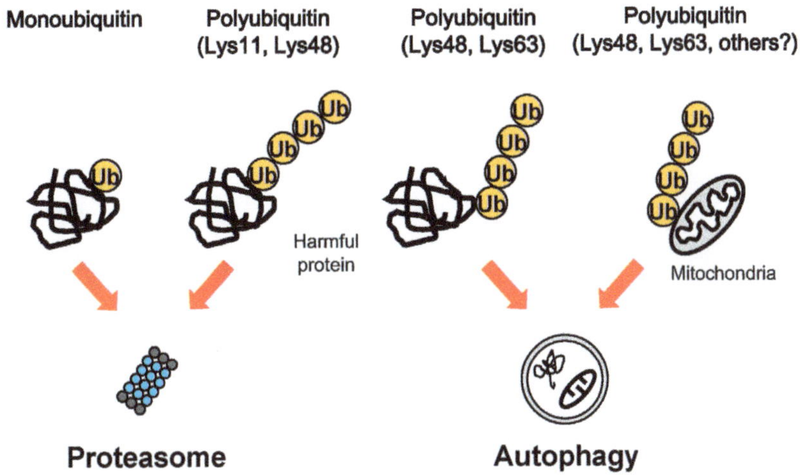

Figure 1. Degradation of ubiquitinated cargoes. Harmful proteins and mitochondria can be modified by various ubiquitin additions, such as mono-ubiquitin and Lys-11-, Lys-48-, and Lys-63-poly-ubiquitin chains. The proteasome preferentially degrades mono-ubiquitinated proteins and Lys11- and Lys-48-linked proteins, whereas autophagy preferentially eliminates Lys-48-, and Lys-63- decorated protein aggregates and mitochondria.

4. Ubiquitination of Neurodegenerative Disease-Associated Proteins

Harmful proteins causing neurodegenerative diseases undergo ubiquitination and pathological analyses using anti-ubiquitin antibodies identify various protein aggregates and inclusions. Paired helical filament-Tau (PHF-Tau) is modified by Lys-6-, Lys-11-, Lys-48-, Lys-63-poly-ubiquitin chains and mono-ubiquitin in AD brains or cultured cells (Figure 2) [83–85]. CHIP, a HSP70 co-chaperone, is an E3 ubiquitin ligase of PHF-Tau [84,86]. Lys-254, 257, 311, and 317 of PHF-Tau are acceptors for ubiquitin [85]. Lys-724, 725, 726, 751, and 763 of APP is intracellularly conjugated with ubiquitin in mouse brain [87] and impairment of this ubiquitination leads to accumulation of both secreted and intracellular Aβ40 [87].

α-Synuclein undergoes ubiquitination by various E3 ubiquitin ligases (Figure 2). Seven in absentia homolog (SIAH), an E3 ubiquitin ligase, mono-ubiquitinates α-synuclein at Lys-12, 21, and 23, resulting in an increase in the aggregation of α-synuclein and apoptotic cell death [88,89]. NEDD4 ubiquitin ligase also targets α-synuclein and mediates Lys-63-poly-ubiquitin [90]. Ubiquitinated α-synuclein is degraded by the endosomal-lysosomal pathway, suggesting that this process might have a protective effect against the pathogenesis of PD and other α-synucleinopathies [90]. Furthermore, CHIP is also involved in α-synuclein mono-ubiquitination or poly-ubiquitination, similarly of PHF-Tau [91,92]. CHIP-mono-ubiquitinated α-synuclein is deubiquitinated by USP9X. USP9X knockdown promotes accumulation of mono-ubiquitinated α-synuclein and enhances the formation of α-synuclein inclusions upon proteolytic inhibition [92]. Ubiquitin ligase E6-AP is localized to Lewy bodies in the PD brain, and is involved in α-synuclein ubiquitination and proteasome-dependent degradation [93].

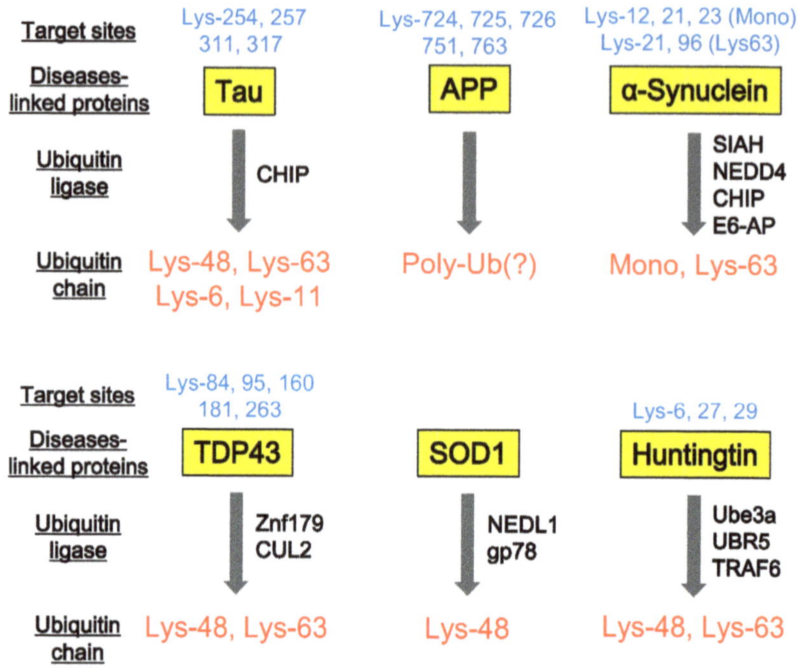

Figure 2. Ubiquitination of neurodegenerative disease-associated proteins. Neurodegenerative disease-associated proteins, such as Tau, APP (β-amyloid precursor protein), α-synuclein, TDP43, SOD1, and Huntingtin are ubiquitinated at individual target sites. Specific ubiquitin ligases involved in this ubiquitination and the pattern of ubiquitin chains can be identified by various biochemical studies.

ALS-causing TDP-43 and SOD1 aggregates are also detected by anti-ubiquitin antibodies. TDP-43 is targeted by Znf179 ubiquitin ligase and is modified by poly-ubiquitin chains [94]. Znf179 knockout suppresses TDP-43 proteosomal turnover, resulting in accumulation of insoluble TDP-43 and cytosolic TDP-43 inclusions in the cortex, hippocampus and midbrain regions [94]. In addition to Znf179, CUL2 ubiquitin ligase can modify misfolded TDP-43 with poly-ubiquitin, coordinately with von Hippel Lindau protein (VHL) [95]. Mass spectrometry analysis identified TDP-43 ubiquitination sites to be Lys-84, Lys-95 Lys-160, Lys-181, and Lys-263 residues and that its poly-ubiquitin chains link via Lys-48 and Lys-63 [96]. SOD1 is targeted by NEDL1 and gp78 ubiquitin ligases. NEDL1 colocalizes with SOD1 inclusions in the spinal cord ventral horn motor neurons of both ALS patients and mutant SOD1 transgenic mice [97]. gp78 ubiquitin ligase is also involved in the ubiquitination of SOD1 [83]. gp78 is a protein with at least five membrane-spanning domains, including a RING finger consensus sequence, and plays an important role in ERAD [98]. Interestingly, this ubiquitin ligase also mediates ubiquitination of spinocerebellar ataxia-associated ataxin-3 [99]. gp78 overexpression promotes the ubiquitination and degradation of SOD1 and ataxin-3 in cultured cells, whereas knockdown of gp78 stabilizes them [99].

Turnover of huntingtin is regulated by ubiquitination, via Lys-48- and Lys-63-poly-ubiquitin. Although aggregated mutant huntingtin mainly includes Lys-63-poly-ubiquitin chains, overexpression of Lys-48-specific ubiquitin ligase, Ube3a, reduces Lys-63-ubiquitination and huntingtin aggregation, enhancing its degradation via the Lys-48 ubiquitin–proteasome system [100]. Similarly, ubiquitin ligase, UBR5, is also involved in Lys-48-proteasomal degradation of both normal and mutant huntingtin [101]. However, tumor necrosis factor receptor-associated factor 6 (TRAF6) can promote the Lys-63-ubiquitin chain on mutant huntingtin and might contribute to autophagic clearance of huntingtin aggregates [102].

Collectively, numerous neurodegenerative disease-associated proteins undergo ubiquitination by a variety of ubiquitin ligases. These ubiquitin signals mainly serve to eliminate pathogenic proteins, although the ubiquitin signal on neurodegenerative disease-associated proteins can be pathogenic.

5. Autophagic Degradation of Neurodegenerative Disease-Associated Proteins

Small protein aggregates are thought to be degraded by ALP. Autophagy receptors recognize ubiquitin chains bound to cargoes and transport them to autophagosomes (Figure 3A). Recent advances in mass spectrometry technology have contributed to the decoding of ubiquitin signals and have revealed the diversity of ubiquitin chains. In addition, the ubiquitin binding-domains of autophagy receptors have been categorized (Figure 3B). p62, Nbr1, and c-Cbl have a UBA domain, a small domain of about 40 residues [103]. The UBA domain of p62 and Nbr1 binds strongly to both Lys-48- and Lys-63-poly-ubiquitin [104]. Moreover, the UBA domain of ubiquilin-1 and yeast Ede1, other ubiquitin binding proteins, have a high affinity for mono-ubiquitin, indicating that autophagy receptors with the UBA domain might bind to mono-ubiquitin [105,106]. OPTN has two ubiquitin binding domains, a UBAN and a zinc finger domain. A UBAN domain can interact not only with Lys-63-linked poly-ubiquitin but also with linear ubiquitin chains, which are generated between the N-terminal methionine of one ubiquitin and the C-terminal glycine of the next in the chain. (Figure 3B) [107,108]. However, the zinc finger domain of OPTN recognizes various protein aggregates in a ubiquitin-independent manner [109], although the same domain of NDP52 can recognize mono-ubiquitin, Lys-48-, and Lys-63-poly-ubiquitin [110]. NDP52 is a selective autophagy receptor for cytosolic bacteria (xenophagy) and damaged mitochondria (mitophagy), which are decorated with ubiquitin [111,112]. Although preferences of ubiquitin binding domains for ubiquitin codes remain unclear, individual autophagy receptors might selectively recognize disease-associated aggregates by the ubiquitin code. Indeed, the UBA domain of p62 can bind both Lys-48-linked and Lys-63-linked ubiquitin chains but has a higher affinity for Lys-63 chains [113]. In addition, the UBA domain of NBR1 is structurally distinct from the p62 UBA domain, resulting in a different interaction with ubiquitin. NBR1 has significantly higher affinity for mono-ubiquitin compared with p62 [114]. Accordingly, substrate preference of autophagy receptors might be dependent on the ubiquitin codes on cargoes (Figure 3A).

Activation of autophagy receptors is mediated by various kinases. For example, p62 is phosphorylated by various kinases, such as mTORC1, casein kinase 1, and TBK1 [64,115,116]. OPTN and NBR1 activities are also regulated by TBK1 and GSK3β, respectively [117,118]. Several phosphorylation sites are located in the ubiquitin binding domain, whose status alters the affinity for ubiquitinated proteins. Indeed, inhibition of several phosphorylation sites reduces cargo-binding potential [119]. This evidence indicates that structural alterations to autophagy receptors by phosphorylation controls autophagic clearance of various cargoes.

Autophagy activation accelerates elimination of neurodegenerative disease-associated protein aggregates and inclusions. The mTORC1 inhibitor, rapamycin, is well known to induce autophagy activity [120]. The effect of rapamycin has been investigated using various neurodegenerative disease models. For example, accumulation of Tau, huntingtin, and α-synuclein aggregates was significantly decreased in cultured cell and Drosophila models of AD, HD, and PD [121,122]. Moreover, the mTORC1-independent autophagy inducer, trehalose, also reduced protein aggregation and neuronal degeneration in ALS and tauopathy model mice [123,124]. Moreover, progression of PD-like pathology was investigated in autophagy suppressor Rubicon-KO mice, in which basal autophagy is constitutively activated. Spread of Lewy body-like α-synuclein aggregates was significantly reduced in the brain of this mouse [125]. These results indicate that autophagy induction may be an effective treatment for various neurodegenerative diseases.

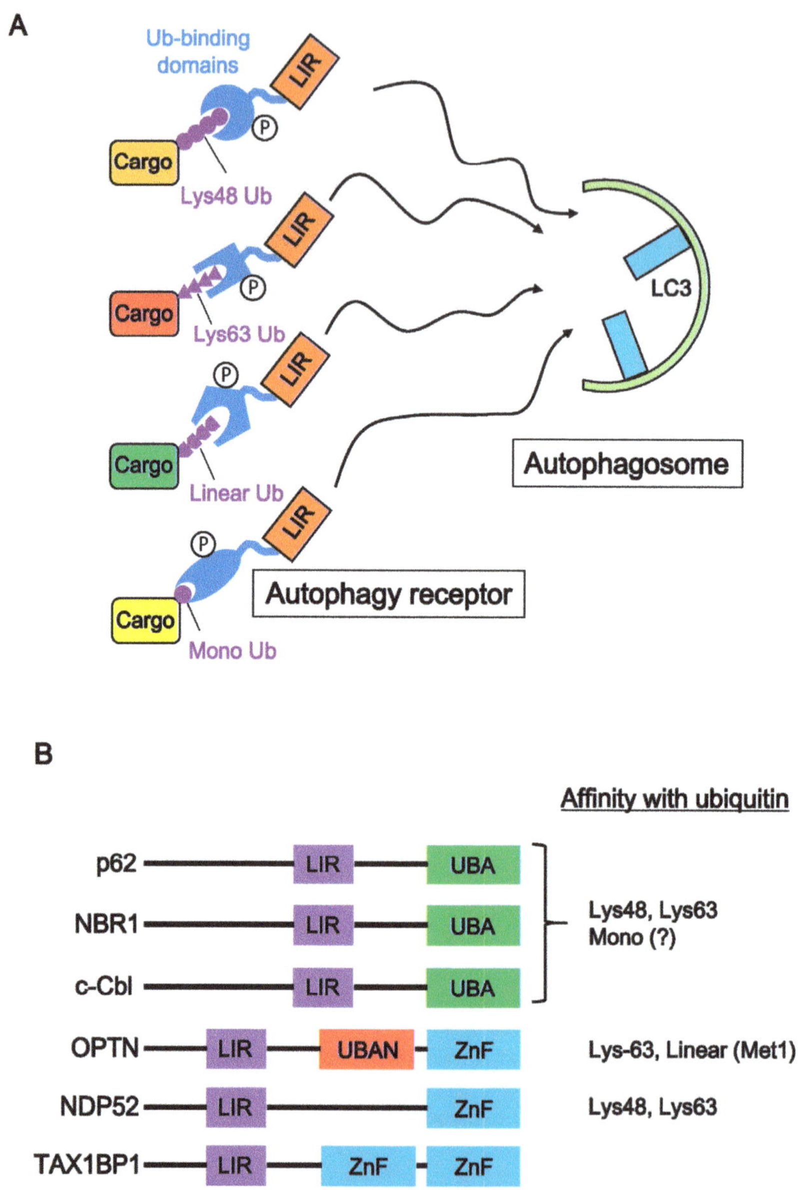

Figure 3. Autophagy receptors and selective autophagy of ubiquitinated cargoes. (**A**) Autophagy receptors play an important role in selective autophagy. Appropriate autophagy receptors bind to various ubiquitinated cargoes (Lys-48-, Lys-63-, linear-poly-ubiquitin chain, and mono-ubiquitin) through their ubiquitin binding domain. Many autophagy receptors are regulated by various kinases, and then autophagy receptor-cargo complexes interact with the autophagosome protein, LC3. (**B**) Structure of major autophagy receptors. Autophagy receptors consist of an LIR domain and a ubiquitin binding domain. Ubiquitin binding domains (UBA, UBAN, and ZnF) are classified based on motif sequences. Biochemical studies show preferences of autophagy receptors for ubiquitin types.

6. Concluding Remarks

Protein quality control systems, such as UPS and ALP, decline with age, which is a leading cause of neurodegenerative diseases. Clinical trials of several autophagy activators have been conducted for AD and ALS patients. For example, resveratrol is a natural polyphenol that induces autophagy activity by directly inhibiting mTOR [126]. In individuals with mild to moderate AD, decline of cerebrospinal fluid and plasma Aβ1–40 levels were observed in a resveratrol-treated group compared with a placebo-treated group [127]. However, improvement of cognitive function was not reported in this trial. The existing drugs, metformin and lithium, are also autophagy inducers, and clinical trials of these drugs for AD patients have also been conducted [128–130]. Currently, clinical trials of rapamycin are planned for ALS patients but not for AD patients, although it is hoped that further evidence warranting trials in AD patients will be forthcoming. While, it has been revealed that deubiquitinating enzymes such as UCL-L1 and ubiquitin-specific proteases are also involved in PD and AD through proteostasis [131]. Various inhibitors of deubiquitinating enzymes might be a new therapeutic target [131]. Collectively, detailed knowledge of ubiquitin chains in neurodegenerative disease-associated proteins and structural analyses of their interactions with ubiquitin binding domains will be beneficial for the development of novel therapies for neurodegenerative diseases.

Author Contributions: Y.W. wrote the paper. M.T. and K.T. edited and revised the manuscript. All authors read and approved the final manuscript.

Funding: This work was funded by Grants-in-Aid for Scientific Research from the Japan Society for the Promotion of Science (Y.W.: 18K07505, K.T.: 18K07371, and M.T.: 17H03553) and by the Shimizu Foundation for Immunology and Neuroscience Grant for 2018.

Acknowledgments: We thank Jeremy Allen, from Edanz Group for editing a draft of this manuscript.

Conflicts of Interest: The authors declare no conflict of interest.

References

1. Galves, M.; Rathi, R.; Prag, G.; Ashkenazi, A. Ubiquitin Signaling and Degradation of Aggregate-Prone Proteins. *Trends Biochem. Sci.* **2019**, *44*, 872–884. [CrossRef] [PubMed]
2. Dugger, B.N.; Dickson, D.W. Pathology of Neurodegenerative Diseases. *Cold Spring Harb. Perspect. Biol.* **2017**, *9*, a028035. [CrossRef] [PubMed]
3. Kwon, Y.T.; Ciechanover, A. The Ubiquitin Code in the Ubiquitin-Proteasome System and Autophagy. *Trends Biochem. Sci.* **2017**, *42*, 873–886. [CrossRef] [PubMed]
4. Ziv, I.; Matiuhin, Y.; Kirkpatrick, D.S.; Erpapazoglou, Z.; Leon, S.; Pantazopoulou, M.; Kim, W.; Gygi, S.P.; Haguenauer-Tsapis, R.; Reis, N.; et al. A perturbed ubiquitin landscape distinguishes between ubiquitin in trafficking and in proteolysis. *Mol. Cell. Proteom.* **2011**, *10*, M111.009753. [CrossRef] [PubMed]
5. Alzheimer, A.; Stelzmann, R.A.; Schnitzlein, H.N.; Murtagh, F.R. An English translation of Alzheimer's 1907 paper, "Uber eine eigenartige Erkankung der Hirnrinde". *Clin. Anat.* **1995**, *8*, 429–431. [CrossRef] [PubMed]
6. Duyckaerts, C.; Delatour, B.; Potier, M.C. Classification and basic pathology of Alzheimer disease. *Acta Neuropathol.* **2009**, *118*, 5–36. [CrossRef] [PubMed]
7. Esler, W.P.; Wolfe, M.S. A portrait of Alzheimer secretases—New features and familiar faces. *Science* **2001**, *293*, 1449–1454. [CrossRef]
8. Scheuner, D.; Eckman, C.; Jensen, M.; Song, X.; Citron, M.; Suzuki, N.; Bird, T.D.; Hardy, J.; Hutton, M.; Kukull, W.; et al. Secreted amyloid beta-protein similar to that in the senile plaques of Alzheimer's disease is increased in vivo by the presenilin 1 and 2 and APP mutations linked to familial Alzheimer's disease. *Nat. Med.* **1996**, *2*, 864–870. [CrossRef]
9. Selkoe, D.J.; Hardy, J. The amyloid hypothesis of Alzheimer's disease at 25 years. *EMBO Mol. Med.* **2016**, *8*, 595–608. [CrossRef]
10. Thal, D.R.; Rub, U.; Orantes, M.; Braak, H. Phases of A beta-deposition in the human brain and its relevance for the development of AD. *Neurology* **2002**, *58*, 1791–1800. [CrossRef]
11. Sinha, S.; Lieberburg, I. Cellular mechanisms of beta-amyloid production and secretion. *Proc. Natl. Acad. Sci. USA* **1999**, *96*, 11049–11053. [CrossRef] [PubMed]

12. Hu, X.; Crick, S.L.; Bu, G.; Frieden, C.; Pappu, R.V.; Lee, J.M. Amyloid seeds formed by cellular uptake, concentration, and aggregation of the amyloid-beta peptide. *Proc. Natl. Acad. Sci. USA* **2009**, *106*, 20324–20329. [CrossRef] [PubMed]
13. Oddo, S.; Caccamo, A.; Shepherd, J.D.; Murphy, M.P.; Golde, T.E.; Kayed, R.; Metherate, R.; Mattson, M.P.; Akbari, Y.; LaFerla, F.M. Triple-transgenic model of Alzheimer's disease with plaques and tangles: Intracellular Abeta and synaptic dysfunction. *Neuron* **2003**, *39*, 409–421. [CrossRef]
14. Gao, Y.L.; Wang, N.; Sun, F.R.; Cao, X.P.; Zhang, W.; Yu, J.T. Tau in neurodegenerative disease. *Ann. Transl. Med.* **2018**, *6*, 175. [CrossRef]
15. Kosik, K.S. The molecular and cellular biology of tau. *Brain Pathol.* **1993**, *3*, 39–43. [CrossRef]
16. Wang, Y.; Mandelkow, E. Tau in physiology and pathology. *Nat. Rev. Neurosci.* **2016**, *17*, 5–21. [CrossRef]
17. Busciglio, J.; Lorenzo, A.; Yeh, J.; Yankner, B.A. Beta-amyloid fibrils induce tau phosphorylation and loss of microtubule binding. *Neuron* **1995**, *14*, 879–888. [CrossRef]
18. Avila, J. Tau phosphorylation and aggregation in Alzheimer's disease pathology. *FEBS Lett.* **2006**, *580*, 2922–2927. [CrossRef]
19. Delacourte, A.; David, J.P.; Sergeant, N.; Buée, L.; Wattez, A.; Vermersch, P.; Ghozali, F.; Fallet-Bianco, C.; Pasquier, F.; Lebert, F.; et al. The biochemical pathway of neurofibrillary degeneration in aging and Alzheimer's disease. *Neurology* **1999**, *52*, 1158–1165. [CrossRef]
20. De Calignon, A.; Polydoro, M.; Suárez-Calvet, M.; William, C.; Adamowicz, D.H.; Kopeikina, K.J.; Pitstick, R.; Sahara, N.; Ashe, K.H.; Carlson, G.A.; et al. Propagation of tau pathology in a model of early Alzheimer's disease. *Neuron* **2012**, *73*, 685–697. [CrossRef]
21. Takeda, A.; Mallory, M.; Sundsmo, M.; Honer, W.; Hansen, L.; Masliah, E. Abnormal accumulation of NACP/alpha-synuclein in neurodegenerative disorders. *Am. J. Pathol.* **1998**, *152*, 367–372. [PubMed]
22. Weinreb, P.H.; Zhen, W.; Poon, A.W.; Conway, K.A.; Lansbury, P.T., Jr. NACP, a protein implicated in Alzheimer's disease and learning, is natively unfolded. *Biochemistry* **1996**, *35*, 13709–13715. [CrossRef] [PubMed]
23. Spillantini, M.G.; Schmidt, M.L.; Lee, V.M.; Trojanowski, J.Q.; Jakes, R.; Goedert, M. Alpha-synuclein in Lewy bodies. *Nature* **1997**, *388*, 839–840. [CrossRef]
24. Forman, M.S.; Lee, V.M.; Trojanowski, J.Q. Nosology of Parkinson's disease: looking for the way out of a quagmire. *Neuron* **2005**, *47*, 479–482. [CrossRef] [PubMed]
25. Taguchi, K.; Watanabe, Y.; Tsujimura, A.; Tanaka, M. Brain region-dependent differential expression of alpha-synuclein. *J. Comp. Neurol.* **2016**, *524*, 1236–1258. [CrossRef] [PubMed]
26. Stefanis, L. α-Synuclein in Parkinson's disease. *Cold Spring Harb. Perspect. Med.* **2012**, *2*, a009399. [CrossRef] [PubMed]
27. Fujiwara, H.; Hasegawa, M.; Dohmae, N.; Kawashima, A.; Masliah, E.; Goldberg, M.S.; Shen, J.; Takio, K.; Iwatsubo, T. alpha-Synuclein is phosphorylated in synucleinopathy lesions. *Nat. Cell Biol.* **2002**, *4*, 160–164. [CrossRef]
28. Giasson, B.I.; Duda, J.E.; Murray, I.V.; Chen, Q.; Souza, J.M.; Hurtig, H.I.; Ischiropoulos, H.; Trojanowski, J.Q.; Lee, V.M. Oxidative damage linked to neurodegeneration by selective alpha-synuclein nitration in synucleinopathy lesions. *Science* **2000**, *290*, 985–989. [CrossRef]
29. Volpicelli-Daley, L.A.; Luk, K.C.; Patel, T.P.; Tanik, S.A.; Riddle, D.M.; Stieber, A.; Meaney, D.F.; Trojanowski, J.Q.; Lee, V.M. Exogenous α-synuclein fibrils induce Lewy body pathology leading to synaptic dysfunction and neuron death. *Neuron* **2011**, *72*, 57–71. [CrossRef]
30. Luk, K.C.; Kehm, V.; Carroll, J.; Zhang, B.; O'Brien, P.; Trojanowski, J.Q.; Lee, V.M. Pathological α-synuclein transmission initiates Parkinson-like neurodegeneration in nontransgenic mice. *Science* **2012**, *338*, 949–953. [CrossRef]
31. Taguchi, K.; Watanabe, Y.; Tsujimura, A.; Tatebe, H.; Miyata, S.; Tokuda, T.; Mizuno, T.; Tanaka, M. Differential expression of alpha-synuclein in hippocampal neurons. *PLoS ONE* **2014**, *9*, e89327. [CrossRef] [PubMed]
32. Kordower, J.H.; Chu, Y.; Hauser, R.A.; Freeman, T.B.; Olanow, C.W. Lewy body-like pathology in long-term embryonic nigral transplants in Parkinson's disease. *Nat. Med.* **2008**, *14*, 504–506. [CrossRef] [PubMed]
33. Li, J.Y.; Englund, E.; Holton, J.L.; Soulet, D.; Hagell, P.; Lees, A.J.; Lashley, T.; Quinn, N.P.; Rehncrona, S.; Björklund, A.; et al. Lewy bodies in grafted neurons in subjects with Parkinson's disease suggest host-to-graft disease propagation. *Nat. Med.* **2008**, *14*, 501–503. [CrossRef] [PubMed]

34. El-Agnaf, O.M.; Salem, S.A.; Paleologou, K.E.; Cooper, L.J.; Fullwood, N.J.; Gibson, M.J.; Curran, M.D.; Court, J.A.; Mann, D.M.; Ikeda, S.; et al. Alpha-synuclein implicated in Parkinson's disease is present in extracellular biological fluids, including human plasma. *FASEB J.* **2003**, *17*, 1945–1947. [CrossRef] [PubMed]
35. Jang, A.; Lee, H.J.; Suk, J.E.; Jung, J.W.; Kim, K.P.; Lee, S.J. Non-classical exocytosis of alpha-synuclein is sensitive to folding states and promoted under stress conditions. *J. Neurochem.* **2010**, *113*, 1263–1274. [CrossRef]
36. Lee, H.J.; Cho, E.D.; Lee, K.W.; Kim, J.H.; Cho, S.G.; Lee, S.J. Autophagic failure promotes the exocytosis and intercellular transfer of α-synuclein. *Exp. Mol. Med.* **2013**, *45*, e22. [CrossRef]
37. Fussi, N.; Höllerhage, M.; Chakroun, T.; Nykänen, N.P.; Rösler, T.W.; Koeglsperger, T.; Wurst, W.; Behrends, C.; Höglinger, G.U. Exosomal secretion of α-synuclein as protective mechanism after upstream blockage of macroautophagy. *Cell Death Dis.* **2018**, *9*, 757. [CrossRef] [PubMed]
38. Holmes, B.B.; DeVos, S.L.; Kfoury, N.; Li, M.; Jacks, R.; Yanamandra, K.; Ouidja, M.O.; Brodsky, F.M.; Marasa, J.; Bagchi, D.P.; et al. Heparan sulfate proteoglycans mediate internalization and propagation of specific proteopathic seeds. *Proc. Natl. Acad. Sci. USA* **2013**, *110*, E3138–E3147. [CrossRef]
39. Mao, X.; Ou, M.T.; Karuppagounder, S.S.; Kam, T.I.; Yin, X.; Xiong, Y.; Ge, P.; Umanah, G.E.; Brahmachari, S.; Shin, J.H.; et al. Pathological α-synuclein transmission initiated by binding lymphocyte-activation gene 3. *Science* **2016**, *353*. [CrossRef]
40. Chou, C.C.; Zhang, Y.; Umoh, M.E.; Vaughan, S.W.; Lorenzini, I.; Liu, F.; Sayegh, M.; Donlin-Asp, P.G.; Chen, Y.H.; Duong, D.M.; et al. TDP-43 pathology disrupts nuclear pore complexes and nucleocytoplasmic transport in ALS/FTD. *Nat. Neurosci.* **2018**, *21*, 228–239. [CrossRef]
41. Gao, F.B.; Almeida, S.; Lopez-Gonzalez, R. Dysregulated molecular pathways in amyotrophic lateral sclerosis-frontotemporal dementia spectrum disorder. *EMBO J.* **2017**, *36*, 2931–2950. [CrossRef] [PubMed]
42. Farrawell, N.E.; Lambert-Smith, I.A.; Warraich, S.T.; Blair, I.P.; Saunders, D.N.; Hatters, D.M.; Yerbury, J.J. Distinct partitioning of ALS associated TDP-43, FUS and SOD1 mutants into cellular inclusions. *Sci. Rep.* **2015**, *5*, 13416. [CrossRef]
43. Saberi, S.; Stauffer, J.E.; Schulte, D.J.; Ravits, J. Neuropathology of Amyotrophic Lateral Sclerosis and Its Variants. *Neurol. Clin.* **2015**, *33*, 855–876. [CrossRef]
44. Wong, P.C.; Pardo, C.A.; Borchelt, D.R.; Lee, M.K.; Copeland, N.G.; Jenkins, N.A.; Sisodia, S.S.; Cleveland, D.W.; Price, D.L. An adverse property of a familial ALS-linked SOD1 mutation causes motor neuron disease characterized by vacuolar degeneration of mitochondria. *Neuron* **1995**, *14*, 1105–1116. [CrossRef]
45. Monahan, Z.; Ryan, V.H.; Janke, A.M.; Burke, K.A.; Rhoads, S.N.; Zerze, G.H.; O'Meally, R.; Dignon, G.L.; Conicella, A.E.; Zheng, W.; et al. Phosphorylation of the FUS low-complexity domain disrupts phase separation, aggregation, and toxicity. *EMBO J.* **2017**, *36*, 2951–2967. [CrossRef] [PubMed]
46. Lee, S.; Kim, H.J. Prion-like Mechanism in Amyotrophic Lateral Sclerosis: are Protein Aggregates the Key? *Exp. Neurobiol.* **2015**, *24*, 1–7. [CrossRef] [PubMed]
47. Grad, L.I.; Guest, W.C.; Yanai, A.; Pokrishevsky, E.; O'Neill, M.A.; Gibbs, E.; Semenchenko, V.; Yousefi, M.; Wishart, D.S.; Plotkin, S.S.; et al. Intermolecular transmission of superoxide dismutase 1 misfolding in living cells. *Proc. Natl. Acad. Sci. USA* **2011**, *108*, 16398–16403. [CrossRef] [PubMed]
48. Nomura, T.; Watanabe, S.; Kaneko, K.; Yamanaka, K.; Nukina, N.; Furukawa, Y. Intranuclear aggregation of mutant FUS/TLS as a molecular pathomechanism of amyotrophic lateral sclerosis. *J. Biol. Chem.* **2014**, *289*, 1192–1202. [CrossRef] [PubMed]
49. Ito, D.; Suzuki, N. Conjoint pathologic cascades mediated by ALS/FTLD-U linked RNA-binding proteins TDP-43 and FUS. *Neurology* **2011**, *77*, 1636–1643. [CrossRef]
50. Wolozin, B.; Ivanov, P. Stress granules and neurodegeneration. *Nat. Rev. Neurosci.* **2019**, *20*, 649–666. [CrossRef]
51. Advani, V.M.; Ivanov, P. Stress granule subtypes: An emerging link to neurodegeneration. *Cell. Mol. Life Sci.* **2020**. [CrossRef] [PubMed]
52. Buchan, J.R.; Parker, R. Eukaryotic stress granules: The ins and outs of translation. *Mol. Cell* **2009**, *36*, 932–941. [CrossRef] [PubMed]
53. Orrù, S.; Coni, P.; Floris, A.; Littera, R.; Carcassi, C.; Sogos, V.; Brancia, C. Reduced stress granule formation and cell death in fibroblasts with the A382T mutation of TARDBP gene: Evidence for loss of TDP-43 nuclear function. *Hum. Mol. Genet.* **2016**, *25*, 4473–4483. [CrossRef] [PubMed]

54. Mackenzie, I.R.; Nicholson, A.M.; Sarkar, M.; Messing, J.; Purice, M.D.; Pottier, C.; Annu, K.; Baker, M.; Perkerson, R.B.; Kurti, A.; et al. TIA1 Mutations in Amyotrophic Lateral Sclerosis and Frontotemporal Dementia Promote Phase Separation and Alter Stress Granule Dynamics. *Neuron* **2017**, *95*, 808–816.e809. [CrossRef]
55. Ajroud-Driss, S.; Siddique, T. Sporadic and hereditary amyotrophic lateral sclerosis (ALS). *Biochim. Biophys. Acta* **2015**, *1852*, 679–684. [CrossRef]
56. Deng, Z.; Lim, J.; Wang, Q.; Purtell, K.; Wu, S.; Palomo, G.M.; Tan, H.; Manfredi, G.; Zhao, Y.; Peng, J.; et al. ALS-FTLD-linked mutations of SQSTM1/p62 disrupt selective autophagy and NFE2L2/NRF2 anti-oxidative stress pathway. *Autophagy* **2020**, *16*, 917–931. [CrossRef]
57. Frontali, M.; Novelletto, A.; Annesi, G.; Jodice, C. CAG repeat instability, cryptic sequence variation and pathogeneticity: evidence from different loci. *Philos. Trans. R. Soc. Lond. Ser. B Biol. Sci.* **1999**, *354*, 1089–1094. [CrossRef]
58. Saudou, F.; Humbert, S. The Biology of Huntingtin. *Neuron* **2016**, *89*, 910–926. [CrossRef]
59. Yushchenko, T.; Deuerling, E.; Hauser, K. Insights into the Aggregation Mechanism of PolyQ Proteins with Different Glutamine Repeat Lengths. *Biophys. J.* **2018**, *114*, 1847–1857. [CrossRef]
60. Mejia, R.O.S.; Friedlander, R.M. Caspases in Huntington's disease. *Neurosci. Rev. J.* **2001**, *7*, 480–489. [CrossRef]
61. Scherzinger, E.; Lurz, R.; Turmaine, M.; Mangiarini, L.; Hollenbach, B.; Hasenbank, R.; Bates, G.P.; Davies, S.W.; Lehrach, H.; Wanker, E.E. Huntingtin-encoded polyglutamine expansions form amyloid-like protein aggregates in vitro and in vivo. *Cell* **1997**, *90*, 549–558. [CrossRef]
62. Cicchetti, F.; Lacroix, S.; Cisbani, G.; Vallières, N.; Saint-Pierre, M.; St-Amour, I.; Tolouei, R.; Skepper, J.N.; Hauser, R.A.; Mantovani, D.; et al. Mutant huntingtin is present in neuronal grafts in Huntington disease patients. *Ann. Neurol.* **2014**, *76*, 31–42. [CrossRef] [PubMed]
63. Pearce, M.M.P.; Kopito, R.R. Prion-Like Characteristics of Polyglutamine-Containing Proteins. *Cold Spring Harb. Perspect. Med.* **2018**, *8*, a024257. [CrossRef]
64. Watanabe, Y.; Tsujimura, A.; Taguchi, K.; Tanaka, M. HSF1 stress response pathway regulates autophagy receptor SQSTM1/p62-associated proteostasis. *Autophagy* **2017**, *13*, 133–148. [CrossRef] [PubMed]
65. Tai, H.C.; Schuman, E.M. Ubiquitin, the proteasome and protein degradation in neuronal function and dysfunction. *Nat. Rev. Neurosci.* **2008**, *9*, 826–838. [CrossRef] [PubMed]
66. Pohl, C.; Dikic, I. Cellular quality control by the ubiquitin–proteasome system and autophagy. *Science* **2019**, *366*, 818–822. [CrossRef] [PubMed]
67. Liu, C.W.; Jacobson, A.D. Functions of the 19S complex in proteasomal degradation. *Trends Biochem. Sci.* **2013**, *38*, 103–110. [CrossRef] [PubMed]
68. Lecker, S.H.; Goldberg, A.L.; Mitch, W.E. Protein degradation by the ubiquitin–proteasome pathway in normal and disease states. *J. Am. Soc. Nephrol.* **2006**, *17*, 1807–1819. [CrossRef]
69. Mizushima, N. Autophagy: Process and function. *Genes Dev.* **2007**, *21*, 2861–2873. [CrossRef]
70. Shabek, N.; Herman-Bachinsky, Y.; Buchsbaum, S.; Lewinson, O.; Haj-Yahya, M.; Hejjaoui, M.; Lashuel, H.A.; Sommer, T.; Brik, A.; Ciechanover, A. The size of the proteasomal substrate determines whether its degradation will be mediated by mono- or polyubiquitylation. *Mol. Cell* **2012**, *48*, 87–97. [CrossRef]
71. Braten, O.; Livneh, I.; Ziv, T.; Admon, A.; Kehat, I.; Caspi, L.H.; Gonen, H.; Bercovich, B.; Godzik, A.; Jahandideh, S.; et al. Numerous proteins with unique characteristics are degraded by the 26S proteasome following monoubiquitination. *Proc. Natl. Acad. Sci. USA* **2016**, *113*, E4639–E4647. [CrossRef] [PubMed]
72. Komander, D.; Reyes-Turcu, F.; Licchesi, J.D.; Odenwaelder, P.; Wilkinson, K.D.; Barford, D. Molecular discrimination of structurally equivalent Lys 63-linked and linear polyubiquitin chains. *EMBO Rep.* **2009**, *10*, 466–473. [CrossRef] [PubMed]
73. Martinez-Fonts, K.; Davis, C.; Tomita, T.; Elsasser, S.; Nager, A.R.; Shi, Y.; Finley, D.; Matouschek, A. The proteasome 19S cap and its ubiquitin receptors provide a versatile recognition platform for substrates. *Nat. Commun.* **2020**, *11*, 477. [CrossRef]
74. Riley, B.E.; Kaiser, S.E.; Shaler, T.A.; Ng, A.C.; Hara, T.; Hipp, M.S.; Lage, K.; Xavier, R.J.; Ryu, K.Y.; Taguchi, K.; et al. Ubiquitin accumulation in autophagy-deficient mice is dependent on the Nrf2-mediated stress response pathway: A potential role for protein aggregation in autophagic substrate selection. *J. Cell Biol.* **2010**, *191*, 537–552. [CrossRef] [PubMed]

75. Watanabe, Y.; Tanaka, M. p62/SQSTM1 in autophagic clearance of a non-ubiquitylated substrate. *J. Cell Sci.* **2011**, *124*, 2692–2701. [CrossRef]
76. Svenning, S.; Johansen, T. Selective autophagy. *Essays Biochem.* **2013**, *55*, 79–92. [CrossRef]
77. Johansen, T.; Lamark, T. Selective Autophagy: ATG8 Family Proteins, LIR Motifs and Cargo Receptors. *J. Mol. Biol.* **2020**, *432*, 80–103. [CrossRef]
78. Gatica, D.; Lahiri, V.; Klionsky, D.J. Cargo recognition and degradation by selective autophagy. *Nat. Cell Biol.* **2018**, *20*, 233–242. [CrossRef]
79. Harper, J.W.; Ordureau, A.; Heo, J.M. Building and decoding ubiquitin chains for mitophagy. *Nat. Rev. Mol. Cell Biol.* **2018**, *19*, 93–108. [CrossRef]
80. Shiba-Fukushima, K.; Inoshita, T.; Hattori, N.; Imai, Y. Lysine 63-linked polyubiquitination is dispensable for Parkin-mediated mitophagy. *J. Biol. Chem.* **2014**, *289*, 33131–33136. [CrossRef]
81. Pickles, S.; Vigié, P.; Youle, R.J. Mitophagy and Quality Control Mechanisms in Mitochondrial Maintenance. *Curr. Biol.* **2018**, *28*, R170–R185. [CrossRef] [PubMed]
82. Koyano, F.; Okatsu, K.; Kosako, H.; Tamura, Y.; Go, E.; Kimura, M.; Kimura, Y.; Tsuchiya, H.; Yoshihara, H.; Hirokawa, T.; et al. Ubiquitin is phosphorylated by PINK1 to activate parkin. *Nature* **2014**, *510*, 162–166. [CrossRef] [PubMed]
83. Cripps, D.; Thomas, S.N.; Jeng, Y.; Yang, F.; Davies, P.; Yang, A.J. Alzheimer disease-specific conformation of hyperphosphorylated paired helical filament-Tau is polyubiquitinated through Lys-48, Lys-11, and Lys-6 ubiquitin conjugation. *J. Biol. Chem.* **2006**, *281*, 10825–10838. [CrossRef] [PubMed]
84. Petrucelli, L.; Dickson, D.; Kehoe, K.; Taylor, J.; Snyder, H.; Grover, A.; De Lucia, M.; McGowan, E.; Lewis, J.; Prihar, G.; et al. CHIP and Hsp70 regulate tau ubiquitination, degradation and aggregation. *Hum. Mol. Genet.* **2004**, *13*, 703–714. [CrossRef]
85. Morishima-Kawashima, M.; Hasegawa, M.; Takio, K.; Suzuki, M.; Titani, K.; Ihara, Y. Ubiquitin is conjugated with amino-terminally processed tau in paired helical filaments. *Neuron* **1993**, *10*, 1151–1160. [CrossRef]
86. Sahara, N.; Murayama, M.; Mizoroki, T.; Urushitani, M.; Imai, Y.; Takahashi, R.; Murata, S.; Tanaka, K.; Takashima, A. In vivo evidence of CHIP up-regulation attenuating tau aggregation. *J. Neurochem.* **2005**, *94*, 1254–1263. [CrossRef]
87. Williamson, R.L.; Laulagnier, K.; Miranda, A.M.; Fernandez, M.A.; Wolfe, M.S.; Sadoul, R.; Di Paolo, G. Disruption of amyloid precursor protein ubiquitination selectively increases amyloid β (Aβ) 40 levels via presenilin 2-mediated cleavage. *J. Biol. Chem.* **2017**, *292*, 19873–19889. [CrossRef]
88. Rott, R.; Szargel, R.; Haskin, J.; Shani, V.; Shainskaya, A.; Manov, I.; Liani, E.; Avraham, E.; Engelender, S. Monoubiquitylation of alpha-synuclein by seven in absentia homolog (SIAH) promotes its aggregation in dopaminergic cells. *J. Biol. Chem.* **2008**, *283*, 3316–3328. [CrossRef]
89. Lee, J.T.; Wheeler, T.C.; Li, L.; Chin, L.S. Ubiquitination of alpha-synuclein by Siah-1 promotes alpha-synuclein aggregation and apoptotic cell death. *Hum. Mol. Genet.* **2008**, *17*, 906–917. [CrossRef]
90. Tofaris, G.K.; Kim, H.T.; Hourez, R.; Jung, J.W.; Kim, K.P.; Goldberg, A.L. Ubiquitin ligase Nedd4 promotes alpha-synuclein degradation by the endosomal-lysosomal pathway. *Proc. Natl. Acad. Sci. USA* **2011**, *108*, 17004–17009. [CrossRef]
91. Shin, Y.; Klucken, J.; Patterson, C.; Hyman, B.T.; McLean, P.J. The co-chaperone carboxyl terminus of Hsp70-interacting protein (CHIP) mediates alpha-synuclein degradation decisions between proteasomal and lysosomal pathways. *J. Biol. Chem.* **2005**, *280*, 23727–23734. [CrossRef] [PubMed]
92. Rott, R.; Szargel, R.; Haskin, J.; Bandopadhyay, R.; Lees, A.J.; Shani, V.; Engelender, S. α-Synuclein fate is determined by USP9X-regulated monoubiquitination. *Proc. Natl. Acad. Sci. USA* **2011**, *108*, 18666–18671. [CrossRef] [PubMed]
93. Mulherkar, S.A.; Sharma, J.; Jana, N.R. The ubiquitin ligase E6-AP promotes degradation of alpha-synuclein. *J. Neurochem.* **2009**, *110*, 1955–1964. [CrossRef] [PubMed]
94. Lee, Y.C.; Huang, W.C.; Lin, J.H.; Kao, T.J.; Lin, H.C.; Lee, K.H.; Shen, C.J.; Chang, W.C.; Huang, C.C. Znf179 E3 ligase-mediated TDP-43 polyubiquitination is involved in TDP-43- ubiquitinated inclusions (UBI) (+)-related neurodegenerative pathology. *J. Biomed. Sci.* **2018**, *25*, 76. [CrossRef]
95. Uchida, T.; Tamaki, Y.; Ayaki, T.; Shodai, A.; Kaji, S.; Morimura, T.; Banno, Y.; Nishitsuji, K.; Sakashita, N.; Maki, T.; et al. CUL2-mediated clearance of misfolded TDP-43 is paradoxically affected by VHL in oligodendrocytes in ALS. *Sci. Rep.* **2016**, *6*, 19118. [CrossRef]

96. Hans, F.; Eckert, M.; von Zweydorf, F.; Gloeckner, C.J.; Kahle, P.J. Identification and characterization of ubiquitinylation sites in TAR DNA-binding protein of 43 kDa (TDP-43). *J. Biol. Chem.* **2018**, *293*, 16083–16099. [CrossRef]
97. Miyazaki, K.; Fujita, T.; Ozaki, T.; Kato, C.; Kurose, Y.; Sakamoto, M.; Kato, S.; Goto, T.; Itoyama, Y.; Aoki, M.; et al. NEDL1, a novel ubiquitin-protein isopeptide ligase for dishevelled-1, targets mutant superoxide dismutase-1. *J. Biol. Chem.* **2004**, *279*, 11327–11335. [CrossRef]
98. Fang, S.; Ferrone, M.; Yang, C.; Jensen, J.P.; Tiwari, S.; Weissman, A.M. The tumor autocrine motility factor receptor, gp78, is a ubiquitin protein ligase implicated in degradation from the endoplasmic reticulum. *Proc. Natl. Acad. Sci. USA* **2001**, *98*, 14422–14427. [CrossRef]
99. Ying, Z.; Wang, H.; Fan, H.; Zhu, X.; Zhou, J.; Fei, E.; Wang, G. Gp78, an ER associated E3, promotes SOD1 and ataxin-3 degradation. *Hum. Mol. Genet.* **2009**, *18*, 4268–4281. [CrossRef]
100. Bhat, K.P.; Yan, S.; Wang, C.E.; Li, S.; Li, X.J. Differential ubiquitination and degradation of huntingtin fragments modulated by ubiquitin-protein ligase E3A. *Proc. Natl. Acad. Sci. USA* **2014**, *111*, 5706–5711. [CrossRef]
101. Koyuncu, S.; Saez, I.; Lee, H.J.; Gutierrez-Garcia, R.; Pokrzywa, W.; Fatima, A.; Hoppe, T.; Vilchez, D. The ubiquitin ligase UBR5 suppresses proteostasis collapse in pluripotent stem cells from Huntington's disease patients. *Nat. Commun.* **2018**, *9*, 2886. [CrossRef]
102. Zucchelli, S.; Marcuzzi, F.; Codrich, M.; Agostoni, E.; Vilotti, S.; Biagioli, M.; Pinto, M.; Carnemolla, A.; Santoro, C.; Gustincich, S.; et al. Tumor necrosis factor receptor-associated factor 6 (TRAF6) associates with huntingtin protein and promotes its atypical ubiquitination to enhance aggregate formation. *J. Biol. Chem.* **2011**, *286*, 25108–25117. [CrossRef]
103. Buchberger, A. From UBA to UBX: New words in the ubiquitin vocabulary. *Trends Cell Biol.* **2002**, *12*, 216–221. [CrossRef]
104. Waters, S.; Marchbank, K.; Solomon, E.; Whitehouse, C.; Gautel, M. Interactions with LC3 and polyubiquitin chains link nbr1 to autophagic protein turnover. *FEBS Lett.* **2009**, *583*, 1846–1852. [CrossRef] [PubMed]
105. Zhang, D.; Raasi, S.; Fushman, D. Affinity makes the difference: nonselective interaction of the UBA domain of Ubiquilin-1 with monomeric ubiquitin and polyubiquitin chains. *J. Mol. Biol.* **2008**, *377*, 162–180. [CrossRef]
106. Swanson, K.A.; Hicke, L.; Radhakrishnan, I. Structural basis for monoubiquitin recognition by the Ede1 UBA domain. *J. Mol. Biol.* **2006**, *358*, 713–724. [CrossRef] [PubMed]
107. Li, F.; Xu, D.; Wang, Y.; Zhou, Z.; Liu, J.; Hu, S.; Gong, Y.; Yuan, J.; Pan, L. Structural insights into the ubiquitin recognition by OPTN (optineurin) and its regulation by TBK1-mediated phosphorylation. *Autophagy* **2018**, *14*, 66–79. [CrossRef] [PubMed]
108. Gleason, C.E.; Ordureau, A.; Gourlay, R.; Arthur, J.S.; Cohen, P. Polyubiquitin binding to optineurin is required for optimal activation of TANK-binding kinase 1 and production of interferon β. *J. Biol. Chem.* **2011**, *286*, 35663–35674. [CrossRef]
109. Korac, J.; Schaeffer, V.; Kovacevic, I.; Clement, A.M.; Jungblut, B.; Behl, C.; Terzic, J.; Dikic, I. Ubiquitin-independent function of optineurin in autophagic clearance of protein aggregates. *J. Cell Sci.* **2013**, *126*, 580–592. [CrossRef]
110. Xie, X.; Li, F.; Wang, Y.; Lin, Z.; Cheng, X.; Liu, J.; Chen, C.; Pan, L. Molecular basis of ubiquitin recognition by the autophagy receptor CALCOCO2. *Autophagy* **2015**, *11*, 1775–1789. [CrossRef]
111. Von Muhlinen, N.; Thurston, T.; Ryzhakov, G.; Bloor, S.; Randow, F. NDP52, a novel autophagy receptor for ubiquitin-decorated cytosolic bacteria. *Autophagy* **2010**, *6*, 288–289. [CrossRef] [PubMed]
112. Lazarou, M.; Sliter, D.A.; Kane, L.A.; Sarraf, S.A.; Wang, C.; Burman, J.L.; Sideris, D.P.; Fogel, A.I.; Youle, R.J. The ubiquitin kinase PINK1 recruits autophagy receptors to induce mitophagy. *Nature* **2015**, *524*, 309–314. [CrossRef] [PubMed]
113. Paul, P.K.; Kumar, A. TRAF6 coordinates the activation of autophagy and ubiquitin–proteasome systems in atrophying skeletal muscle. *Autophagy* **2011**, *7*, 555–556. [CrossRef]
114. Walinda, E.; Morimoto, D.; Sugase, K.; Konuma, T.; Tochio, H.; Shirakawa, M. Solution structure of the ubiquitin-associated (UBA) domain of human autophagy receptor NBR1 and its interaction with ubiquitin and polyubiquitin. *J. Biol. Chem.* **2014**, *289*, 13890–13902. [CrossRef]
115. Sánchez-Martín, P.; Komatsu, M. p62/SQSTM1—Steering the cell through health and disease. *J. Cell Sci.* **2018**, *131*. [CrossRef]

116. Ichimura, Y.; Waguri, S.; Sou, Y.S.; Kageyama, S.; Hasegawa, J.; Ishimura, R.; Saito, T.; Yang, Y.; Kouno, T.; Fukutomi, T.; et al. Phosphorylation of p62 activates the Keap1-Nrf2 pathway during selective autophagy. *Mol. Cell* **2013**, *51*, 618–631. [CrossRef] [PubMed]
117. Wild, P.; Farhan, H.; McEwan, D.G.; Wagner, S.; Rogov, V.V.; Brady, N.R.; Richter, B.; Korac, J.; Waidmann, O.; Choudhary, C.; et al. Phosphorylation of the autophagy receptor optineurin restricts Salmonella growth. *Science* **2011**, *333*, 228–233. [CrossRef]
118. Nicot, A.S.; Lo Verso, F.; Ratti, F.; Pilot-Storck, F.; Streichenberger, N.; Sandri, M.; Schaeffer, L.; Goillot, E. Phosphorylation of NBR1 by GSK3 modulates protein aggregation. *Autophagy* **2014**, *10*, 1036–1053. [CrossRef]
119. Deng, Z.; Purtell, K.; Lachance, V.; Wold, M.S.; Chen, S.; Yue, Z. Autophagy Receptors and Neurodegenerative Diseases. *Trends Cell Biol.* **2017**, *27*, 491–504. [CrossRef]
120. Rubinsztein, D.C.; Nixon, R.A. Rapamycin induces autophagic flux in neurons. *Proc. Natl. Acad. Sci. USA* **2010**, *107*, E181. [CrossRef]
121. Berger, Z.; Ravikumar, B.; Menzies, F.M.; Oroz, L.G.; Underwood, B.R.; Pangalos, M.N.; Schmitt, I.; Wullner, U.; Evert, B.O.; O'Kane, C.J.; et al. Rapamycin alleviates toxicity of different aggregate-prone proteins. *Hum. Mol. Genet.* **2006**, *15*, 433–442. [CrossRef] [PubMed]
122. Watanabe, Y.; Tatebe, H.; Taguchi, K.; Endo, Y.; Tokuda, T.; Mizuno, T.; Nakagawa, M.; Tanaka, M. p62/SQSTM1-dependent autophagy of Lewy body-like α-synuclein inclusions. *PLoS ONE* **2012**, *7*, e52868. [CrossRef] [PubMed]
123. Zhang, X.; Chen, S.; Song, L.; Tang, Y.; Shen, Y.; Jia, L.; Le, W. MTOR-independent, autophagic enhancer trehalose prolongs motor neuron survival and ameliorates the autophagic flux defect in a mouse model of amyotrophic lateral sclerosis. *Autophagy* **2014**, *10*, 588–602. [CrossRef] [PubMed]
124. Schaeffer, V.; Goedert, M. Stimulation of autophagy is neuroprotective in a mouse model of human tauopathy. *Autophagy* **2012**, *8*, 1686–1687. [CrossRef] [PubMed]
125. Nakamura, S.; Oba, M.; Suzuki, M.; Takahashi, A.; Yamamuro, T.; Fujiwara, M.; Ikenaka, K.; Minami, S.; Tabata, N.; Yamamoto, K.; et al. Suppression of autophagic activity by Rubicon is a signature of aging. *Nat. Commun.* **2019**, *10*, 847. [CrossRef]
126. Park, D.; Jeong, H.; Lee, M.N.; Koh, A.; Kwon, O.; Yang, Y.R.; Noh, J.; Suh, P.G.; Park, H.; Ryu, S.H. Resveratrol induces autophagy by directly inhibiting mTOR through ATP competition. *Sci. Rep.* **2016**, *6*, 21772. [CrossRef]
127. Turner, R.S.; Thomas, R.G.; Craft, S.; van Dyck, C.H.; Mintzer, J.; Reynolds, B.A.; Brewer, J.B.; Rissman, R.A.; Raman, R.; Aisen, P.S. A randomized, double-blind, placebo-controlled trial of resveratrol for Alzheimer disease. *Neurology* **2015**, *85*, 1383–1391. [CrossRef]
128. Cano-Cuenca, N.; Solís-García del Pozo, J.E.; Jordán, J. Evidence for the efficacy of latrepirdine (Dimebon) treatment for improvement of cognitive function: A meta-analysis. *J. Alzheimer's Dis.* **2014**, *38*, 155–164. [CrossRef]
129. Koenig, A.M.; Mechanic-Hamilton, D.; Xie, S.X.; Combs, M.F.; Cappola, A.R.; Xie, L.; Detre, J.A.; Wolk, D.A.; Arnold, S.E. Effects of the Insulin Sensitizer Metformin in Alzheimer Disease: Pilot Data From a Randomized Placebo-controlled Crossover Study. *Alzheimer Dis. Assoc. Disord.* **2017**, *31*, 107–113. [CrossRef]
130. Suresh, S.N.; Chakravorty, A.; Giridharan, M.; Garimella, L.; Manjithaya, R. Pharmacological Tools to Modulate Autophagy in Neurodegenerative Diseases. *J. Mol. Biol.* **2020**, *432*, 2822–2842. [CrossRef]
131. Lim, K.H.; Joo, J.Y.; Baek, K.H. The potential roles of deubiquitinating enzymes in brain diseases. *Ageing Res. Rev.* **2020**, *61*, 101088. [CrossRef] [PubMed]

cells

MDPI

Review

Sorting Nexins in Protein Homeostasis

Sara E. Hanley and Katrina F. Cooper *

Department of Molecular Biology, Graduate School of Biomedical Sciences, Rowan University, Stratford, NJ 08084, USA; hanleys2@rowan.edu
* Correspondence: cooperka@rowan.edu; Tel.: +1-(856)-566-2887

Abstract: Protein homeostasis is maintained by removing misfolded, damaged, or excess proteins and damaged organelles from the cell by three major pathways; the ubiquitin-proteasome system, the autophagy-lysosomal pathway, and the endo-lysosomal pathway. The requirement for ubiquitin provides a link between all three pathways. Sorting nexins are a highly conserved and diverse family of membrane-associated proteins that not only traffic proteins throughout the cells but also provide a second common thread between protein homeostasis pathways. In this review, we will discuss the connections between sorting nexins, ubiquitin, and the interconnected roles they play in maintaining protein quality control mechanisms. Underlying their importance, genetic defects in sorting nexins are linked with a variety of human diseases including neurodegenerative, cardiovascular diseases, viral infections, and cancer. This serves to emphasize the critical roles sorting nexins play in many aspects of cellular function.

Keywords: sorting nexins; retromer; endosome; autophagy; ubiquitin; lysosome; proteasome

Citation: Hanley, S.E.; Cooper, K.F. Sorting Nexins in Protein Homeostasis. *Cells* **2021**, *10*, 17. https://dx.doi.org/10.3390/cells10010017

Received: 4 November 2020
Accepted: 18 December 2020
Published: 24 December 2020

1. Introduction

The integrity of the proteome is essential for maintaining homeostasis as well as coordinating stress response mechanisms. Maintenance of homeostasis is handled by sophisticated quality control systems that ensure that malformed or excess proteins are degraded at the appropriate time and location. Selective proteolysis is largely mediated by the ubiquitin-proteasomal system (UPS) and the autophagy-lysosomal pathway (ALP). In general, the UPS is the primary proteolytic route for misfolded or damaged proteins, and short-lived proteins, having essential functions in many critical cellular pathways, including cell cycle progression and transcriptional regulation [1]. The ALP tackles long-lived proteins, dysfunctional or superfluous organelles, and protein aggregates. As the ALP is upregulated in response to cellular stress (nutrient deprivation, hypoxia, oxidative stress), it is considered a major adaptive mechanism, critical for cell survival following unfavorable environmental onslaughts [2,3].

There are three major types of autophagy: macroautophagy, microautophagy, and chaperon-mediated autophagy (CMA) [4]. Macroautophagy (herein autophagy) is also further classified as being selective or non-selective. In non-selective bulk autophagy, autophagic vesicles randomly engulf portions of the cytoplasm and various cytoplasmic components, primarily in response to starvation signals. Selective mechanisms are predominantly utilized to maintain homeostasis under physiological conditions. Cargos include defective organelles such as mitochondria, endoplasmic reticulum, peroxisomes [5], as well as cytoplasmic protein aggregates [6] and pathogenic intracellular invaders including RNA viruses like SARS-CoV-2 (COVID-19) [7–9]. In mammalian cells, ubiquitylation of the cargo is critical for recognition by the autophagic machinery, thereby linking UPS and ALP pathways [10,11].

The third protein quality control system, the endo-lysosomal pathway, is intricately linked to UPS and ALP [12]. Cargos are transported from the plasma membrane to sorting endosomes where their fate is decided. In a ubiquitin-dependent pathway, proteins can be selected for inclusion into intraluminal vesicles (ILVs), that by budding away from sorting endosomes, are ultimately delivered to lysosomes for degradation [13]. Alternatively, proteins can be retrieved from this degradative fate and selected for enrichment in endosomal "retrieval" subdomains, namely the *trans*-Golgi network (TGN), or recycling endosomes. From here they are recycled back to the plasma membrane by the secretory pathway [12,14]. Retrieval of cargos is mediated either by the retromer or retriever complex paired with a sorting nexin, and aided by several complexes (CCC (CCDC22, CCDC93, and COMMD), WASH (Wiskott–Aldrich syndrome protein and SCAR homolog)) and branched actin [15–17].

Although initially reported as independent pathways, the UPS and ALP are now known to be interconnected, linked by their common requirement for ubiquitin in substrate targeting. Another common thread is that they all utilize the evolutionarily conserved sorting nexin (SNX) family of proteins to move substrates to different destinations (Figure 1). Underlying their importance, genetic defects in SNXs are linked with a variety of human diseases including neurodegenerative, cardiovascular diseases, and cancer [18,19]. In this review, we will discuss the interplay between sorting nexins and protein homeostasis.

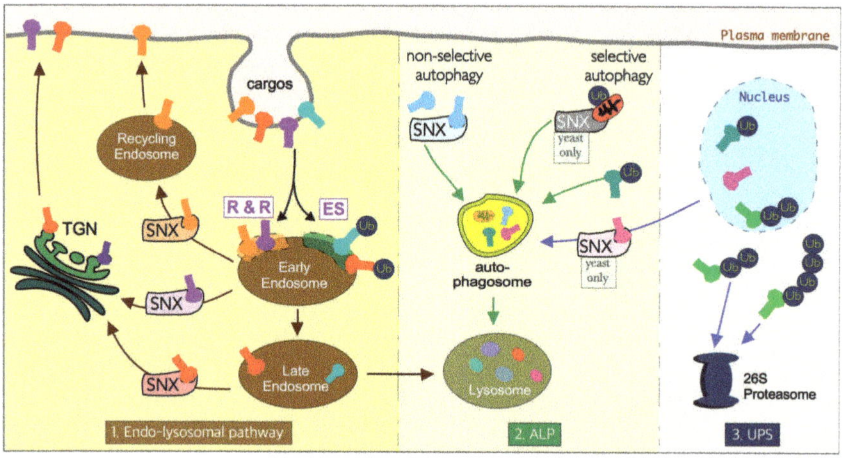

Figure 1. Sorting nexins and ubiquitin coordinate the three distinct but interconnected protein proteolysis pathways. In the endo-lysosomal pathway, membrane proteins are sorted at the early endosome. Cargos destined for lysosomal degradation are marked by ubiquitination and internalized using endosomal sorting complexes required for transport pathways (ES). Cargos destined for recycling are retrieved either by the retromer or retrieval complex (R & R) coupled with various subclasses of sorting nexins (SNX- see text for details). In the autophagy-lysosome pathway (ALP) cargos are sequestered to the vacuole by double-membraned vesicles called autophagosomes by selective or non-selective mechanisms. Selective pathways in yeast are mediated by SNX-Bar heterodimers. In mammalian cells, the recognition of selective autophagy cargos is dependent upon ubiquitination (Ub). The ubiquitin-proteasomal system (UPS) targets short-lived regulatory proteins that are selectively targeted and degraded. TGN- *trans*-Golgi network, SNX-sorting nexin. In cells, the TGN and nucleus are in close proximity, whereas here they are drawn apart for clarity.

2. Classification of SNX Proteins

Sorting nexins are a large group of diverse cellular trafficking proteins, with 10 and 33 members identified in yeast and mammals, respectively [20]. The members are classified into different subfamilies based on the structural arrangements of their scaffolding, enzymatic, and regulatory domains [21] (Figure 2). They all share a common phox homology (PX) domain that binds to phosphoinositide (PI) lipids found in organelle membranes [20,22]. Many SNX family members also contain various other conserved structural domains, BAR and FERM domains being the most prevalent. This modularity confers a wide variety of functions, from signal transduction to membrane deformation and cargo binding. Importantly, sorting nexins are crucial modulators of endosome dynamics as well as autophagic functions.

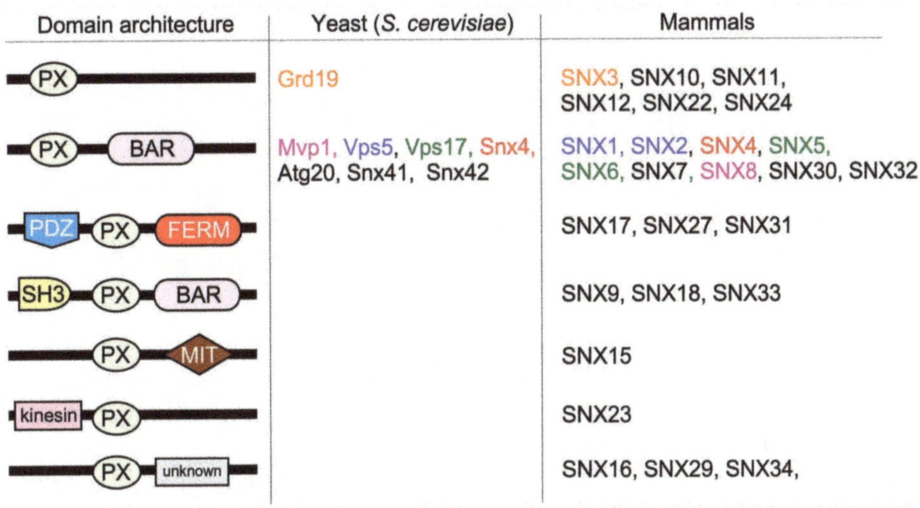

Figure 2. Structural classification of SNX subfamilies. The domain architecture describes the functional domains within different SNX subfamilies. The phox homology (PX) domain denotes the highly conserved lipid-binding domain that unifies the SNX protein family. It enables SNXs to bind to specific phosphoinositides (PtdIns) which mark different membrane surfaces. SNX-BAR (Bin/Amphiphysin/Rvs) proteins contain coiled-coil regions that enhance membrane binding, membrane remodeling, and protein-protein interactions. PDZ (postsynaptic density 95/discs large/zonula occludens) domains, FERM (protein 4.1/ezrin/radixin/moesin) domains, SH3 (SRC homology 3) domains, MIT (microtubule interacting and trafficking) domains, and Kinesin motor domains play a role in membrane binding, substrate recognition, kinase activity regulation, protein trafficking, and binding/movement along microtubules. The color code signifies yeast and mammalian homologs.

2.1. Lipid Binding PX Domain of SNX Proteins

The canonical 100-130 amino acid phox homology (PX) domain was first identified in the NADPH phagocyte oxidase complex subunits p40phox and p47phox [23]. It is highly conserved and predominantly occurs in sorting nexins. Despite this, the PX domains show little sequence conservation across SNX family members [24]. PX domains all possess the same core fold, consisting of three antiparallel β-strands (β1–β3), followed by three α-helices (α1–α3). Crystal structure analysis has shown that these helices form a loop structure that is required for PtdIns3P binding. Mutations in the loop structure invariably result in the dissociation of PX domain proteins from endosomal compartments [25].

It is well established that different PtdIns decorate different membranes [26]. This has resulted in the concept of a phosphoinositide code, that provides membrane identity within the endocytic system [27]. The PX domains of SNX family members predominantly bind phosphatidylinositol 3-monophosphate (PtdIns3P), a signaling lipid enriched in the early endosome membrane [28,29]. However, SNX proteins can also bind to the other PtdIns phospholipids such as PtdIns(3,4)P2, PtdIns(3,5)P2, PtdIns(4,5)P2 and PtdIns(3,4,5)P3 [24].

2.2. SNX-PX Proteins

The SNX-PX subfamily consists only of a PX domain. This family includes SNX3 which binds to a multi-protein complex called the retromer that consists of three conserved subunits (VPS26, VPS29, and VPS35) [30–32]. Cargo recognition is dependent on SNX3 *binding to the retromer. This* exposes a binding site at the interface between SNX3 and

VPS26 for cargos containing the Øx/L/M/V recycling motif (where Ø is a bulky aromatic residue) [33,34]. The PX domain associates with the cytosolic face of early endosomes and recycles its various cargos to the *trans*-Golgi network (TGN) [35]. This is aided by the Ankyrin-repeat protein ANKRD50 [36] and WASH mediated actin polymerization (see Figure 3) [33,34]. Multiple SNX3-retromer cargos have been identified including the Wnt sorting receptor Wntless [30,37], the transferrin receptor [38], and the divalent metal ion transporter Dmt1-II [39]. As such, retromer dysfunction impairs many cellular processes and underlies the pathogenesis of various neurodegenerative disorders. For example, a mutation in VPS35, causes late-onset Parkinson's disease [40], and microarray studies have implicated the retromer complex in Alzheimer's disease [18,41].

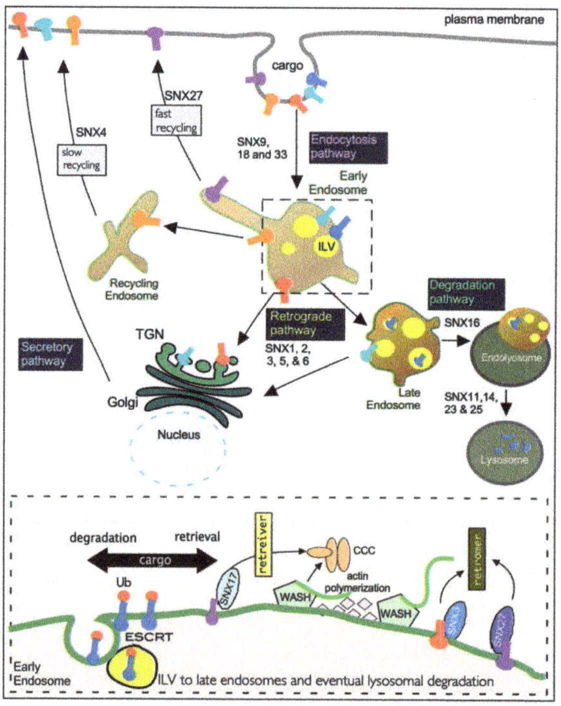

Figure 3. The role of SNXs in mammalian endocytosis. Transmembrane proteins are internalized into early endosomes from the plasma membrane by the endocytosis pathway. From here they are sorted through the complex and dynamic endomembrane network that consists of three different endosome vesicles (early, late, and recycling). Retrograde transport of membrane proteins requires recycling endosomes and the *trans*-Golgi network (TGN) for delivery back to the plasma membrane by the secretory pathway. Degradation of membrane proteins requires multiple rounds of cargo sorting and intra-luminal vesicle (ILV) biogenesis and fusion of the late endosome with the lysosome to form endo-lysosomes. Here ILVs and their accompanying cargos are degraded. The traced box at the bottom of the diagram is a zoomed-in schematic of sorting at endosomal membranes. It indicates that ubiquitinated membrane proteins destined for lysosomal proteolysis mediated by the ESCRT pathway. Transmembrane proteins are sorted by retromer and retrieval complexes aided by the actin remodeling WASH complex. Only SNX17 associates with the retriever and cooperates with the CCC complex to mediate endosomal trafficking. The retromer complex controls the recycling of a wide range of different cargos in cooperation with multiple SNX proteins including SNX3 and SNX27. Ub—Ubiquitin; CCC—CCDC22, CCDC93, and COMMD; WASH—Wiskott–Aldrich syndrome protein and SCAR homolog; ESCRT—endosomal sorting complexes required for transport.

SNX16 is a unique SNX family protein containing a coiled-coil (CC) domain downstream of the PX domain that deserves a special mention. The PX domain binds specifically to PI3P which is found on early and late endosomes. Recently the recycling of E-Cadherin, which mediates cell-cell adhesion in epithelial tissues, was found to be mediated by SNX16 in a novel mechanism. Here the PX and CC domains form a novel pear-shaped homodimer that interacts with E-Cadherin while simultaneously binding to endosomal membranes [42]. Further studies also discovered that higher order assembly of SNX16 via its CC domain drives membrane tubulation and controls neuronal endosomal maturation [43].

2.3. SNX-FERM Proteins (SNX17, SNX27, and SNX31)

The PX-FERM sorting nexins are a sub-group of the PX superfamily. This subfamily has a PX domain and a C-terminal 4.1, ezrin radixin, moesin (FERM) domain with an atypical tertiary structure [44]. They are required for endosomal-to-cell-surface recycling of diverse transmembrane cargos [45]. PX-FERM nexins are further divided into two groups: SNX17 and SNX31. These SNXs are cargo adaptor proteins for a retromer-independent complex called the retriever complex, whereas SNX27 associates with the retromer [46]. The retriever complex localizes to early endosomes and recycles its NPxY/NxxY motif-containing cargo proteins to the cell surface [46]. Through quantitative proteomic analysis, over 120 cell surface proteins, including numerous integrins, signaling receptors, and solute transporters, that require SNX17–retriever to maintain their surface levels have been identified [46]. These include the LDL receptor, amyloid precursor protein, and integrins [44,45,47,48]. SNX17 also recognizes all Human Papillomavirus (HPV) L2 proteins. This interaction aids lysosomal viral escape thereby being crucial for HPV infection [49–51].

SNX27 is a unique SNX-FERM protein that contains an N-terminal density 95/discs large/zonula occludens-1 (PDZ) domain. This domain binds PDZ-binding motif (PDZbm)-containing cargo such as the β2-adrenergic receptor [21,52,53]. Quantitative proteomics of the SNX27 interactome has provided an unbiased global view of SNX27-mediated sorting. Here over 100 cell surface proteins, including the glucose transporter GLUT1, the Menkes disease copper transporter ATP7A, various zinc and amino acid transporters, and numerous signaling receptors, require the SNX27–retromer [54]. The FERM domain recognizes Asn-Pro-Xaa-Tyr–sorting signals in transmembrane cargos. Some of these cargo proteins need to be phosphorylated to facilitate binding to SNX27 [44,45,47,48]. SNX27 is highly enriched in the brain. Consequently, cargos include proteins involved in neuronal signaling, such, AMPA receptors [55]. Deficiencies in SNX27 function are associated with Down syndrome [56] and epilepsy [57]. More recently, SNX27-mediated recycling of neuroligin-2 (NL2), a protein required for stabilization of synaptic inhibitory receptors contributes to the regulation of inhibitory synapse composition [22].

2.4. SNX-BAR Proteins

The SNX-BAR proteins contain an additional BAR (Bin/Amphiphysin/Rvs) domain that can sense membrane curvature and induce membrane tubulation. The BAR domain is a dimerization motif that forms a rigid cup-shaped structure. This permits the BAR regions to induce membrane deformation, transitioning flat membranes to tubular membrane surfaces [28]. Current models propose that both PX and BAR domains have to be engaged with the membrane to ensure specificity and efficient binding. Mammalian cells possess twelve SNX-BAR family members (SNX1, SNX2, SNX4, SNX5, SNX7, SNX8, SNX9, SNX18, SNX30, SNX32, and SNX33) [58]. Many SNX-BAR proteins (SNX1, SNX2, SNX5, SNX6, and the neuronal SNX32) form heterodimeric complexes. These are critical for endosome-to-TGN retrieval and endosome–to–plasma membrane recycling [59]. Current models suggest that endosome–to–plasma membrane recycling is aided by SNX-BAR association with SNX27 and the retromer, with the PDZ domain of SNX27 as the predominant cargo recognition module [54]. Lastly, SNX-BAR proteins, have retromer-independent roles in autophagic processes, which is discussed in more detail below (Section 6) [60,61].

Until recently the molecular details of how SNX-BAR proteins recognize their cargos have remained elusive. Most of the mechanistic insight was gained from studying the retrograde transport of cation-independent mannose 6-phosphate receptor (CI-MPR) [62]. Interestingly, a wealth of earlier studies, report that CI-MPR recycles to the TGN through direct binding of the CI-MPR tail to the VPS retromer [63,64]. The role of the retromer in CI-MPR recycling has recently been reappraised by two independent studies [65,66]. Surprisingly, both groups concluded that SNX-BAR proteins can directly bind to CI-MPR independently of the core retromer trimer. Furthermore, other proteins that are recycled to the TGN e.g., Insulin-like growth factor 1 receptor (IGF1R) also directly bind to SNX-BAR proteins [65]. Supporting these studies, Simonetti et al. used a SILAC-based proteomic approach to identify a recycling motif, (WLM) [65]. Yong et al extended this work and both identified an additional upstream hydrophobic stretch to the recycling motif and over 70 putative SNX-BAR cargos [58]. Based upon these results, an alternative model has been proposed in which SNX-BARs function as a direct cargo selecting module for a large set of transmembrane proteins transiting the endosome.

2.5. Other Domains

Some SNX proteins also contain additional domains including SH3 (Src homology 3), RA (RasGTP effector), and RGS (regulator of G-protein signaling) domains [20,67]. These additional protein-protein binding domains enable SNXs to form homo- or heterodimers and associate with larger protein complexes such as the retromer or autophagy vesicles. More recently, the tetratricopeptide repeat (TPR) protein-protein motif found in SNX21 (a member of the PX-associated B subfamily [21]) was identified as a scaffold for the endosomal recruitment of the Huntington's disease protein huntingtin (Htt) [68].

3. Sorting Nexins and Endocytosis Pathways

3.1. The Endocytic Network

Endocytic recycling (outlined in Figure 3) has historically been considered a relatively passive process. Now it is understood to be a highly orchestrated program that plays a major role in cellular homeostasis. How this is conducted at a molecular level is critical as thousands of integral membrane proteins, that regulate many cell functions, journey through this network [69]. Here the SNXs play prominent roles in many aspects of plasma membrane remodeling especially in response to physiological conditions.

Although several distinct endocytic pathways are known, clathrin-mediated endocytosis (CME) is the key process in vesicular trafficking for internalization of transmembrane cargos and their ligands. The process requires the coordinated actions of over 60 different proteins [70]. The arrival and departure of these proteins define the different stages of the pathway that eventually results in the fusion of the internalized vesicles with early/sorting RAB5 GTPase positive endosomes. The pathway has been described in detail in other reviews [71,72].

Early/sorting positive endosomes are formed from primary endocytic vesicles that have undergone homotypic fusion or fused with a pre-existing endosome. These mature into late RAB7 positive endosomes by gradually acidifying the fluid within the endosomal lumen, ending with a pH of 5.5. Both early and late endosomes are characterized by a vacuolar domain that contains ILVs, formed by ESCRT (endosomal sorting complexes required for transport) complexes and enriched in proteins earmarked for lysosomal degradation (see [73] for an excellent review). Early endosomes contain significantly fewer ILV's than late endosomes and are characterized by a tubular domain that buds from the endosome and ferries their contents to recycling pathways. The formation of late endosomes is dependent upon multiple rounds of cargo sorting and ILV biogenesis, coupled with maturation of early endosomes (also known as the multivesicular endosome or multivesicular body). Upon fusion of the late endosome with lysosomes in a structure named an endo-lysosome, ILV's and their accompanying cargos are degraded by lysosomal enzymes.

3.2. SNX Proteins in Recycling Pathways

The fate of endocytosed proteins is decided upon reaching early endosomes [13]. Cargos destined for lysosomal degradation by ESCRT are earmarked by ubiquitin [73]. Other internalized cargos are recycled by retromer or retriever complexes, aided by the actin-remodeling WASH complex and sorting nexins [74]. There are three distinct forms of retromer, SNX-BAR-retromer, SNX27-retromer, and the SNX3-retromer which have been discussed in detail in Section 2 of this review. By associating with endosomal membranes through interaction with the GTPase RAB7 [75] retromer complexes recycle a wide range of internalized transmembrane cargos from early and maturing endosomes (see Figure 3). The cargos recycle back to the cell surface either by fast or slow pathways mediated by SNX27 and SNX4 respectively. Alternatively, cargo-enriched tubules bud from endosomes and passage through the TGN to the cell surface via the secretory pathway in a process called retrograde transport [76,77]. Importantly, as mentioned above it has now been shown that SNX-BAR proteins mediate retromer-independent retrograde transport of various cargos [65,66]. The retriever complex recycles plasma membrane cargos including integrins and lipoprotein receptors. Here the interaction of the retriever complex with SNX17 is essential for cargo selection. The CCC and WASH complexes aid in the recruitment of the retriever to endosomes [46].

3.3. SNX Proteins in Promote Endo-Lysosomal Degradation

The trafficking of cargos by recycling pathways to the plasma membrane is a critical physiological role of sorting nexins. In contrast, some sorting nexins prevent promote lysosomal degradation of their cargo. For example, SNX11 promotes the trafficking of TRPV3 (transient receptor potential vanilloid 3) ion channel from the plasma membrane to lysosomes for degradation via protein-protein interactions [78]. Likewise, SNX1 promotes retromer independent trafficking of protease-activated receptor-1 (PAR1), to the lysosomes [79]. Also, SNX1 and SNX6 facilitate the fate of epidermal growth factor receptor (EGFR) and the tumor suppression p27Kip1 [80,81]. Similarly, SNX4 also regulates the accumulation of BACE1, (ß-site amyloid precursor protein-cleaving enzyme), preventing it from trafficking to the lysosome. This is important as BACE1 is an enzyme involved in proteolytic processing of the amyloid precursor protein, which leads to the formation of the pathological amyloid-β (Aβ) peptide in Alzheimer's disease [82].

3.4. SNX Proteins in Other Pathways

The SNX9 subfamily (SNX9, SN18, and SNX33) are also involved in endocytosis. This subfamily is characterized by containing an SH3 domain at the N terminus, a low complexity domain, and a BAR domain at the C terminus. By regulating dynamin polymerization SNX9 is required for efficient clathrin-mediated endocytosis and SNX18 and SNX9 can compensate for each other [83]. Sorting nexins also interact with cargos outside of endo-lysosomal-TGN pathways (see Figure 1 and Section 6 of this review). Here they play significant roles in autophagy pathways in yeast and mammalian cells [61,84–86]. Taken together, this serves to emphasize the diverse roles sorting nexins play in maintaining protein homeostasis.

4. Sorting Nexin Cargo Recognition in Yeast

4.1. The Yeast Endosome System

In *Saccharomyces cerevisiae*, the makeup of the plasma membrane is adjusted in response to different physiological conditions to maintain homeostasis. This is achieved by the internalization of plasma membrane proteins by endocytosis through clathrin-dependent or independent mechanisms [87]. Some proteins are recycled back to the cell surface either via the TGN or by a recycling pathway originating from the endosome [88]. Others, tagged by ubiquitination, are degraded in the vacuole (the yeast equivalent of lysosomes) [89].

Recently it has been proposed that unlike other eukaryotic species, budding yeast lack early endosomes [90,91]. Instead, cargo-carrying vesicles are initially targeted to the

TGN. From here, cargos are either recycled or transferred to late endosomes (also known as multivesicular bodies (MVBs) or pre-vacuolar endosome (PVE) compartments [60]). Late endosomes contain cargo-laden intraluminal vesicles which require ESCRT pathways for their formation. Here ubiquitin plays a key role, as the transport of cargos to the vacuole depends on ubiquitin linkages. Thus, ubiquitination serves both as a signal for endocytosis from the plasma membrane and a specific sorting signal for entry into the vacuolar lumen [13]. In the final step, late endosomes fuse with the highly acidic vacuoles that contain proteases for degradation of the endosomal contents. Recently, very elegant experiments using an engineered fluorescent vacuolar cargo and 4D microscopy have suggested that transfer of material from late endosomes to the vacuole most likely involves "*kiss-and-run*" fusion events [90,92].

4.2. The Yeast Retromer Architecture

The yeast retromer (Vps35, Vps26, and Vps29), is an evolutionarily conserved protein coat complex. Significantly, pioneering genetic studies in *S. cerevisiae* initially led to the identification of this trimeric structure, and its cargo Vps10 [32,93]. Vsp10 (vacuolar protein sorting 10), sorts the transmembrane protein receptor carboxypeptidase (CPY) into vesicles at the Golgi [94]. Thereafter CPY-containing vesicles plus Vsp10 are transported to the endosome, which upon maturation, fuses with the vacuole, delivering soluble CPY to the vacuole lumen. Vps10 escapes this fate, being recycled back to the Golgi by the retromer complex, making Vps10 available for additional rounds of CPY sorting. Further studies revealed that the yeast retromer forms a pentameric structure with the Vsp1-Vsp17, SNX-BAR sorting nexin [35]. Within this pentameric complex, the retromer recognizes and mediates the packaging of cargos into endosome-derived transport carriers whereas the sorting nexin mediates endosome recruitment [59,95].

How the retromer coats tubulovesicle carriers remains unclear and is somewhat controversial. This is a critical mechanism to understand as retromer disruption is associated with major neurodegenerative disorders [18]. Studies in yeast by five independent groups using different methodologies show that the retromer interacts with the SNX-BAR dimer through Vps29 and Vps35 independent of Vsp26 [34,96–98]. However, Cryo-EM studies of the thermophilic yeast *Chaetomium thermophilum* SNX-BAR protein Vps5, and the retromer report different results. Here Koevtun et al. [99] propose a model in which the SNX-BAR protein, Vps5 interacts only with Vps26. Moreover, they suggest that retromer binding to membranes is dependent on Vps5 from which arches of retromer extend away from the membrane surface. Taken together, retromer assembly in *S. cerevisiae* and *C. thermophilum* may be different.

4.3. Yeast Cargo Recognition

It is well understood that retromer complexes selectively recognize their cargos through a recycling sequence. In mammalian cells, this has been defined as (ØX[L/M/V], where Ø is F/Y/W,) [73]. However, in yeast, almost all proteins contain this sequence suggesting a different mechanism for cargo identification. By using mutational analysis of two different retromer cargos, Vsp10 and Ear1, Scott Emr's group discovered that different sites in the retromer subunit Vps26 are required for their recognition [34]. This suggests a model in which a bipartite recycling signal sequence ensures precise cargo recognition by the retromer complex. These striking results show that the retromer utilizes different binding sites depending on the cargo, allowing this complex to recycle different proteins.

Other sorting nexins also contribute to retromer function [60]. Snx3 is an accessory protein that binds the retromer and recycles cargos from endosomes to the TGN [100]. In yeast, it recognizes relatively few cargos, though a recent systematic genome-wide screen expanded its repertoire [60,86]. One cargo, Neo1, deserves a special mention as its discovery uncovered a previously unknown role for the Snx3-retromer [101,102]. Neo1 is an aminophospholipid flippase, that contributes to the phosphatidylethanolamine asymmetry of endosomal membranes [103]. As anticipated, deletion of the canonical Snx3 recycling motif inhibited the sorting of Neo1 [86]. Moreover, it was also discovered that the sorting of Neo1 by Snx3 is required to recycle other Snx3 cargos. This suggests a model in which Neo1-driven lipid flippase activity promotes vesicle or tubule formation [101]. Similarly, the packaging of human SNX3-retromer cargo, Wntless, also requires Neo1 [102]. Taken together, this suggests that the incorporation of Neo1 into recycling tubules is highly conserved and may influence tubule formation.

Another less well understood sorting nexin that contributes to retromer function is the SNX-BAR protein Mvp1 [60,104]. Mvp1 shares conservation with the mammalian SNX–BAR SNX8, whose function, is involved in endosomal sorting [105,106]. Likewise, *mvp1Δ* cells exhibit defects in retromer-dependent retrograde trafficking [107,108]. Recent structural studies have revealed that the Mvp1 SNX-BAR protein exists as an autoinhibited tetramer in which the PX lipid-binding sites are occluded. The Mvp1 dimer retains membrane-remodeling activity and exhibits enhanced membrane binding. This suggests a model in which unmasking of PX and BAR domains is required for Mvp1 function. As most SNX–BAR proteins are invariably dimeric, this finding adds an intriguing layer of complexity to the regulation of SNX–BAR function.

4.4. Retromer-Independent Sorting Nexin Function in Yeast

Snx4, Snx41, Atg20 form two distinct retromer-independent complexes (Snx4-Snx41 and Snx4-Atg20 (Snx42)) that are required for endocytic recycling and selective autophagy. Consistent with these roles they co-localize to the endosome and the pre-autophagosomal structure (PAS) [85,109,110]. Moreover, we and others have shown that after nitrogen starvation, they sequester to the perinucleus where they transport nuclear cargos to the vacuole [84,111] (see Section 6.1).

The most studied cargo of the Snx4-Atg20 complex is Snc1, a plasma membrane-directed v-SNARE, required for the fusion of secretory vesicles with the plasma membrane [112]. It is subsequently retrieved from the plasma membrane by endocytosis and recycled to the Golgi apparatus [85,113]. Two distinct pathways move Snc1 within the cell. Plasma membrane recycling is Snx4 independent and requires F-box protein Rcy1 and the aminophospholipid flippase, Drs2, [114]. Retrograde recycling is Snx4-Atg20-dependent and delivers Snc1 back to the TGN from late endosomes [115]. Taken together, this indicates that multiple pathways can regulate a single SNARE as it cycles through the endo-lysosomal system.

A critical responsibility of Snx4-Atg20 heterodimer is to mediate the endosome-to-Golgi transport of Atg9 [116]. Atg9 is an essential integral membrane protein required for autophagosome biogenesis. Atg27 maintains this Golgi-localized pool of Atg9 [117]. In turn, Atg27 recycling and trafficking are regulated by the retromer and Snx4 [118,119]. More recently, it has been shown that Atg27 is recycled from the vacuole membrane using a 2-step recycling process. First, the Snx4 complex recycles Atg27 from the vacuole to the endosome. Then, the retromer complex mediates endosome-to-Golgi retrograde transport [118]. This is exciting as it represents the first physiological substrate for the vacuole-to-endosome retrograde trafficking pathway.

5. Interplay between the Ubiquitin Proteolytic System and SNXs

Ubiquitin (Ub) is a small molecule that covalently attaches to lysine resides on its targets. Ub itself can be conjugated to a second Ub molecule resulting in ubiquitin chains differing in linkage types and lengths [11,120]. This wide variety of Ub modifications can have pleiotropic effects on its substrates [1]. K48-linked ubiquitin chains typically target proteins for degradation by the 26S proteasome [121]. On the other hand, K63-linked ubiquitination typically acts as a signaling event to modify function, such as DNA repair, altering protein-protein interactions, and protein trafficking [122].

During endocytosis, membrane proteins are identified as cargo either as part of a programmed biological response (such as ligand-mediated receptor down-regulation) or as a way to remove aberrantly folded or damaged proteins from the cell surface as a quality control mechanism. Membrane proteins are decorated with Ub on the cell surface and early endosomes to trigger their internalization and endosomal sorting respectively [123,124]. A functional ESCRT pathway is also required. In short, cargos that are tagged with the ubiquitin sorting signal are recognized by ESCRT-0. These are then sequentially handed to ESCRT-I and -II or recruited to the ESCRT-I-II supercomplex before being incorporated into ILVs for delivery to lysosomes [89,123].

5.1. Sorting Nexins Regulate UPS Activity

Sorting nexins influence the regulation of proteasome activity and substrate degradation by a variety of different mechanisms. These include blocking ubiquitination of protein substrates, inhibiting ubiquitin specificity factors, regulating protein stability of E3 ligases by either enhancing their recycling or degradation pathways, and degrading inactive or excess proteasome complexes. These are summarized in Table 1. Intriguingly, there are several examples of the relationship of sorting nexins with E3 ligases. In yeast, the E3 ligase specificity factor for Rsp5-dependent ubiquitination, Ear1, is recycled by Snx3 [86]. In mammalian cells, SNX18 is regulated by the E3 ligase Mib1, which indirectly promotes Notch signaling [125]. Likewise, Itch (atrophin-1 interacting protein 4), a member of the NEDD4 family of E3 ubiquitin ligases, ubiquitylates SNX9, thereby regulating intracellular SNX9 levels [126]. In a seminal discovery, the E3 ubiquitin ligase partner of MAGE-L2, a protein that enhances E3 ubiquitin activity [127], was found to be K48 E3 TRIM27 [127]. The MAGE-L2-TRIM27 complex localizes to endosomes through interactions with the retromer. The outcome of this interaction is K63 ubiquitination of the WASH complex, a known regulator of retromer-mediated transport. This action permits WASH to nucleate endosomal F-actin (see Figure 3). Moreover, this pathway is regulated by the deubiquitinating enzyme USP7 [128].

Table 1. List of mammalian sorting nexins and associated proteins (WASH) that play a role in the UPS.

SNX	UPS Activity	Role	Ref
SNX3	ubiquitin-specific protease 10 (USP10)	Deubiquitylates and stabilizes SNX3	[129]
SNX5	FBW7	Interacts with FBW7 and blocks FBW7-mediated ubiquitination of oncoproteins such as c-Myc, NOTCH1, and Cyclin E1	[130]
SNX9	Itch (atrophin-1 interacting protein 4, Nedd family member)	Itch regulates intracellular levels of SNX9	[126]
SNX16	indirect	Postulated that SNX16 interacts with and inhibits proteasome-dependent ubiquitination of eukaryotic translation elongation factor 1 A2 (eEF1A2), thereby activating c-myc signaling.	[131]
SNX18	MIBd1 E3 ligase	Promote the endocytosis of Delta-like protein 1 (Dll1) which is the transmembrane ligand protein for the Notch proteins.	[125]
SNX27	Non-catalytic role of the deubiquitinase OTULIN	OTULIN antagonizes SNX27-dependent cargo loading and binding to the VPS26A and affects endosome-to-plasma membrane trafficking.	[132]
retromer	TRIM27 E3 ubiquitin ligase (non-catalytic role)	Mediates the phosphorylation and activation of STAT	[133]
retromer	MAGE-L2-TRIM27 E3 ubiquitin ligase	The MAGE-L2-TRIM27 E3 ubiquitin ligase localizes to retromer-positive endosomes.	[134]
WASH	K63-linked ubiquitination and deubiquitinase USP7	WASH is activated by K63-linked ubiquitination of WASH K220 by MAGE-L2-TRIM27. USP7 regulates this activity.	[128,134]

Less is known about sorting nexins and E2 activity. In *Drosophila*, UBC-13, the E2 ubiquitin-conjugating enzyme that generates K63-linked ubiquitin chains, is essential for retrograde transport of multiple retromer-dependent cargos, including MIG-14/Wntless. Here UBC-13 function is required for retrograde transport of SNX1 retromer-dependent cargos [135].

In the budding yeast, nitrogen starvation triggers the disassembly of nuclear 26S proteasomes into 19S and 20S subcomplexes. The subcomplexes are consequently transported through the nuclear pore complex (NPC) and targeted to autophagosomes for degradation [84,136,137]. Intriguingly, nitrogen starvation triggers cytosolic Snx4-Atg20 and Snx4-Snx41 heterodimers to relocate to the perinucleus. Moreover, both these heterodimers and the core autophagic machinery are required for 19S and 20S vacuolar proteolysis [84]. This has led to the model that these heterodimers mediate the transport of the proteasome subcomplexes to the phagosomes. How Snx4-Atg20 and Snx4-Snx41 interact with the proteasome subunits remains unknown, but it adds another example of how sorting nexins regulate the UPS machinery.

5.2. Sorting Nexins Can Be Regulated by the UPS

In addition to regulating UPS activity, sorting nexins themselves are regulated by the UPS. Although there are significantly fewer examples of this activity reported it still serves to demonstrate the reciprocal nature of the relationship between sorting nexins and the UPS. In short, sorting nexins have been reported to be stabilized by interacting deubiquitinating enzymes (DUBs) [138]. The hormone vasopressin increases the expression of the DUB USP10 that deubiquitinates and stabilizes SNX3 [129]. SNX27 interacts with OTULIN (OTU Deubiquitinase With Linear Linkage Specificity) that specifically hydrolyzes methionine1 (Met1)-linked ubiquitin chains. This antagonizes SNX27-dependent cargo loading and binding of SNX27 to VPS26A and affects endosome-to-plasma membrane trafficking. Moreover, these findings define a non-catalytic function of deubiquitinases in sorting nexin function [132].

6. Sorting Nexins in the Autophagy-Lysosomal Pathway (ALP)

Autophagy is considered the first line of defense in response to many forms of extracellular stress [11,139]. Sorting nexins engage in non-selective autophagy pathways following unfavorable physiological cues in both yeast and mammalian cells. In yeast, SNX-BAR heterodimers also are required for efficient selective autophagy pathways.

6.1. Sorting Nexins in Non-Selective Autophagy in Yeast

In physiological conditions, the primary function of sorting nexins within the endosomal pathway is to maintain steady-state levels of membrane proteins. Following different stress cues such as nutrient depletion or starvation, SNXs engage in stress-induced regulatory roles that require sorting nexin cellular relocalization. The role of sorting nexins in autophagy was first identified in yeast where it was found to be a component of the Atg1 initiation complex [109]. Here Snx4 binds the Atg17 scaffold complex which is required to localize the PAS to vacuole outer membranes [140,141]. Consistent with this, deletion of Snx4 results in insufficient PAS formation coupled with a delayed autophagic response [142]. In a more recent phosphoproteomics study, Snx4 was identified as a direct substrate for Atg1 [143]. This phosphorylation event may direct Snx4 away from its physiological functions in endosomal sorting and towards its role in starvation-induced autophagy. Interestingly, during selective autophagy, Atg20 interacts with the scaffold protein Atg11 that replaces Atg17 and initiates autophagosome assembly at the cargo site [140]. SNXs also play a role in autophagosome and vacuolar membrane fusion. Here the Snx4-Atg20 heterodimer promotes non-selective autophagy by exporting lipids from the vacuole which maintains its fusion competence [144].

6.2. Sorting Nexins in Non-Selective Autophagy in Mammalian Systems

The biogenesis of the phagophore is not fully understood. It is believed to originate from the ER and expand by receiving membranes from different sources including the plasma membrane and Golgi [145]. Two essential proteins needed to build phagophores, ATG16L1 and ATG9A, traffic from the plasma membrane to recycling endosomes on their way to sites of autophagosome formation [146–148]. New insight into how this is achieved has come from Anne Simonsen's group. They revealed that the PX-BAR-containing protein SNX18 recruits Dynamin-2 to induce budding of ATG9A and ATG16L1 containing membranes from recycling endosomes to sites of autophagosome formation (Figure 4A) [149,150]. More recently, the SNX4-SNX7 heterodimer has been shown to play a role in phagophore biogenesis and ATG9A distribution, although the precise molecular mechanisms remain unknown [61]. Given the dynamic nature between organelles during autophagosome biogenesis, it is likely that future studies may reveal similar roles for other sorting nexins.

6.3. Sorting Nexins in Selective Autophagy in Yeast

In yeast, Snx4 and Atg20 are required for several selective autophagy pathways. Cargos include mitochondria [151], peroxisomes [152,153], proteasomes [84], ribosomes [84], fatty acid synthase complexes [154] and transcription factors [111] (Figure 4B). A role for sorting nexins in selective autophagy was first identified while studying the cytoplasm-to-vacuole targeting pathway (CVT).

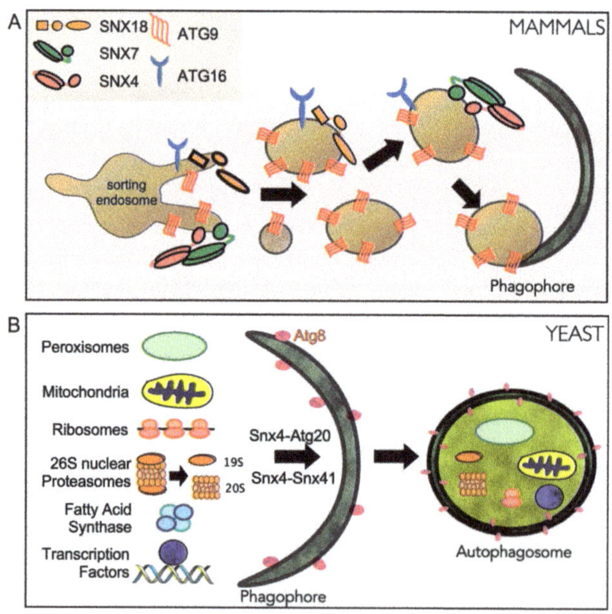

Figure 4. Sorting nexins play critical roles in autophagy following starvation stress. (**A**) In mammalian cells, SNX4/SNX7 and SNX18 are required for autophagy-dependent localization of ATG16 and ATG9 from recycling endosomes to the pre-autophagosomal site (PAS). ATG9 containing vesicles are required for PAS formation and autophagosome biogenesis. (**B**) In yeast, Snx4-Atg20 (Snx42) and Snx4-Snx41 heterodimers regulate different forms of selective autophagy.

This biosynthetic pathway functions in physiological conditions to transport the aminopeptidase, Ape1, to the vacuole. The Snx4-Atg20 heterodimer is required for the recruitment of proteins to the site of CVT formation [109]. The molecular details of Snx4's role in this pathway remain unclear. Lastly, a role for sorting nexins in mammalian selective autophagy pathways has not yet been reported.

6.4. SNX5 and Viral Autophagy Induction

It is well established that inactivation of the target of rapamycin complex 1 (TORC1) induces autophagy [155]. In pioneering studies, Xiaonan Dong and colleagues have demonstrated that viral-induced autophagy is distinct from previously described selective autophagy and basal autophagy activated by nutrient deprivation and mTOR suppression. Following viral infection, SNX5 initiates autophagosome biogenesis by localizing to virion-containing early endosomes. Mechanistically SNX5 interacts with beclin1 and ATG14-containing Class III phosphatidylinositol 3-kinase (PI3KC3) complex 1 (PI3KC3-C1). SNX5 is also required for increasing the kinase activity of PI3KC3-C1, generation of endosomal PI3P, and recruitment of WIPI2 to virion-containing endosomes. Deletion of SNX5 in mice models results in increased susceptibility to multiple human viral infections and enhances lethality after infection. These exciting results reveal that SNX5 thereby plays an important role in the immune response following viral infections [156].

7. The Interplay between Sorting Nexins, Lysosomal Degradation, and UPS-Mediated Degradation

In recent years, it has become apparent that the UPS and autophagy pathways are functionally interconnected [157–159]. Key findings from these studies have revealed that when the UPS is overwhelmed, autophagy is upregulated to eliminate aberrant proteins [11]. Furthermore, ubiquitination is utilized as a degradation signal by autophagy pathways, being critical for removing damaged mitochondria by mitophagy in mammalian cells [160,161]. Ubiquitin is also required for autophagic degradation of protein aggregates [162,163], peroxisomes [164] pathogens [165] and ribosomes [166,167]. Here the crosstalk between ubiquitination and autophagy is provided by autophagic adaptor proteins (or autophagy receptors), which bind both ubiquitin and autophagy-specific ubiquitin-like modifiers of Atg8 and its homologs [168,169]. This has led to the more current hypothesis that UPS and autophagy pathways constitute a single integrated degradation system [140]. Consistent with this, following TORC1 inhibition, in yeast, nuclear proteasomes are disassembled and then destroyed by Snx4-Atg20 and Snx4-Atg41 mediated autophagy [84,137,170].

7.1. Environmental Cues Dictate the Degradative Fate of Med13

There are a limited group of proteins that are both ALP and UPS substrates. Our group has discovered that in *S. cerevisiae*, Med13, a member of the conserved Cdk8 kinase module (CKM) of the mediator complex, is degraded either by a novel Snx4-mediated autophagy pathway or by the UPS in response to cell survival and death signals respectively (Figure 5) [111,171,172]. The CKM interacts with the mediator complex of RNA pol. II to predominantly repress genes induced by environmental stress [173–175]. Activation of these genes is achieved by disrupting the CKM association with the mediator [174,176]. We have shown that Med13 and cyclin C are both targets of the UPS system following oxidative stress [171,172,174,177]. Importantly, before cyclin C is destroyed it translocates to the mitochondria where it mediates stress-induced mitochondrial fission and promotes cell death in both yeast and mammalian cells [178–180]. In mammalian cells, mitochondrial located cyclin C also associates with Bax to promote its activation [179].

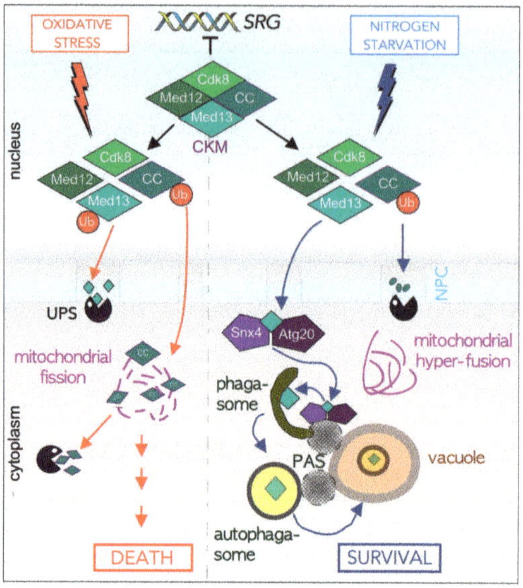

Figure 5. Med13 is destroyed either by the UPS or Snx4-Atg20 mediated autophagic degradation following cell death (left) or survival signals (right). Cartoon outlining stress-dependent fates of cyclin C and Med13, two members of the Cdk8 Kinase module (CKM). Here the subcellular address of cyclin C following stress mediates cell fate decisions by affecting mitochondria morphology. See text for details and [111,171,172,181].

In contrast, following a survival cue (nitrogen starvation), cyclin C is rapidly destroyed by the UPS before its nuclear release. This promotes cell survival by preventing mitochondrial fission and upregulating AuTophaGy (*ATG*) genes [181]. To our surprise, we discovered that instead of being targeted by the UPS, here Med13 is destroyed by vacuolar proteolysis. After transitioning through the nuclear pore complex (NPC), Med13 is transported by the Snx4-Atg20 heterodimer to Atg17-initiated phagophores at the vacuole [111]. Moreover, two transcriptional activators (Rim15 and Msn2) that positively regulate *ATG* expression, are degraded upon nitrogen starvation by this mechanism. Taken together, this suggests a model in which Snx4-mediated autophagy of *ATG* transcriptional regulators allows fine-tuning of the autophagic response. Moreover, it outlines a new autophagy pathway by which transcription factors are selectively targeted for degradation.

7.2. p27 Is Regulated by Proteasome Degradation and SNX6-Mediated Endo-Lysosomal Pathways

It is well established that proteasomal degradation of the growth suppressor p27 regulates cell cycle progression. Proteolytic degradation occurs via two different pathways. In G1 this cyclin-CDK (cyclin-dependent kinase) inhibitor is degraded in the cytoplasm by the E3 ligase Kip1 whereas its degradation at the G1/S transition and in G2 occurs in the nucleus, mediated by the E3 ligase SCF^{Skp2} [182,183]. Cytoplasmic p27 can also be recognized by SNX6 for endo-lysosomal degradation. This is also important for cell cycle progression as silencing SNX6 delays S-phase entry in starvation-synchronized NIH-3T3 cells [81]. It remains unknown if the SNX-mediated endo-lysosomal pathway interacts with p27 proteasomal degradation but it is important to determine as this cyclin-CDK inhibitor not only inhibits the catalytic activity of cyclin D-, E-, A-, and B-CDK complexes but also regulates other processes including cell migration and development independent of its CDK inhibitory action [184].

8. Sorting Nexins in Disease

The SNX family consists of a diverse group of proteins involved in various aspects of protein trafficking. Underpinning their importance in protein homeostasis, the etiology of many diseases such as cancer, cardiovascular and neurodegenerative diseases is linked to the dysregulation of sorting nexin function [185]. Moreover, viruses and bacteria can hijack sorting nexin pathways to both invade and replicate in host cells.

8.1. The Role of Sorting Nexins in Cardiovascular Disease

Sorting nexins are implemented in the development of cardiovascular diseases such as hypertension, coronary heart disease, and heart failure [19]. Here SNXs influence the maintenance of blood pressure by regulating the expression and function of G-protein coupled receptors (GPCRs) such as dopamine receptors, ion channels, and transporters [19,129,186]. Consistent with this, knockdown/knockout animal models of SNX1, SNX5, and SNX19 correlate with hypertension [187,188]. This has led to addressing if SNXs could potentially be therapeutic targets for hypertension. Therapeutic strategies have focused on expressing specific SNX subtypes within the kidney to decrease blood pressure [19].

SNXs also influence the pathogenesis of coronary artery disease by regulating lipid metabolism. SNXs interact with the leptin receptor and the low-density lipoprotein (LDL) receptor [22,189] and decreasing SNX1 levels results in increased levels of triglycerides and cholesterol [22,189]. SNXs may also influence coronary artery disease by regulating inflammation, which is linked to the etiology of vascular diseases [190]. SNX13 deficiencies correlate with decreased heart function associated with cardiomyocyte apoptosis. SNX13 mediates the recycling of the apoptotic repressor, ARC. Thereby, loss of SNX3 results in the degradation of ARC and promotes cardiomyocyte apoptosis and heart failure [191].

Insulin insensitivity is a major hallmark of type 2 diabetes mellitus which is a characteristic feature of heart failure [192]. Sorting nexins are linked to this pathophysiology as SNX5, SNX19 and SNX27 regulate insulin degradation, secretion, and signaling. Silencing SNX5 in animal models increases blood insulin, decreases insulin excretion, and causes insulin resistance [54,193,194].

8.2. The Role of Sorting Nexins in Neurogenerative Diseases

SNX dysregulation has been linked to several neurodegenerative diseases such as Alzheimer's disease (AD), Parkinson's disease, and Down syndrome [195]. In neuronal cells, the composition of the cellular membrane is essential for responding to extracellular stimuli and neuroplasticity. SNX-mediated regulation of the cellular membrane composition influences several processes such as neuronal excitability, plasticity, neural development, signaling, psychostimulant response, and cellular drug resistance [196].

Best described is the role of sorting nexins in the pathogenesis of Alzheimer's disease (for details see [197]). AD is characterized by brain accumulation of extracellular neuritic plaques containing deposits of β-amyloid peptide (Aβ) and neurofibrillary tangles compromised of the microtubule-associated protein tau. One of the proteins involved in regulating the β-amyloid peptide is the SNX33 retromer. This complex inhibits endocytosis of amyloid precursor protein (APP) which in turn leads to retention of APP at the plasma membrane which promotes plaque formation [198]. SNX15 and SNX17 also regulate APP processing [199]. SNX4, SNX6, and SNX12 have been shown to regulate BACE1 trafficking which also controls Aβ peptide generation [200,201]. In addition, SNX27 binds and inhibits γ-secretases thereby decreasing Aβ peptide [202].

Parkinson's disease (PD) is defined by the loss of dopaminergic neurons and the accumulation of α-synuclein-enriched Lewy bodies. Genome-wide association studies have identified various mutations that increase Parkinson's disease susceptibility such as PINK1 and Parkin whose gene products regulate mitophagy [203,204]. As such mitochondrial defects such as disruptions in mitochondrial fission and mitophagy are hallmarks of PD. It can be speculated that dysregulation of SNXs may perturb autophagy pathways that are necessary to clear α-synuclein aggregates and damaged mitochondria. In support of this, the pathophysiology of Parkinson's disease is linked with a mutation in the retromer subunit VPS35 [195]. This is relevant as VPS35 mutations result in decreased association with the WASH complex which perturbs ATG9 transport, ultimately compromising autophagosome biogenesis [205,206]. Lastly, VPS35 also interacts with the mitochondrial fusion regulator, Drp1 [207]. VPS35 mutations are linked to increased mitochondrial fragmentation and cell death [208]. In neurons, this is particularly devastating as mitochondrial fission directs mitochondrial transport to their potential docking sites in axons and dendrites [209]. Lastly, the etiology of PD has also been linked with Pink1 and Parkin function and SNX9 mediated degradation of mitochondria in a vacuolar pathway distinct from mitophagy [210].

SNX deficiencies have been implemented in Down syndrome [56] as well as associated with epilepsy, developmental delays, and subcortical brain abnormalities [196]. SNX27 knockdown/knockout animal models or human patients with non-functional SNX27 variants exhibit a wide range of neurological aberrations that may be associated with defects in cell surface receptors [57]. Some of these receptors include neuroreceptors (AMPA, NMDA), ATPase copper transporters, glucose transporters, disintegrin, metalloproteinase, and adhesion proteins (NLGN2) [196]. For example, SNX27 expression is downregulated in human Down's syndrome brains. Mechanistically, SNX27 may regulate the retention of cell surface membrane proteins such as the myelination-related protein, GRP17 which plays an important role in oligodendrocyte development [211].

8.3. Oncogenic Roles of Sorting Nexins and the UPS

In recent years oncogenic roles of sorting nexins have been reported. Therefore, it comes as no surprise that many of these roles lead to the activation of well-characterized oncoproteins. Recently, TRIM27 has been classified as an oncoprotein. Consistent with this role, it is overexpressed in many cancers, including breast, endometrial, ovarian, lung, and colon [212]. TRIM27 associates with the retromer and activates the cytoplasmic transcription factor, STAT3 [133]. This is an important discovery as STAT3 plays a central role in various physiological processes and its aberrant and persistent activation results in serious diseases, including cancer [213]. It is a cytoplasmic transcription factor as its activation and translocation to the nucleus is dependent upon its passage through the endosome system [214]. In response to several cytokines or growth factors including interleukin-6 (IL-6), STAT 3 is phosphorylated [215]. This promotes its release from the endosome and translocation to the nucleus, resulting in the induction of downstream effector genes. Intriguingly, the E3 ubiquitin ligase activity of TRIM27 is dispensable for its ability to mediate STAT3 activation. Confirming a retromer linked role, knockdown of each of the retromer components significantly inhibits IL-6-induced transcription of STAT-dependent genes [133]. It is well established that endocytosis is an effective mechanism to downregulate cellular signaling events by internalizing receptors or ligand-receptor complexes [216]. Further studies are needed to address if other signaling proteins that are imprisoned by endocytosis, e.g., the promiscuous kinase glycogen synthase 3 beta (GSK3-β), are similarly regulated. This is important as this kinase has numerous phosphorylation targets in distinct pathways, including WNT, Hedgehog, and MAPK signaling [217].

Sorting nexins interaction with E3 ligases plays a role in oncogenesis in other cancers. In head and neck squamous cell carcinoma (HNSCC) SNX5 interacts with the E3 ligase F box proteins, thereby blocking FBW7 mediated ubiquitination of oncoproteins including c-Myc, NOTCH, and cyclin E1 [130]. SNX16 also has oncogenic properties in colorectal cancer, where it is significantly upregulated. This affects eEF1A2/c-Myc signaling, possibly by inhibiting proteasome-dependent ubiquitination of eukaryotic translation elongation factor 1 A2 (eEF1A2) [131]. As such, SNX16 has been implemented in the development of other tumors such as bladder and ovarian cancer [218,219]. SNX10 also may be a tumor suppressor in mouse models of colorectal cancer. Here SNX10 deficiency prevents the degradation of LAMP-2A, the essential CMA lysosomal receptor [220,221]. Given the key role of soring nexins in many biological processes, there is no doubt that future work will reveal more links to cancer and other diseases.

8.4. Viruses Can Hijack Sorting Nexin Pathways

In the last few years, it has become apparent that viruses can exploit retromer-mediated trafficking for their replication. Although many details remain unclear it has emerged that the strategies used are diverse. For example, some viral effectors recruit retromer components to viral replication sites to promote infection. An excellent example is the NS5A protein from the hepatitis C virus which interacts with VPS35 [222]. Others may mimic retromer cargo to travel i.e., hitchhike, from endosomes to the TGN, either to escape lysosomal degradation or to gain access to the nucleus. One of the best examples is the interaction of human papillomavirus (HPV) with SNX17 [49]. Here the PDZ domain of SNX17 and SNX27 interacts with the viral capsid protein L2 and enhances HPV infection by trafficking L2 and the bound viral DNA from the late endosomes to the TGN and subsequently to the nucleus [223]. In a similar way, SNX2 traffics the human respiratory syncytial virus (HRSV) structural proteins to enhance viral infection [224]. Consistent with this, various studies have shown that deletion of specific retromer components inhibits specific steps in the intracellular life cycle of the vaccinia virus, hepatitis C virus (HCV), and human papillomavirus (HPV) [225]. Other pathogens have also evolved elegant mechanisms to inhibit the innate immune response roles of SNXs. SNX5 and SNX6 are inhibited by *Legionella* RidL and *Chlamydia* IncE to evade the lysosomal mediated degradation [226]. It would be of great interest to see if SARS-CoV-2 and other pathogenic viruses regulate SNX trafficking activity to enhance viral progeny production or evade the innate immune response.

9. Conclusions

Cells sense and respond to various internal and external stimuli to regulate processes such as gene expression, cell cycle progression, metabolism, and protein homeostasis. In the cell, there are severe quality control mechanisms held in place to regulate protein degradation. The mode of protein degradation depends on several factors including size, localization, and timing of substrate proteolysis. For example, the large size of organelles and multiple subunit complexes requires lysosomal degradation. Localization of proteins such as transmembrane proteins requires lysosomal degradation because these proteins are embedded within the membrane making proteasomal-mediated degradation unfavorable. For the cell to quickly turn genes on and off transcription factors are degraded via nuclear 26S proteasomes. This mode of degradation, therefore, relies on spatiotemporal factors because degradation needs to happen rapidly, and proteasomes are localized in close proximity within the nucleus. Understanding the molecular details behind SNX cargo recognition, membrane binding, and protein degradation will provide insight into the diverse roles of SNXs in various biological processes. The growing evidence of SNXs in the three quality control protein homeostasis pathways will shed light on pathologies associated with perturbed proteolysis and provide innovative targets for therapeutics.

Author Contributions: S.E.H. and K.F.C. both wrote and edited the text. All authors have read and agreed to the published version of the manuscript.

Funding: This work was supported by a grant from the National Institutes of Health awarded to K.F.C. (GM113196).

Acknowledgments: We thank Taras Nazarko for his support and invitation to write this review. We especially thank the members of the Cooper laboratory for critically reading this manuscript.

Conflicts of Interest: The authors declare no competing or financial interests.

References

1. Finley, D. Recognition and processing of ubiquitin-protein conjugates by the proteasome. *Annu. Rev. Biochem.* **2009**, *78*, 477–513. [CrossRef] [PubMed]
2. Klionsky, D.J. Autophagy: From phenomenology to molecular understanding in less than a decade. *Nat. Rev. Mol. Cell Biol.* **2007**, *8*, 931–937. [CrossRef] [PubMed]

3. Lamb, C.A.; Yoshimori, T.; Tooze, S.A. The autophagosome: Origins unknown, biogenesis complex. *Nat. Rev. Mol. Cell Biol.* **2013**, *14*, 759–774. [CrossRef] [PubMed]
4. Khandia, R.; Dadar, M.; Munjal, A.; Dhama, K.; Karthik, K.; Tiwari, R.; Yatoo, M.I.; Iqbal, H.M.N.; Singh, K.P.; Joshi, S.K.; et al. A Comprehensive Review of Autophagy and Its Various Roles in Infectious, Non-Infectious, and Lifestyle Diseases: Current Knowledge and Prospects for Disease Prevention, Novel Drug Design, and Therapy. *Cells* **2019**, *8*, 674. [CrossRef] [PubMed]
5. Anding, A.L.; Baehrecke, E.H. Cleaning House: Selective Autophagy of Organelles. *Dev. Cell* **2017**, *41*, 10–22. [CrossRef]
6. Lamark, T.; Johansen, T. Aggrephagy: Selective disposal of protein aggregates by macroautophagy. *Int. J. Cell Biol.* **2012**, *2012*, 736905. [CrossRef]
7. Hu, W.; Chan, H.; Lu, L.; Wong, K.T.; Wong, S.H.; Li, M.X.; Xiao, Z.G.; Cho, C.H.; Gin, T.; Chan, M.T.V.; et al. Autophagy in intracellular bacterial infection. *Semin. Cell Dev. Biol.* **2020**, *101*, 41–50. [CrossRef]
8. Wong, H.H.; Sanyal, S. Manipulation of autophagy by (+) RNA viruses. *Semin. Cell Dev. Biol.* **2020**, *101*, 3–11. [CrossRef]
9. Bonam, S.R.; Muller, S.; Bayry, J.; Klionsky, D.J. Autophagy as an emerging target for COVID-19: Lessons from an old friend, chloroquine. *Autophagy* **2020**, 2260–2266. [CrossRef]
10. Grumati, P.; Dikic, I. Ubiquitin signaling and autophagy. *J. Biol. Chem.* **2018**, *293*, 5404–5413. [CrossRef]
11. Dikic, I. Proteasomal and Autophagic Degradation Systems. *Annu. Rev. Biochem.* **2017**, *86*, 193–224. [CrossRef] [PubMed]
12. Repnik, U.; Cesen, M.H.; Turk, B. The endolysosomal system in cell death and survival. *Cold Spring Harb. Perspect. Biol.* **2013**, *5*, a008755. [CrossRef] [PubMed]
13. Schoneberg, J.; Lee, I.H.; Iwasa, J.H.; Hurley, J.H. Reverse-topology membrane scission by the ESCRT proteins. *Nat. Rev. Mol. Cell Biol.* **2017**, *18*, 5–17. [CrossRef] [PubMed]
14. Mukherjee, S.; Ghosh, R.N.; Maxfield, F.R. Endocytosis. *Physiol. Rev.* **1997**, *77*, 759–803. [CrossRef]
15. Puthenveedu, M.A.; Lauffer, B.; Temkin, P.; Vistein, R.; Carlton, P.; Thorn, K.; Taunton, J.; Weiner, O.D.; Parton, R.G.; von Zastrow, M. Sequence-dependent sorting of recycling proteins by actin-stabilized endosomal microdomains. *Cell* **2010**, *143*, 761–773. [CrossRef] [PubMed]
16. Lee, P.P.; Lobato-Marquez, D.; Pramanik, N.; Sirianni, A.; Daza-Cajigal, V.; Rivers, E.; Cavazza, A.; Bouma, G.; Moulding, D.; Hultenby, K.; et al. Wiskott-Aldrich syndrome protein regulates autophagy and inflammasome activity in innate immune cells. *Nat. Commun.* **2017**, *8*, 1576. [CrossRef]
17. McNally, K.E.; Cullen, P.J. Endosomal Retrieval of Cargo: Retromer Is Not Alone. *Trends Cell Biol.* **2018**, *28*, 807–822. [CrossRef]
18. McMillan, K.J.; Korswagen, H.C.; Cullen, P.J. The emerging role of retromer in neuroprotection. *Curr. Opin. Cell Biol.* **2017**, *47*, 72–82. [CrossRef]
19. Yang, J.; Villar, V.A.M.; Rozyyev, S.; Jose, P.A.; Zeng, C. The emerging role of sorting nexins in cardiovascular diseases. *Clin. Sci. (Lond.)* **2019**, *133*, 723–737. [CrossRef]
20. Cullen, P.J. Endosomal sorting and signalling: An emerging role for sorting nexins. *Nat. Rev. Mol. Cell Biol.* **2008**, *9*, 574–582. [CrossRef]
21. Teasdale, R.D.; Collins, B.M. Insights into the PX (phox-homology) domain and SNX (sorting nexin) protein families: Structures, functions and roles in disease. *Biochem. J.* **2012**, *441*, 39–59. [CrossRef] [PubMed]
22. Haft, C.R.; de la Luz Sierra, M.; Barr, V.A.; Haft, D.H.; Taylor, S.I. Identification of a family of sorting nexin molecules and characterization of their association with receptors. *Mol. Cell. Biol.* **1998**, *18*, 7278–7287. [CrossRef]
23. Ponting, C.P. Novel domains in NADPH oxidase subunits, sorting nexins, and PtdIns 3-kinases: Binding partners of SH3 domains? *Protein Sci.* **1996**, *5*, 2353–2357. [CrossRef]
24. Chandra, M.; Chin, Y.K.; Mas, C.; Feathers, J.R.; Paul, B.; Datta, S.; Chen, K.E.; Jia, X.; Yang, Z.; Norwood, S.J.; et al. Classification of the human phox homology (PX) domains based on their phosphoinositide binding specificities. *Nat. Commun.* **2019**, *10*, 1528. [CrossRef]
25. Chandra, M.; Collins, B.M. The Phox Homology (PX) Domain. *Adv. Exp. Med. Biol.* **2019**, *1111*, 1–17. [CrossRef] [PubMed]
26. Shewan, A.; Eastburn, D.J.; Mostov, K. Phosphoinositides in cell architecture. *Cold Spring Harb. Perspect. Biol.* **2011**, *3*, a004796. [CrossRef]
27. Kutateladze, T.G. Translation of the phosphoinositide code by PI effectors. *Nat. Chem. Biol.* **2010**, *6*, 507–513. [CrossRef] [PubMed]
28. van Weering, J.R.; Verkade, P.; Cullen, P.J. SNX-BAR proteins in phosphoinositide-mediated, tubular-based endosomal sorting. *Semin. Cell Dev. Biol.* **2010**, *21*, 371–380. [CrossRef] [PubMed]
29. Carlton, J.; Bujny, M.; Peter, B.J.; Oorschot, V.M.; Rutherford, A.; Mellor, H.; Klumperman, J.; McMahon, H.T.; Cullen, P.J. Sorting nexin-1 mediates tubular endosome-to-TGN transport through coincidence sensing of high- curvature membranes and 3-phosphoinositides. *Curr. Biol.* **2004**, *14*, 1791–1800. [CrossRef] [PubMed]
30. Harterink, M.; Port, F.; Lorenowicz, M.J.; McGough, I.J.; Silhankova, M.; Betist, M.C.; van Weering, J.R.T.; van Heesbeen, R.; Middelkoop, T.C.; Basler, K.; et al. A SNX3-dependent retromer pathway mediates retrograde transport of the Wnt sorting receptor Wntless and is required for Wnt secretion. *Nat. Cell Biol.* **2011**, *13*, 914–923. [CrossRef]
31. Zhang, D.; Isack, N.R.; Glodowski, D.R.; Liu, J.; Chen, C.C.; Xu, X.Z.; Grant, B.D.; Rongo, C. RAB-6.2 and the retromer regulate glutamate receptor recycling through a retrograde pathway. *J. Cell Biol.* **2012**, *196*, 85–101. [CrossRef] [PubMed]
32. Seaman, M.N.; McCaffery, J.M.; Emr, S.D. A membrane coat complex essential for endosome-to-Golgi retrograde transport in yeast. *J. Cell Biol.* **1998**, *142*, 665–681. [CrossRef] [PubMed]

33. Lucas, M.; Gershlick, D.C.; Vidaurrazaga, A.; Rojas, A.L.; Bonifacino, J.S.; Hierro, A. Structural Mechanism for Cargo Recognition by the Retromer Complex. *Cell* **2016**, *167*, 1623–1635.e14. [CrossRef] [PubMed]
34. Suzuki, S.W.; Chuang, Y.S.; Li, M.; Seaman, M.N.J.; Emr, S.D. A bipartite sorting signal ensures specificity of retromer complex in membrane protein recycling. *J. Cell Biol.* **2019**, *218*, 2876–2886. [CrossRef] [PubMed]
35. Burd, C.; Cullen, P.J. Retromer: A master conductor of endosome sorting. *Cold Spring Harb. Perspect. Biol.* **2014**, *6*. [CrossRef]
36. Kvainickas, A.; Orgaz, A.J.; Nagele, H.; Diedrich, B.; Heesom, K.J.; Dengjel, J.; Cullen, P.J.; Steinberg, F. Retromer- and WASH-dependent sorting of nutrient transporters requires a multivalent interaction network with ANKRD50. *J. Cell Sci.* **2017**, *130*, 382–395. [CrossRef]
37. Zhang, P.; Wu, Y.; Belenkaya, T.Y.; Lin, X. SNX3 controls Wingless/Wnt secretion through regulating retromer-dependent recycling of Wntless. *Cell Res.* **2011**, *21*, 1677–1690. [CrossRef]
38. Chen, C.; Garcia-Santos, D.; Ishikawa, Y.; Seguin, A.; Li, L.; Fegan, K.H.; Hildick-Smith, G.J.; Shah, D.I.; Cooney, J.D.; Chen, W.; et al. Snx3 regulates recycling of the transferrin receptor and iron assimilation. *Cell Metab.* **2013**, *17*, 343–352. [CrossRef]
39. Tabuchi, M.; Yanatori, I.; Kawai, Y.; Kishi, F. Retromer-mediated direct sorting is required for proper endosomal recycling of the mammalian iron transporter DMT1. *J. Cell Sci.* **2010**, *123*, 756–766. [CrossRef]
40. Zimprich, A.; Benet-Pages, A.; Struhal, W.; Graf, E.; Eck, S.H.; Offman, M.N.; Haubenberger, D.; Spielberger, S.; Schulte, E.C.; Lichtner, P.; et al. A mutation in VPS35, encoding a subunit of the retromer complex, causes late-onset Parkinson disease. *Am. J. Hum. Genet.* **2011**, *89*, 168–175. [CrossRef]
41. Small, S.A.; Kent, K.; Pierce, A.; Leung, C.; Kang, M.S.; Okada, H.; Honig, L.; Vonsattel, J.P.; Kim, T.W. Model-guided microarray implicates the retromer complex in Alzheimer's disease. *Ann. Neurol.* **2005**, *58*, 909–919. [CrossRef] [PubMed]
42. Xu, J.; Zhang, L.; Ye, Y.; Shan, Y.; Wan, C.; Wang, J.; Pei, D.; Shu, X.; Liu, J. SNX16 Regulates the Recycling of E-Cadherin through a Unique Mechanism of Coordinated Membrane and Cargo Binding. *Structure* **2017**, *25*, 1251–1263.e5. [CrossRef] [PubMed]
43. Wang, S.; Zhao, Z.; Rodal, A.A. Higher-order assembly of Sorting Nexin 16 controls tubulation and distribution of neuronal endosomes. *J. Cell Biol.* **2019**, *218*, 2600–2618. [CrossRef] [PubMed]
44. Ghai, R.; Mobli, M.; Norwood, S.J.; Bugarcic, A.; Teasdale, R.D.; King, G.F.; Collins, B.M. Phox homology band 4.1/ezrin/radixin/moesin-like proteins function as molecular scaffolds that interact with cargo receptors and Ras GTPases. *Proc. Natl. Acad. Sci. USA* **2011**, *108*, 7763–7768. [CrossRef]
45. Ghai, R.; Bugarcic, A.; Liu, H.; Norwood, S.J.; Skeldal, S.; Coulson, E.J.; Li, S.S.; Teasdale, R.D.; Collins, B.M. Structural basis for endosomal trafficking of diverse transmembrane cargos by PX-FERM proteins. *Proc. Natl. Acad. Sci. USA* **2013**, *110*, E643–E652. [CrossRef]
46. McNally, K.E.; Faulkner, R.; Steinberg, F.; Gallon, M.; Ghai, R.; Pim, D.; Langton, P.; Pearson, N.; Danson, C.M.; Nagele, H.; et al. Retriever is a multiprotein complex for retromer-independent endosomal cargo recycling. *Nat. Cell Biol.* **2017**, *19*, 1214–1225. [CrossRef]
47. Bottcher, R.T.; Stremmel, C.; Meves, A.; Meyer, H.; Widmaier, M.; Tseng, H.Y.; Fassler, R. Sorting nexin 17 prevents lysosomal degradation of beta1 integrins by binding to the beta1-integrin tail. *Nat. Cell Biol.* **2012**, *14*, 584–592. [CrossRef]
48. Stockinger, W.; Sailler, B.; Strasser, V.; Recheis, B.; Fasching, D.; Kahr, L.; Schneider, W.J.; Nimpf, J. The PX-domain protein SNX17 interacts with members of the LDL receptor family and modulates endocytosis of the LDL receptor. *EMBO J.* **2002**, *21*, 4259–4267. [CrossRef]
49. Bergant Marusic, M.; Ozbun, M.A.; Campos, S.K.; Myers, M.P.; Banks, L. Human papillomavirus L2 facilitates viral escape from late endosomes via sorting nexin 17. *Traffic* **2012**, *13*, 455–467. [CrossRef]
50. Bergant, M.; Banks, L. SNX17 facilitates infection with diverse papillomavirus types. *J. Virol.* **2013**, *87*, 1270–1273. [CrossRef]
51. Day, P.M.; Lowy, D.R.; Schiller, J.T. Papillomaviruses infect cells via a clathrin-dependent pathway. *Virology* **2003**, *307*, 1–11. [CrossRef]
52. Temkin, P.; Lauffer, B.; Jager, S.; Cimermancic, P.; Krogan, N.J.; von Zastrow, M. SNX27 mediates retromer tubule entry and endosome-to-plasma membrane trafficking of signalling receptors. *Nat. Cell Biol.* **2011**, *13*, 715–721. [CrossRef] [PubMed]
53. Lauffer, B.E.; Melero, C.; Temkin, P.; Lei, C.; Hong, W.; Kortemme, T.; von Zastrow, M. SNX27 mediates PDZ-directed sorting from endosomes to the plasma membrane. *J. Cell Biol.* **2010**, *190*, 565–574. [CrossRef] [PubMed]
54. Steinberg, F.; Gallon, M.; Winfield, M.; Thomas, E.C.; Bell, A.J.; Heesom, K.J.; Tavare, J.M.; Cullen, P.J. A global analysis of SNX27-retromer assembly and cargo specificity reveals a function in glucose and metal ion transport. *Nat. Cell Biol.* **2013**, *15*, 461–471. [CrossRef]
55. Temkin, P.; Morishita, W.; Goswami, D.; Arendt, K.; Chen, L.; Malenka, R. The Retromer Supports AMPA Receptor Trafficking During LTP. *Neuron* **2017**, *94*, 74–82.e5. [CrossRef] [PubMed]
56. Wang, X.; Zhao, Y.; Zhang, X.; Badie, H.; Zhou, Y.; Mu, Y.; Loo, L.S.; Cai, L.; Thompson, R.C.; Yang, B.; et al. Loss of sorting nexin 27 contributes to excitatory synaptic dysfunction by modulating glutamate receptor recycling in Down's syndrome. *Nat. Med.* **2013**, *19*, 473–480. [CrossRef]
57. Damseh, N.; Danson, C.M.; Al-Ashhab, M.; Abu-Libdeh, B.; Gallon, M.; Sharma, K.; Yaacov, B.; Coulthard, E.; Caldwell, M.A.; Edvardson, S.; et al. A defect in the retromer accessory protein, SNX27, manifests by infantile myoclonic epilepsy and neurodegeneration. *Neurogenetics* **2015**, *16*, 215–221. [CrossRef]
58. Yong, X.; Zhao, L.; Deng, W.; Sun, H.; Zhou, X.; Mao, L.; Hu, W.; Shen, X.; Sun, Q.; Billadeau, D.D.; et al. Mechanism of cargo recognition by retromer-linked SNX-BAR proteins. *PLoS Biol.* **2020**, *18*, e3000631. [CrossRef]

59. van Weering, J.R.; Sessions, R.B.; Traer, C.J.; Kloer, D.P.; Bhatia, V.K.; Stamou, D.; Carlsson, S.R.; Hurley, J.H.; Cullen, P.J. Molecular basis for SNX-BAR-mediated assembly of distinct endosomal sorting tubules. *EMBO J.* **2012**, *31*, 4466–4480. [CrossRef]
60. Ma, M.; Burd, C.G. Retrograde trafficking and plasma membrane recycling pathways of the budding yeast Saccharomyces cerevisiae. *Traffic* **2020**, *21*, 45–59. [CrossRef]
61. Anton, Z.; Betin, V.M.S.; Simonetti, B.; Traer, C.J.; Attar, N.; Cullen, P.J.; Lane, J.D. A heterodimeric SNX4–SNX7 SNX-BAR autophagy complex coordinates ATG9A trafficking for efficient autophagosome assembly. *J. Cell Sci.* **2020**, *133*. [CrossRef] [PubMed]
62. Ghosh, P.; Dahms, N.M.; Kornfeld, S. Mannose 6-phosphate receptors: New twists in the tale. *Nat. Rev. Mol. Cell Biol.* **2003**, *4*, 202–212. [CrossRef] [PubMed]
63. Arighi, C.N.; Hartnell, L.M.; Aguilar, R.C.; Haft, C.R.; Bonifacino, J.S. Role of the mammalian retromer in sorting of the cation-independent mannose 6-phosphate receptor. *J. Cell Biol.* **2004**, *165*, 123–133. [CrossRef] [PubMed]
64. Seaman, M.N. Cargo-selective endosomal sorting for retrieval to the Golgi requires retromer. *J. Cell Biol.* **2004**, *165*, 111–122. [CrossRef]
65. Simonetti, B.; Danson, C.M.; Heesom, K.J.; Cullen, P.J. Sequence-dependent cargo recognition by SNX-BARs mediates retromer-independent transport of CI-MPR. *J. Cell Biol.* **2017**, *216*, 3695–3712. [CrossRef] [PubMed]
66. Kvainickas, A.; Jimenez-Orgaz, A.; Nagele, H.; Hu, Z.; Dengjel, J.; Steinberg, F. Cargo-selective SNX-BAR proteins mediate retromer trimer independent retrograde transport. *J. Cell Biol.* **2017**, *216*, 3677–3693. [CrossRef] [PubMed]
67. Worby, C.A.; Dixon, J.E. Sorting out the cellular functions of sorting nexins. *Nat. Rev. Mol. Cell Biol.* **2002**, *3*, 919–931. [CrossRef] [PubMed]
68. Danson, C.M.; Pearson, N.; Heesom, K.J.; Cullen, P.J. Sorting nexin-21 is a scaffold for the endosomal recruitment of huntingtin. *J. Cell Sci.* **2018**, *131*. [CrossRef] [PubMed]
69. Uhlen, M.; Fagerberg, L.; Hallstrom, B.M.; Lindskog, C.; Oksvold, P.; Mardinoglu, A.; Sivertsson, A.; Kampf, C.; Sjostedt, E.; Asplund, A.; et al. Proteomics. Tissue-based map of the human proteome. *Science* **2015**, *347*, 1260419. [CrossRef]
70. Taylor, M.J.; Perrais, D.; Merrifield, C.J. A high precision survey of the molecular dynamics of mammalian clathrin-mediated endocytosis. *PLoS Biol.* **2011**, *9*, e1000604. [CrossRef]
71. Kaksonen, M.; Roux, A. Mechanisms of clathrin-mediated endocytosis. *Nat. Rev. Mol. Cell Biol.* **2018**, *19*, 313–326. [CrossRef] [PubMed]
72. Godlee, C.; Kaksonen, M. Review series: From uncertain beginnings: Initiation mechanisms of clathrin-mediated endocytosis. *J. Cell Biol.* **2013**, *203*, 717–725. [CrossRef] [PubMed]
73. Cullen, P.J.; Steinberg, F. To degrade or not to degrade: Mechanisms and significance of endocytic recycling. *Nat. Rev. Mol. Cell Biol.* **2018**, *19*, 679–696. [CrossRef] [PubMed]
74. Chen, K.E.; Healy, M.D.; Collins, B.M. Towards a molecular understanding of endosomal trafficking by Retromer and Retriever. *Traffic* **2019**, *20*, 465–478. [CrossRef]
75. Seaman, M.N.; Harbour, M.E.; Tattersall, D.; Read, E.; Bright, N. Membrane recruitment of the cargo-selective retromer subcomplex is catalysed by the small GTPase Rab7 and inhibited by the Rab-GAP TBC1D5. *J. Cell Sci.* **2009**, *122*, 2371–2382. [CrossRef]
76. Mallet, W.G.; Maxfield, F.R. Chimeric forms of furin and TGN38 are transported with the plasma membrane in the trans-Golgi network via distinct endosomal pathways. *J. Cell Biol.* **1999**, *146*, 345–359. [CrossRef]
77. Goldenring, J.R. Recycling endosomes. *Curr. Opin. Cell Biol.* **2015**, *35*, 117–122. [CrossRef]
78. Li, C.; Ma, W.; Yin, S.; Liang, X.; Shu, X.; Pei, D.; Egan, T.M.; Huang, J.; Pan, A.; Li, Z. Sorting Nexin 11 Regulates Lysosomal Degradation of Plasma Membrane TRPV3. *Traffic* **2016**, *17*, 500–514. [CrossRef]
79. Gullapalli, A.; Wolfe, B.L.; Griffin, C.T.; Magnuson, T.; Trejo, J. An essential role for SNX1 in lysosomal sorting of protease-activated receptor-1: Evidence for retromer-, Hrs-, and Tsg101-independent functions of sorting nexins. *Mol. Biol. Cell* **2006**, *17*, 1228–1238. [CrossRef]
80. Kurten, R.C.; Cadena, D.L.; Gill, G.N. Enhanced degradation of EGF receptors by a sorting nexin, SNX1. *Science* **1996**, *272*, 1008–1010. [CrossRef]
81. Fuster, J.J.; Gonzalez, J.M.; Edo, M.D.; Viana, R.; Boya, P.; Cervera, J.; Verges, M.; Rivera, J.; Andres, V. Tumor suppressor p27(Kip1) undergoes endolysosomal degradation through its interaction with sorting nexin 6. *FASEB J.* **2010**, *24*, 2998–3009. [CrossRef] [PubMed]
82. Muller, U.C.; Deller, T.; Korte, M. Not just amyloid: Physiological functions of the amyloid precursor protein family. *Nat. Rev. Neurosci.* **2017**, *18*, 281–298. [CrossRef] [PubMed]
83. Bendris, N.; Schmid, S.L. Endocytosis, Metastasis and Beyond: Multiple Facets of SNX9. *Trends Cell Biol.* **2017**, *27*, 189–200. [CrossRef] [PubMed]
84. Nemec, A.A.; Howell, L.A.; Peterson, A.K.; Murray, M.A.; Tomko, R.J., Jr. Autophagic clearance of proteasomes in yeast requires the conserved sorting nexin Snx4. *J. Biol. Chem.* **2017**, *292*, 21466–21480. [CrossRef] [PubMed]
85. Ma, M.; Burd, C.G.; Chi, R.J. Distinct complexes of yeast Snx4 family SNX-BARs mediate retrograde trafficking of Snc1 and Atg27. *Traffic* **2017**, *18*, 134–144. [CrossRef]
86. Bean, B.D.; Davey, M.; Conibear, E. Cargo selectivity of yeast sorting nexins. *Traffic* **2017**, *18*, 110–122. [CrossRef]
87. Robinson, M.S. Forty Years of Clathrin-coated Vesicles. *Traffic* **2015**, *16*, 1210–1238. [CrossRef]

88. MacDonald, C.; Piper, R.C. Genetic dissection of early endosomal recycling highlights a TORC1-independent role for Rag GTPases. *J. Cell Biol.* **2017**, *216*, 3275–3290. [CrossRef]
89. Haglund, K.; Dikic, I. The role of ubiquitylation in receptor endocytosis and endosomal sorting. *J. Cell Sci.* **2012**, *125*, 265–275. [CrossRef]
90. Day, K.J.; Casler, J.C.; Glick, B.S. Budding Yeast Has a Minimal Endomembrane System. *Dev. Cell* **2018**, *44*, 56–72.e4. [CrossRef]
91. Grissom, J.H.; Segarra, V.A.; Chi, R.J. New Perspectives on SNARE Function in the Yeast Minimal Endomembrane System. *Genes* **2020**, *11*, 899. [CrossRef] [PubMed]
92. Casler, J.C.; Glick, B.S. A microscopy-based kinetic analysis of yeast vacuolar protein sorting. *Elife* **2020**, *9*. [CrossRef]
93. Seaman, M.N.; Marcusson, E.G.; Cereghino, J.L.; Emr, S.D. Endosome to Golgi retrieval of the vacuolar protein sorting receptor, Vps10p, requires the function of the VPS29, VPS30, and VPS35 gene products. *J. Cell Biol.* **1997**, *137*, 79–92. [CrossRef] [PubMed]
94. Stevens, T.; Esmon, B.; Schekman, R. Early stages in the yeast secretory pathway are required for transport of carboxypeptidase Y to the vacuole. *Cell* **1982**, *30*, 439–448. [CrossRef]
95. Liu, T.T.; Gomez, T.S.; Sackey, B.K.; Billadeau, D.D.; Burd, C.G. Rab GTPase regulation of retromer-mediated cargo export during endosome maturation. *Mol. Biol. Cell* **2012**, *23*, 2505–2515. [CrossRef]
96. Collins, B.M.; Skinner, C.F.; Watson, P.J.; Seaman, M.N.; Owen, D.J. Vps29 has a phosphoesterase fold that acts as a protein interaction scaffold for retromer assembly. *Nat. Struct. Mol. Biol.* **2005**, *12*, 594–602. [CrossRef]
97. Reddy, J.V.; Seaman, M.N. Vps26p, a component of retromer, directs the interactions of Vps35p in endosome-to-Golgi retrieval. *Mol. Biol. Cell* **2001**, *12*, 3242–3256. [CrossRef]
98. Seaman, M.N.; Williams, H.P. Identification of the functional domains of yeast sorting nexins Vps5p and Vps17p. *Mol. Biol. Cell* **2002**, *13*, 2826–2840. [CrossRef]
99. Kovtun, O.; Leneva, N.; Bykov, Y.S.; Ariotti, N.; Teasdale, R.D.; Schaffer, M.; Engel, B.D.; Owen, D.J.; Briggs, J.A.G.; Collins, B.M. Structure of the membrane-assembled retromer coat determined by cryo-electron tomography. *Nature* **2018**, *561*, 561–564. [CrossRef]
100. Strochlic, T.I.; Setty, T.G.; Sitaram, A.; Burd, C.G. Grd19/Snx3p functions as a cargo-specific adapter for retromer-dependent endocytic recycling. *J. Cell Biol.* **2007**, *177*, 115–125. [CrossRef]
101. Dalton, L.E.; Bean, B.D.M.; Davey, M.; Conibear, E. Quantitative high-content imaging identifies novel regulators of Neo1 trafficking at endosomes. *Mol. Biol. Cell* **2017**, *28*, 1539–1550. [CrossRef] [PubMed]
102. McGough, I.J.; de Groot, R.E.A.; Jellett, A.P.; Betist, M.C.; Varandas, K.C.; Danson, C.M.; Heesom, K.J.; Korswagen, H.C.; Cullen, P.J. SNX3-retromer requires an evolutionary conserved MON2:DOPEY2:ATP9A complex to mediate Wntless sorting and Wnt secretion. *Nat. Commun.* **2018**, *9*, 3737. [CrossRef] [PubMed]
103. Takar, M.; Wu, Y.; Graham, T.R. The Essential Neo1 Protein from Budding Yeast Plays a Role in Establishing Aminophospholipid Asymmetry of the Plasma Membrane. *J. Biol. Chem.* **2016**, *291*, 15727–15739. [CrossRef] [PubMed]
104. Ekena, K.; Stevens, T.H. The Saccharomyces cerevisiae MVP1 gene interacts with VPS1 and is required for vacuolar protein sorting. *Mol. Cell. Biol.* **1995**, *15*, 1671–1678. [CrossRef]
105. Dyve, A.B.; Bergan, J.; Utskarpen, A.; Sandvig, K. Sorting nexin 8 regulates endosome-to-Golgi transport. *Biochem. Biophys. Res. Commun.* **2009**, *390*, 109–114. [CrossRef] [PubMed]
106. Chi, R.J.; Liu, J.; West, M.; Wang, J.; Odorizzi, G.; Burd, C.G. Fission of SNX-BAR-coated endosomal retrograde transport carriers is promoted by the dynamin-related protein Vps1. *J. Cell Biol.* **2014**, *204*, 793–806. [CrossRef]
107. Bonangelino, C.J.; Chavez, E.M.; Bonifacino, J.S. Genomic screen for vacuolar protein sorting genes in Saccharomyces cerevisiae. *Mol. Biol. Cell* **2002**, *13*, 2486–2501. [CrossRef]
108. Sun, D.; Varlakhanova, N.V.; Tornabene, B.A.; Ramachandran, R.; Zhang, P.; Ford, M.G.J. The cryo-EM structure of the SNX-BAR Mvp1 tetramer. *Nat. Commun.* **2020**, *11*, 1506. [CrossRef]
109. Nice, D.C.; Sato, T.K.; Stromhaug, P.E.; Emr, S.D.; Klionsky, D.J. Cooperative binding of the cytoplasm to vacuole targeting pathway proteins, Cvt13 and Cvt20, to phosphatidylinositol 3-phosphate at the pre-autophagosomal structure is required for selective autophagy. *J. Biol. Chem.* **2002**, *277*, 30198–30207. [CrossRef]
110. Reggiori, F.; Klionsky, D.J. Autophagic processes in yeast: Mechanism, machinery and regulation. *Genetics* **2013**, *194*, 341–361. [CrossRef]
111. Hanley, S.E.; SD, W.; Cooper, K. Snx4 is required for the destruction of the transcription factors by a novel autophgay pathway. *Autophagy* **2021**, in press.
112. Lewis, M.J.; Nichols, B.J.; Prescianotto-Baschong, C.; Riezman, H.; Pelham, H.R. Specific retrieval of the exocytic SNARE Snc1p from early yeast endosomes. *Mol. Biol. Cell* **2000**, *11*, 23–38. [CrossRef] [PubMed]
113. Hettema, E.H.; Lewis, M.J.; Black, M.W.; Pelham, H.R. Retromer and the sorting nexins Snx4/41/42 mediate distinct retrieval pathways from yeast endosomes. *EMBO J.* **2003**, *22*, 548–557. [CrossRef] [PubMed]
114. Hanamatsu, H.; Fujimura-Kamada, K.; Yamamoto, T.; Furuta, N.; Tanaka, K. Interaction of the phospholipid flippase Drs2p with the F-box protein Rcy1p plays an important role in early endosome to trans-Golgi network vesicle transport in yeast. *J. Biochem.* **2014**, *155*, 51–62. [CrossRef] [PubMed]
115. Ma, M.; Burd, C.G. Retrograde trafficking and quality control of yeast synaptobrevin, Snc1, are conferred by its transmembrane domain. *Mol. Biol. Cell* **2019**, *30*, 1729–1742. [CrossRef] [PubMed]

116. Shirahama-Noda, K.; Kira, S.; Yoshimori, T.; Noda, T. TRAPPIII is responsible for vesicular transport from early endosomes to Golgi, facilitating Atg9 cycling in autophagy. *J. Cell Sci.* **2013**, *126*, 4963–4973. [CrossRef]
117. Reggiori, F.; Tooze, S.A. Autophagy regulation through Atg9 traffic. *J. Cell Biol.* **2012**, *198*, 151–153. [CrossRef]
118. Suzuki, S.W.; Emr, S.D. Membrane protein recycling from the vacuole/lysosome membrane. *J. Cell Biol.* **2018**, *217*, 1623–1632. [CrossRef]
119. Matscheko, N.; Mayrhofer, P.; Rao, Y.; Beier, V.; Wollert, T. Atg11 tethers Atg9 vesicles to initiate selective autophagy. *PLoS Biol.* **2019**, *17*, e3000377. [CrossRef]
120. Budenholzer, L.; Cheng, C.L.; Li, Y.; Hochstrasser, M. Proteasome Structure and Assembly. *J. Mol. Biol.* **2017**, *429*, 3500–3524. [CrossRef]
121. Bochtler, M.; Ditzel, L.; Groll, M.; Hartmann, C.; Huber, R. The proteasome. *Annu. Rev. Biophys. Biomol. Struct.* **1999**, *28*, 295–317. [CrossRef] [PubMed]
122. Sun, L.; Chen, Z.J. The novel functions of ubiquitination in signaling. *Curr. Opin. Cell Biol.* **2004**, *16*, 119–126. [CrossRef]
123. Piper, R.C.; Dikic, I.; Lukacs, G.L. Ubiquitin-dependent sorting in endocytosis. *Cold Spring Harb. Perspect. Biol.* **2014**, *6*. [CrossRef]
124. Polo, S. Signaling-mediated control of ubiquitin ligases in endocytosis. *BMC Biol.* **2012**, *10*, 25. [CrossRef] [PubMed]
125. Okano, M.; Matsuo, H.; Nishimura, Y.; Hozumi, K.; Yoshioka, S.; Tonoki, A.; Itoh, M. Mib1 modulates dynamin 2 recruitment via Snx18 to promote Dll1 endocytosis for efficient Notch signaling. *Genes Cells* **2016**, *21*, 425–441. [CrossRef]
126. Baumann, C.; Lindholm, C.K.; Rimoldi, D.; Levy, F. The E3 ubiquitin ligase Itch regulates sorting nexin 9 through an unconventional substrate recognition domain. *FEBS J.* **2010**, *277*, 2803–2814. [CrossRef] [PubMed]
127. Doyle, J.M.; Gao, J.; Wang, J.; Yang, M.; Potts, P.R. MAGE-RING protein complexes comprise a family of E3 ubiquitin ligases. *Mol. Cell* **2010**, *39*, 963–974. [CrossRef]
128. Hao, Y.H.; Fountain, M.D., Jr.; Fon Tacer, K.; Xia, F.; Bi, W.; Kang, S.H.; Patel, A.; Rosenfeld, J.A.; Le Caignec, C.; Isidor, B.; et al. USP7 Acts as a Molecular Rheostat to Promote WASH-Dependent Endosomal Protein Recycling and Is Mutated in a Human Neurodevelopmental Disorder. *Mol. Cell* **2015**, *59*, 956–969. [CrossRef]
129. Boulkroun, S.; Ruffieux-Daidie, D.; Vitagliano, J.J.; Poirot, O.; Charles, R.P.; Lagnaz, D.; Firsov, D.; Kellenberger, S.; Staub, O. Vasopressin-inducible ubiquitin-specific protease 10 increases ENaC cell surface expression by deubiquitylating and stabilizing sorting nexin 3. *Am. J. Physiol. Ren. Physiol.* **2008**, *295*, F889–F900. [CrossRef]
130. Cai, J.; Sun, M.; Hu, B.; Windle, B.; Ge, X.; Li, G.; Sun, Y. Sorting Nexin 5 Controls Head and Neck Squamous Cell Carcinoma Progression by Modulating FBW7. *J. Cancer* **2019**, *10*, 2942–2952. [CrossRef]
131. Shen, Z.; Li, Y.; Fang, Y.; Lin, M.; Feng, X.; Li, Z.; Zhan, Y.; Liu, Y.; Mou, T.; Lan, X.; et al. SNX16 activates c-Myc signaling by inhibiting ubiquitin-mediated proteasomal degradation of eEF1A2 in colorectal cancer development. *Mol. Oncol.* **2020**, *14*, 387–406. [CrossRef] [PubMed]
132. Stangl, A.; Elliott, P.R.; Pinto-Fernandez, A.; Bonham, S.; Harrison, L.; Schaub, A.; Kutzner, K.; Keusekotten, K.; Pfluger, P.T.; El Oualid, F.; et al. Regulation of the endosomal SNX27-retromer by OTULIN. *Nat. Commun.* **2019**, *10*, 4320. [CrossRef] [PubMed]
133. Zhang, H.X.; Xu, Z.S.; Lin, H.; Li, M.; Xia, T.; Cui, K.; Wang, S.Y.; Li, Y.; Shu, H.B.; Wang, Y.Y. TRIM27 mediates STAT3 activation at retromer-positive structures to promote colitis and colitis-associated carcinogenesis. *Nat. Commun.* **2018**, *9*, 3441. [CrossRef] [PubMed]
134. Hao, Y.H.; Doyle, J.M.; Ramanathan, S.; Gomez, T.S.; Jia, D.; Xu, M.; Chen, Z.J.; Billadeau, D.D.; Rosen, M.K.; Potts, P.R. Regulation of WASH-dependent actin polymerization and protein trafficking by ubiquitination. *Cell* **2013**, *152*, 1051–1064. [CrossRef]
135. Sato, M.; Konuma, R.; Sato, K.; Tomura, K.; Sato, K. Fertilization-induced K63-linked ubiquitylation mediates clearance of maternal membrane proteins. *Development* **2014**, *141*, 1324–1331. [CrossRef]
136. Marshall, R.S.; Li, F.; Gemperline, D.C.; Book, A.J.; Vierstra, R.D. Autophagic Degradation of the 26S Proteasome Is Mediated by the Dual ATG8/Ubiquitin Receptor RPN10 in Arabidopsis. *Mol. Cell* **2015**, *58*, 1053–1066. [CrossRef]
137. Waite, K.A.; De-La Mota-Peynado, A.; Vontz, G.; Roelofs, J. Starvation Induces Proteasome Autophagy with Different Pathways for Core and Regulatory Particles. *J. Biol. Chem.* **2016**, *291*, 3239–3253. [CrossRef]
138. Sowa, M.E.; Bennett, E.J.; Gygi, S.P.; Harper, J.W. Defining the human deubiquitinating enzyme interaction landscape. *Cell* **2009**, *138*, 389–403. [CrossRef]
139. Galluzzi, L.; Baehrecke, E.H.; Ballabio, A.; Boya, P.; Bravo-San Pedro, J.M.; Cecconi, F.; Choi, A.M.; Chu, C.T.; Codogno, P.; Colombo, M.I.; et al. Molecular definitions of autophagy and related processes. *EMBO J.* **2017**, *36*, 1811–1836. [CrossRef]
140. Zientara-Rytter, K.; Subramani, S. The Roles of Ubiquitin-Binding Protein Shuttles in the Degradative Fate of Ubiquitinated Proteins in the Ubiquitin-Proteasome System and Autophagy. *Cells* **2019**, *8*, 40. [CrossRef]
141. Hollenstein, D.M.; Kraft, C. Autophagosomes are formed at a distinct cellular structure. *Curr. Opin. Cell Biol.* **2020**, *65*, 50–57. [CrossRef] [PubMed]
142. Popelka, H.; Damasio, A.; Hinshaw, J.E.; Klionsky, D.J.; Ragusa, M.J. Structure and function of yeast Atg20, a sorting nexin that facilitates autophagy induction. *Proc. Natl. Acad. Sci. USA* **2017**, *114*, E10112–E10121. [CrossRef] [PubMed]
143. Hu, Z.; Raucci, S.; Jaquenoud, M.; Hatakeyama, R.; Stumpe, M.; Rohr, R.; Reggiori, F.; De Virgilio, C.; Dengjel, J. Multilayered Control of Protein Turnover by TORC1 and Atg1. *Cell Rep.* **2019**, *28*, 3486–3496.e6. [CrossRef] [PubMed]
144. Ma, M.; Kumar, S.; Purushothaman, L.; Babst, M.; Ungermann, C.; Chi, R.J.; Burd, C.G. Lipid trafficking by yeast Snx4 family SNX-BAR proteins promotes autophagy and vacuole membrane fusion. *Mol. Biol. Cell* **2018**, *29*, 2190–2200. [CrossRef]

145. Guo, Y.; Chang, C.; Huang, R.; Liu, B.; Bao, L.; Liu, W. AP1 is essential for generation of autophagosomes from the trans-Golgi network. *J. Cell Sci.* **2012**, *125*, 1706–1715. [CrossRef]
146. Imai, K.; Hao, F.; Fujita, N.; Tsuji, Y.; Oe, Y.; Araki, Y.; Hamasaki, M.; Noda, T.; Yoshimori, T. Atg9A trafficking through the recycling endosomes is required for autophagosome formation. *J. Cell Sci.* **2016**, *129*, 3781–3791. [CrossRef]
147. Longatti, A.; Lamb, C.A.; Razi, M.; Yoshimura, S.; Barr, F.A.; Tooze, S.A. TBC1D14 regulates autophagosome formation via Rab11- and ULK1-positive recycling endosomes. *J. Cell Biol.* **2012**, *197*, 659–675. [CrossRef]
148. Ravikumar, B.; Moreau, K.; Jahreiss, L.; Puri, C.; Rubinsztein, D.C. Plasma membrane contributes to the formation of pre-autophagosomal structures. *Nat. Cell Biol.* **2010**, *12*, 747–757. [CrossRef]
149. Knaevelsrud, H.; Soreng, K.; Raiborg, C.; Haberg, K.; Rasmuson, F.; Brech, A.; Liestol, K.; Rusten, T.E.; Stenmark, H.; Neufeld, T.P.; et al. Membrane remodeling by the PX-BAR protein SNX18 promotes autophagosome formation. *J. Cell Biol.* **2013**, *202*, 331–349. [CrossRef]
150. Soreng, K.; Munson, M.J.; Lamb, C.A.; Bjorndal, G.T.; Pankiv, S.; Carlsson, S.R.; Tooze, S.A.; Simonsen, A. SNX18 regulates ATG9A trafficking from recycling endosomes by recruiting Dynamin-2. *EMBO Rep.* **2018**, *19*. [CrossRef]
151. Kanki, T.; Wang, K.; Cao, Y.; Baba, M.; Klionsky, D.J. Atg32 is a mitochondrial protein that confers selectivity during mitophagy. *Dev. Cell* **2009**, *17*, 98–109. [CrossRef] [PubMed]
152. Ano, Y.; Hattori, T.; Oku, M.; Mukaiyama, H.; Baba, M.; Ohsumi, Y.; Kato, N.; Sakai, Y. A sorting nexin PpAtg24 regulates vacuolar membrane dynamics during pexophagy via binding to phosphatidylinositol-3-phosphate. *Mol. Biol. Cell* **2005**, *16*, 446–457. [CrossRef] [PubMed]
153. Germain, K.; Kim, P.K. Pexophagy: A Model for Selective Autophagy. *Int. J. Mol. Sci.* **2020**, *21*, 578. [CrossRef] [PubMed]
154. Shpilka, T.; Welter, E.; Borovsky, N.; Amar, N.; Shimron, F.; Peleg, Y.; Elazar, Z. Fatty acid synthase is preferentially degraded by autophagy upon nitrogen starvation in yeast. *Proc. Natl. Acad. Sci. USA* **2015**, *112*, 1434–1439. [CrossRef] [PubMed]
155. Loewith, R.; Hall, M.N. Target of rapamycin (TOR) in nutrient signaling and growth control. *Genetics* **2011**, *189*, 1177–1201. [CrossRef] [PubMed]
156. Dong, X.; Yang, Y.; Zou, Y.; Zhao, Y.; Ci, B.; Zhong, L.; Bhave, M.; Wang, L.; Kuo, Y.-C.; Zang, X.; et al. Sorting Nexin 5 Mediates Virus-Induced Autophagy and Immunity. *Nature* **2020**, in press. [CrossRef]
157. Balch, W.E.; Morimoto, R.I.; Dillin, A.; Kelly, J.W. Adapting proteostasis for disease intervention. *Science* **2008**, *319*, 916–919. [CrossRef]
158. Korolchuk, V.I.; Menzies, F.M.; Rubinsztein, D.C. Mechanisms of cross-talk between the ubiquitin-proteasome and autophagy-lysosome systems. *FEBS Lett.* **2010**, *584*, 1393–1398. [CrossRef]
159. Ji, C.H.; Kwon, Y.T. Crosstalk and Interplay between the Ubiquitin-Proteasome System and Autophagy. *Mol. Cells* **2017**, *40*, 441–449. [CrossRef]
160. Pickles, S.; Vigie, P.; Youle, R.J. Mitophagy and Quality Control Mechanisms in Mitochondrial Maintenance. *Curr. Biol.* **2018**, *28*, R170–R185. [CrossRef]
161. Springer, M.Z.; Macleod, K.F. In Brief: Mitophagy: Mechanisms and role in human disease. *J. Pathol.* **2016**, *240*, 253–255. [CrossRef] [PubMed]
162. Fortun, J.; Verrier, J.D.; Go, J.C.; Madorsky, I.; Dunn, W.A.; Notterpek, L. The formation of peripheral myelin protein 22 aggregates is hindered by the enhancement of autophagy and expression of cytoplasmic chaperones. *Neurobiol. Dis.* **2007**, *25*, 252–265. [CrossRef] [PubMed]
163. Matsumoto, G.; Inobe, T.; Amano, T.; Murai, K.; Nukina, N.; Mori, N. N-Acyldopamine induces aggresome formation without proteasome inhibition and enhances protein aggregation via p62/SQSTM1 expression. *Sci. Rep.* **2018**, *8*, 9585. [CrossRef] [PubMed]
164. Nazarko, T.Y. Atg37 regulates the assembly of the pexophagic receptor protein complex. *Autophagy* **2014**, *10*, 1348–1349. [CrossRef] [PubMed]
165. Siqueira, M.D.S.; Ribeiro, R.M.; Travassos, L.H. Autophagy and Its Interaction with Intracellular Bacterial Pathogens. *Front. Immunol.* **2018**, *9*, 935. [CrossRef] [PubMed]
166. Kraft, C.; Deplazes, A.; Sohrmann, M.; Peter, M. Mature ribosomes are selectively degraded upon starvation by an autophagy pathway requiring the Ubp3p/Bre5p ubiquitin protease. *Nat. Cell Biol.* **2008**, *10*, 602–610. [CrossRef]
167. Nakatogawa, H. Spoon-Feeding Ribosomes to Autophagy. *Mol. Cell* **2018**, *71*, 197–199. [CrossRef]
168. Shaid, S.; Brandts, C.H.; Serve, H.; Dikic, I. Ubiquitination and selective autophagy. *Cell Death Differ.* **2013**, *20*, 21–30. [CrossRef]
169. Liebl, M.P.; Hoppe, T. It's all about talking: Two-way communication between proteasomal and lysosomal degradation pathways via ubiquitin. *Am. J. Physiol. Cell Physiol.* **2016**, *311*, C166–C178. [CrossRef]
170. Marshall, R.S.; McLoughlin, F.; Vierstra, R.D. Autophagic Turnover of Inactive 26S Proteasomes in Yeast Is Directed by the Ubiquitin Receptor Cue5 and the Hsp42 Chaperone. *Cell Rep.* **2016**, *16*, 1717–1732. [CrossRef]
171. Stieg, D.C.; Willis, S.D.; Ganesan, V.; Ong, K.L.; Scuorzo, J.; Song, M.; Grose, J.; Strich, R.; Cooper, K.F. A complex molecular switch directs stress-induced cyclin C nuclear release through SCF(Grr1)-mediated degradation of Med13. *Mol. Biol. Cell* **2018**, *29*, 363–375. [CrossRef] [PubMed]
172. Khakhina, S.; Cooper, K.F.; Strich, R. Med13p prevents mitochondrial fission and programmed cell death in yeast through nuclear retention of cyclin C. *Mol. Biol. Cell* **2014**, *25*, 2807–2816. [CrossRef] [PubMed]

173. Allen, B.L.; Taatjes, D.J. The Mediator complex: A central integrator of transcription. *Nat. Rev. Mol. Cell Biol.* **2015**, *16*, 155–166. [CrossRef] [PubMed]
174. Cooper, K.F.; Mallory, M.J.; Smith, J.B.; Strich, R. Stress and developmental regulation of the yeast C-type cyclin Ume3p (Srb11p/Ssn8p). *EMBO J.* **1997**, *16*, 4665–4675. [CrossRef]
175. van de Peppel, J.; Kettelarij, N.; van Bakel, H.; Kockelkorn, T.T.; van Leenen, D.; Holstege, F.C. Mediator expression profiling epistasis reveals a signal transduction pathway with antagonistic submodules and highly specific downstream targets. *Mol. Cell* **2005**, *19*, 511–522. [CrossRef] [PubMed]
176. Jeronimo, C.; Robert, F. The Mediator Complex: At the Nexus of RNA Polymerase II Transcription. *Trends Cell Biol.* **2017**, *27*, 765–783. [CrossRef]
177. Cooper, K.F.; Mallory, M.J.; Strich, R. Oxidative stress-induced destruction of the yeast C-type cyclin Ume3p requires phosphatidylinositol-specific phospholipase C and the 26S proteasome. *Mol. Cell. Biol.* **1999**, *19*, 3338–3348. [CrossRef] [PubMed]
178. Ganesan, V.; Willis, S.D.; Chang, K.T.; Beluch, S.; Cooper, K.F.; Strich, R. Cyclin C directly stimulates Drp1 GTP affinity to mediate stress-induced mitochondrial hyperfission. *Mol. Biol. Cell* **2019**, *30*, 302–311. [CrossRef]
179. Jezek, J.; Chang, K.T.; Joshi, A.M.; Strich, R. Mitochondrial translocation of cyclin C stimulates intrinsic apoptosis through Bax recruitment. *EMBO Rep.* **2019**, *20*, e47425. [CrossRef]
180. Wang, K.; Yan, R.; Cooper, K.F.; Strich, R. Cyclin C mediates stress-induced mitochondrial fission and apoptosis. *Mol. Biol. Cell* **2015**, *26*, 1030–1043. [CrossRef]
181. Willis, S.D.; Hanley, S.E.; Beishke, T.; Tati, P.D.; Cooper, K.F. Ubiquitin-proteasome-mediated cyclin C degradation promotes cell survival following nitrogen starvation. *Mol. Biol. Cell* **2020**, *31*, 1015–1031. [CrossRef]
182. Borriello, A.; Cucciolla, V.; Oliva, A.; Zappia, V.; Della Ragione, F. p27Kip1 metabolism: A fascinating labyrinth. *Cell Cycle* **2007**, *6*, 1053–1061. [CrossRef] [PubMed]
183. Abbastabar, M.; Kheyrollah, M.; Azizian, K.; Bagherlou, N.; Tehrani, S.S.; Maniati, M.; Karimian, A. Multiple functions of p27 in cell cycle, apoptosis, epigenetic modification and transcriptional regulation for the control of cell growth: A double-edged sword protein. *DNA Repair (Amst)* **2018**, *69*, 63–72. [CrossRef]
184. Hydbring, P.; Malumbres, M.; Sicinski, P. Non-canonical functions of cell cycle cyclins and cyclin-dependent kinases. *Nat. Rev. Mol. Cell Biol.* **2016**, *17*, 280–292. [CrossRef] [PubMed]
185. Schreij, A.M.; Fon, E.A.; McPherson, P.S. Endocytic membrane trafficking and neurodegenerative disease. *Cell. Mol. Life Sci.* **2016**, *73*, 1529–1545. [CrossRef] [PubMed]
186. Singh, V.; Yang, J.; Cha, B.; Chen, T.E.; Sarker, R.; Yin, J.; Avula, L.R.; Tse, M.; Donowitz, M. Sorting nexin 27 regulates basal and stimulated brush border trafficking of NHE3. *Mol. Biol. Cell* **2015**, *26*, 2030–2043. [CrossRef]
187. Villar, V.A.; Armando, I.; Sanada, H.; Frazer, L.C.; Russo, C.M.; Notario, P.M.; Lee, H.; Comisky, L.; Russell, H.A.; Yang, Y.; et al. Novel role of sorting nexin 5 in renal D(1) dopamine receptor trafficking and function: Implications for hypertension. *FASEB J.* **2013**, *27*, 1808–1819. [CrossRef]
188. Villar, V.A.; Jones, J.E.; Armando, I.; Asico, L.D.; Escano, C.S., Jr.; Lee, H.; Wang, X.; Yang, Y.; Pascua-Crusan, A.M.; Palmes-Saloma, C.P.; et al. Sorting nexin 1 loss results in D5 dopamine receptor dysfunction in human renal proximal tubule cells and hypertension in mice. *J. Biol. Chem.* **2013**, *288*, 152–163. [CrossRef] [PubMed]
189. Burden, J.J.; Sun, X.M.; Garcia, A.B.; Soutar, A.K. Sorting motifs in the intracellular domain of the low density lipoprotein receptor interact with a novel domain of sorting nexin-17. *J. Biol. Chem.* **2004**, *279*, 16237–16245. [CrossRef]
190. Williams, F.M.; Cherkas, L.F.; Spector, T.D.; MacGregor, A.J. A common genetic factor underlies hypertension and other cardiovascular disorders. *BMC Cardiovasc. Disord.* **2004**, *4*, 20. [CrossRef]
191. Li, J.; Li, C.; Zhang, D.; Shi, D.; Qi, M.; Feng, J.; Yuan, T.; Xu, X.; Liang, D.; Xu, L.; et al. SNX13 reduction mediates heart failure through degradative sorting of apoptosis repressor with caspase recruitment domain. *Nat. Commun.* **2014**, *5*, 5177. [CrossRef] [PubMed]
192. Valera Mora, M.E.; Scarfone, A.; Calvani, M.; Greco, A.V.; Mingrone, G. Insulin clearance in obesity. *J. Am. Coll Nutr.* **2003**, *22*, 487–493. [CrossRef] [PubMed]
193. Li, F.; Yang, J.; Villar, V.A.M.; Asico, L.D.; Ma, X.; Armando, I.; Sanada, H.; Yoneda, M.; Felder, R.A.; Jose, P.A.; et al. Loss of renal SNX5 results in impaired IDE activity and insulin resistance in mice. *Diabetologia* **2018**, *61*, 727–737. [CrossRef] [PubMed]
194. Harashima, S.; Horiuchi, T.; Wang, Y.; Notkins, A.L.; Seino, Y.; Inagaki, N. Sorting nexin 19 regulates the number of dense core vesicles in pancreatic beta-cells. *J. Diabetes Investig.* **2012**, *3*, 52–61. [CrossRef]
195. Zhang, H.; Huang, T.; Hong, Y.; Yang, W.; Zhang, X.; Luo, H.; Xu, H.; Wang, X. The Retromer Complex and Sorting Nexins in Neurodegenerative Diseases. *Front. Aging Neurosci.* **2018**, *10*, 79. [CrossRef]
196. Parente, D.J.; Morris, S.M.; McKinstry, R.C.; Brandt, T.; Gabau, E.; Ruiz, A.; Shinawi, M. Sorting nexin 27 (SNX27) variants associated with seizures, developmental delay, behavioral disturbance, and subcortical brain abnormalities. *Clin. Genet.* **2020**, *97*, 437–446. [CrossRef]
197. Zhang, J.; Liu, J.; Norris, A.; Grant, B.D.; Wang, X. A novel requirement for ubiquitin-conjugating enzyme UBC-13 in retrograde recycling of MIG-14/Wntless and Wnt signaling. *Mol. Biol. Cell* **2018**, *29*, 2098–2112. [CrossRef]
198. Schobel, S.; Neumann, S.; Hertweck, M.; Dislich, B.; Kuhn, P.H.; Kremmer, E.; Seed, B.; Baumeister, R.; Haass, C.; Lichtenthaler, S.F. A novel sorting nexin modulates endocytic trafficking and alpha-secretase cleavage of the amyloid precursor protein. *J. Biol. Chem.* **2008**, *283*, 14257–14268. [CrossRef]

199. Lee, J.; Kim, C.H.; Simon, D.K.; Aminova, L.R.; Andreyev, A.Y.; Kushnareva, Y.E.; Murphy, A.N.; Lonze, B.E.; Kim, K.S.; Ginty, D.D.; et al. Mitochondrial cyclic AMP response element-binding protein (CREB) mediates mitochondrial gene expression and neuronal survival. *J. Biol. Chem.* **2005**, *280*, 40398–40401. [CrossRef]
200. Kim, N.Y.; Cho, M.H.; Won, S.H.; Kang, H.J.; Yoon, S.Y.; Kim, D.H. Sorting nexin-4 regulates beta-amyloid production by modulating beta-site-activating cleavage enzyme-1. *Alzheimers Res.* **2017**, *9*, 4. [CrossRef]
201. Zhao, C.; Deng, W.; Gage, F.H. Mechanisms and functional implications of adult neurogenesis. *Cell* **2008**, *132*, 645–660. [CrossRef] [PubMed]
202. Wang, X.; Huang, T.; Zhao, Y.; Zheng, Q.; Thompson, R.C.; Bu, G.; Zhang, Y.W.; Hong, W.; Xu, H. Sorting nexin 27 regulates Abeta production through modulating gamma-secretase activity. *Cell Rep.* **2014**, *9*, 1023–1033. [CrossRef] [PubMed]
203. Li, X.; Huang, L.; Lan, J.; Feng, X.; Li, P.; Wu, L.; Peng, Y. Molecular mechanisms of mitophagy and its roles in neurodegenerative diseases. *Pharm. Res.* **2020**. [CrossRef] [PubMed]
204. Wang, Y.; Liu, N.; Lu, B. Mechanisms and roles of mitophagy in neurodegenerative diseases. *CNS Neurosci.* **2019**, *25*, 859–875. [CrossRef] [PubMed]
205. McGough, I.J.; Steinberg, F.; Jia, D.; Barbuti, P.A.; McMillan, K.J.; Heesom, K.J.; Whone, A.L.; Caldwell, M.A.; Billadeau, D.D.; Rosen, M.K.; et al. Retromer binding to FAM21 and the WASH complex is perturbed by the Parkinson disease-linked VPS35(D620N) mutation. *Curr. Biol.* **2014**, *24*, 1670–1676. [CrossRef] [PubMed]
206. Zavodszky, E.; Seaman, M.N.; Rubinsztein, D.C. VPS35 Parkinson mutation impairs autophagy via WASH. *Cell Cycle* **2014**, *13*, 2155–2156. [CrossRef]
207. Kandimalla, R.; Reddy, P.H. Multiple faces of dynamin-related protein 1 and its role in Alzheimer's disease pathogenesis. *Biochim. Biophys. Acta* **2016**, *1862*, 814–828. [CrossRef]
208. Wang, W.; Wang, X.; Fujioka, H.; Hoppel, C.; Whone, A.L.; Caldwell, M.A.; Cullen, P.J.; Liu, J.; Zhu, X. Parkinson's disease-associated mutant VPS35 causes mitochondrial dysfunction by recycling DLP1 complexes. *Nat. Med.* **2016**, *22*, 54–63. [CrossRef]
209. Kodavati, M.; Wang, H.; Hegde, M.L. Altered Mitochondrial Dynamics in Motor Neuron Disease: An Emerging Perspective. *Cells* **2020**, *9*, 65. [CrossRef]
210. Matheoud, D.; Sugiura, A.; Bellemare-Pelletier, A.; Laplante, A.; Rondeau, C.; Chemali, M.; Fazel, A.; Bergeron, J.J.; Trudeau, L.E.; Burelle, Y.; et al. Parkinson's Disease-Related Proteins PINK1 and Parkin Repress Mitochondrial Antigen Presentation. *Cell* **2016**, *166*, 314–327. [CrossRef]
211. Meraviglia, V.; Ulivi, A.F.; Boccazzi, M.; Valenza, F.; Fratangeli, A.; Passafaro, M.; Lecca, D.; Stagni, F.; Giacomini, A.; Bartesaghi, R.; et al. SNX27, a protein involved in down syndrome, regulates GPR17 trafficking and oligodendrocyte differentiation. *Glia* **2016**, *64*, 1437–1460. [CrossRef] [PubMed]
212. Mandell, M.A.; Saha, B.; Thompson, T.A. The Tripartite Nexus: Autophagy, Cancer, and Tripartite Motif-Containing Protein Family Members. *Front. Pharm.* **2020**, *11*, 308. [CrossRef] [PubMed]
213. Yu, H.; Jove, R. The STATs of cancer–new molecular targets come of age. *Nat. Rev. Cancer* **2004**, *4*, 97–105. [CrossRef] [PubMed]
214. Bild, A.H.; Turkson, J.; Jove, R. Cytoplasmic transport of Stat3 by receptor-mediated endocytosis. *EMBO J.* **2002**, *21*, 3255–3263. [CrossRef]
215. Wang, Y.; van Boxel-Dezaire, A.H.; Cheon, H.; Yang, J.; Stark, G.R. STAT3 activation in response to IL-6 is prolonged by the binding of IL-6 receptor to EGF receptor. *Proc. Natl. Acad. Sci. USA* **2013**, *110*, 16975–16980. [CrossRef]
216. Palfy, M.; Remenyi, A.; Korcsmaros, T. Endosomal crosstalk: Meeting points for signaling pathways. *Trends Cell Biol.* **2012**, *22*, 447–456. [CrossRef]
217. Dobrowolski, R.; De Robertis, E.M. Endocytic control of growth factor signalling: Multivesicular bodies as signalling organelles. *Nat. Rev. Mol. Cell Biol.* **2011**, *13*, 53–60. [CrossRef]
218. Osman, I.; Bajorin, D.F.; Sun, T.T.; Zhong, H.; Douglas, D.; Scattergood, J.; Zheng, R.; Han, M.; Marshall, K.W.; Liew, C.C. Novel blood biomarkers of human urinary bladder cancer. *Clin. Cancer Res.* **2006**, *12*, 3374–3380. [CrossRef]
219. Pharoah, P.D.; Tsai, Y.Y.; Ramus, S.J.; Phelan, C.M.; Goode, E.L.; Lawrenson, K.; Buckley, M.; Fridley, B.L.; Tyrer, J.P.; Shen, H.; et al. GWAS meta-analysis and replication identifies three new susceptibility loci for ovarian cancer. *Nat. Genet.* **2013**, *45*, 362–370. [CrossRef]
220. Zhang, S.; Hu, B.; You, Y.; Yang, Z.; Liu, L.; Tang, H.; Bao, W.; Guan, Y.; Shen, X. Sorting nexin 10 acts as a tumor suppressor in tumorigenesis and progression of colorectal cancer through regulating chaperone mediated autophagy degradation of p21(Cip1/WAF1). *Cancer Lett.* **2018**, *419*, 116–127. [CrossRef]
221. Alfaro, I.E.; Albornoz, A.; Molina, A.; Moreno, J.; Cordero, K.; Criollo, A.; Budini, M. Chaperone Mediated Autophagy in the Crosstalk of Neurodegenerative Diseases and Metabolic Disorders. *Front. Endocrinol. (Lausanne)* **2018**, *9*, 778. [CrossRef] [PubMed]
222. Yin, P.; Hong, Z.; Yang, X.; Chung, R.T.; Zhang, L. A role for retromer in hepatitis C virus replication. *Cell. Mol. Life Sci.* **2016**, *73*, 869–881. [CrossRef] [PubMed]
223. Pim, D.; Broniarczyk, J.; Bergant, M.; Playford, M.P.; Banks, L. A Novel PDZ Domain Interaction Mediates the Binding between Human Papillomavirus 16 L2 and Sorting Nexin 27 and Modulates Virion Trafficking. *J. Virol.* **2015**, *89*, 10145–10155. [CrossRef] [PubMed]

224. Cardoso, R.S.; Tavares, L.A.; Jesus, B.L.S.; Criado, M.F.; de Carvalho, A.N.; Souza, J.P.; Bedi, S.; de Souza, M.M.; Silva, M.L.; Lanfredi, G.P.; et al. Host Retromer Protein Sorting Nexin 2 Interacts with Human Respiratory Syncytial Virus Structural Proteins and is Required for Efficient Viral Production. *mBio* **2020**, *11*. [CrossRef] [PubMed]
225. Elwell, C.; Engel, J. Emerging Role of Retromer in Modulating Pathogen Growth. *Trends Microbiol.* **2018**, *26*, 769–780. [CrossRef]
226. Elwell, C.A.; Czudnochowski, N.; von Dollen, J.; Johnson, J.R.; Nakagawa, R.; Mirrashidi, K.; Krogan, N.J.; Engel, J.N.; Rosenberg, O.S. Chlamydia interfere with an interaction between the mannose-6-phosphate receptor and sorting nexins to counteract host restriction. *Elife* **2017**, *6*. [CrossRef]

cells

MDPI

Review

Interplay between the Ubiquitin Proteasome System and Ubiquitin-Mediated Autophagy in Plants

Tong Su, Mingyue Yang, Pingping Wang, Yanxiu Zhao and Changle Ma *

Shandong Provincial Key Laboratory of Plant Stress, College of Life Sciences, Shandong Normal University, Jinan 250014, China; sut0229@sdnu.edu.cn (T.S.); yangmingyue1@stu.sdnu.edu.cn (M.Y.); pingping.wang@sdnu.edu.cn (P.W.); zhaoyx@sdnu.edu.cn (Y.Z.)
* Correspondence: machangle@sdnu.edu.cn; Tel.: +86-0531-86180792

Received: 21 August 2020; Accepted: 25 September 2020; Published: 1 October 2020

check for updates

Abstract: All eukaryotes rely on the ubiquitin-proteasome system (UPS) and autophagy to control the abundance of key regulatory proteins and maintain a healthy intracellular environment. In the UPS, damaged or superfluous proteins are ubiquitinated and degraded in the proteasome, mediated by three types of ubiquitin enzymes: E1s (ubiquitin activating enzymes), E2s (ubiquitin conjugating enzymes), and E3s (ubiquitin protein ligases). Conversely, in autophagy, a vesicular autophagosome is formed that transfers damaged proteins and organelles to the vacuole, mediated by a series of ATGs (autophagy related genes). Despite the use of two completely different componential systems, the UPS and autophagy are closely interconnected and mutually regulated. During autophagy, ATG8 proteins, which are autophagosome markers, decorate the autophagosome membrane similarly to ubiquitination of damaged proteins. Ubiquitin is also involved in many selective autophagy processes and is thus a common factor of the UPS and autophagy. Additionally, the components of the UPS, such as the 26S proteasome, can be degraded via autophagy, and conversely, ATGs can be degraded by the UPS, indicating cross regulation between the two pathways. The UPS and autophagy cooperate and jointly regulate homeostasis of cellular components during plant development and stress response.

Keywords: autophagy; degradation; the ubiquitin-proteasome system; ubiquitin; plants

1. Introduction

Over the course of plant development from germination to senescence, plant morphology and metabolism change dramatically, resulting in accumulation of many obsolete proteins and cellular structures that need to be degraded. Moreover, as sessile organisms, plants must adapt to constantly changing environmental conditions [1].

In all eukaryotes, the ubiquitin-proteasome system (UPS) and autophagy are the two major protein quality control pathways utilized to degrade misfolded or redundant proteins to maintain the homeostasis of the plant proteome [2]. The UPS and autophagy pathways also play crucial roles in regulating a wide range of cytological and physiological processes by selectively removing regulatory proteins after they are no longer needed, such as those involved in plant development and stress response [3]. In the UPS, proteins to be degraded are labeled with ubiquitin molecules. This ubiquitin conjugation process requires three types of ubiquitin enzymes: E1 (ubiquitin activating enzyme), E2 (ubiquitin conjugating enzyme), and E3 (ubiquitin protein ligase) [4]. Subsequently, the poly-ubiquitinated proteins are degraded in the proteasomes, which are composed of a variety of subunits and regulatory proteins [5]. In eukaryotes, autophagy, mediated by a set of core autophagy-related (ATG) proteins, is a membrane trafficking system used to degrade protein aggregates and cytoplasmic components in the vacuole (yeast and plant) or lysosome (mammals) [6,7]. Recycling of nutrients through autophagy is essential for plant survival under starvation conditions [8]. Although both the UPS and autophagy pathways play important roles in plant growth, development, and stress response, they were long viewed as independent, parallel systems with no point of intersection. Recently, growing evidence has demonstrated that these two degradation pathways are interconnected. In this manuscript, we review the recent advances regarding the roles of the UPS and autophagy in stress response and their interconnectedness.

2. Ubiquitin Proteasome System

The UPS consists of two successive processes: (1) the attachment of a chain of ubiquitin molecules to the targeted protein and (2) degradation of the targeted protein by the 26S proteasome with recycling of the ubiquitin molecules [9]. During the first step, the enzymatic cascade consisting of E1, E2, and E3 is essential for ubiquitin attachment [3].

2.1. Ubiquitin-Proteasome System

Ubiquitin was discovered in the early 1980′s as a reusable recognition signal for selective protein degradation and is a highly conserved protein found in all eukaryotic organisms and tissues [10,11]. Ubiquitin is encoded by multiple genes as a precursor protein and undergoes proteolytic processing to produce mature ubiquitin consisting of 76 amino acids and a diglycine (Gly75-Gly76) sequence in the carboxyl-terminus [3].

Ubiquitination of proteins is a multistep process in all eukaryotes. First, the C-terminal carboxyl group of ubiquitin is connected to the conserved active-site cysteine (Cys) residue on the E1, forming an E1-ubiquitin thioester in an energy-dependent manner. Then, ubiquitin is transferred from E1 to an active cysteine residue on the E2, forming an E2-ubiquitin thioester-linked intermediate. Finally, the ubiquitin is transferred from E2 to a lysine (Lys) residue on the substrate protein by catalysis of E3 (monoubiquitination) [12]. Generally, monoubiquitination has been considered to be non-proteolytic and plays roles in the regulation of interactions and activation of substrates. However, there is evidence in yeast that monoubiquitination can also lead to the degradation of some substrates and monoubiquitination-dependent degradation is more widespread than previously assumed [13]. In addition, in plants, a single ubiquitin molecule can be attached to multiple Lys residues (multimonoubiqutination) or generate a polyubiquitin chain on a single Lys (polyubiquitination) on the target protein [14,15]. It has been shown that ubiquitinated Lys residues tend to reside within intrinsically disordered regions of the substrate proteins in plants [14]. There are seven conserved Lys residues on ubiquitin which are used to form an isopeptide bond with Gly76 of another ubiquitin molecule. In *Arabidopsis*, Lys48 is used the most frequently to form polyubiquitin chains [14,15]. Ubiquitin-modified proteins can be deubiquitinated by deubiquitinating enzymes (DUBs), which hydrolyze the isopeptide bond between ubiquitin and the substrate Lys residues [16]. Therefore, ubiquitination is reversible, making it a controllable process.

2.2. Enzymes of the UPS

In the *Arabidopsis* genome, there are two E1 encoding genes, 37 E2 encoding genes, and more than 1400 genes encoding E3 ubiquitin ligases or the components of the E3 complex [3,17,18]. E1s are encoded by *Ubiquitin Activating 1* (*UBA1*) and *UBA2*, which share similar sequences and expression patterns [17]. However, their function may not be completely redundant, as evidenced by the finding that mutation of *Arabidopsis UBA1*, but not *UBA2*, results in defects in innate immunity [19].

E2 is characterized by the UBC (ubiquitin conjugating) domain containing an active cysteine to form thioester linkages with E1 and E3 [20]. The 37 ubiquitin E2s, AtUBC1 to AtUBC37, are classified into 14 subgroups according to the amino acid sequences of the UBC domain. The largest subgroup, V1, contains 8 members, in which UBC8, UBC10, and UBC11 are the most active [18]. The *Arabidopsis* genome also encodes another 11 UBC-domain containing proteins. Some of these proteins are ubiquitin-conjugating enzyme variants (UEVs) that can form complexes with ubiquitin E2s to promote substrate ubiquitination [18,21]. For example, the UEV COP10 can enhance the activities of UBC8 and UBC9 to promote the formation of Lys48- and Lys63-linked ubiquitin chains [21].

The substrate specificity of ubiquitination is primarily determined by E3; thus, the E3 ubiquitin ligases are diverse. E3s are classified into three types based on the domain they use to interact with the E2-ubiquitin intermediate: Really Interesting New Gene (RING) proteins, Homology to E6-Associated Carboxyl-Terminus (HECT) proteins, and U-box domain proteins [9]. The *Arabidopsis* genome encodes about 490 RING proteins, making it the largest E3 subfamily [22]. The E3s work in two ways: by facilitating the direct transfer of ubiquitin from E2 to the substrate such as for the RING and U-box E3s, or by accepting ubiquitin from E2, forming a thioester-linked E3-ubiquitin intermediate, and then transferring the ubiquitin to the substrate such as for the HECT E3s [9]. The E3 ubiquitin ligases can interact with themselves or with other E3s to form homomeric or heteromeric multimeric E3s [9].

3. The Role of the UPS in Plant Stress Response

The UPS is a key regulator of plant proteome homeostasis. In response to environmental stress, the UPS can regulate gene expression and cellular responses via ubiquitination of regulatory proteins such as transcription factors involved in environmental stress responses [23]. Many genes encoding components involved in the UPS are induced by abiotic stress [3].

The role of various E3s in regulating stress signaling is largely dependent on the function of its substrate and is influenced by its ability to bind to the substrate proteins [9]. For example, interaction of an E3 and its substrate may only occur under normal conditions or conversely only during abiotic stresses, leading to different physiological outcomes. Dehydration-responsive element-binding protein 2A (DREB2A) is a key transcription factor that regulates the expression of numerous genes which are responsive to heat, drought, and salt stresses in plants [24–26]. Under non-stress conditions, DREB2A is ubiquitinated by RING E3s, DRIP1 (DREB2A-INTERACTING PROTEIN1), and DRIP2, and subsequently degraded [25]. Whereas under stress conditions, an A20/AN1-type zinc finger protein, AtSAP5 (Stress Associated Protein 5), functions as a ubiquitin ligase that mediates the ubiquitination and degradation of DRIP1/2 to release and stabilize DREB2A and subsequently induce the expression of downstream responsive genes [27]. Additionally, INDUCER OF CBF EXPRESSION1 (ICE1) can positively regulate the expression of *C-REPEAT BINDING FACTOR/DRE BINDING FACTOR1* (*CBF/DREB1*), which encodes a transcription factor that activates a series of Cold-Regulated (*COR*) genes in response to cold stress [28,29]. The stability of ICE1 is crucial for its function in cold stress and is regulated through the UPS. In *Arabidopsis*, a RING protein HIGH EXPRESSION OF OSMOTICALLY RESPONSIVE GENES1 (HOS1) acts as an E3 ligase for the ubiquitination of ICE1, thus negatively regulating cold response. The interaction between ICE1 and HOS1 is blocked by the phosphorylation of ICE1, which is catalyzed by a member of the SNF1-related protein kinase family, OPEN STOMATA1 (OST1), thus, enhancing the stability of ICE1 and inducing cold tolerance [30].

The UPS can also play important roles is regulating hormone pathways in plants. The ABA (abscisic acid) signal pathway includes multiple receptors PYR1 (pyrabactin resistance 1)/PYL (PYR1-like)/RCAR (regulatory components of ABA receptors), protein phosphatase type 2Cs (PP2Cs), SnRKs (Sucrose non-fermenting related kinase), and transcription factors ABI5 (ABA INSENSITIVE 5) and ABF (ABRE-binding factor). The PP2Cs, SnRKs, and transcription factors are activated successively by the binding of ABA to the receptors under abiotic stress conditions, including drought, salt, and cold stresses [31]. Under normal conditions, the UPS plays a crucial role in inhibiting the ABA signal pathway by degrading some of its components. Deetiolated 1 (DET1) and DDB1-associated protein 1 (DDA1) and a F-box protein RIFP1 (RCAR3 INTERACTING F-BOX PROTEIN 1) act as the substrate receptor and recruiting component for the RING E3 Clu4 to mediate the ubiquitination and degradation of PYL8, PYL4, and PYL9 [32,33]. PYL9 can also interact with PUB (Plant U-Box) E3s, PUB22 and PUB23, and is ubiquitinated for degradation [34]. The members of the SnRK superfamily, SnRK1, SnRK2 and SnRK3, play crucial roles in activating ABA responsive transcription factors. Under normal conditions, in addition to being inactivated by phosphatase PP2C, SnRKs can interact with pleiotropic regulatory locus 1 (PRL1), which may function as the substrate-binding component of a Cul4 (CULLIN 4)-based RING E3, to be ubiquitinated for degradation [35]. The ubiquitination and proteasomal degradation of SnRK3.26 can also be promoted by interaction with RING E3 Keep on Going (KEG) [36]. Some transcription factors involved in ABA signaling pathway, such as ABI5, ABF1, and ABF3, can be ubiquitinated via a RING E3, KEG to be proteasomally degraded [37–39]. The ABA signaling pathway is activated by the binding of ABA to the PYR1/PLY/RCAR receptors, which interact with and inhibit the phosphatase PP2C [40]. PP2C activity is also inhibited by the RING E3s RGLG1 (RING DOMAIN LIGASE 1) and RGLG5 via the UPS [41]. Meanwhile, the proteasome-dependent degradation of KEG requires its own ubiquitin ligase activity and subsequently leads to the accumulation of SnRK3.26. The self-ubiquitination of KEG is phosphorylation-dependent, which is catalyzed by CBL-Interacting Protein Kinase 26 [42]. Thus, in plants, the UPS is involved throughout the ABA signaling pathway to ensure the correct timing of the response (Figure 1).

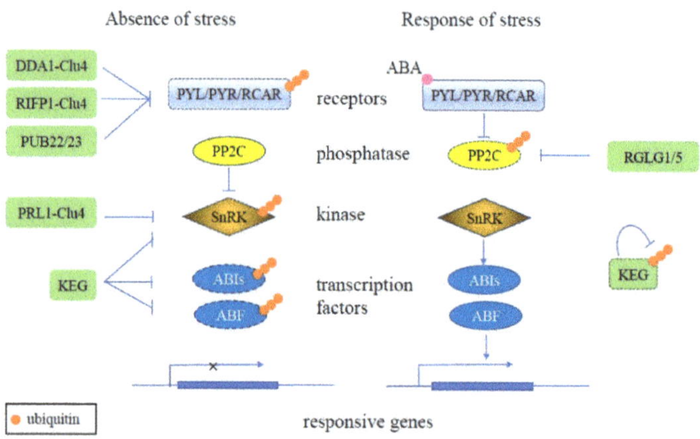

Figure 1. Schematic of ubiquitination regulating the plant ABA signaling pathway. Under normal conditions (left), the redundant ABA receptors PYL, PYR, and RCAR are ubiquitinated (orange circles) and degraded (dashed outline) by E3 ligases, such as DDA1-Clu4, RIFP1-Clu4, and PUB22/23. Kinase SnRK and transcription factors ABI5 and ABF also are degraded via the UPS, mediated by RING E3 KEG. Together, these result in the ABA pathway being inactivated in the absence of stress. During stress conditions (right), the ABA receptors are activated by the binding of ABA, whereby they then interact with PP2C to inhibit its activity. PP2C activity is also inhibited by the RING E3s RGLG1 and RGLG5 and degraded via the UPS. KEG also undergoes proteasome-dependent degradation dependent on its own ligase activity, resulting in stabilization of SnRK, ABI5, and ABF and activation of the ABA signaling pathway throughinduction of responsive gene expression.

Ethylene signaling is important for regulating abiotic stress responses and numerous plant developmental processes, including germination, soil emergence, fruit ripening, and senescence [43]. The expression of ethylene signaling responsive genes are largely regulated by the transcription factors ETHYLENE INSENSITIVE3 (EIN3) and EIN3-Like (EIL) [44]. The stability of EIN3 and EIL1 are regulated by the coaction of ubiquitin-ligases EIN3 BINDING F-BOX1 (EBF1) and EBF2 [45]. After seed germination, EIN3/EIL1 are stabilized by both light and ethylene signaling to promote the underground growth of seedlings. An E3 ubiquitin ligase CONSTITUTIVE PHOTOMORPHOGENIC 1 (COP1) functions as a central repressor of light signaling to mediate the ubiquitination and degradation of EBF1/2 [46]. As seedlings grow toward the soil surface, COP1 activity, which is negatively regulated by photoreceptors, gradually decreases, resulting in attenuation of the ethylene response [47]. Moreover, another key regulator of the ethylene signaling pathway, ETHYLENE-INSENSITIVE2 (EIN2), can also be regulated via ubiquitination and degradation. The ethylene response can be repressed by ubiquitination of the integral membrane protein EIN2 by EIN2 TARGETING PROTEIN1 (ETP1) and ETP2. However, ETP abundance is negatively regulated by ethylene, contributing to activation of the ethylene response in the presence of ethylene [48] (Figure 2).

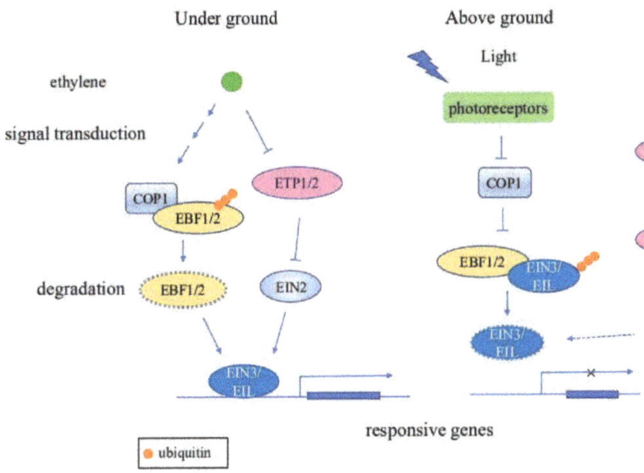

Figure 2. Schematic diagram of ubiquitination regulating ethylene signaling pathways. When seedlings grow underground, ethylene signaling promotes the ubiquitination of EBF1/2 by the E3 ligase COP1. In the absence of EBF1/2, EIN3/EIL is stabilized, resulting in induction of expression of ethylene responsive genes. Ethylene reduces the protein levels of ETP1/2, thus stabilizing EIN2, which in turn promotes the expression of ethylene responsive genes. As seedlings grow toward the soil surface, regulation by photoreceptors causes COP1 activity to decrease gradually, resulting in ubiquitination and subsequent degradation of EIN3/EIL by EBF1/2, causing termination of the ethylene response. Moreover, EIN2 is ubiquitinated and degraded by ETP1/2 to repress the ethylene response.

In addition, the plant auxin signaling pathway is also regulated by the UPS. In the presence of auxin, the transcriptional repressors AUXIN/INDOLE-3-ACETIC ACID (AUX/IAA) are targeted for degradation by SKP1-CULLIN1-F-BOX (SCF) ubiquitin-protein ligases to turn on the auxin signaling pathway [49]. However, further study is needed to uncover the intricacies of the relationship between ubiquitination and auxin signaling.

4. Ubiquitin-Mediated Regulation of Autophagy in Plants

Found in all eukaryotic cells, autophagy is a pathway to degrade cytoplasmic content. This is achieved by forming specialized autophagic vesicles which are delivered to the vacuole for degradation [50]. There are three types of autophagy which have been described in plants: microautophagy, macroautophagy, and mega-autophagy [51]. During microautophagy, cellular contents congregate at the vacuole surface and are trapped by invagination of the vacuole for degradation. Mega-autophagy is an extreme form of autophagy in which the vacuole ruptures to release hydrolases into the cytoplasm, ultimately resulting in programmed cell death [8,51]. The mechanism of macroautophagy, or autophagy for short hereafter, is well characterized in yeast and occurs in several distinct steps, requiring a series of ATGs [8]: (1) The dephosphorylation of ATG13 caused by the inactivation of TOR (Target of Rapamycin) and the hyperphosphorylation of ATG1 catalyzed by other kinases lead to the formation of the ATG1-ATG13-ATG11-ATG101 complex to initiate autophagy; (2) Activated ATG1 kinase promotes ATG9-mediated delivery of lipids to the developing phagophore; (3) ATG8s, a marker of autophagy, are attached to the autophagosome membrane by the ATG5-ATG12-ATG16 E3 ligase complex via the ubiquitin-like pathway; and (4) the autophagosome is delivered to the vacuole for degradation. The orthologs of most yeast ATG proteins have been identified in plants, indicating that the mechanism of autophagy may be well conserved in plants and yeast.

The UPS and autophagy, the two major cellular degradative pathways controlling the stability of the cellular proteome, employ separate molecular machinery. The UPS specifically degrades substrates via the E1-E2-E3 system which conjugates ubiquitin on to the substrates, whereas autophagy is a vesicular trafficking pathway that transfers damaged or redundant proteins and organelles to the vacuole or lysosome [52–54]. Because of this, the UPS and autophagy pathways have long been thought of as parallel processes with no intersection. However, it now appears that autophagy and the UPS are interconnected and may mutually affect each other (Figure 3).

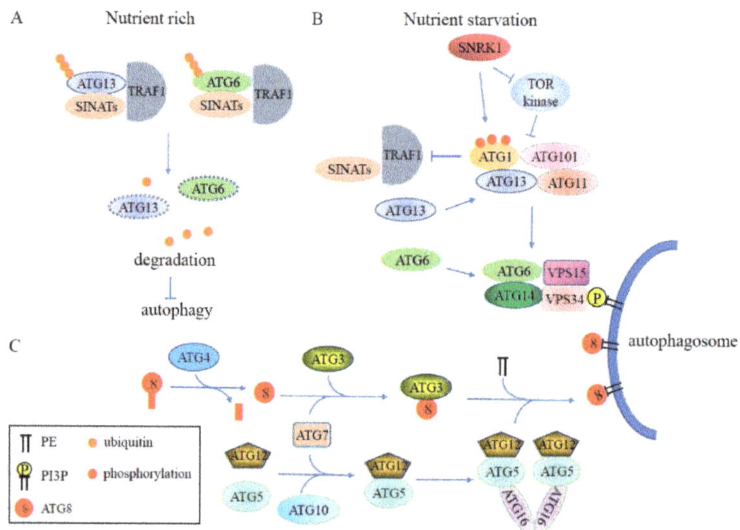

Figure 3. Schematic diagram of ubiquitination involved in the regulation of plant autophagy. (**A**) Under nutrient rich conditions, ATG13 and ATG6 are ubiquitinated by RING-type E3 ligases SINATs with the assistance of TRAF1 and then degraded via the 26S proteasome, leading to inhibition of autophagy. (**B**) During nutrient starvation, TRAF1 stability is impaired by activated ATG1, resulting in accumulation of ATG13 and ATG6 and formation of autophagosomes. (**C**) ATG8 is labeled into the autophagosome membrane in a similar way as the ubiquitination pathway. The ubiquitin-like protein ATG8 is cleaved by Cys protease ATG4 to expose a conserved C-terminal glycine, then ATG8 is transferred to the E2-like enzyme ATG3 through E1-like ATG7, and ultimately, labeled into the autophagosome membrane as ATG8-PE by E3 ligase complex ATG5-ATG12-ATG16. The decoration of the phagophore with PI3P is catalyzed by a complex containing ATG6/ATG14/VPS15/VPS34. PE: phosphatidylethanolamine; PI3P: phosphatidylinositol-3-phosphate.

4.1. Regulation of Autophagy by Ubiquitination

In plants, the assembly of the ATG1-ATG13 complex promotes autophagy initiation. The activity of the ATG1-ATG13 complex is negatively regulated by TOR kinase in response to nutrient conditions. Under starvation conditions, the levels of phosphorylated ATG1a and ATG13a drop dramatically, but this turnover is abolished when the ATG system is inhibited, indicating that the ATG1-ATG13 complex is degraded by autophagy [55]. Recent studies revealed that the stability of the ATG1-ATG13 complex is modulated by the ubiquitination pathway [56]. In *Arabidopsis* wild-type plants, ATG1a and ATG13a accumulate under carbon or nitrogen starvation, but are significantly reduced under prolonged starvation treatment. Degradation of ATG1a and ATG13a is repressed by treatment with the proteasome inhibitor MG132, indicating that the 26S proteasome modulates the stability of ATG1a and ATG13a. Further studies have shown that the ubiquitination of ATG13a is enhanced by overexpression of the RING-type E3 ligase SEVEN IN ABSENTIA OF ARABIDOPSIS THALIANA1 (SINAT1) in wild-type plants. However, this enhanced ubiquitination is impaired in the *traf1a/b* (tumor necrosis factor receptor-associated factor1a/b) mutant, demonstrating that TRAF1a and TRAF1b act as adaptors to mediate the ubiquitylation and degradation of ATG13a by SINAT1. Although ATG1 protein levels are also significantly reduced during prolonged starvation treatment, the mechanism of ATG1 degradation is still elusive. It was shown that the stability of TRAF1 can be regulated by ATG1, indicating feedback regulation between the ATG1-ATG13 kinase complex and TRAF1 proteins [56,57].

In addition, in *Arabidopsis*, TRAF and SINAT proteins are also involved in regulating ATG6 stability, which subsequently mediates autophagosome formation [57,58]. TRAF1a and TRAF1b can interact directly with ATG6. Loss of *TRAF1a* and *TRAF1b* causes stabilization of ATG6 protein levels by reducing its ubiquitination level. It was also found that SINAT1 and SINAT2 can interact with and ubiquitinate ATG6. In the *traf1a/b* mutant, SINAT1- and SINAT2-induced degradation of ATG6 is impaired, indicating that TRAF1a and TRAF1b function as adaptors required for SINAT1/SINAT2-mediated ubiquitination and degradation of ATG6 [58].

SINATs (SINAT1, SINAT2, SINAT3, and SINAT4) are primarily localized to the endosomal and autophagic vesicles and are involved in the ubiquitination and proteasomal degradation of FYVE DOMAIN PROTEIN REQUIRED FOR ENDOSOMAL SORTING 1 (FREE1) and VACUOLAR PROTEIN SORTING 23A (VPS23A). FREE1 and VPS23A are key components of the plant specific ESCRT (endosomal sorting complex required for transport), which is involved in endosomal sorting and autophagic degradation [59,60]. Furthermore, the SINATs can be co-degraded with FREE1 and VPS23A via the vacuolar pathway and expression of SINATs promotes increased sensitivity to ABA and induction of ABA-responsive gene expression. Together, these findings indicate that SINATs-mediated degradation of FREE1 and VPS23A may be involved in the regulation of ABA signaling [59]. In addition, the protein levels of SINAT1-4 are induced by iron deficiency, promoting the ubiquitination and degradation of FREE1 and subsequently relieving the repression of FREE1 on iron absorption [60]. These studies suggest that SINAT-mediated ubiquitination plays important roles in autophagy and participates in the regulation of important physiological processes in plants.

4.2. Ubiquitin-Like Systems in Autophagy

The ubiquitin-like conjugation system is comprised of E1-like ATG7, E2-like ATG3, and the E3-like ATG5-ATG12-ATG16 complex, and is conserved in all eukaryotes [8,61,62]. It is involved in the formation of autophagic vesicles via mediation of ATG8 lipidation. This process starts with the cleavage of the ubiquitin-like protein ATG8 by the Cys protease ATG4 to expose a conserved C-terminal glycine, which is essential for the subsequent conjugation reaction [61,63]. Subsequently, ATG8 is activated at this glycine by the ATP-dependent activating enzyme, E1-like ATG7, to form a thioester bond with the Cys residue of ATG7 [64]. Then, ATG8 is transferred by the E2-like enzyme ATG3 to its Cys residue [62]. Finally, ATG8 is covalently conjugated to the lipid molecule phosphatidylethanolamine (PE) and attached to the phagophore membrane, mediated by the E3 ligase complex ATG5-ATG12-ATG16, via an amide bond between the C-terminal Gly residue of ATG8 and the amino group of PE [8,65].

To form the ATG5-ATG12-ATG16 E3 ligase complex, the Ub-fold protein ATG12 is first adenylated at its C-terminal glycine by E1-like activating enzyme ATG7 to bind via a thioester bond to the active-site Cys in ATG7 [64,66]. Then, the activated ATG12 is transferred to the active-site Cys of E2 conjugating enzyme ATG10 by transesterification and subsequently conjugated to a conserved lysine residue of ATG5 via a peptide bond [67]. The ATG12-ATG5 conjugate then assembles with the dimeric scaffold protein ATG16 to function as an E3-like ligase in ATG8 lipidation [68,69]. ATG8s decorate the autophagosome membrane and serve as an anchor for autophagy receptors bringing cargo to the autophagosome, similar to how the UPS labels substrates to be degraded.

4.3. Ubiquitination in Selective Autophagy

Many studies have demonstrated that cellular substrates and organelles—including chloroplasts, peroxisomes, mitochondria, the proteasome, nucleus, protein aggregates, and pathogens—can be selectively degraded via autophagy [8]. Some selective autophagy receptors have been found to be ubiquitinated during the course of degrading their respective selective substrates or organelles [8], indicating that the UPS plays an important role in selective autophagy (Figure 4).

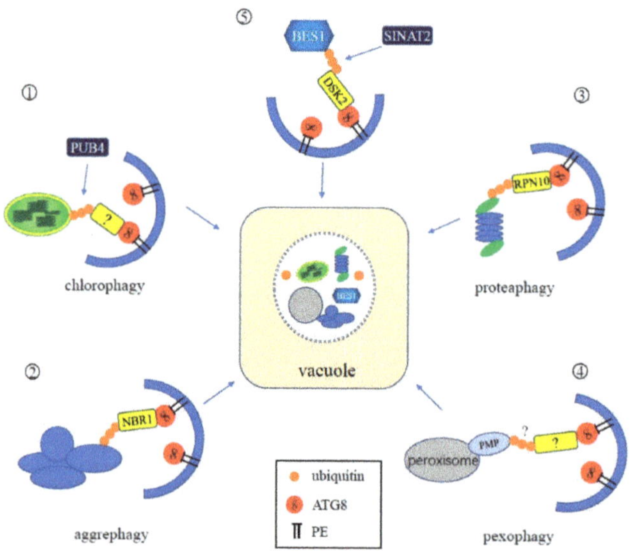

Figure 4. Schematic representation of the interplay between ubiquitination and selective autophagy. ①: In chlorophagy, the surface of an aberrant chloroplast is polyubiquitinated by cytoplasmic-localized E3 ligase Plant U-Box 4 (PUB4). An unknown adapter mediates the binding of ubiquitinated chloroplasts and ATG8 for degradation. ②: NBR1 is the adaptor of aggrephagy in plants. It can simultaneously interact with ubiquitin attached to the aggregates and ATG8. ③: The proteasome itself can be ubiquitinated and degraded by autophagy, during which RPN10 functions as a ubiquitin receptor of proteaphagy in plants. ④: In plants, an unknown pexophagy receptor likely interacts with ubiquitinated peroxisomal membrane proteins, mediating the degradation of peroxisomes. ⑤: In *Arabidopsis*, BES1 is ubiquitinated by E3 ligase SINAT2 and is subsequently recognized by selective autophagy receptor protein DSK2 for degradation through autophagy.

4.3.1. Chlorophagy

Chloroplasts are specialized plastid organelles found exclusively in plants and play an important role in photosynthesis and other metabolic functions necessary for plant growth and development [70]. Damaged chloroplasts are degraded via selective autophagy, termed chlorophagy, for quality control and remobilization of nitrogen and fixed carbon [71,72]. Depending on the status of the chloroplasts and the nutritional needs of the plant, chlorophagy can occur through several different mechanisms, including packaging of whole chloroplasts, budding of stromal proteins into autophagic vesicles called Rubisco-containing bodies (RCBs), or via ATI1 (ATG8-INTERACTING PROTEIN1)-decorated plastid bodies [8].

Ubiquitination is involved in the degradation of whole chloroplasts, in which entire chloroplasts are encapsulated into ATG8-decorated autophagic vesicles [71]. This process is promoted by strong photo-oxidative damage of chloroplasts caused by UV-B and high-intensity visible light [71]. Recent studies reveal that photo-damaged chloroplasts are associated with excess reactive oxygen species (ROS). Damaged proteins on the envelope membranes of aberrant chloroplasts are polyubiquitinated and the ubiquitylated chloroplasts are recognized by unknown autophagic receptors to form autophagy vesicles and delivered to the vacuole for degradation [73]. It has been reported that a cytoplasmic-localized E3 ligase PUB4 (Plant U-Box 4) is involved in the ubiquitination of chloroplast membrane proteins, but the specific targets of PUB4 have not been identified, nor is it clear if they are damaged [73]. Interestingly, PUB4 is not necessary for the induction of chlorophagy since chlorophagy can be induced by high light and low temperature treatment in the *pub4* mutant [74]. However, the double *atg5 pub4* mutant results in accelerated leaf chlorosis during senescence and reduced seed production, suggesting that both autophagy and the ubiquitination activity of PUB4 contribute to protein degradation in senescing leaves. Thus, PUB4 and autophagy may function in two semi-redundant pathways to remove chloroplasts [74]. The TOC (translocon at the outer envelope of chloroplasts) complexes in the plastid outer membrane have been shown to be ubiquitinated by a RING-type ubiquitin E3 ligase, SP1 (suppressor of ppi1 locus1), during chloroplast development [75]. However, recent evidence suggest that SP1-mediated ubiquitination may be not involved in chlorophagy [75].

4.3.2. Aggrephagy

Cellular proteins must fold into their correct conformation to function properly. However, under various stresses, these proteins can become misfolded or damaged and become toxic to cell. The UPS is considered the primary degradation process to remove ubiquitinated misfolded and damaged proteins, but recent studies have uncovered that a form of selective autophagy called aggrephagy plays an important role in degrading ubiquitinated protein aggregates for quality control [8].

In mammals, the autophagy receptors p62 and NBR1 (NEIGHBOR OF BRCA 1) can bind ubiquitinated aggregates via their ubiquitin binding domains (UBA) and tether microtubule-associated protein 1 light chain 3 (LC3), the mammalian ATG8 homolog, through their LC3-interacting region (LIR) motifs to facilitate selective sequestration of ubiquitin condensates to the autophagosome for clearance [76,77]. Loss of function of *Arabidopsis* NBR1, a homolog of mammalian p62 and NBR1, causes increased sensitivity to abiotic stresses and accumulation of insoluble protein aggregates. *Arabidopsis* NBR1 binds ubiquitin through a C-terminal UBA-domain and interacts with *Arabidopsis* ATG8, via the conserved LIR motif, indicating that NBR1 functions as the selective autophagy cargo adaptor of the aggrephagy pathway in plants [78,79].

In *Arabidopsis*, a chaperone-associated E3 ubiquitin ligase, CHIP (carboxyl terminus of Hsc70- interacting protein), which mediates degradation of nonnative proteins by 26S proteasomes, participates in aggrephagy during stress response. While *chip* mutants are similarly sensitive to abiotic stresses as *nbr1* mutants, *chip nbr1* double mutants display increased sensitivity, suggesting cumulative roles for CHIP and NBR1 [80]. However, protein aggregates induced by stress are still ubiquitinated in *chip* mutants. After a relatively short period of heat stress, different types of protein aggregates accumulate in *nbr1* and *chip* single mutants [80]. For example, rubisco activase and catalases are preferentially accumulated in the *nbr1* mutant while light-harvesting complex proteins accumulate at high levels in the *chip* mutant. These findings indicate that CHIP and NBR1 mediate two distinct but complementary anti-proteotoxic pathways under stress conditions [80]. Additionally, NBR1 can also interact with the ABA signaling regulatory proteins ABI3, ABI4, and ABI5, indicating that the ubiquitinated ABIs may be degraded by autophagy [81].

4.3.3. Proteaphagy

In eukaryotes, the proteasome contains a core particle (CP) and a regulatory particle (RP) [82]. The 20S CP consists of four heteroheptameric rings and is responsible for the entrance of substrates and removal of degradation products. The 19S RP is a multifunctional complex that functions as the proteasome cap and mediates substrate recognition, binding, unfolding, and translocation to the CP [82]. Despite the ability to degrade ubiquitylated proteins via the UPS, the proteasome itself can be degraded by autophagy, termed as proteaphagy.

Proteaphagy was first discovered in *Arabidopsis*, but is conserved in yeast and mammals [83–85]. In *Arabidopsis atg* mutants, such as *atg1*, *atg4*, *atg5*, *atg7*, *atg12*, and *atg13*, the protein levels of the 26S proteasome subunits are significantly increased [85]. Fluorescently labeled CP subunit PAG1 (proteasome α subunit G1) and the RP subunit RPN5 were found to colocalize with autophagic vesicles and eventually appear in the vacuole under nitrogen starvation. Moreover, ubiquitination of the proteasome has been detected and the deposition of proteasomes into the vacuole is induced when plants are treated with the proteasome inhibitor MG132, indicating that whole proteasomes are ubiquitylated and eliminated through autophagy [85]. The ubiquitin receptors of proteaphagy, Cue5, p62, and RPN10, have been identified in yeast, mammals, and plants, respectively [83–85]. RPN10 contains two UIM (ubiquitin-interacting motif) domains at its C-terminus, allowing RPN10 to simultaneously bind to ubiquitylated proteasomes through one UIM and ATG8 via another [85]. In *Arabidopsis* mutants lacking the C-terminal domain of RPN10, MG132-induced selective proteaphagy is blocked, demonstrating that RPN10 functions as a ubiquitin receptor of proteaphagy [85].

4.3.4. Pexophagy

Peroxisomes universally exist in all eukaryotes and function in fatty acid β-oxidation, glyoxylate cycles, photorespiration, and other metabolic processes which are involved in plant growth, development, and response to various stresses [86–88]. Damaged, obsolete, or redundant peroxisomes need to be eliminated by a selective autophagy pathway termed pexophagy to maintain cellular redox homeostasis [89,90]. In yeast, the selective receptors of pexophagy are Atg30 (*Pichia pastoris*) and Atg36 (*Saccharomyces cerevisiae*), which recognize peroxisomes to be degraded by direct interaction with peroxisomal membrane proteins, such as Pex3 (peroxin 3) and Pex14 [91,92]. However, no evidence has demonstrated that ubiquitination is required for yeast pexophagy. Interestingly, in mammals, the pexophagy receptors NBR1 and p62 have the ability to recognize ubiquitylated peroxisomal proteins, suggesting that ubiquitination is involved in mammalian pexophagy [93]. In mammals, overexpression of PEX3 or PEX2 induces peroxisome ubiquitination and degradation through NBR1-mediated pexophagy [94,95]. These findings demonstrate that ubiquitination of peroxisomal membrane proteins can trigger pexophagy in mammals. Moreover, recent evidence suggests that ubiquitinated mammalian Pex5 serves as a potential peroxisome degradation signal, which is recognized by autophagy receptor, p62 [96]. Similar to yeast pexophagy, the role of ubiquitination has not yet been established in plant pexophagy. Nevertheless, in *Arabidopsis*, the ubiquitin receptor protein DOMINANT SUPPRESSOR OF KAR2 (DSK2) was found to interact with PEX2 and PEX12 and has been suggested to participate in the peroxisomal membrane-associated protein degradation pathway [97]. As a selective autophagy receptor, DSK2 recognizes and targets ubiquitinated BRI1-EMS SUPPRESSOR1 (BES1) for autophagy in response to drought and starvation stresses (see Section 4.3.5). However, whether DSK2 serves as a potential pexophagy receptor and whether ubiquitination is involved in plant pexophagy are still not clear and need further investigation.

4.3.5. Others

In *Arabidopsis*, ubiquitinated BES1, which is a key transcription factor regulating gene expression in the brassinosteroid signal pathway, accumulates under treatment with proteasome and autophagy inhibitors, indicating that BES1 can be degraded through the UPS and autophagy [98,99]. Further exploration revealed that ubiquitin-binding and selective autophagy receptor protein DSK2 can interact with both BES1 and ATG8 to mediate BES1 degradation through autophagy [99]. In addition, the E3 ligase SINAT2 can interact with DSK2 to form a complex to target ubiquitinated BES1 for the UPS under starvation conditions [99]. In summary, ubiquitin modifications can regulate plant selective autophagy. Hence, the UPS and autophagy cooperate and complement each other during plant development and stress response to jointly regulate cellular homeostasis.

5. Future Perspectives

As complex and well-regulated pathways, the UPS and autophagy act together to form an integrated quality control network in response to general cellular stress. Although their mechanisms vary widely, there is growing evidence that they are closely related. Ubiquitin, as a molecular marker, not only plays a role in the UPS pathway, but also mediates the initiation of many autophagic processes. Furthermore, the components of the UPS can be degraded by autophagy and vice versa. While this area of study has been well characterized in mammals, it is just beginning to be revealed in plants, leaving many mysteries to be solved. For example, under stress, the UPS and autophagy pathways may both need to be activated to withstand adverse environments. Thus, how do plants coordinate these different, interconnected pathways? What determines whether ubiquitinated protein substrates are degraded by UPS or by autophagy? As proteolysis plays a central role in various biological processes of plants, our understanding on the relationship between the UPS and autophagy will aid in the understanding of the molecular mechanisms of plant development and stress response.

Author Contributions: T.S. and C.M. conceived and wrote the review. M.Y., P.W. and Y.Z. contributed to revision of the manuscript. All authors have read and agreed to the published version of the manuscript.

Funding: This review was funded by National Natural Science Foundation of China (31770290 and 31970301), the Key Technology Research and Development Program of Shandong (2018GNC113010), and the China Postdoctoral Science Foundation (2017M612333).

Conflicts of Interest: The authors declare no conflict of interest.

References

1. Smalle, J.; Vierstra, R.D. The ubiquitin 26S proteasome proteolytic pathway. *Annu. Rev. Plant Biol.* **2004**, *55*, 555–590. [CrossRef] [PubMed]
2. Balchin, D.; Hayer-Hartl, M.; Hartl, F.U. In vivo aspects of protein folding and quality control. *Science* **2016**, *353*, aac4354. [CrossRef] [PubMed]
3. Xu, F.Q.; Xue, H.W. The ubiquitin-proteasome system in plant responses to environments. *Plant Cell Environ.* **2019**, *42*, 2931–2944. [CrossRef] [PubMed]
4. Zientara-Rytter, K.; Subramani, S. The roles of ubiquitin-binding protein shuttles in the degradative fate of ubiquitinated proteins in the ubiquitin-proteasome system and autophagy. *Cells* **2019**, *8*, 40. [CrossRef] [PubMed]
5. Dikic, I. Proteasomal and autophagic degradation systems. *Annu. Rev. Biochem.* **2017**, *86*, 193–224. [CrossRef] [PubMed]
6. Avin-Wittenberg, T. Autophagy and its role in plant abiotic stress management. *Plant Cell Environ.* **2019**, *42*, 1045–1053. [CrossRef]
7. Rabinowitz, J.D.; White, E. Autophagy and metabolism. *Science* **2010**, *330*, 1344–1348. [CrossRef]
8. Marshall, R.S.; Vierstra, R.D. Autophagy: The master of bulk and selective recycling. *Annu. Rev. Plant Biol.* **2018**, *69*, 173–208. [CrossRef]
9. Stone, S.L. Role of the ubiquitin proteasome system in plant response to abiotic stress. *Int. Rev. Cell Mol. Biol.* **2019**, *343*, 65–110.
10. Ciechanover, A.; Heller, H.; Elias, S.; Haas, A.L.; Hershko, A. ATP-dependent conjugation of reticulocyte proteins with the polypeptide required for protein degradation. *Proc. Natl. Acad. Sci. USA* **1980**, *77*, 1365–1368. [CrossRef]
11. Hershko, A.; Ciechanover, A.; Heller, H.; Haas, A.L.; Rose, I.A. Proposed role of ATP in protein breakdown: Conjugation of protein with multiple chains of the polypeptide of ATP-dependent proteolysis. *Proc. Natl. Acad. Sci. USA* **1980**, *77*, 1783–1786. [CrossRef] [PubMed]
12. Kwon, Y.T.; Ciechanover, A. The ubiquitin code in the ubiquitin-proteasome system and autophagy. *Trends Biochem. Sci.* **2017**, *42*, 873–886. [CrossRef] [PubMed]

13. Braten, O.; Livneh, I.; Ziv, T.; Admon, A.; Kehat, I.; Caspi, L.H.; Gonen, H.; Bercovich, B.; Godzik, A.; Jahandideh, S.; et al. Numerous proteins with unique characteristics are degraded by the 26S proteasome following monoubiquitination. *Proc. Natl. Acad. Sci. USA* **2016**, *113*, E4639–E4647. [CrossRef] [PubMed]
14. Kim, D.Y.; Scalf, M.; Smith, L.M.; Vierstra, R.D. Advanced proteomic analyses yield a deep catalog of ubiquitylation targets in *Arabidopsis*. *Plant Cell* **2013**, *25*, 1523–1540. [CrossRef] [PubMed]
15. Maor, R.; Jones, A.; Nühse, T.S.; Studholme, D.J.; Peck, S.C.; Shirasu, K. Multidimensional protein identification technology (MudPIT) analysis of ubiquitinated proteins in plants. *Mol. Cell Proteom.* **2007**, *6*, 601–610. [CrossRef] [PubMed]
16. Komander, D. Mechanism, specificity and structure of the deubiquitinases. *Subcell Biochem.* **2010**, *54*, 69–87.
17. Hatfield, P.M.; Gosink, M.M.; Carpenter, T.B.; Vierstra, R.D. The ubiquitin-activating enzyme (E1) gene family in *Arabidopsis thaliana*. *Plant J.* **1997**, *11*, 213–226. [CrossRef]
18. Kraft, E.; Stone, S.L.; Ma, L.; Su, N.; Gao, Y.; Lau, O.S.; Deng, X.W.; Callis, J. Genome analysis and functional characterization of the E2 and RING-type E3 ligase ubiquitination enzymes of *Arabidopsis*. *Plant Physiol.* **2005**, *139*, 1597–1611. [CrossRef]
19. Goritschnig, S.; Zhang, Y.; Li, X. The ubiquitin pathway is required for innate immunity in *Arabidopsis*. *Plant J.* **2007**, *49*, 540–551. [CrossRef]
20. Wenzel, D.M.; Stoll, K.E.; Klevit, R.E. E2s: Structurally economical and functionally replete. *Biochem. J.* **2011**, *433*, 31–42. [CrossRef]
21. Yanagawa, Y.; Sullivan, J.A.; Komatsu, S.; Gusmaroli, G.; Suzuki, G.; Yin, J.; Ishibashi, T.; Saijo, Y.; Rubio, V.; Kimura, S.; et al. *Arabidopsis* COP10 forms a complex with DDB1 and DET1 in vivo and enhances the activity of ubiquitin conjugating enzymes. *Genes Dev.* **2004**, *18*, 2172–2181. [CrossRef] [PubMed]
22. Stone, S.L.; Hauksdóttir, H.; Troy, A.; Herschleb, J.; Kraft, E.; Callis, J. Functional analysis of the RING-type ubiquitin ligase family of Arabidopsis. *Plant Physiol.* **2005**, *137*, 13–30. [CrossRef] [PubMed]
23. Lee, J.H.; Kim, W.T. Regulation of abiotic stress signal transduction by E3 ubiquitin ligases in Arabidopsis. *Mol. Cells* **2011**, *31*, 201–208. [CrossRef]
24. Mizoi, J.; Kanazawa, N.; Kidokoro, S.; Takahashi, F.; Qin, F.; Morimoto, K.; Shinozaki, K.; Yamaguchi-Shinozaki, K. Heat-induced inhibition of phosphorylation of the stress-protective transcription factor DREB2A promotes thermotolerance of *Arabidopsis thaliana*. *J. Biol. Chem.* **2019**, *294*, 902–917. [CrossRef]
25. Qin, F.; Sakuma, Y.; Tran, L.S.; Maruyama, K.; Kidokoro, S.; Fujita, Y.; Fujita, M.; Umezawa, T.; Sawano, Y.; Miyazono, K.; et al. *Arabidopsis* DREB2A-interacting proteins function as RING E3 ligases and negatively regulate plant drought stress-responsive gene expression. *Plant Cell* **2008**, *20*, 1693–1707. [CrossRef] [PubMed]
26. Sakuma, Y.; Maruyama, K.; Qin, F.; Osakabe, Y.; Shinozaki, K.; Yamaguchi-Shinozaki, K. Dual function of an *Arabidopsis* transcription factor DREB2A in water-stress-responsive and heat-stress-responsive gene expression. *Proc. Natl. Acad. Sci. USA* **2006**, *103*, 18822–18827. [CrossRef] [PubMed]
27. Kang, M.; Fokar, M.; Abdelmageed, H.; Allen, R.D. *Arabidopsis* SAP5 functions as a positive regulator of stress responses and exhibits E3 ubiquitin ligase activity. *Plant Mol. Biol.* **2011**, *75*, 451–466. [CrossRef]
28. Chinnusamy, V.; Ohta, M.; Kanrar, S.; Lee, B.H.; Hong, X.; Agarwal, M.; Zhu, J.K. ICE1: A regulator of cold-induced transcriptome and freezing tolerance in *Arabidopsis*. *Genes Dev.* **2003**, *17*, 1043–1054. [CrossRef]
29. Zhao, C.; Zhang, Z.; Xie, S.; Si, T.; Li, Y.; Zhu, J.K. Mutational evidence for the critical role of CBF transcription factors in cold acclimation in *Arabidopsis*. *Plant Physiol.* **2016**, *171*, 2744–2759.
30. Ding, Y.; Li, H.; Zhang, X.; Xie, Q.; Gong, Z.; Yang, S. OST1 kinase modulates freezing tolerance by enhancing ICE1 stability in *Arabidopsis*. *Dev. Cell* **2015**, *32*, 278–289. [CrossRef]
31. Cutler, S.R.; Rodriguez, P.L.; Finkelstein, R.R.; Abrams, S.R. Abscisic acid: Emergence of a core signaling network. *Annu. Rev. Plant Biol.* **2010**, *61*, 651–679. [CrossRef] [PubMed]
32. Irigoyen, M.L.; Iniesto, E.; Rodriguez, L.; Puga, M.I.; Yanagawa, Y.; Pick, E.; Strickland, E.; Paz-Ares, J.; Wei, N.; De Jaeger, G.; et al. Targeted degradation of abscisic acid receptors is mediated by the ubiquitin ligase substrate adaptor DDA1 in *Arabidopsis*. *Plant Cell* **2014**, *26*, 712–728. [CrossRef] [PubMed]
33. Li, Y.; Zhang, L.; Li, D.; Liu, Z.; Wang, J.; Li, X.; Yang, Y. The *Arabidopsis* F-box E3 ligase RIFP1 plays a negative role in abscisic acid signalling by facilitating ABA receptor RCAR3 degradation. *Plant Cell Environ.* **2016**, *39*, 571–582. [CrossRef] [PubMed]
34. Zhao, J.; Zhao, L.; Zhang, M.; Zafar, S.A.; Fang, J.; Li, M.; Zhang, W.; Li, X. *Arabidopsis* E3 ubiquitin ligases PUB22 and PUB23 negatively regulate drought tolerance by targeting ABA receptor PYL9 for degradation. *Int. J. Mol. Sci.* **2017**, *18*, 1841. [CrossRef] [PubMed]

35. Lee, J.H.; Terzaghi, W.; Gusmaroli, G.; Charron, J.B.; Yoon, H.J.; Chen, H.; He, Y.J.; Xiong, Y.; Deng, X.W. Characterization of *Arabidopsis* and rice DWD proteins and their roles as substrate receptors for CUL4-RING E3 ubiquitin ligases. *Plant Cell* **2008**, *20*, 152–167. [CrossRef] [PubMed]
36. Lyzenga, W.J.; Liu, H.; Schofield, A.; Muise-Hennessey, A.; Stone, S.L. *Arabidopsis* CIPK26 interacts with KEG, components of the ABA signalling network and is degraded by the ubiquitin-proteasome system. *J. Exp. Bot.* **2013**, *64*, 2779–2791. [CrossRef] [PubMed]
37. Liu, H.; Stone, S.L. Cytoplasmic degradation of the *Arabidopsis* transcription factor abscisic acid insensitive 5 is mediated by the RING-type E3 ligase KEEP ON GOING. *J. Biol. Chem.* **2013**, *288*, 20267–20279. [CrossRef]
38. Stone, S.L.; Williams, L.A.; Farmer, L.M.; Vierstra, R.D.; Callis, J. KEEP ON GOING, a RING E3 ligase essential for *Arabidopsis* growth and development, is involved in abscisic acid signaling. *Plant Cell* **2006**, *18*, 3415–3428. [CrossRef]
39. Chen, Y.T.; Liu, H.; Stone, S.; Callis, J. ABA and the ubiquitin E3 ligase KEEP ON GOING affect proteolysis of the *Arabidopsis thaliana* transcription factors ABF1 and ABF3. *Plant J.* **2013**, *75*, 965–976. [CrossRef]
40. Ma, Y.; Szostkiewicz, I.; Korte, A.; Moes, D.; Yang, Y.; Christmann, A.; Grill, E. Regulators of PP2C phosphatase activity function as abscisic acid sensors. *Science* **2009**, *324*, 1064–1068. [CrossRef]
41. Wu, Q.; Zhang, X.; Peirats-Llobet, M.; Belda-Palazon, B.; Wang, X.; Cui, S.; Yu, X.; Rodriguez, P.L.; An, C. Ubiquitin ligases RGLG1 and RGLG5 regulate abscisic acid signaling by controlling the turnover of phosphatase PP2CA. *Plant Cell* **2016**, *28*, 2178–2196. [CrossRef] [PubMed]
42. Lyzenga, W.J.; Sullivan, V.; Liu, H.; Stone, S.L. The kinase activity of calcineurin B-like interacting protein kinase 26 (CIPK26) influences its own stability and that of the ABA-regulated ubiquitin ligase, Keep on Going (KEG). *Front. Plant Sci.* **2017**, *8*, 502. [CrossRef] [PubMed]
43. Yang, C.; Lu, X.; Ma, B.; Chen, S.-Y.; Zhang, J.-S. Ethylene signaling in rice and Arabidopsis: Conserved and diverged aspects. *Mol. Plant* **2015**, *8*, 495–505. [CrossRef] [PubMed]
44. Dolgikh, V.A.; Pukhovaya, E.M.; Zemlyanskaya, E.V. Shaping ethylene response: The role of EIN3/EIL1 transcription factors. *Front. Plant Sci.* **2019**, *10*, 1030. [CrossRef]
45. An, F.; Zhao, Q.; Ji, Y.; Li, W.; Jiang, Z.; Yu, X.; Zhang, C.; Han, Y.; He, W.; Liu, Y.; et al. Ethylene-induced stabilization of ETHYLENE INSENSITIVE3 and EIN3-LIKE1 is mediated by proteasomal degradation of EIN3 binding F-box 1 and 2 that requires EIN2 in *Arabidopsis*. *Plant Cell* **2010**, *22*, 2384–2401. [CrossRef]
46. Shi, H.; Liu, R.; Xue, C.; Shen, X.; Wei, N.; Deng, X.W.; Zhong, S. Seedlings transduce the depth and mechanical pressure of covering soil using COP1 and ethylene to regulate EBF1/EBF2 for soil emergence. *Curr. Biol.* **2016**, *26*, 139–149. [CrossRef]
47. Podolec, R.; Ulm, R. Photoreceptor-mediated regulation of the COP1/SPA E3 ubiquitin ligase. *Curr. Opin. Plant Biol.* **2018**, *45*, 18–25. [CrossRef]
48. Qiao, H.; Chang, K.N.; Yazaki, J.; Ecker, J.R. Interplay between ethylene, ETP1/ETP2 F-box proteins, and degradation of EIN2 triggers ethylene responses in *Arabidopsis*. *Genes Dev.* **2009**, *23*, 512–521. [CrossRef]
49. Vain, T.; Raggi, S.; Ferro, N.; Barange, D.K.; Kieffer, M.; Ma, Q.; Doyle, S.M.; Thelander, M.; Pařízková, B.; Novák, O.; et al. Selective auxin agonists induce specific AUX/IAA protein degradation to modulate plant development. *Proc. Natl. Acad. Sci. USA* **2019**, *116*, 6463–6472. [CrossRef]
50. Su, T.; Li, X.; Yang, M.; Shao, Q.; Zhao, Y.; Ma, C.; Wang, P. Autophagy: An intracellular degradation pathway regulating plant survival and stress response. *Front. Plant Sci.* **2020**, *11*, 164. [CrossRef]
51. van Doorn, W.G.; Papini, A. Ultrastructure of autophagy in plant cells. *Autophagy* **2013**, *9*, 1922–1936. [CrossRef] [PubMed]
52. Klionsky, D.J.; Abeliovich, H.; Agostinis, P.; Agrawal, D.K.; Aliev, G.; Askew, D.S.; Baba, M.; Baehrecke, E.H.; Bahr, B.A.; Ballabio, A.; et al. Guidelines for the use and interpretation of assays for monitoring autophagy in higher eukaryotes. *Autophagy* **2008**, *4*, 151–175. [CrossRef] [PubMed]
53. Korolchuk, V.I.; Menzies, F.M.; Rubinsztein, D.C. Mechanisms of cross-talk between the ubiquitin-proteasome and autophagy-lysosome systems. *FEBS Lett.* **2010**, *584*, 1393–1398. [CrossRef]
54. Streich, F.C., Jr.; Lima, C.D. Structural and functional insights to ubiquitin-like protein conjugation. *Annu. Rev. Biophys.* **2014**, *43*, 357–379. [CrossRef] [PubMed]
55. Suttangkakul, A.; Li, F.; Chung, T.; Vierstra, R.D. The ATG1/ATG13 protein kinase complex is both a regulator and a target of autophagic recycling in *Arabidopsis*. *Plant Cell* **2011**, *23*, 3761–3779. [CrossRef] [PubMed]

56. Qi, H.; Li, J.; Xia, F.N.; Chen, J.Y.; Lei, X.; Han, M.Q.; Xie, L.J.; Zhou, Q.M.; Xiao, S. *Arabidopsis* SINAT proteins control autophagy by mediating ubiquitylation and degradation of ATG13. *Plant Cell* **2020**, *32*, 263–284. [CrossRef]
57. Qi, H.; Xia, F.-N.; Xiao, S. Autophagy in plants: Physiological roles and post-translational regulation. *J. Integr. Plant Biol.* **2020**. [CrossRef]
58. Qi, H.; Xia, F.N.; Xie, L.J.; Yu, L.J.; Chen, Q.F.; Zhuang, X.H.; Wang, Q.; Li, F.; Jiang, L.; Xie, Q.; et al. TRAF family proteins regulate autophagy dynamics by modulating AUTOPHAGY PROTEIN6 stability in *Arabidopsis*. *Plant Cell* **2017**, *29*, 890–911. [CrossRef]
59. Xia, F.N.; Zeng, B.; Liu, H.S.; Qi, H.; Xie, L.J.; Yu, L.J.; Chen, Q.F.; Li, J.F.; Chen, Y.Q.; Jiang, L.; et al. SINAT E3 ubiquitin ligases mediate FREE1 and VPS23A degradation to modulate abscisic acid signalling. *Plant Cell* **2020**. [CrossRef]
60. Xiao, Z.; Yang, C.; Liu, C.; Yang, L.; Yang, S.; Zhou, J.; Li, F.; Jiang, L.; Xiao, S.; Gao, C.; et al. SINAT E3 ligases regulate the stability of the ESCRT component FREE1 in response to iron deficiency in plants. *J. Integr. Plant Biol.* **2020**, *62*, 1399–1417. [CrossRef]
61. Ohsumi, Y. Molecular dissection of autophagy: Two ubiquitin-like systems. *Nat. Rev. Mol. Cell Biol.* **2001**, *2*, 211–216. [CrossRef] [PubMed]
62. Chung, T.; Phillips, A.R.; Vierstra, R.D. ATG8 lipidation and ATG8-mediated autophagy in *Arabidopsis* require ATG12 expressed from the differentially controlled ATG12A and ATG12B loci. *Plant J.* **2010**, *62*, 483–493. [CrossRef] [PubMed]
63. Yoshimoto, K.; Hanaoka, H.; Sato, S.; Kato, T.; Tabata, S.; Noda, T.; Ohsumi, Y. Processing of ATG8s, ubiquitin-like proteins, and their deconjugation by ATG4s are essential for plant autophagy. *Plant Cell* **2004**, *16*, 2967–2983. [CrossRef] [PubMed]
64. Doelling, J.H.; Walker, J.M.; Friedman, E.M.; Thompson, A.R.; Vierstra, R.D. The APG8/12-activating enzyme APG7 is required for proper nutrient recycling and senescence in *Arabidopsis thaliana*. *J. Biol. Chem.* **2002**, *277*, 33105–33114. [CrossRef] [PubMed]
65. Ichimura, Y.; Kirisako, T.; Takao, T.; Satomi, Y.; Shimonishi, Y.; Ishihara, N.; Mizushima, N.; Tanida, I.; Kominami, E.; Ohsumi, M.; et al. A ubiquitin-like system mediates protein lipidation. *Nature* **2000**, *408*, 488–492. [CrossRef] [PubMed]
66. Klionsky, D.J.; Schulman, B.A. Dynamic regulation of macroautophagy by distinctive ubiquitin-like proteins. *Nat. Struct. Mol. Biol.* **2014**, *21*, 336–345. [CrossRef]
67. Phillips, A.R.; Suttangkakul, A.; Vierstra, R.D. The ATG12-conjugating enzyme ATG10 is essential for autophagic vesicle formation in *Arabidopsis thaliana*. *Genetics* **2008**, *178*, 1339–1353. [CrossRef]
68. Thompson, A.R.; Doelling, J.H.; Suttangkakul, A.; Vierstra, R.D. Autophagic nutrient recycling in *Arabidopsis* directed by the ATG8 and ATG12 conjugation pathways. *Plant Physiol.* **2005**, *138*, 2097–2110. [CrossRef]
69. Bu, F.; Yang, M.; Guo, X.; Huang, W.; Chen, L. Multiple functions of ATG8 family proteins in plant autophagy. *Front. Cell Dev. Biol.* **2020**, *8*, 466. [CrossRef]
70. Jarvis, P.; López-Juez, E. Biogenesis and homeostasis of chloroplasts and other plastids. *Nat. Rev. Mol. Cell Biol.* **2013**, *14*, 787–802. [CrossRef]
71. Izumi, M.; Ishida, H.; Nakamura, S.; Hidema, J. Entire photodamaged chloroplasts are transported to the central vacuole by autophagy. *Plant Cell* **2017**, *29*, 377–394. [CrossRef] [PubMed]
72. Havé, M.; Marmagne, A.; Chardon, F.; Masclaux-Daubresse, C. Nitrogen remobilization during leaf senescence: Lessons from *Arabidopsis* to crops. *J. Exp. Bot.* **2017**, *68*, 2513–2529. [PubMed]
73. Woodson, J.D.; Joens, M.S.; Sinson, A.B.; Gilkerson, J.; Salomé, P.A.; Weigel, D.; Fitzpatrick, J.A.; Chory, J. Ubiquitin facilitates a quality-control pathway that removes damaged chloroplasts. *Science* **2015**, *350*, 450–454. [CrossRef]
74. Kikuchi, Y.; Nakamura, S.; Woodson, J.D.; Ishida, H.; Ling, Q.; Hidema, J.; Jarvis, R.P.; Hagihara, S.; Izumi, M. Chloroplast autophagy and ubiquitination combine to manage oxidative damage and starvation responses. *Plant Physiol.* **2020**, *183*, 1531–1544. [CrossRef] [PubMed]
75. Ling, Q.; Huang, W.; Baldwin, A.; Jarvis, P. Chloroplast biogenesis is regulated by direct action of the ubiquitin-proteasome system. *Science* **2012**, *338*, 655–659. [CrossRef]
76. Johansen, T.; Lamark, T. Selective autophagy mediated by autophagic adapter proteins. *Autophagy* **2011**, *7*, 279–296. [CrossRef] [PubMed]

77. Shaid, S.; Brandts, C.H.; Serve, H.; Dikic, I. Ubiquitination and selective autophagy. *Cell Death Differ.* **2013**, *20*, 21–30. [CrossRef]
78. Svenning, S.; Lamark, T.; Krause, K.; Johansen, T. Plant NBR1 is a selective autophagy substrate and a functional hybrid of the mammalian autophagic adapters NBR1 and p62/SQSTM1. *Autophagy* **2011**, *7*, 993–1010. [CrossRef]
79. Jung, H.; Lee, H.N.; Marshall, R.S.; Lomax, A.W.; Yoon, M.J.; Kim, J.; Kim, J.H.; Vierstra, R.D.; Chung, T. *Arabidopsis* cargo receptor NBR1 mediates selective autophagy of defective proteins. *J. Exp. Bot.* **2020**, *71*, 73–89. [CrossRef]
80. Zhou, J.; Zhang, Y.; Qi, J.; Chi, Y.; Fan, B.; Yu, J.Q.; Chen, Z. E3 ubiquitin ligase CHIP and NBR1-mediated selective autophagy protect additively against proteotoxicity in plant stress responses. *PLoS Genet.* **2014**, *10*, e1004116. [CrossRef]
81. Tarnowski, L.; Rodriguez, M.C.; Brzywczy, J.; Piecho-Kabacik, M.; Krčkova, Z.; Martinec, J.; Wawrzynska, A.; Sirko, A. A selective autophagy cargo receptor NBR1 modulates abscisic acid signalling in *Arabidopsis thaliana*. *Sci. Rep.* **2020**, *10*, 7778. [CrossRef] [PubMed]
82. Livneh, I.; Cohen-Kaplan, V.; Cohen-Rosenzweig, C.; Avni, N.; Ciechanover, A. The life cycle of the 26S proteasome: From birth, through regulation and function, and onto its death. *Cell Res.* **2016**, *26*, 869–885. [CrossRef] [PubMed]
83. Marshall, R.S.; McLoughlin, F.; Vierstra, R.D. Autophagic turnover of inactive 26S proteasomes in yeast is directed by the ubiquitin receptor Cue5 and the Hsp42 chaperone. *Cell Rep.* **2016**, *16*, 1717–1732. [CrossRef] [PubMed]
84. Cohen-Kaplan, V.; Livneh, I.; Avni, N.; Fabre, B.; Ziv, T.; Kwon, Y.T.; Ciechanover, A. p62- and ubiquitin-dependent stress-induced autophagy of the mammalian 26S proteasome. *Proc. Natl. Acad. Sci. USA* **2016**, *113*, E7490–E7499. [CrossRef]
85. Marshall, R.S.; Li, F.; Gemperline, D.C.; Book, A.J.; Vierstra, R.D. Autophagic degradation of the 26S proteasome is mediated by the dual ATG8/ubiquitin receptor RPN10 in *Arabidopsis*. *Mol. Cell* **2015**, *58*, 1053–1066. [CrossRef]
86. Pan, R.; Liu, J.; Hu, J. Peroxisomes in plant reproduction and seed-related development. *Integr. Plant Biol.* **2019**, *61*, 784–802. [CrossRef]
87. Su, T.; Wang, P.; Li, H.; Zhao, Y.; Lu, Y.; Dai, P.; Ren, T.; Wang, X.; Li, X.; Shao, Q.; et al. The *Arabidopsis* catalase triple mutant reveals important roles of catalases and peroxisome-derived signaling in plant development. *J. Integr. Plant Biol.* **2018**, *60*, 591–607. [CrossRef]
88. Corpas, F.J.; Del Río, L.A.; Palma, J.M. Plant peroxisomes at the crossroad of NO and H_2O_2 metabolism. *J. Integr. Plant Biol.* **2019**, *61*, 803–816.
89. Su, T.; Li, W.; Wang, P.; Ma, C. Dynamics of peroxisome homeostasis and its role in stress response and signaling in plants. *Front. Plant Sci.* **2019**, *10*, 705. [CrossRef]
90. Till, A.; Lakhani, R.; Burnett, S.F.; Subramani, S. Pexophagy: The selective degradation of peroxisomes. *Int. J. Cell Biol.* **2012**, *2012*, 512721. [CrossRef]
91. Farré, J.C.; Manjithaya, R.; Mathewson, R.D.; Subramani, S. PpAtg30 tags peroxisomes for turnover by selective autophagy. *Dev. Cell* **2008**, *14*, 365–376. [CrossRef] [PubMed]
92. Motley, A.M.; Nuttall, J.M.; Hettema, E.H. Atg36: The *Saccharomyces cerevisiae* receptor for pexophagy. *Autophagy* **2012**, *8*, 1680–1681. [CrossRef]
93. Zientara-Rytter, K.; Sirko, A. To deliver or to degrade—An interplay of the ubiquitin-proteasome system, autophagy and vesicular transport in plants. *FEBS J.* **2016**, *283*, 3534–3555. [CrossRef] [PubMed]
94. Yamashita, S.; Abe, K.; Tatemichi, Y.; Fujiki, Y. The membrane peroxin PEX3 induces peroxisome-ubiquitination-linked pexophagy. *Autophagy* **2014**, *10*, 1549–1564. [CrossRef]
95. Sargent, G.; van Zutphen, T.; Shatseva, T.; Zhang, L.; Di Giovanni, V.; Bandsma, R.; Kim, P.K. PEX2 is the E3 ubiquitin ligase required for pexophagy during starvation. *J. Cell Biol.* **2016**, *214*, 677–690. [CrossRef]
96. Zhang, J.; Tripathi, D.N.; Jing, J.; Alexander, A.; Kim, J.; Powell, R.T.; Dere, R.; Tait-Mulder, J.; Lee, J.H.; Paull, T.T.; et al. ATM functions at the peroxisome to induce pexophagy in response to ROS. *Nat. Cell Biol.* **2015**, *17*, 1259–1269. [CrossRef] [PubMed]
97. Kaur, N.; Zhao, Q.; Xie, Q.; Hu, J. *Arabidopsis* RING peroxins are E3 ubiquitin ligases that interact with two homologous ubiquitin receptor proteins (F). *J. Integr. Plant Biol.* **2013**, *55*, 108–120. [CrossRef]

98. Yin, Y.; Wang, Z.Y.; Mora-Garcia, S.; Li, J.; Yoshida, S.; Asami, T.; Chory, J. BES1 accumulates in the nucleus in response to brassinosteroids to regulate gene expression and promote stem elongation. *Cell* **2002**, *109*, 181–191. [CrossRef]
99. Nolan, T.M.; Brennan, B.; Yang, M.; Chen, J.; Zhang, M.; Li, Z.; Wang, X.; Bassham, D.C.; Walley, J.; Yin, Y. Selective autophagy of BES1 mediated by DSK2 balances plant growth and survival. *Dev. Cell* **2017**, *41*, 33–46.e7. [CrossRef]

cells

MDPI

Article

TonEBP Promotes β-Cell Survival under ER Stress by Enhancing Autophagy

Hyun Je Kang †, Eun Jin Yoo †, Hwan Hee Lee, Seung Min An, Hyun Park, Whaseon Lee-Kwon, Soo Youn Choi * and Hyug Moo Kwon *

School of Life Sciences, Ulsan National Institute of Science and Technology, Ulsan 44919, Korea; hjkang90@unist.ac.kr (H.J.K.); ejyoo89@unist.ac.kr (E.J.Y.); hheelee89@unist.ac.kr (H.H.L.); asm7615@unist.ac.kr (S.M.A.); skyline@unist.ac.kr (H.P.); wlee@unist.ac.kr (W.L.-K.)

* Correspondence: sychoi@unist.ac.kr (S.Y.C.); hmkwon@unist.ac.kr (H.M.K.)

† These authors contributed equally to this work.

Received: 9 June 2020; Accepted: 19 August 2020; Published: 20 August 2020

check for updates

Abstract: The endoplasmic reticulum (ER) stress response and autophagy are important cellular responses that determine cell fate and whose dysregulation is implicated in the perturbation of homeostasis and diseases. Tonicity-responsive enhancer-binding protein (TonEBP, also called NFAT5) is a pleiotropic stress protein that mediates both protective and pathological cellular responses. Here, we examined the role of TonEBP in β-cell survival under ER stress. We found that TonEBP increases β-cell survival under ER stress by enhancing autophagy. The level of TonEBP protein increased under ER stress due to a reduction in its degradation via the ubiquitin–proteasome pathway. In response to ER stress, TonEBP increased autophagosome formations and suppressed the accumulation of protein aggregates and β-cell death. The Rel-homology domain of TonEBP interacted with FIP200, which is essential for the initiation of autophagy, and was required for autophagy and cell survival upon exposure to ER stress. Mice in which *TonEBP* was specifically deleted in pancreatic endocrine progenitor cells exhibited defective glucose homeostasis and a loss of islet mass. Taken together, these findings demonstrate that TonEBP protects against ER stress-induced β-cell death by enhancing autophagy.

Keywords: NFAT5; autophagy initiation; islet; FIP200; unfolded protein response; UPR

1. Introduction

The endoplasmic reticulum (ER) is an important intracellular organelle for the synthesis, folding, and assembly of secreted and transmembrane proteins. ER function is disturbed in several physiological and pathological conditions, and this leads to ER stress, which is characterized by the accumulation and aggregation of unfolded and/or misfolded proteins in the ER [1,2]. Cells react to ER stress by initiating the unfolded protein response (UPR), which is an adaptive and cellular protective response that aims to reduce the accumulation of unfolded proteins and restore ER homeostasis. However, an insufficient UPR and/or persistent ER stress trigger cellular dysfunction and cell death, leading to human diseases [2–4]. Importantly, β-cells in pancreatic islets contain a highly developed ER to produce insulin and thus are vulnerable to ER stress [5]. ER stress and the UPR are being increasingly implicated in the dysfunction and loss of pancreatic β-cells associated with the development of type 1 and type 2 diabetes mellitus (DM) [6,7].

Autophagy is a conserved lysosomal degradation pathway that involves recognizing the material for autophagic degradation, isolating the material via autophagosome formation and fusing autophagosomes with lysosomes (autolysosome) to degrade the cargo, and is essential for cellular homeostasis and adaptations to stress [8,9]. Autophagy is reciprocally linked to ER stress in eukaryotic cells. In response to ER stress, autophagy is induced to remove misfolded/aggregated proteins and damaged organelles, thereby improving cellular function and cell survival [10]. Conversely, blockade of autophagy increases ER stress and cell death [10–12]. The interplay between ER stress and autophagy is implicated in the physiology of β-cells. Transgenic mice in which autophagy is perturbed in β-cells exhibit increased β-cell death, decreased β-cell proliferation, and thus a reduced β-cell mass [13]. Autophagy-deficient β-cells display a distention of the ER, accumulation of polyubiquitinated proteins, and increased formation of large intracellular aggregates, which are related to the susceptibility of β-cells to ER stress [1], indicating that autophagy is essential for the survival and function of β-cells. However, the cellular mechanism underlying the interplay between autophagy and ER stress remains to be fully elucidated.

Tonicity-responsive enhancer-binding protein (TonEBP), which is also known as nuclear factor of activated T-cells 5 (NFAT5), was initially identified as a transcriptional regulator of the cellular response to hypertonic stress in the renal medulla [14–16]. Numerous studies have revealed that TonEBP is a pleiotropic stress protein that is involved in the response not only to hypertonicity but also to various types of stress, and leads to physiological or pathological consequences depending on the context [17]. The induction and activation of TonEBP in response to autoimmune and metabolic stresses are implicated in immunometabolic diseases such as rheumatoid arthritis [18,19], atherosclerosis [20], hepatocellular carcinoma [21], obesity [22], and DM [22,23]. By contrast, TonEBP-mediated responses to hypertonicity [16,24–27], bacterial infection [28–30], and genotoxic stress [31] have protective or homeostatic functions. Although the function of TonEBP in the responses to a range of cellular stresses is well-established, its role in the determination of cell fate under ER stress remains to be elucidated. Here, we explored the potential role of TonEBP in β-cell survival under ER stress. We found that TonEBP enhances ER stress-induced autophagy in β-cells and thereby increases β-cell survival. These findings suggest that TonEBP protects against ER stress-induced β-cell death.

2. Materials and Methods

2.1. Cells and Reagents

The MIN6-M9 mouse pancreatic β-cell line was provided by Prof. Seino (Kobe University, Kobe, Japan). MIN6-M9 and human embryonic kidney 293 (HEK293; ATCC CRL-1573) cells were cultured in Dulbecco's Modified Eagle's medium supplemented with 10% fetal bovine serum (Thermo Fisher Scientific, Waltham, MA, USA) and penicillin/streptomycin (100 U/mL and 100 μg/mL, respectively; GE Healthcare Life Sciences, Boston, UT, USA). Cells were maintained at 37 °C in an incubator containing 5% CO_2. The antibodies used for immunoblotting or immunoprecipitation were obtained from various companies. Antibodies against ubiquitin (SantaCruz Biotechnology, Santa Cruz, CA, USA), BiP (Abcam, Cambridge, UK), LC3 (Cell Signaling Technology, Beverly, MA, USA), Hsc70 (Rockland, Gilbertsville, PA, USA), FIP200 (Abcam), Flag (Sigma-Aldrich, Saint Louis, MO, USA), Myc (Cell Signaling Technology), and HA (Sigma-Aldrich) were used. The primary antibodies were detected with horseradish peroxidase-conjugated mouse, rabbit, or goat secondary antibodies (Thermo Fisher Scientific) and Alexa Fluor 488-, 568-, or 633-conjugated secondary antibodies (Invitrogen, Carlsbad, CA, USA).

2.2. Transfection

Cells were transfected with the same concentrations of scrambled or gene-targeted siRNAs for 24 h using Lipofectamine RNAimax (Invitrogen) according to the manufacturer's instructions. All siRNA duplexes were purchased from Integrated DNA Technologies (Coralville, IA, USA). All plasmids were purified using an endotoxin-free purification system (Qiagen, Valencia, CA, USA) and transfected into cells using Lipofectamine 2000 (Invitrogen) for 24 h. Transfected cells were then cultured in fresh complete medium and were analyzed as indicated in the figure legends.

2.3. Cell Viability Assay

MIN6-M9 cells were plated in triplicate in 96-well plates, and treated with brefeldin A (20 μM; Sigma-Aldrich), or tunicamycin (1 μg/mL; Sigma-Aldrich) for 24 h. Lactate dehydrogenase (LDH) release (Clontech, Mountain View, CA, USA) and 3-(4,5-dimethylthiazol-2-yl)-2,5-diphenyltetrazolium bromide (MTT) reduction (Sigma-Aldrich) were used to measure cell viability according to the manufacturer's protocol. The LDH release was calculated as a percentage using the following formula: percentage = (sample–spontaneous release/maximum release–spontaneous release) × 100. MTT reduction was calculated as a percentage of the respective controls.

2.4. Immunofluorescence Staining

Cells were plated on LabTek chamber slides (Thermo Fisher Scientific), incubated for 24 h, and treated with brefeldin A (20 μM) or tunicamycin (1 μg/mL) with or without a pretreatment of LY294002 (10 μM; Sigma-Aldrich) or chloroquine (10 μM; Sigma-Aldrich) for 1 h. Cells were treated for 4 h or 6 h to detect LC3 puncta and 24 h to detect ubiquitination and BiP, and fixed with 100% methanol at 20 °C for 30 min. Fixed cells were stained with anti-ubiquitin, anti-BiP, and anti-LC3 primary antibodies overnight at 4 °C, washed with 0.05% Triton X-100, and then stained with Alexa Fluor-conjugated secondary antibodies for 1 h. The stained cells were mounted. Pancreatic tissues were fixed overnight in 4% paraformaldehyde and embedded in paraffin. The paraffin sections were deparaffinized and dehydrated. Immunohistochemistry was performed using anti-BiP antibodies under optimized conditions. Images were acquired using an Olympus FV1000 confocal fluorescence microscope.

2.5. Immunoblotting

Cells were treated with brefeldin A (20 μM) or tunicamycin (1 μg/mL) for 1 h to detect ubiquitination and 6 h to detect the levels of proteins. Cells were washed twice with cold phosphate-buffered saline (PBS) and lysed in RIPA buffer (0.01 M Tris, pH 7.4, 0.15 M NaCl, 0.001 M EDTA, 0.001 M EGTA, and 1% Triton X-100; all from Sigma-Aldrich) containing 0.002 M phenylmethylsulfonyl fluoride (PMSF; Sigma-Aldrich) and protease inhibitors (Roche, Rotkreuz, Switzerland). After the centrifugation of the lysate, the supernatant was used for an immunoblot analysis. The protein concentration was measured with the BCA Protein Assay System (Pierce Biotechnology, Rockford, IL, USA). Proteins were denatured in Laemmli buffer. Equal amounts of each sample were separated on SDS-polyacrylamide gels and then transferred to polyvinylidene difluoride (PVDF) membranes. The membranes were blocked, incubated with primary antibodies, and washed using PBS supplemented with 0.05% (*v*/*v*) Tween-20 and 5% (*w*/*v*) nonfat dry milk. Anti-TonEBP [14], anti-LC3, anti-FIP200, anti-Flag, anti-Myc, anti-HA, and anti-BiP primary antibodies were used for immunoblotting. An anti-Hsc70 primary antibody was used as a loading control. Horseradish peroxidase-conjugated secondary antibodies were used for detection. Reactive bands were detected by chemiluminescence using the ImageQuant LAS 4000 imaging system (GE Healthcare Life Sciences).

2.6. Immunoprecipitation

Cells were washed three times with ice-cold PBS and then incubated with RIPA buffer in a tube on ice to prepare the total cell lysates. The lysates were incubated with anti-TonEBP, anti-Myc, anti-Flag, and anti-HA antibodies overnight at 4 °C under rotary agitation and then with Protein A/G Sepharose beads (GE Healthcare Sciences) for 2 h at 4 °C under rotary agitation. Bead–antibody–antigen complexes were pelleted by centrifugation at 4 °C for 1 min and the supernatant was removed. Complexes were washed three times for 10 min with RIPA buffer at 4 °C, supplemented with a sample buffer, and boiled at 95 °C for 5 min. The samples were analyzed by immunoblotting.

2.7. Mice

All procedures involving live mice were carried out in accordance with the approved guidelines of the Institutional Animal Care and Use Committee of the Ulsan National Institute of Science and Technology (UNISTACUC-16-08). All experiments used male C57BL/6 J mice. Mice carrying the loxP-targeted *TonEBP* gene (*TonEBP$^{fl/fl}$*) have been described previously [32] and were provided by Dr. Neuhofer (Division of Nephrology and Rheumatology, Clinical Center Traunstein, D-83278 Traunstein, Germany). Neurog3-cre knock-in mice, known as Ngn3-cre mice, were obtained from Jackson Laboratories (Bar Harbor, ME, USA). The *TonEBP$^{fl/fl}$* mice were crossed with Ngn3-cre mice to generate mice that lacked TonEBP in pancreatic endocrine progenitor cells. Age- and sex-matched littermates were used as controls in all experiments.

2.8. Statistical Analysis

Data are expressed as the mean + standard deviation or standard error of the mean. The statistical significance of the differences between the two conditions was estimated using an unpaired *t*-test. More than two conditions were compared using a one-way ANOVA and Tukey's post-hoc test. A p-value < 0.05 was deemed significant. All statistical analyses were performed using GraphPad Prism 8.2 software (GraphPad, San Jose, CA, USA).

3. Results

3.1. TonEBP Suppresses the ER Stress-Induced Accumulation of Unfolded Proteins and β-Cell Death

To investigate the role of TonEBP in β-cell survival under ER stress, we first examined whether the siRNA-mediated depletion of TonEBP affects cell death triggered by agents that induce ER stress, namely, brefeldin A (BFA) and tunicamycin (TM). MIN6-M9 mouse β-cells transfected with scrambled siRNA or TonEBP-targeted siRNA were incubated with 20 μM BFA or 1 μg/mL TM for 24 h followed by an LDH or MTT assay. Treatment with these agents reduced the viability of β-cells, and the TonEBP depletion dramatically increased the cell death induced by ER stress inducers, BFA and TM (Figure 1A,B). Conversely, TonEBP overexpression attenuated cell death induced by BFA, but not by TM (Figure S1A). These results suggest that TonEBP increases β-cell survival under ER stress.

We next investigated the mechanism by which TonEBP determines β-cell fate under ER stress. Ubiquitinated and unfolded proteins commonly accumulate in response to ER stress [33,34] and this triggers the activation of the UPR, which leads to the removal of these proteins [2]. However, a prolonged activation of the UPR leads to cell death under persistent ER stress [35]. We first examined whether TonEBP modulates the accumulation of ubiquitinated proteins by performing an immunofluorescence analysis of ubiquitin. The formation of ubiquitin foci markedly increased under ER stress, and TonEBP depletion enhanced the formation of ubiquitin foci in response to ER stress inducers (Figure 1C,D). We next examined the induction of BiP (also called GRP78), which is a molecular indicator of ER stress and UPR activation [36]. TonEBP depletion increased the number of BiP-positive cells observed after 24 h treatment with ER stressors (Figure 1E,F). Additionally, TonEBP depletion did not affect the protein and mRNA expression of BiP (Figure S1B,C), and the mRNA expression of ER stress-related genes *dit3*, *Atf4*, and *Ire1a* (Figure S1D–F) in a 4 h treatment with ER stressors.

These findings suggest that TonEBP is required for the clearance of unfolded protein aggregates and thereby increases cell survival in response to ER stress.

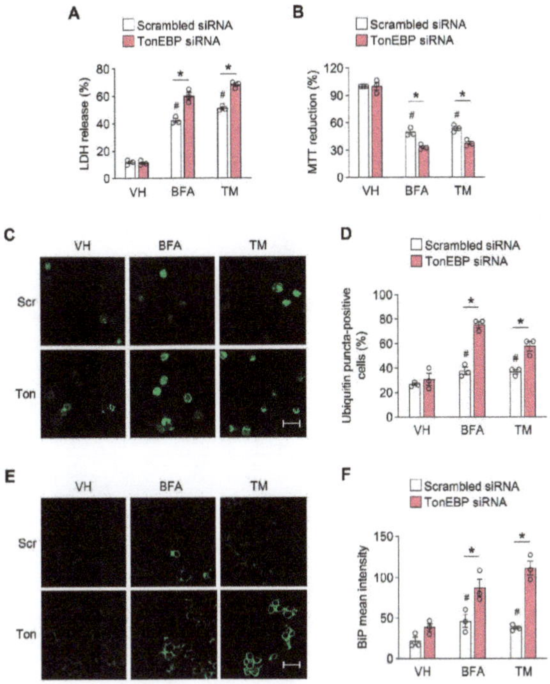

Figure 1. Tonicity-responsive enhancer-binding protein (TonEBP) prevents the accumulation of unfolded proteins. (**A**,**B**) MIN6-M9 cells were transfected with scrambled siRNA or TonEBP-targeted siRNA, and then treated with vehicle (VH), brefeldin A (BFA; 20 μM), or tunicamycin (TM; 1 μg/mL). Cell viability was assessed by the LDH release (**A**) and MTT reduction (**B**) after 24 h. (**C**) Cells were transfected with scrambled siRNA (Scr) or TonEBP-targeting siRNA (Ton) and then treated as above. Ubiquitin was visualized with an anti-ubiquitin antibody by immunostaining. (**D**) Percent of ubiquitin puncta positive cells were counted from 100 cells in each group. (**E**) Cells were transfected and treated as in (**C**). BiP was detected with an anti-BiP antibody by immunostaining. (**F**) Percent of BiP positive cells were counted from 100 cells in each group. VH, vehicle. Data (mean + SD) were from three independent experiments (n = 3) each with more than three replicates. # $p < 0.05$ vs. scrambled siRNA-VH. * $p < 0.05$ ((**A**,**B**); unpaired t-test, (**D**,**F**); one-way ANOVA). Scale bars, 50 μm (**C**,**E**).

3.2. TonEBP Is Required for ER Stress-Induced Autophagosome Formation

The induction of autophagy increases β-cell survival under ER stress by mediating the clearance of protein aggregates [37]. To elucidate the mechanism by which TonEBP increases β-cell survival under ER stress, we examined whether it stimulates autophagy in response to ER stress. During autophagosome formation, microtubule-associated protein 1 light chain 3 (LC3)-I is converted to LC3-II, which is then incorporated into the autophagosomal membrane [38]. Thus, the levels of LC3-II and LC3 correlate with the number of autophagosomes and are reliable markers of autophagosome formation [39]. A six hour treatment with ER stress inducers (20 μM BFA, and 1 μg/mL TM) markedly increased the level of LC3-II proteins in β-cells; however, this increase was markedly smaller in TonEBP-depleted cells than in control cells (Figure 2A). Furthermore, the number and intensity of LC3 puncta were higher in cells treated with ER stress inducers than in control cells, and TonEBP depletion markedly suppressed the accumulation of LC3 in response to ER stress inducers (Figure 2B,C). To further clarify the role of TonEBP during the autophagy process, we examined the effect of TonEBP depletion at the early stage (autophagosome formation) and the late stage (autophagosome-lysosome fusion) of autophagy using pharmaceutical inhibitors [40]. LY294002 (LY; 10 μM), an inhibitor of autophagosome formation, markedly suppressed TM-induced LC3 puncta (Figure 2D). On the other hand, chloroquine (CQ; 10 μM), an inhibitor of autolysosome formation, increased the accumulation of LC3, as expected from the blockade of autolysosome formation (Figure 2D). Notably, TonEBP depletion showed a similar inhibition on the accumulation of LC3 under both CQ-treated and untreated conditions (Figure 2D) indicating that TonEBP is involved in the early stage of autophagy formation. TonEBP depletion did not obviously affect the mRNA expression of the autophagy-related genes *Atg7*, *Atg14*, *p62*, and *Ulk1* (Figure S2A–D). Collectively, these data suggest that TonEBP is necessary for the induction of autophagy in β-cells.

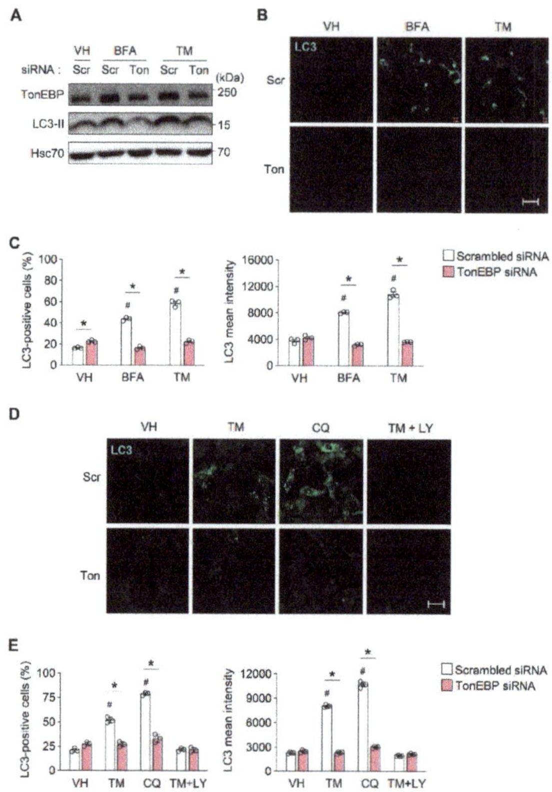

Figure 2. TonEBP promotes autophagy in pancreatic β cells. (**A**) MIN6-M9 cells transfected with scrambled siRNA (scr) or TonEBP-targeted siRNA (Ton) were treated for 6 h with vehicle (VH), brefeldin A (BFA; 20 μM), or tunicamycin (TM; 1 μg/mL). TonEBP, LC3-II, and Hsc70 were immunoblotted. (**B**) Cells transfected and treated as above were immunostained for LC3. (**C**) Percent of LC3 positive cells and LC3 signal intensity was measured in 150 cells from each group from (**B**). (**D**,**E**) Cells transfected with siRNA as above were pre-treated for 1 h with chloroquine (CQ; 10 μM) or LY294002 (LY; 10 μM) followed by a 4 h treatment with TM (1 μg/mL). (**D**) Cells were immunostained for LC3. (**E**) Percent of LC3 positive cells and LC3 signal intensity was measured in 50 cells from each group. Mean + SD. # $p < 0.05$ vs. scrambled siRNA-VH. * $p < 0.05$. Scale bars, 50 μm (**B**,**D**).

We asked whether TonEBP mediated other forms of stress for autophagy induction. To answer this question, we examined autophagy induction by rapamycin which is a potent inducer of autophagy via the suppression of mTOR [1]. As expected, rapamycin increased the level of LC3 protein in β-cells. TonEBP depletion markedly suppressed the accumulation of LC3 in response to rapamycin (Figure S2E) indicating that TonEBP contributes to autophagy induced by multiple forms of cellular stress including, but not limited to, ER stress.

3.3. TonEBP Interacts with FIP200 through Its Rel-Homology Domain (RHD)

Next, we investigated the mechanism by which TonEBP functions in ER stress-induced autophagy. To this end, we analyzed proteins that interacted with an N-terminal truncated form of TonEBP containing the intact RHD (Yc1) (Figure 3A) by performing a tandem affinity purification [31]. FIP200, a ULK-interacting protein that is required for autophagosome formation in mammalian cells [41], was a top hit (Figure S3). We performed reciprocal co-immunoprecipitation experiments to confirm the interaction between TonEBP and FIP200. Experiments using both endogenous (Figure 3B) and overexpressed (Figure 3C) proteins revealed that TonEBP pulled down FIP200 and vice versa.

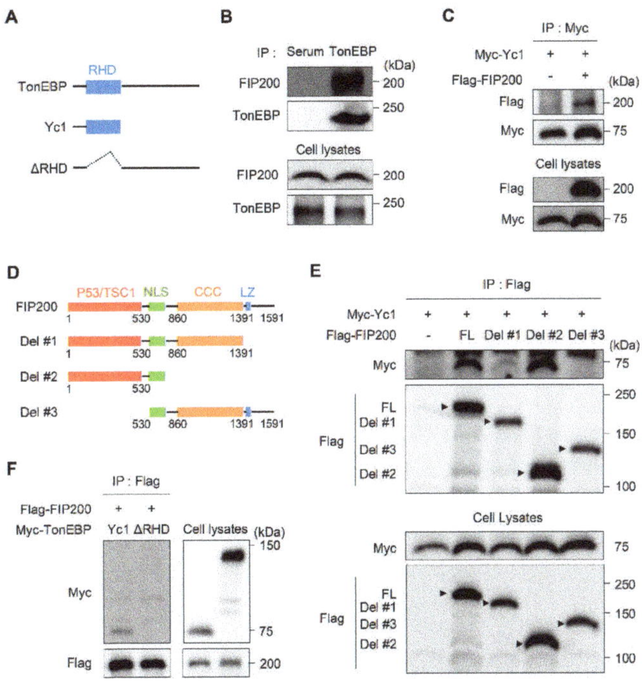

Figure 3. TonEBP interacts with FIP200 through the Rel-homology domain. (**A**) Domain structures of human TonEBP and deletion constructs Yc1 and ΔRel-homology domain (RHD). (**B**) MIN6-M9 cell lysates were immunoprecipitated (IP) with TonEBP antibody or normal serum as indicated. Cell lysates and precipitated proteins were immunoblotted for FIP200 and TonEBP. (**C**) Cells were transfected with plasmids expressing Myc–Yc1 without or with Flag–FIP200 as indicated. Cell lysates were immunoprecipitated with an anti-Myc antibody. Cell lysates and precipitated proteins were immunoblotted for Myc and Flag. (**D**) Domain structures of human FIP200 and deletion constructs Del #1, #2 and #3. (**E**) HEK293 cells were transfected with plasmids expressing Myc-Yc1 together with Flag-tagged FIP200 (FL), Del #1, Del #2 or Del #3 as indicated. Cell lysates were immunoprecipitated with an anti-Flag antibody. Cell lysates and precipitated proteins were immunoblotted for Myc and Flag. (**F**) HEK293 cells were transfected with a plasmid expressing Flag–FIP200 together with a plasmid expressing Myc–Yc1 or Myc–ΔRHD as indicated.

To define which sites mediate the TonEBP–FIP200 interaction, we generated constructs that expressed several TonEBP (Figure 3A) and FIP200 (Figure 3D) mutant proteins. To identify which structural elements of FIP200 are important for its interaction with TonEBP, cells were transfected with constructs that expressed Yc1 and full-length FIP200 or a deletion mutant. Yc1 was co-immunoprecipitated by full-length FIP200, Del #1 (ΔLz), and Del #2 (ΔCCC and ΔLz). However, the deletion of the p53/TSC1 domain

(Del #3) abolished the interaction with Yc1, demonstrating that this domain of FIP200 is required for its interaction with TonEBP (Figure 3E). In addition, a TonEBP mutant lacking the RHD (ΔRHD) did not interact with FIP200 (Figure 3F), indicating that this domain of TonEBP is essential for its interaction with FIP200. Collectively, these data suggest that the RHD of TonEBP and the p53/TSC1 domain of FIP200 mediate the interaction of these two proteins.

Based on these results, we hypothesized that the RHD of TonEBP plays an important role in cell viability and the activation of autophagy under ER stress. To investigate this, we performed rescue experiments in which wild-type TonEBP or a mutant lacking the RHD was expressed. The siRNA-mediated knockdown of TonEBP decreased cell viability over a 24 h treatment with the ER stress inducers BFA (20 μM) and TM (1 μg/mL). The reduction in cell viability by TonEBP depletion upon treatment with each of the two ER stress inducers was rescued by the expression of wild-type TonEBP, but not by the expression of the TonEBP mutant lacking the RHD (Figure 4A). Furthermore, the accumulation of LC3 in TonEBP-depleted cells under ER stress was enhanced by the expression of wild-type TonEBP but was unaffected by the expression of the TonEBP mutant lacking the RHD (Figure 4B–D).

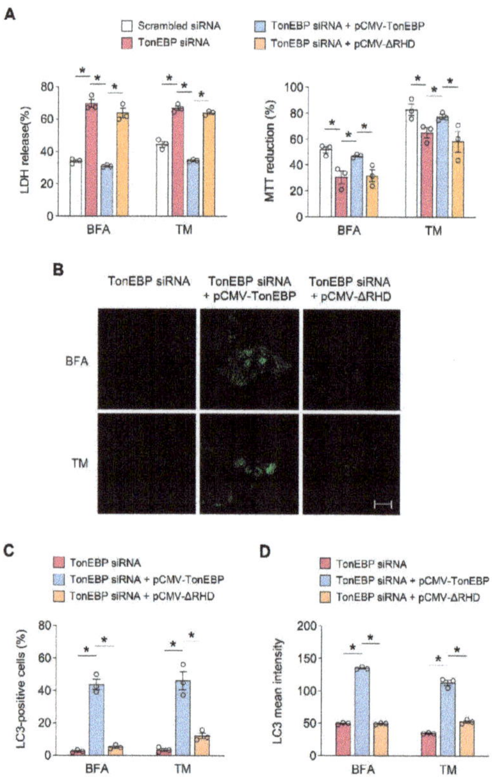

Figure 4. RHD of TonEBP is required for endoplasmic reticulum (ER) stress-induced autophagy. (**A–D**) MIN6-M9 cells were transfected with scrambled or TonEBP-targeted siRNA followed by a second transfection with a plasmid expressing TonEBP or ΔRHD as indicated. (**A**) Cell viability was assessed by the LDH release (**A**) and MTT reduction (**B**) after a 24 h treatment of brefeldin A (BFA; 20 μM) or tunicamycin (TM; 1 μg/mL). (**B**) LC3 was detected with immunostaining. (**C**) Percent of LC3 positive cells was measured in 150 cells from each group. (**D**) LC3 signal intensity was measured in 150 cells from each group. Data (mean + SD) were from three independent experiments ($n = 3$) each with more than three replicates. * $p < 0.05$ ((**A**,**C**,**D**); Two-way ANOVA with Tukey's post-hoc test). Scale bars, 50 μm (**B**).

Taken together, these data suggest that the RHD of TonEBP is required for its interaction with FIP200, activation of autophagy, and cell survival under ER stress.

3.4. ER Stress Enhances the Stability of TonEBP Proteins

The contribution of TonEBP to cell survival under ER stress (Figure 1) led us to examine whether ER stress influences its expression. A treatment of of four hours with the ER stress inducers (20 μM of BFA and 1 μg/mL of TM) increased the protein expression of TonEBP in β-cells (Figure 5A). However, the mRNA expression of TonEBP was unaffected (Figure 5B), demonstrating that TonEBP is regulated post-translationally in response to ER stress. Consistently, the treatment with MG132, a potent proteasome inhibitor, dose-dependently (10–100 nM) increased the level of TonEBP proteins in β-cells (Figure 5C), suggesting that the ubiquitin–proteasome pathway contributes to the stability of TonEBP. To investigate the ubiquitination of TonEBP, we transfected HEK293 cells with constructs expressing various deletion mutants of TonEBP (Figure 5D). Only cells expressing Yc1 displayed ubiquitination and this was enhanced by the ectopic expression of ubiquitin (Figure 5E), suggesting that the RHD of TonEBP is the main ubiquitination target. We next examined the ubiquitination of TonEBP under ER stress. A one-hour treatment with the ER stress inducers (20 μM of BFA and 1 μg/mL of TM) reduced the ubiquitination of TonEBP, indicating that the stability of TonEBP proteins is enhanced under ER stress (Figure 5F). Recent studies have shown that different linkage types of the ubiquitin chain elicit different effects on substrates [42,43]. The K48-linked ubiquitin chain mediates the proteasomal degradation of substrates, whereas the K63-linked polyubiquitin chain is involved in the regulation of the activities, localizations, and binding partners of substrates. Yc1 was ubiquitinated when wild-type ubiquitin was expressed; however, its ubiquitination was markedly decreased when the K48R or K63R ubiquitin mutant was expressed (Figure 4G). Collectively, these data suggest that the protein stability of TonEBP is enhanced under ER stress due to a reduction in its degradation via the ubiquitin–proteasome pathway and that this increases cell survival.

3.5. Deletion of TonEBP in Pancreatic Endocrine Progenitor Cells Perturbs Glucose Homeostasis

We next sought to determine the impact of TonEBP deficiency on pancreatic homeostasis in vivo. To this end, we generated *TonEBP$^{fl/fl}$ NGN3-cre* mice in which TonEBP was deleted in pancreatic endocrine progenitor cells using the Cre-lox system (*TonEBP$^{fl/fl}$*; *neurogenin 3 promoter driven-Cre*). Floxed TonEBP mice that did not express Cre recombinase (*TonEBP$^{fl/f}$* alone) were used as a control. The deletion of TonEBP in pancreatic endocrine progenitor cells significantly decreased the size of islets (Figure 6A), number of islets (Figure 6B), and pancreas weight (Figure 6C), but did not affect body weight (Figure 6D). Consistently, the serum glucose level was higher in *TonEBP$^{fl/fl}$ NGN3-cre* mice than in control mice (Figure 6E), but the serum insulin level was unchanged (Figure 6F), suggesting that TonEBP deletion in pancreatic endocrine progenitor cells perturbs glucose homeostasis. We also examined ER stress and autophagy in the pancreas by performing immunofluorescence staining for BiP, respectively. The accumulation of BiP was greater in *TonEBP$^{fl/fl}$ NGN3-cre* mice than in control mice (Figure 6G). Taken together, these data suggest that TonEBP is required for homeostasis in the pancreas via the modulation of ER stress and autophagy. However, these findings do not exclude the possibility that TonEBP is involved in pancreas development and this requires further investigation.

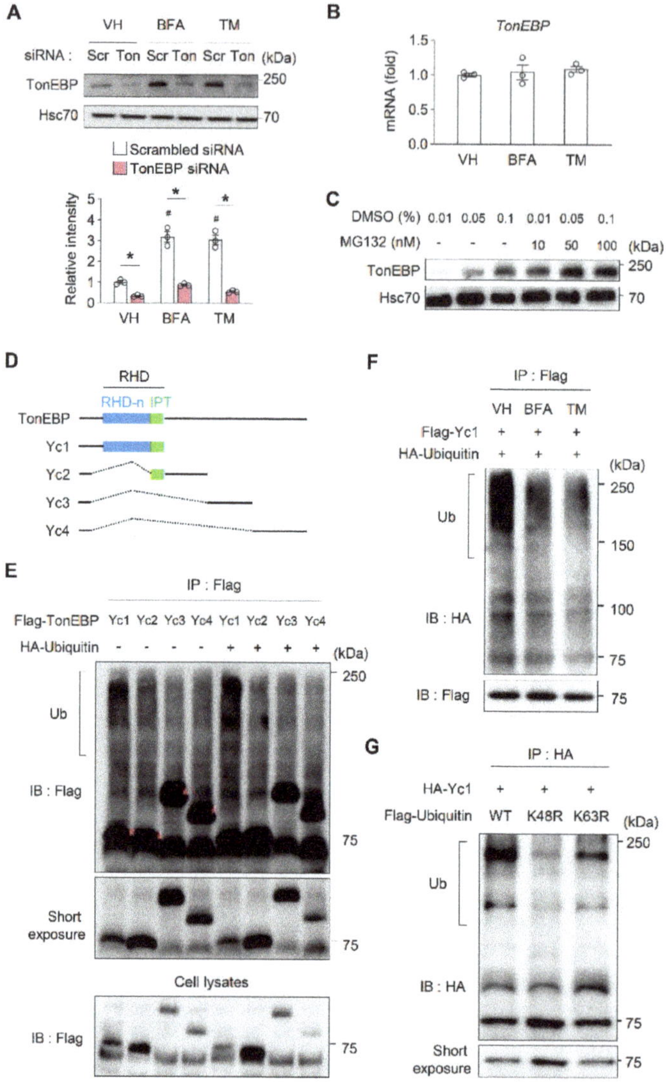

Figure 5. ER stress dramatically increases TonEBP protein stability. (**A**) MIN6-M9 cells transfected with scrambled siRNA (scr) or TonEBP-targeted siRNA (Ton) were treated for 4 h with vehicle (VH), brefeldin A (BFA; 20 µM), or tunicamycin (TM; 1 µg/mL) as indicated. TonEBP and Hsc70 were immunoblotted. Data (mean + SD) were from three independent experiments ($n = 3$) each with more than three replicates. # $p < 0.05$ vs. scrambled siRNA-VH. * $p < 0.05$ (one-way ANOVA). (**B**) Cells were treated for 6 h with the same agents as in (**A**). TonEBP mRNA was measured by RT Q-PCR. Mean + SD, $n = 4$. (**C**) Cells were treated for 4 h with 10–100 nM MG132 or 0.01–0.1% of DMSO (vehicle) as indicated. (**D**) Structures of human TonEBP and their serial deletion constructs Yc1, Yc2, Yc3 and Yc4. (**E**) HEK293 cells were transfected with plasmid expressing Flag-tagged Yc1, Yc2, Yc3 or Yc4 alone or in combination with another plasmid expressing HA-ubiquitin as indicated. Proteins immunoprecipitated with an anti-Flag antibody were immunoblotted with the same antibody. (**F**) Cells were transfected with

two plasmids expressing Flag–Yc1 and HA–ubiquitin. Cells were the treated for 1 h with agents shown in (**A**). Proteins immunoprecipitated with an anti-Flag antibody were immunoblotted with an anti-HA antibody. (**G**) HEK293 cells were transfected with HA–Yc1 together with Flag–ubiquitin (WT), Flag–ubiquitin K48R mutant (K48R) or Flag–ubiquitin K48R mutant (K63R). Proteins immunoprecipitated with an anti-HA antibody were immunoblotted with the same antibody. (**E–G**) Ub, ubiquitinated proteins.

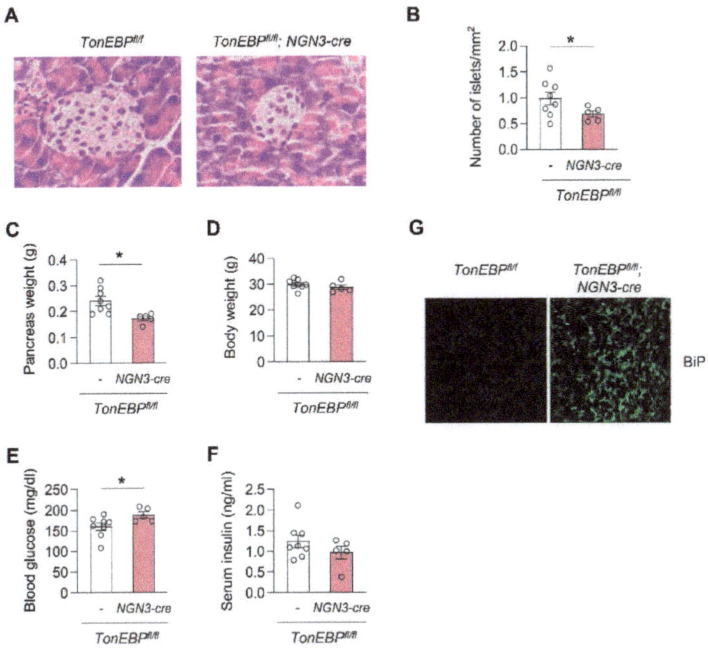

Figure 6. Smaller and fewer pancreatic islets in association with hyperglycemia in animals with islet-specific TonEBP deficiency. (**A**) Representative hematoxylin and eosin (H&E) stained images of pancreatic islets in *TonEBP*$^{fl/fl}$; *NGN3 cre*$^{+/-}$ mice and their *TonEBP*$^{fl/fl}$ littermates. (**B**) Number of pancreatic islets per mm^2 of tissues sections are shown in (**A**). Pancreas weight (**C**), body weight (**D**), plasma glucose (**E**) and plasma insulin (**F**) were analyzed from 14 week old animals: *TonEBP*$^{fl/fl}$; *NGN3 cre*$^{+/-}$ mice (blue, n = 5) and their *TonEBP*$^{fl/f}$ littermates (red, n = 8). (**G**) Representative fluorescence microscopic images showing BiP from *TonEBP*$^{fl/fl}$; *NGN3 cre*$^{+/-}$ mice and their *TonEBP*$^{fl/f}$ littermates. n represents the number of biologically independent animals. Mean + SE, * $p < 0.05$ ((**B–F**); unpaired t-test).

4. Discussion

TonEBP is a pleiotropic stress protein that mediates both protective and pathological cellular responses in a stress- and cell type-dependent manner [17]. The primary finding of this study is that TonEBP protects pancreatic β-cells against ER stress. We showed that (a) the protein expression of TonEBP in β-cells is elevated in response to ER stress due to its increased protein stability, (b) TonEBP increases β-cell survival under ER stress, (c) TonEBP inhibits the accumulation of ER stress-related proteins, and (d) TonEBP attenuates ER stress-associated cell death via the regulation of autophagy (Figure 7). These findings provide new insights into the role of TonEBP in the context of ER stress and autophagy.

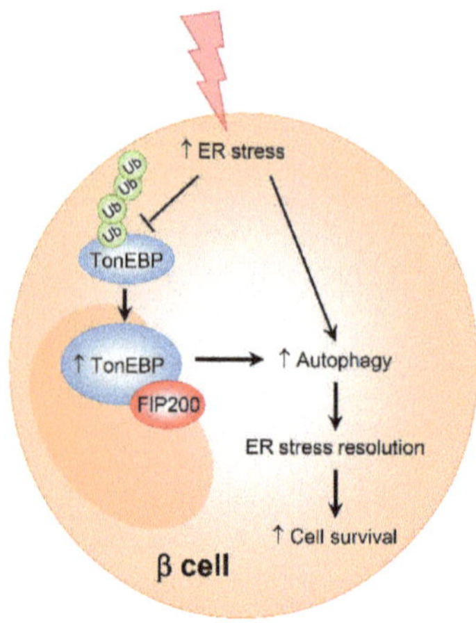

Figure 7. Model for ER stress-induced stabilization of TonEBP and autophagy formation in pancreatic β cells. TonEBP is deubiquitinated in response to ER stress. As a result, TonEBP is stabilized and interacts with FIP200 leading to the initiation of autophagy formation which prevents cell death.

The ER stress response and autophagy are important cellular responses that determine cell fate (survival and death) [1–3,8,9] and whose dysregulation is implicated in various human diseases [4,44–46]. Persistent ER stress results in the accumulation of misfolded and/or aggregated proteins in the ER that are a hallmark of protein conformational disorders [1,2]. Thus, the cellular protein quality control machinery is critical for the cellular and organismal physiology. The aggregated proteins and dysfunctional organelles are removed by autophagy-mediated lysosomal degradation. Misfolded proteins without aggregation can be restored by molecular chaperones [47] or removed by proteasomal degradation [10]. Emerging evidence indicates that there is crosstalk and coordination between the three protein degradation systems [47,48]. Autophagy can be stimulated as a secondary response to multiple types of cellular stress, including ER stress, in order to alleviate stress [11,12,49]. Autophagy functions in physiology and pathophysiology, and thus the finding that it is regulated by TonEBP, are particularly important. Autophagy occurs at a basal rate in a range of normal human physiological processes to maintain cellular homeostasis [50], and is essential for development and differentiation [50,51]. More importantly, autophagy is involved in various human disorders. The induction of autophagy protects against aging [52], metabolic syndrome [53–55], neurodegenerative diseases [56–58], infectious diseases [59–61], and some cancers [62,63]. Conversely, autophagy acts as a pro-survival pathway in certain cancers [64–66]. Given these observations, approaches to activate or inhibit autophagy are currently receiving considerable attention as potential therapeutic strategies for diverse diseases. The identification of TonEBP as a novel regulator of autophagy provides a significant insight into the mechanisms underlying the regulation of autophagy and autophagy-modulating strategies.

Although the molecular mechanism linking ER stress and autophagy remains to be fully elucidated, multiple signaling pathways are reportedly involved in the crosstalk between autophagy and ER stress [3,67]. Autophagy is tightly regulated by mTOR, the ULK1–ATG13–FIP200 complex, and ATG, indicating that significant crosstalk occurs between signaling pathways [44,45]. The ULK1–ATG13–FIP200 complex is essential for the initiation of autophagy [68]. Here, we showed that TonEBP interacts with FIP200, which is essential for autophagosome formation [69], via its RHD; this domain of TonEBP is necessary for its function in autophagy and β-cell survival under ER stress (Figure 4A–D). We hypothesize that TonEBP functions in ER stress-induced autophagy via its interaction with FIP200; however, further studies are required to verify this. The finding that TonEBP interacts with FIP200 is of great interest because FIP200 has distinct roles in cellular homeostasis and disease pathogenesis in different cell types. The depletion of FIP200 in neurons [70] and hematopoietic stem cells [71] leads to phenotypic defects associated with the suppression of autophagy. The non-autophagic functions of FIP200 are crucial during embryogenesis [72]. FIP200 also inhibits the progression of several types of cancer [73]. By contrast, another study showed that the function of FIP200 in autophagy supports tumor cell growth [74], suggesting that FIP200 is a potential target for cancer therapy. FIP200 regulates several intracellular signaling pathways via interactions with other proteins [75–78], and we identified TonEBP as an intracellular binding partner of FIP200. It will be interesting to investigate the impact of the TonEBP–FIP200 interaction on autophagy and other cellular responses in future studies.

The primary finding of this study is that TonEBP protects β-cells against ER stress. Although TonEBP has both protective and pathological effects in a stress- and cell type-dependent manner, the interpretation of this result is complicated by the previous findings that islet autoimmunity in humans is associated with an increased expression of TonEBP [23] and that the blood glucose level in mice with global TonEBP haplo-deficiencies is comparable with that in their wild-type littermates [79]. However, there is a possible explanation for these discrepant findings. Pancreatic islets contain numerous other cell types (e.g., immune cells, vascular cells, and stromal cells) in addition to endocrine cells [80]. The survival and death of β-cells might involve cell–cell crosstalk through the coordination of multiple processes as well as intrinsic pathways. Pro-inflammatory M1 macrophages, $CD4^+$ T lymphocytes, and $CD8^+$ cytotoxic T-cells are considered to be the major cell types that promote the development of type 1 DM [81,82], while M2 macrophages, which function in wound healing and tissue remodeling, promote β-cell proliferation by inducing crosstalk among different cell types [83]. Notably, a depletion of TonEBP in macrophages suppresses the polarization of M1 macrophages [30,84] and activates the polarization of M2 macrophages [29,85]. More importantly, the downregulation of TonEBP attenuates pathological $CD4^+$ T cell differentiation and autoimmunity [19,86]. Based on these findings, we believe that the loss of β-cells induced by a TonEBP deficiency can be rescued by enhanced protective or homeostatic functions in immune cells lacking TonEBP.

The level of TonEBP is elevated in response to various stresses [17]. The upregulation of TonEBP protein under stress is paralleled by an increase in *TonEBP* mRNA [17]. The *TonEBP* gene promoter has not been defined and thus it remains unknown whether it is regulated by stress. In many cases, the downregulation of microRNAs targeting *TonEBP* mRNA leads to the upregulation of this mRNA [17]. Here, we demonstrated that the level of TonEBP is regulated by its protein stability. We showed that ER stress increases the level of TonEBP in β-cells (Figure 5A) and that this is due to an increase in the protein stability of TonEBP owing to a reduction in its ubiquitin-mediated proteasomal degradation. Additionally, we showed that both K48- and K63-linked ubiquitin chains, which are the two most abundant chain types, are involved in the ubiquitination of TonEBP. To the best of our knowledge, this is the first study to show that the level of TonEBP protein is influenced by its stability. Our previous study demonstrated that TonEBP interacts with the E3 ubiquitin ligase SHPRH and the deubiquitinase USP1, and that this correlates with PCNA polyubiquitination in response to DNA damage [31]. Based on these findings, we speculate that SHPRH and USP1 mediate the ubiquitination and deubiquitination of TonEBP. However, further studies focusing on the mechanism that regulates TonEBP ubiquitination are needed to verify this hypothesis.

In summary, our results reveal that TonEBP is upregulated and increases β-cell survival by enhancing autophagy under ER stress. These findings demonstrate the previously unknown role of TonEBP in the ER stress response and autophagy.

Supplementary Materials: The following are available online at http://www.mdpi.com/2073-4409/9/9/1928/s1, Figure S1: TonEBP promotes autophagy without changes in ER stress-related protein and mRNA expression, Figure S2: TonEBP promotes autophagy without changes in ATG7, ATG14, p62 and ULK1 mRNA, Figure S3: TonEBP-interacting proteins related to autophagy initiation, Table S1: Primers used for real time PCR.

Author Contributions: H.J.K., E.J.Y., S.Y.C., and H.M.K. made substantial contributions to the conception and design, acquisition of data, analysis and interpretation of data, drafting of the article, revisions of the article critical for important intellectual content, and provided the final approval of the version to be submitted. H.H.L., S.M.A.,

H.P., and W.L.-K. made substantial contributions to the acquisition of data, and the analysis and interpretation of data. All authors have read and agreed to the published version of the manuscript.

Funding: This research was funded by the National Research Foundation grants (NRF-2018R1A5A1024340, NRF-2017R1E1A1A01074673, NRF-2019R1A2C1089260, and NRF-2020R1I1A1A01072535) of Korea. This work was also supported by UNIST funds (1.200037.01).

Conflicts of Interest: The authors declare no conflict of interest.

References

1. Wu, J.; Kaufman, R.J. From acute ER stress to physiological roles of the Unfolded Protein Response. *Cell Death Differ.* **2006**, *13*, 374–384. [CrossRef]
2. Ron, D.; Walter, P. Signal integration in the endoplasmic reticulum unfolded protein response. *Nat. Rev. Mol. Cell Biol.* **2007**, *8*, 519–529. [CrossRef]
3. Sano, R.; Reed, J.C. ER stress-induced cell death mechanisms. *Biochim. Biophys. Acta* **2013**, *1833*, 3460–3470. [CrossRef]
4. Schönthal, A.H. Endoplasmic reticulum stress: Its role in disease and novel prospects for therapy. *Scientifica* **2012**, *2012*, 857516. [CrossRef]
5. Scheuner, D.; Kaufman, R.J. The unfolded protein response: A pathway that links insulin demand with beta-cell failure and diabetes. *Endocr. Rev.* **2008**, *29*, 317–333. [CrossRef] [PubMed]
6. Cnop, M.; Welsh, N.; Jonas, J.C.; Jörns, A.; Lenzen, S.; Eizirik, D.L. Mechanisms of pancreatic beta-cell death in type 1 and type 2 diabetes: Many differences, few similarities. *Diabetes* **2005**, *54* (Suppl. 2), S97–S107. [CrossRef] [PubMed]
7. Fonseca, S.G.; Gromada, J.; Urano, F. Endoplasmic reticulum stress and pancreatic β-cell death. *Trends Endocrinol. Metab.* **2011**, *22*, 266–274. [CrossRef] [PubMed]
8. Mizushima, N. Autophagy: Process and function. *Genes Dev.* **2007**, *21*, 2861–2873. [CrossRef] [PubMed]
9. Das, G.; Shravage, B.V.; Baehrecke, E.H. Regulation and function of autophagy during cell survival and cell death. *Cold Spring Harb. Perspect. Biol.* **2012**, *4*. [CrossRef]
10. Ogata, M.; Hino, S.; Saito, A.; Morikawa, K.; Kondo, S.; Kanemoto, S.; Murakami, T.; Taniguchi, M.; Tanii, I.; Yoshinaga, K.; et al. Autophagy is activated for cell survival after endoplasmic reticulum stress. *Mol. Cell. Biol.* **2006**, *26*, 9220–9231. [CrossRef]
11. Ding, W.X.; Ni, H.M.; Gao, W.; Hou, Y.F.; Melan, M.A.; Chen, X.; Stolz, D.B.; Shao, Z.M.; Yin, X.M. Differential effects of endoplasmic reticulum stress-induced autophagy on cell survival. *J. Biol. Chem.* **2007**, *282*, 4702–4710. [CrossRef] [PubMed]
12. Yorimitsu, T.; Nair, U.; Yang, Z.; Klionsky, D.J. Endoplasmic reticulum stress triggers autophagy. *J. Biol. Chem.* **2006**, *281*, 30299–30304. [CrossRef] [PubMed]
13. Jung, H.S.; Lee, M.S. Macroautophagy in homeostasis of pancreatic beta-cell. *Autophagy* **2009**, *5*, 241–243. [CrossRef] [PubMed]
14. Miyakawa, H.; Woo, S.K.; Dahl, S.C.; Handler, J.S.; Kwon, H.M. Tonicity-responsive enhancer binding protein, a rel-like protein that stimulates transcription in response to hypertonicity. *Proc. Natl. Acad. Sci. USA* **1999**, *96*, 2538–2542. [CrossRef] [PubMed]
15. Go, W.Y.; Liu, X.; Roti, M.A.; Liu, F.; Ho, S.N. NFAT5/TonEBP mutant mice define osmotic stress as a critical feature of the lymphoid microenvironment. *Proc. Natl. Acad. Sci. USA* **2004**, *101*, 10673–10678. [CrossRef] [PubMed]
16. Lee, S.D.; Choi, S.Y.; Lim, S.W.; Lamitina, S.T.; Ho, S.N.; Go, W.Y.; Kwon, H.M. TonEBP stimulates multiple cellular pathways for adaptation to hypertonic stress: Organic osmolyte-dependent and -independent pathways. *Am. J. Physiol. Ren. Physiol.* **2011**, *300*, F707–F715. [CrossRef] [PubMed]
17. Choi, S.Y.; Lee-Kwon, W.; Kwon, H.M. The evolving role of TonEBP as an immunometabolic stress protein. *Nat. Rev. Nephrol.* **2020**, *16*, 352–364. [CrossRef]
18. Choi, S.; You, S.; Kim, D.; Choi, S.Y.; Kwon, H.M.; Kim, H.S.; Hwang, D.; Park, Y.J.; Cho, C.S.; Kim, W.U. Transcription factor NFAT5 promotes macrophage survival in rheumatoid arthritis. *J. Clin. Investig.* **2017**, *127*, 954–969. [CrossRef]

19. Ye, B.J.; Lee, H.H.; Yoo, E.J.; Lee, C.Y.; Lee, J.H.; Kang, H.J.; Jung, G.W.; Park, H.; Lee-Kwon, W.; Choi, S.Y.; et al. TonEBP in dendritic cells mediates pro-inflammatory maturation and Th1/Th17 responses. *Cell Death Dis.* **2020**, *11*, 421. [CrossRef]
20. Halterman, J.A.; Kwon, H.M.; Leitinger, N.; Wamhoff, B.R. NFAT5 expression in bone marrow-derived cells enhances atherosclerosis and drives macrophage migration. *Front. Physiol.* **2012**, *3*, 313. [CrossRef]
21. Lee, J.H.; Suh, J.H.; Choi, S.Y.; Kang, H.J.; Lee, H.H.; Ye, B.J.; Lee, G.R.; Jung, S.W.; Kim, C.J.; Lee-Kwon, W.; et al. Tonicity-responsive enhancer-binding protein promotes hepatocellular carcinogenesis, recurrence and metastasis. *Gut* **2019**, *68*, 347–358. [CrossRef] [PubMed]
22. Lee, H.H.; An, S.M.; Ye, B.J.; Lee, J.H.; Yoo, E.J.; Jeong, G.W.; Kang, H.J.; Alfadda, A.A.; Lim, S.W.; Park, J.; et al. TonEBP/NFAT5 promotes obesity and insulin resistance by epigenetic suppression of white adipose tissue beiging. *Nat. Commun.* **2019**, *10*, 3536. [CrossRef] [PubMed]
23. Serr, I.; Scherm, M.G.; Zahm, A.M.; Schug, J.; Flynn, V.K.; Hippich, M.; Kälin, S.; Becker, M.; Achenbach, P.; Nikolaev, A.; et al. A miRNA181a/NFAT5 axis links impaired T cell tolerance induction with autoimmune type 1 diabetes. *Sci. Transl. Med.* **2018**, *10*. [CrossRef] [PubMed]
24. López-Rodríguez, C.; Antos, C.L.; Shelton, J.M.; Richardson, J.A.; Lin, F.; Novobrantseva, T.I.; Bronson, R.T.; Igarashi, P.; Rao, A.; Olson, E.N. Loss of NFAT5 results in renal atrophy and lack of tonicity-responsive gene expression. *Proc. Natl. Acad. Sci. USA* **2004**, *101*, 2392–2397. [CrossRef]
25. Nakayama, Y.; Peng, T.; Sands, J.M.; Bagnasco, S.M. The TonE/TonEBP pathway mediates tonicity-responsive regulation of UT-A urea transporter expression. *J. Biol. Chem.* **2000**, *275*, 38275–38280. [CrossRef]
26. Machnik, A.; Neuhofer, W.; Jantsch, J.; Dahlmann, A.; Tammela, T.; Machura, K.; Park, J.K.; Beck, F.X.; Müller, D.N.; Derer, W.; et al. Macrophages regulate salt-dependent volume and blood pressure by a vascular endothelial growth factor-C-dependent buffering mechanism. *Nat. Med.* **2009**, *15*, 545–552. [CrossRef]
27. Jantsch, J.; Schatz, V.; Friedrich, D.; Schröder, A.; Kopp, C.; Siegert, I.; Maronna, A.; Wendelborn, D.; Linz, P.; Binger, K.J.; et al. Cutaneous Na+ storage strengthens the antimicrobial barrier function of the skin and boosts macrophage-driven host defense. *Cell Metab.* **2015**, *21*, 493–501. [CrossRef]
28. Buxadé, M.; Huerga Encabo, H.; Riera-Borrull, M.; Quintana-Gallardo, L.; López-Cotarelo, P.; Tellechea, M.; Martínez-Martínez, S.; Redondo, J.M.; Martín-Caballero, J.; Flores, J.M.; et al. Macrophage-specific MHCII expression is regulated by a remote Ciita enhancer controlled by NFAT5. *J. Exp. Med.* **2018**, *215*, 2901–2918. [CrossRef]
29. Yoo, E.J.; Lee, H.H.; Ye, B.J.; Lee, J.H.; Lee, C.Y.; Kang, H.J.; Jeong, G.W.; Park, H.; Lim, S.W.; Lee-Kwon, W.; et al. TonEBP Suppresses the HO-1 Gene by Blocking Recruitment of Nrf2 to Its Promoter. *Front. Immunol.* **2019**, *10*, 850. [CrossRef]
30. Buxadé, M.; Lunazzi, G.; Minguillón, J.; Iborra, S.; Berga-Bolaños, R.; Del Val, M.; Aramburu, J.; López-Rodríguez, C. Gene expression induced by Toll-like receptors in macrophages requires the transcription factor NFAT5. *J. Exp. Med.* **2012**, *209*, 379–393. [CrossRef]
31. Kang, H.J.; Park, H.; Yoo, E.J.; Lee, J.H.; Choi, S.Y.; Lee-Kwon, W.; Lee, K.Y.; Hur, J.H.; Seo, J.K.; Ra, J.S.; et al. TonEBP Regulates PCNA Polyubiquitination in Response to DNA Damage through Interaction with SHPRH and USP1. *Iscience* **2019**, *19*, 177–190. [CrossRef]
32. Küper, C.; Beck, F.X.; Neuhofer, W. Generation of a conditional knockout allele for the NFAT5 gene in mice. *Front. Physiol.* **2014**, *5*, 507. [CrossRef]
33. Hamdan, N.; Kritsiligkou, P.; Grant, C.M. ER stress causes widespread protein aggregation and prion formation. *J. Cell Biol.* **2017**, *216*, 2295–2304. [CrossRef] [PubMed]
34. Berke, S.J.; Paulson, H.L. Protein aggregation and the ubiquitin proteasome pathway: Gaining the UPPer hand on neurodegeneration. *Curr. Opin. Genet. Dev.* **2003**, *13*, 253–261. [CrossRef]
35. Tabas, I.; Ron, D. Integrating the mechanisms of apoptosis induced by endoplasmic reticulum stress. *Nat. Cell Biol.* **2011**, *13*, 184–190. [CrossRef] [PubMed]
36. Bertolotti, A.; Zhang, Y.; Hendershot, L.M.; Harding, H.P.; Ron, D. Dynamic interaction of BiP and ER stress transducers in the unfolded-protein response. *Nat. Cell Biol.* **2000**, *2*, 326–332. [CrossRef]
37. Bachar-Wikstrom, E.; Wikstrom, J.D.; Kaiser, N.; Cerasi, E.; Leibowitz, G. Improvement of ER stress-induced diabetes by stimulating autophagy. *Autophagy* **2013**, *9*, 626–628. [CrossRef]
38. Kabeya, Y.; Mizushima, N.; Ueno, T.; Yamamoto, A.; Kirisako, T.; Noda, T.; Kominami, E.; Ohsumi, Y.; Yoshimori, T. LC3, a mammalian homologue of yeast Apg8p, is localized in autophagosome membranes after processing. *EMBO J.* **2000**, *19*, 5720–5728. [CrossRef]

39. Mizushima, N.; Yoshimori, T. How to interpret LC3 immunoblotting. *Autophagy* **2007**, *3*, 542–545. [CrossRef]
40. Pasquier, B. Autophagy inhibitors. *Cell. Mol. Life Sci. Cmls* **2016**, *73*, 985–1001. [CrossRef]
41. Jung, C.H.; Jun, C.B.; Ro, S.H.; Kim, Y.M.; Otto, N.M.; Cao, J.; Kundu, M.; Kim, D.H. ULK-Atg13-FIP200 complexes mediate mTOR signaling to the autophagy machinery. *Mol. Biol. Cell* **2009**, *20*, 1992–2003. [CrossRef] [PubMed]
42. Pickart, C.M.; Fushman, D. Polyubiquitin chains: Polymeric protein signals. *Curr. Opin. Chem. Biol.* **2004**, *8*, 610–616. [CrossRef] [PubMed]
43. Mukhopadhyay, D.; Riezman, H. Proteasome-independent functions of ubiquitin in endocytosis and signaling. *Science* **2007**, *315*, 201–205. [CrossRef] [PubMed]
44. Eskelinen, E.L.; Saftig, P. Autophagy: A lysosomal degradation pathway with a central role in health and disease. *Biochim. Biophys. Acta* **2009**, *1793*, 664–673. [CrossRef]
45. Rubinsztein, D.C.; Codogno, P.; Levine, B. Autophagy modulation as a potential therapeutic target for diverse diseases. *Nat. Rev. Drug Discov.* **2012**, *11*, 709–730. [CrossRef]
46. Dikic, I.; Elazar, Z. Mechanism and medical implications of mammalian autophagy. *Nat. Rev. Mol. Cell Biol.* **2018**, *19*, 349–364. [CrossRef]
47. Hartl, F.U.; Bracher, A.; Hayer-Hartl, M. Molecular chaperones in protein folding and proteostasis. *Nature* **2011**, *475*, 324–332. [CrossRef]
48. Lamark, T.; Johansen, T. Aggrephagy: Selective Disposal of Protein Aggregates by Macroautophagy. *Int. J. Cell Biol.* **2012**, *2012*, 736905. [CrossRef]
49. Kroemer, G.; Mariño, G.; Levine, B. Autophagy and the integrated stress response. *Mol. Cell* **2010**, *40*, 280–293. [CrossRef]
50. Ravikumar, B.; Sarkar, S.; Davies, J.E.; Futter, M.; Garcia-Arencibia, M.; Green-Thompson, Z.W.; Jimenez-Sanchez, M.; Korolchuk, V.I.; Lichtenberg, M.; Luo, S.; et al. Regulation of mammalian autophagy in physiology and pathophysiology. *Physiol. Rev.* **2010**, *90*, 1383–1435. [CrossRef]
51. Mizushima, N.; Levine, B. Autophagy in mammalian development and differentiation. *Nat. Cell Biol.* **2010**, *12*, 823–830. [CrossRef] [PubMed]
52. Rubinsztein, D.C.; Mariño, G.; Kroemer, G. Autophagy and aging. *Cell* **2011**, *146*, 682–695. [CrossRef] [PubMed]
53. He, C.; Bassik, M.C.; Moresi, V.; Sun, K.; Wei, Y.; Zou, Z.; An, Z.; Loh, J.; Fisher, J.; Sun, Q.; et al. Exercise-induced BCL2-regulated autophagy is required for muscle glucose homeostasis. *Nature* **2012**, *481*, 511–515. [CrossRef] [PubMed]
54. Yang, L.; Li, P.; Fu, S.; Calay, E.S.; Hotamisligil, G.S. Defective hepatic autophagy in obesity promotes ER stress and causes insulin resistance. *Cell Metab.* **2010**, *11*, 467–478. [CrossRef] [PubMed]
55. Codogno, P.; Meijer, A.J. Autophagy: A potential link between obesity and insulin resistance. *Cell Metab.* **2010**, *11*, 449–451. [CrossRef] [PubMed]
56. Williams, A.; Sarkar, S.; Cuddon, P.; Ttofi, E.K.; Saiki, S.; Siddiqi, F.H.; Jahreiss, L.; Fleming, A.; Pask, D.; Goldsmith, P.; et al. Novel targets for Huntington's disease in an mTOR-independent autophagy pathway. *Nat. Chem. Biol.* **2008**, *4*, 295–305. [CrossRef]
57. Hara, T.; Nakamura, K.; Matsui, M.; Yamamoto, A.; Nakahara, Y.; Suzuki-Migishima, R.; Yokoyama, M.; Mishima, K.; Saito, I.; Okano, H.; et al. Suppression of basal autophagy in neural cells causes neurodegenerative disease in mice. *Nature* **2006**, *441*, 885–889. [CrossRef]
58. Komatsu, M.; Waguri, S.; Chiba, T.; Murata, S.; Iwata, J.; Tanida, I.; Ueno, T.; Koike, M.; Uchiyama, Y.; Kominami, E.; et al. Loss of autophagy in the central nervous system causes neurodegeneration in mice. *Nature* **2006**, *441*, 880–884. [CrossRef]
59. Gutierrez, M.G.; Master, S.S.; Singh, S.B.; Taylor, G.A.; Colombo, M.I.; Deretic, V. Autophagy is a defense mechanism inhibiting BCG and Mycobacterium tuberculosis survival in infected macrophages. *Cell* **2004**, *119*, 753–766. [CrossRef]
60. Virgin, H.W.; Levine, B. Autophagy genes in immunity. *Nat. Immunol.* **2009**, *10*, 461–470. [CrossRef]
61. Levine, B.; Mizushima, N.; Virgin, H.W. Autophagy in immunity and inflammation. *Nature* **2011**, *469*, 323–335. [CrossRef] [PubMed]
62. Liang, X.H.; Jackson, S.; Seaman, M.; Brown, K.; Kempkes, B.; Hibshoosh, H.; Levine, B. Induction of autophagy and inhibition of tumorigenesis by beclin 1. *Nature* **1999**, *402*, 672–676. [CrossRef] [PubMed]

63. Mah, L.Y.; Ryan, K.M. Autophagy and cancer. *Cold Spring Harb. Perspect. Biol.* **2012**, *4*, a008821. [CrossRef] [PubMed]
64. Inami, Y.; Waguri, S.; Sakamoto, A.; Kouno, T.; Nakada, K.; Hino, O.; Watanabe, S.; Ando, J.; Iwadate, M.; Yamamoto, M.; et al. Persistent activation of Nrf2 through p62 in hepatocellular carcinoma cells. *J. Cell Biol.* **2011**, *193*, 275–284. [CrossRef] [PubMed]
65. Duran, A.; Linares, J.F.; Galvez, A.S.; Wikenheiser, K.; Flores, J.M.; Diaz-Meco, M.T.; Moscat, J. The signaling adaptor p62 is an important NF-kappaB mediator in tumorigenesis. *Cancer Cell* **2008**, *13*, 343–354. [CrossRef] [PubMed]
66. Yang, Z.J.; Chee, C.E.; Huang, S.; Sinicrope, F.A. The role of autophagy in cancer: Therapeutic implications. *Mol. Cancer Ther.* **2011**, *10*, 1533–1541. [CrossRef]
67. Qi, Z.; Chen, L. Endoplasmic Reticulum Stress and Autophagy. *Adv. Exp. Med. Biol.* **2019**, *1206*, 167–177. [CrossRef]
68. Ganley, I.G.; Lam, D.H.; Wang, J.; Ding, X.; Chen, S.; Jiang, X. ULK1.ATG13.FIP200 complex mediates mTOR signaling and is essential for autophagy. *J. Biol. Chem.* **2009**, *284*, 12297–12305. [CrossRef]
69. Hara, T.; Takamura, A.; Kishi, C.; Iemura, S.; Natsume, T.; Guan, J.L.; Mizushima, N. FIP200, a ULK-interacting protein, is required for autophagosome formation in mammalian cells. *J. Cell Biol.* **2008**, *181*, 497–510. [CrossRef]
70. Liang, C.C.; Wang, C.; Peng, X.; Gan, B.; Guan, J.L. Neural-specific deletion of FIP200 leads to cerebellar degeneration caused by increased neuronal death and axon degeneration. *J. Biol. Chem.* **2010**, *285*, 3499–3509. [CrossRef]
71. Liu, F.; Lee, J.Y.; Wei, H.; Tanabe, O.; Engel, J.D.; Morrison, S.J.; Guan, J.L. FIP200 is required for the cell-autonomous maintenance of fetal hematopoietic stem cells. *Blood* **2010**, *116*, 4806–4814. [CrossRef] [PubMed]
72. Chen, S.; Wang, C.; Yeo, S.; Liang, C.C.; Okamoto, T.; Sun, S.; Wen, J.; Guan, J.L. Distinct roles of autophagy-dependent and -independent functions of FIP200 revealed by generation and analysis of a mutant knock-in mouse model. *Genes Dev.* **2016**, *30*, 856–869. [CrossRef] [PubMed]
73. Melkoumian, Z.K.; Peng, X.; Gan, B.; Wu, X.; Guan, J.L. Mechanism of cell cycle regulation by FIP200 in human breast cancer cells. *Cancer Res.* **2005**, *65*, 6676–6684. [CrossRef] [PubMed]
74. Wei, H.; Wei, S.; Gan, B.; Peng, X.; Zou, W.; Guan, J.L. Suppression of autophagy by FIP200 deletion inhibits mammary tumorigenesis. *Genes Dev.* **2011**, *25*, 1510–1527. [CrossRef]
75. Abbi, S.; Ueda, H.; Zheng, C.; Cooper, L.A.; Zhao, J.; Christopher, R.; Guan, J.L. Regulation of focal adhesion kinase by a novel protein inhibitor FIP200. *Mol. Biol. Cell* **2002**, *13*, 3178–3191. [CrossRef]
76. Ueda, H.; Abbi, S.; Zheng, C.; Guan, J.L. Suppression of Pyk2 kinase and cellular activities by FIP200. *J. Cell Biol.* **2000**, *149*, 423–430. [CrossRef]
77. Gan, B.; Guan, J.L. FIP200, a key signaling node to coordinately regulate various cellular processes. *Cell. Signal.* **2008**, *20*, 787–794. [CrossRef]
78. Martin, N.; Schwamborn, K.; Urlaub, H.; Gan, B.; Guan, J.L.; Dejean, A. Spatial interplay between PIASy and FIP200 in the regulation of signal transduction and transcriptional activity. *Mol. Cell. Biol.* **2008**, *28*, 2771–2781. [CrossRef]
79. Choi, S.Y.; Lim, S.W.; Salimi, S.; Yoo, E.J.; Lee-Kwon, W.; Lee, H.H.; Lee, J.H.; Mitchell, B.D.; Sanada, S.; Parsa, A.; et al. Tonicity-Responsive Enhancer-Binding Protein Mediates Hyperglycemia-Induced Inflammation and Vascular and Renal Injury. *J. Am. Soc. Nephrol. JASN* **2018**, *29*, 492–504. [CrossRef]
80. In't Veld, P.; Marichal, M. Microscopic anatomy of the human islet of Langerhans. In *The Islets of Langerhans*; Advances in Experimental Medicine and Biology; Springer: Dordrecht, The Netherlands, 2010; Volume 654, pp. 1–19. [CrossRef]
81. Rojas, J.; Bermudez, V.; Palmar, J.; Martínez, M.S.; Olivar, L.C.; Nava, M.; Tomey, D.; Rojas, M.; Salazar, J.; Garicano, C.; et al. Pancreatic Beta Cell Death: Novel Potential Mechanisms in Diabetes Therapy. *J. Diabetes Res.* **2018**, *2018*, 9601801. [CrossRef]
82. Burrack, A.L.; Martinov, T.; Fife, B.T. T Cell-Mediated Beta Cell Destruction: Autoimmunity and Alloimmunity in the Context of Type 1 Diabetes. *Front. Endocrinol.* **2017**, *8*, 343. [CrossRef] [PubMed]
83. Espinoza-Jiménez, A.; Peón, A.N.; Terrazas, L.I. Alternatively activated macrophages in types 1 and 2 diabetes. *Mediat. Inflamm.* **2012**, *2012*, 815953. [CrossRef] [PubMed]

84. Lee, H.H.; Sanada, S.; An, S.M.; Ye, B.J.; Lee, J.H.; Seo, Y.K.; Lee, C.; Lee-Kwon, W.; Küper, C.; Neuhofer, W.; et al. LPS-induced NFκB enhanceosome requires TonEBP/NFAT5 without DNA binding. *Sci. Rep.* **2016**, *6*, 24921. [CrossRef] [PubMed]
85. Choi, S.Y.; Lee, H.H.; Lee, J.H.; Ye, B.J.; Yoo, E.J.; Kang, H.J.; Jung, G.W.; An, S.M.; Lee-Kwon, W.; Chiong, M.; et al. TonEBP suppresses IL-10-mediated immunomodulation. *Sci. Rep.* **2016**, *6*, 25726. [CrossRef]
86. Kleinewietfeld, M.; Manzel, A.; Titze, J.; Kvakan, H.; Yosef, N.; Linker, R.A.; Muller, D.N.; Hafler, D.A. Sodium chloride drives autoimmune disease by the induction of pathogenic TH17 cells. *Nature* **2013**, *496*, 518–522. [CrossRef]

cells

MDPI

Article

The Proteasomal Deubiquitinating Enzyme PSMD14 Regulates Macroautophagy by Controlling Golgi-to-ER Retrograde Transport

Hianara A Bustamante [1,†], Karina Cereceda [2], Alexis E González [1,‡], Guillermo E Valenzuela [3,4], Yorka Cheuquemilla [4], Sergio Hernández [2], Eloisa Arias-Muñoz [2], Cristóbal Cerda-Troncoso [2], Susanne Bandau [5], Andrea Soza [2], Gudrun Kausel [3], Bredford Kerr [2], Gonzalo A Mardones [1,6], Jorge Cancino [2], Ronald T Hay [5], Alejandro Rojas-Fernandez [4,5,*] and Patricia V Burgos [2,7,*]

1. Instituto de Fisiología, Facultad de Medicina, Universidad Austral de Chile, Valdivia 5110566, Chile; lahianara@gmail.com (H.A.B.); alexisgonzalez003@gmail.com (A.E.G.); gonzalo.mardones@uach.cl (G.A.M.)
2. Centro de Biología Celular y Biomedicina (CEBICEM), Facultad de Medicina y Ciencia, Universidad San Sebastián, Lota 2465, Santiago 7510157, Chile; cereceda.karina@gmail.com (K.C.); felipehg1991@gmail.com (S.H.); eloisa.arias.m@gmail.com (E.A.-M.); cct.troncoso@gmail.com (C.C.-T.); andrea.soza@uss.cl (A.S.); bredford.kerr@uss.cl (B.K.); jorge.cancino@uss.cl (J.C.)
3. Instituto de Bioquímica y Microbiología, Facultad de Ciencias, Universidad Austral de Chile, Valdivia 5110566, Chile; guillervn7@gmail.com (G.E.V.); gkausel@gmail.com (G.K.)
4. Instituto de Medicina & Centro Interdisciplinario de Estudios del Sistema Nervioso (CISNe), Universidad Austral de Chile, Valdivia 5110566, Chile; ycheuquemilla@gmail.com
5. Centre for Gene Regulation and Expression, College of Life Sciences, University of Dundee, DD1 4HN, Dundee DD1 4HN UK; s.bandau@dundee.ac.uk (S.B.); R.T.Hay@dundee.ac.uk (R.T.H.)
6. Centro Interdisciplinario de Estudios del Sistema Nervioso (CISNe), Universidad Austral de Chile, Valdivia 5110566, Chile
7. Centro de Envejecimiento y Regeneración (CARE-UC), Facultad de Ciencias Biológicas, Pontificia Universidad Católica de Chile, Santiago 83330023, Chile

* Correspondence: alejandro.rojas@uach.cl (A.R.-F.); patricia.burgos@uss.cl (P.V.B.); Tel.: +56-63-2444315 (A.R.-F.); +56-2-22606309 (P.V.B.)

† Current address: Instituto de Microbiología Clínica, Facultad de Medicina, Universidad Austral de Chile, Valdivia 5110566, Chile.

‡ Current address: Institute of Biochemistry II, School of Medicine, Goethe University Frankfurt, Theoder-Stern-Kai 7, 60590 Frankfurt am Main, Germany.

Received: 25 February 2020; Accepted: 19 March 2020; Published: 23 March 2020

Abstract: Ubiquitination regulates several biological processes, however the role of specific members of the ubiquitinome on intracellular membrane trafficking is not yet fully understood. Here, we search for ubiquitin-related genes implicated in protein membrane trafficking performing a High-Content siRNA Screening including 1187 genes of the human "ubiquitinome" using amyloid precursor protein (APP) as a reporter. We identified the deubiquitinating enzyme PSMD14, a subunit of the 19S regulatory particle of the proteasome, specific for K63-Ub chains in cells, as a novel regulator of Golgi-to-endoplasmic reticulum (ER) retrograde transport. Silencing or pharmacological inhibition of PSMD14 with Capzimin (CZM) caused a robust increase in APP levels at the Golgi apparatus and the swelling of this organelle. We showed that this phenotype is the result of rapid inhibition of Golgi-to-ER retrograde transport, a pathway implicated in the early steps of the autophagosomal formation. Indeed, we observed that inhibition of PSMD14 with CZM acts as a potent blocker of macroautophagy by a mechanism related to the retention of Atg9A and Rab1A at the Golgi apparatus. As pharmacological inhibition of the proteolytic core of the 20S proteasome did not recapitulate these effects, we concluded that PSMD14, and the K63-Ub chains, act as a crucial regulatory factor for macroautophagy by controlling Golgi-to-ER retrograde transport.

Keywords: PSMD14; ubiquitin; retrograde; trafficking; APP

1. Introduction

Several post-translational modifications (PTMs) contribute to membrane transport [1,2]. Among these, ubiquitination acts as a relevant player in protein membrane trafficking [3–5] participating as an important protein localization signal [3,6–8]. However, to date, most studies have focused on its role in endocytosis, particularly in the inclusion of cargoes in intraluminal vesicles (ILVs) of multivesicular bodies (MVBs) [9–11] such as cell-surface receptors [12]. Ubiquitination is a regulated post-translational modification that conjugates ubiquitin (Ub) to lysine (K) residues and is involved in many cellular pathways including the degradation of target proteins through the proteasomes and autophagy [13,14]. Ubiquitin modification is a cascade of reactions catalyzed by three classes of enzymes (E1, E2 and E3). An ATP-dependent E1 activating enzymes first forms a covalent intermediate with ubiquitin, followed by the transfer of Ub to E2 conjugating enzymes and finally the covalent attachment of Ub to a K residue in the target protein by E3 Ub ligases [15]. Ubiquitination is a reversible reaction, with specific deubiquitinating enzymes (DUBs) that catalyze the removal of Ub-moieties for their recycling. DUBs serve to counterbalance ubiquitination reactions within the cell, thus dynamically contributing to the regulation of various cellular processes, such as endosomal sorting [11,16–18].

Ub increases proteome complexity, providing an additional surface for protein-protein interactions and functional regulation [19,20]. Ub-modified proteins and Ub chains are recognized by several Ub receptors including a family of specialized proteins carrying Ub binding domains (UBDs). To date, over 20 UBD families have been identified and characterized in mammals participating in the recognition of Ub monomers or Ub chains on specific substrates [4,14,21]. Together, the Ub network offers a variety of choices to modulate cellular processes including protein membrane trafficking. However, the contribution of the ubiquitinome regarding its impact on intracellular membrane trafficking is not yet fully understood.

To investigate the contribution of most of the members of the ubiquitinome in protein membrane trafficking, we performed high-content siRNA screening (HCS) including 1187 genes of the human "ubiquitinome". For this screening we used a stable cell line expressing amyloid precursor protein (APP) fused to the enhanced green fluorescent protein (EGFP). APP was the protein selected as target because it traffics dynamically through the secretory and endocytic pathways containing specific sorting signal motifs [22–24] and specific lysine residues targets of ubiquitination [25–28]. We thus identified the deubiquitinating (DUB) enzyme PSMD14, a subunit of the 19S regulatory particle (RP) of the proteasome, as a crucial player of Golgi-to-Endoplasmic reticulum (ER) retrograde transport. The DUB enzyme PSMD14 has been shown to be specific for K63-Ub chains in cells [29]. Here, we found that the inhibition of its activity blocks Golgi-to-ER retrograde transport, causing the swelling of the Golgi apparatus. We also found that PSMD14 inhibition acts as a potent blocker of macroautophagy as a result of its negative impact in Golgi-to-ER retrograde transport. Inhibition of the proteolytic core of the 20S proteasome by MG132 did not recapitulate these effects, indicating that the 19S RP plays a role in macroautophagy by controlling membrane trafficking at the early secretory pathway.

2. Materials and Methods

2.1. Chemical Reagents

Earle's balanced salt solution (EBSS) and the cocktail of protease inhibitors were purchased from Sigma-Aldrich (St. Louis, MO, USA). MG132 was purchased from Millipore (Burlington, MA, USA). Torin-1 was purchased from Tocris Bioscience (Bristol, UK). Dr. Raymond Deshaies from California Institute of Technology (Caltech, CA, USA) and Dr. Yuyong Ma from The University of California, CA, USA kindly donated Capzimin (CZM).

2.2. Antibodies

The following monoclonal antibodies were used: mouse anti-ubiquitin clone P4D1 (Cytoskeleton, Inc, Denver, CO, USA), mouse anti-β-actin clone BA3R (Thermo Fisher Scientific, Waltham, MA, USA), rabbit anti-RAB1A clone D3X9S, mouse anti-GM130 clone 35/GM130, rabbit anti-PSMD14 clone D18C7 (Cell Signaling Technology, Danvers, MA, USA) and rabbit anti-ATG9A clone EPR2450(2) (Abcam, Cambridge, UK). We used the following polyclonal antibodies: rabbit anti-ubiquitin (cat: Z0458, Dako, Carpintería, CA, USA), rabbit anti-giantin (cat: AB24586, Abcam, Cambridge, UK), rabbit anti-LC3 (cat: 2775S, Cell Signaling Technology, Danvers, MA, USA), rabbit anti-APP CT695 (cat: 51-2700, Thermo Fisher Scientific). Horseradish peroxidase-conjugated secondary antibodies were purchased from Jackson ImmunoResearch Laboratories (West Grove, PA, USA), and DAPI probe, Alexa and Dylight fluorophore-conjugated secondary antibodies were purchased from Thermo Fisher Scientific.

2.3. Cell Culture

H4 ATCC ® HTB-148™ Homo sapiens brain neuroglioma cells, referred to here as H4 human neuroglioma cells and HeLa ATCC ® CCL-2™ cells were obtained from the American Type Culture Collection (Manassas, VA, USA). Cell lines were cultured in Dulbecco's modified Eagle's medium (DMEM; Thermo Fisher Scientific) supplemented with 10% (*vol/vol*) heat-inactivated fetal bovine serum (FBS; Thermo Fisher Scientific), and penicillin/streptomycin (Thermo Fisher Scientific), in a 5% CO_2 atmosphere at 37 °C. The generation of the H4 stable cell line expressing hemagglutinin-tagged APP_{695}-F/P-D/A-EGFP (APP-EGFP) and the HeLa stable cell line expressing KDELR1-GFP were previously reported [30–32]. Stably transfected cells were maintained in culture medium supplemented with 100 µg/mL Cells were grown to sub confluence and then treated with drugs or transfected with siRNAs for further western blot and immunofluorescence analyses. Nutrient starvation assays were performed in the presence of EBSS. Assays to detect Mycoplasm were performed periodically.

2.4. High Content siRNA Transfection and Imaging

The primary siRNA screen for levels of APP-EGFP was performed in duplicates with our custom-assembled "Ubiquitinome" siRNA library, which consists of 1187 SMARTpools siRNAs targeting all known and assumed components of the ubiquitin and ubiquitin-like systems in 96-well format, as previously reported [33,34]. The H4 cells stably expressing APP-EGFP were reverse transfected [33,34] in µClear bottom 96-well plates (Greiner Bio-One, Kremsmünster, Austria) with ON-TARGETplus SMARTpools siRNAs (GE Dharmacon, Lafayette, CO, USA). Additionally, ON-TARGETplus Non-targeting pool (NT siRNA) was used together with untransfected cells (Mock) as neutral control; siRNA against GFP (GFP siRNA) was used as a positive control for APP downregulation. Briefly, 10 µL siRNA (200 nM) were stamped from 96-well plates of the library on to µClear bottom 96 well plates. Thereafter, 10 µL of Opti-MEM I Reduced Serum Medium (Thermo Fisher Scientific) containing Lipofectamine RNAiMAX transfection reagent (Thermo Fisher Scientific) was added to the siRNA (dilution 1:50). Plates were shaken at 900 rpm for 1 min followed by incubation for 20 min at room temperature. During this incubation, cells were resuspended in DMEM supplemented with 10% (*v*/*v*) FBS and 1× Normocin (InvivoGen, San Diego, CA, USA) and 6000 cells were dispensed onto each well loaded with medium for a final siRNA concentration of 20 nM. Plates were incubated for 72 h and further prepared for high content measurement.

After transfection, cells were stained using 100 µL of reagent per step, dispensed by the automated reagent dispenser XPP-721 (fluidX, Manchester, UK), according to the following protocol: two washes in phosphate buffered saline (PBS); fixation in 3.7% (*v*/*v*) paraformaldehyde for 10 min; one wash in PBS; incubation with 0.2% (*v*/*v*) Triton X-100 in PBS for 10 min; two washes in PBS; incubation with 0.1 µg/mL DAPI for 5 min; two washes with PBS. Finally, 100 µL of PBS was left in each well. Images were acquired and analyzed using the automated microscope IN Cell 2000 Analyzer (GE Healthcare, Little Chalfont, UK). Images of six randomized fields per well with ~600 cells each were acquired, and further analyzed the total fluorescence intensity in $n > 2000$ cells per condition. A secondary siRNA screening was performed in triplicate targeting the 35 most responsive hits, using each single siRNA duplex derived from the SMARTpools used in the primary siRNA screening.

2.5. siRNA Transfection for the siRNA Screening Validation Stage

Four single siRNA sequences targeting human PSMD14 (Accession number: NM_005805), derived from the ON-TARGETplus SMARTpool used in the siRNA Screening (Figure S1) were purchased from GE Dharmacon (Lafayette, CO, USA). siRNA transfections were carried out in 60 mm tissue culture plates using the Lipofectamine RNAiMax transfection reagent (Thermo Fisher Scientific) according to the manufacturer's protocol, and after 72 h cells were collected for further analysis.

2.6. RNA isolation and RT-qPCR Analysis

Total RNA extraction from H4 cells was carried out using the E.Z.N.A.® Total RNA Kit I (Omega Biotek, Norcross, GA, USA), and either purity (260/280 nm ratio and 260/230 nm ratio) or quantity (260 nm absorbance) were determined by spectrophotometry using NanoVue Spectrophotometer (GE Healthcare). The cDNA synthesis was performed from 2.5 μg of total RNA and oligo-dT and MMLV reverse transcriptase (Promega, Madison, WI, USA) according to supplier instructions. Specific primer pairs for tbp (NM_003194), psmd14 (NM_005805) and app (NM_000484) human genes were designed for quantitative reverse transcription PCR on cDNA template (RT-qPCR) (Figure S2). First, the specificity of amplicons was verified by cloning and sequencing, including tbp (223 bp), psmd14 (150 bp) and app (247 bp). mRNA levels were quantified in cDNA by qPCR with GoTaq qPCR Master Mix (Promega) according to supplier's instructions in a Mx3000 Real-Time Thermocycler (Stratagene, San Diego, CA, USA). In a 40-cycle PCR reaction, each cycle consisted of 20 s at 94 °C, 15 s at 55 °C and 15 s at 72 °C, followed by a final heating at 95 °C, revealing melting curves that confirmed single amplification products. All analyses were performed in triplicate. The expression level of each gene was normalized to tbp expression as reference gene using exon-spanning primers to control for genomic DNA contamination since no DNAse treatment of total RNA was included. RT-qPCR assays were analyzed with 2(-ΔΔCt) method [35] via MxPro software (Stratagene) and expressed as relative quantity to normalizer [36].

2.7. Preparation of Protein Extracts, Electrophoresis, SDS-PAGE and Western Blot Analysis

Cells were washed in ice-cold phosphate buffered saline (PBS) and lysed at 4 °C in lysis buffer (50 mM Tris-HCl pH 7.4, 150 mM NaCl, 1 mM EDTA, 1% (*v*/*v*) Triton X-100) supplemented with a cocktail of protease inhibitors (Sigma-Aldrich). All lysates were cleared by centrifugation at 16,000× *g* for 20 min at 4 °C, and protein concentration was determined with a protein assay dye reagent (Bio-Rad Laboratories, Hercules, CA, USA). Samples with an equivalent amount of protein were boiled for 5 min with Laemmli SDS-PAGE sample buffer, and then analyzed by SDS-PAGE. Proteins were electroblotted onto nitrocellulose membranes, blocked by incubation for 30 min in PBS containing 5% (*wt*/*vol*) free-fat dry milk, and incubated sequentially with primary and secondary antibodies, both diluted in blocking solution, for 1 h at room temperature, or overnight at 4 °C. Chemiluminescence protein detection was performed using SuperSignal West Pico (Thermo Fisher Scientific). β-actin was used as an internal loading control.

2.8. In vitro Proteasomal Activity Assay

Proteasome activity was quantitatively assessed in H4 cell extracts using the β5-selective fluorogenic substrate succinyl-leucine-leucine-valine-tyrosine-4-methyl-7-courmarylamide (Suc-LLVY-AMC, Calbiochem, Burlington, MA, USA) using an adapted protocol [37]. Briefly, cells were left untreated or treated for 4 h with different amounts of CZM (between 2 and 10 μM) or with 10 μM MG132. Cells were lysed in lysis buffer (20 mM Tris-HCl, pH 7.2, 1 mM EDTA, 1 mM dithiothreitol (DTT) and 0.1% (*v*/*v*) Nonidet P-40) supplemented with a cocktail of protease inhibitors (Sigma-Aldrich). Lysates were cleared by centrifugation at 16,000 × g for 20 min at 4 °C, and protein concentration was determined with a protein assay dye reagent (Bio-Rad Laboratories, Hercules, CA, USA). Soluble extracts (20 μg) were incubated for 60 min at 37 °C in proteasome assay buffer (50 mM Tris-HCl, pH 7.2, 1 mM DTT, 0,5 mM EDTA and 100 μM Suc-LLVY-AMC) all dispensed in triplicate into a 96-well black opaque plate (Nunc, Thermo Fisher Scientific). Proteasome activity was measured by monitoring the production of free AMC using a Synergy HT Multi-detection Microplate Reader (BioTek Instruments, VT, USA) with excitation and emission wavelengths of 360 and 460 nm, respectively, and the data were obtained by Gen5 Version 2.09.1 data analysis software (BioTek Instruments, VT, USA). Statistical significance was determined by One-Way ANOVA, followed by Dunnett's test. Value of $p < 0.01$(**) and $p < 0.001$(***) were regarded as statistically significant and are indicated in the figure.

2.9. Immunofluorescence

Cells grown on glass coverslips were washed with PBS and fixed in 4% (*v*/*v*) paraformaldehyde for 30 min at room temperature. After fixation, cells were washed in PBS and permeabilized with 0.2% (*v*/*v*) Triton X-100 in PBS for 10 min at room temperature. Cells were incubated with the indicated primary antibodies diluted in immunofluorescence buffer (PBS containing 10% (*v*/*v*) FBS and 0.1% (*w*/*v*) saponin for 30 min at 37 °C. Coverslips were washed in PBS and incubated with the corresponding Alexa-conjugated secondary antibody diluted in immunofluorescence buffer for 30 min at 37 °C. For nuclei staining, cells were washed with PBS and incubated for 10 min at room temperature with 0.1 mg/mL DAPI. After the final wash, coverslips were mounted onto glass slides with Fluoromount-G (SouthernBiotech, Birmingham, AL, USA).

2.10. Fluorescence Microscopy

Images of fixed cells were acquired by using a TCS SP8 laser-scanning confocal microscope (Leica Microsystems, Wetzlar, Germany) equipped with a 63× oil immersion objective (1.4 NA), 405 nm, 488 nm and 561 nm laser lines, with Photomultiplier (PMT), a hybrid detector system (Leica HyD) and the Leica Application Suite LAS X software. For quantification of fluorescent signals, 8-bit images were acquired under identical settings avoiding signal saturation and corrected for background signal on each image. The corrected fluorescent signal in each cell of each image was used in Image J (version 1.44o; Wayne Rasband, NIH, http://imagej.nih.gov) to determine the total integrated pixel intensity per cell area. Colocalization analyses were performed with sets of immunofluorescence images (Z-stack, with 0.2 μm intervals) of the same cells for each marker. Quantification of the acquired images was performed with the ICY software (Quantitative Image Analysis Unit, Institut Pasteur, http://icy.bioimageanalysis.org/) using the protocols plugin to create a pipeline to analyze the images in batch, the active contours plugin to perform the cell segmentation, the hk-means plugin for threshold detection, the wavelet spot detector plugin for spot detection and the colocalization studio plugin for colocalization analysis.

For live cell imaging assays, H4 cells were grown in glass bottom culture dishes (MatTek Corporation, Ashland, MA, USA) and transiently transfected for 48-h with KDELR-VSVG-YFP, using TransIT-LT1 Transfection Reagent (Mirus Bio LLC, Madison, Wi, USA) according to the manufacturer's protocol. Before the live cell imaging assay, the culture medium was replaced with phenol red-free DMEM supplemented with HEPES (10 mM, pH 7.4), and the cells were treated with CZM (10 μM) at 32 °C in a controlled temperature chamber in the TCS SP8 laser-scanning confocal microscope. Cells were kept at 32 °C to allow KDELR-VSVG-YFP localization on the Golgi, followed by a shift in temperature to 40 °C. Imaging was done with a 63x oil immersion objective (1.4 NA), running the Leica Application Suite LAS X software, acquiring 8-bit images at 1-min interval for 15 min at 40 °C (488 laser for excitation; HyD: 510–550 nm; 1024 × 1024 pixels; frame average 1). Quantification of the acquired images was performed with the MetaMorph Software version 7.0.

2.11. D Golgi Reconstruction and Golgi Volume and Area Measurements

H4 and HeLa cells stably expressing KDELR1-GFP, under the specified conditions, were fixed and immunostained with anti-GM130 or anti-Giantin, respectively. The immunofluorescence protocol was performed as described above to visualize the Golgi structure. For Golgi volume and area measurements, Z-stack (250 nm) fluorescence images were acquired by using a TCS SP8 laser-scanning confocal microscope (Leica Microsystems, Wetzlar, Germany) equipped with a 63x oil immersion objective (1.4 NA) running the Leica Application Suite LAS X software. Images were then processed with ImageJ software version FIJI to remove the background by using a threshold. The threshold images were then visualized using the "3D Viewer" plugin. The Golgi volume was quantified using ImageJ software version FIJI by setting a threshold region to select only the Golgi fluorescence. Then, the Golgi structures were separated by ROI (Regions of Interest), and the individual Golgi volume was measured with the plugin "Voxel Counting". The Golgi volume in μm^3 was determined by the number of voxels contained in the stack of images (voxel $0.2 \times 0.2 \times 0.25$ nm). The Golgi area was quantified using ICY software, and the Golgi marker GM130 was used to determine the ROI. To separate the specific signal from the background, the "k means threshold" plugin was used, using the same threshold level to all images. Data analysis was performed using GraphPad Prism 6 (GraphPad Software, La Jolla, CA, USA), and the results are represented in graphs depicting the mean ± SEM of at least 20 cells. The statistical significance of the data was determined with Student's T-test. The value of $p < 0.001$(***) was regarded as statistically significant and is indicated in the respective figures.

2.12. Densitometric Quantification and Statistical Analysis

The amount of immunoblot signal was estimated using Image J software version 1.48v (Wayne Rasband, NIH, http://imagej.nih.gov). For each condition, protein bands were quantified from at least three independent experiments in order to ensure adequate statistical power. Data analysis was performed using Microsoft Excel 2013 for Windows (Redmond, WA, USA) or GraphPad Prism Results are represented in graphs depicting the mean ± standard deviation. Statistical significance of data comparisons from two groups comparisons was determined with Student's T-test for parametric data. Values of $p < 0.05$ (*), $p < 0.01$ (**), $p < 0.001$ (***) were regarded as statistically significant and are indicated in the figures. The statistical significance of data from many groups was analyzed using One-Way ANOVA, followed by Tukey's test in order to evaluate pair-wise comparisons. The value of $p < 0.05$ was regarded as statistically significant and indicated in the figure in different letters above bars mean.

3. Results

3.1. High-Content siRNA Screening Revealed PSMD14 Deubiquitinating Enzyme as a Novel Regulator of Protein Trafficking

To investigate the contribution of the ubiquitinome in protein membrane trafficking, an automated HCS using a human small interfering RNA (siRNA) "ubiquitinome" library was performed consisting of 1187 siRNA duplex pools targeting all known and predicted Ub-genes (Supplementary Materials). This library was used previously in mammalian screening assays [33,34], and includes E1 activating enzymes (0.8%), E2 conjugating enzymes (4.7%), E3 Ub ligases (61.2%), UBD-containing proteins (12.8%), DUBs (8.8%), SUMO-specific proteases (SENPs) (0.2%) and others (11.4%) (Figure 1A). This library was used to identify novel regulators of protein membrane trafficking using an H4 neuroglioma cell line stably expressing APP-EGFP, a cell line previously characterized and used in siRNAs knockdown experiments [24,30,38]. We standardized the basal (background) and maximum fluorescence intensity by analyzing the total fluorescence of silenced and non-silenced APP-EGFP, respectively. A reduction in total fluorescence intensity from 100% to 6% was observed with the EGFP siRNA, compared to the non-target (NT) siRNA with a Z factor of 0.69 (Figure 1B). Representative images of basal and maximum levels of total fluorescence intensity in reporter cells are shown in Figure 1C. Cells were further assessed with a primary siRNA screening consisting of a pool of four-different siRNA duplexes for each of the 1187 genes selected. Each siRNA targeted a distinct region to minimize their off-targets effects [39]. Figure 1D shows a summary of the results with all siRNA pools tested and normalized in respect to the expression of the NT siRNA. Among the 1187 tested genes, we found that the highest total fluorescence increase (4.15-fold increase) corresponded to protein PSMD14 (POH1, also known as

Rpn11/MPR1/SS13/CepP1) (Figure 1D), a subunit of the 19S regulatory particle (RP) of the proteasome, which has DUB activity [40,41]. Representative images of reporter cells knock-down (KD) of PSMD14 in comparison to cells transfected with NT siRNA are shown in Figure 1E.

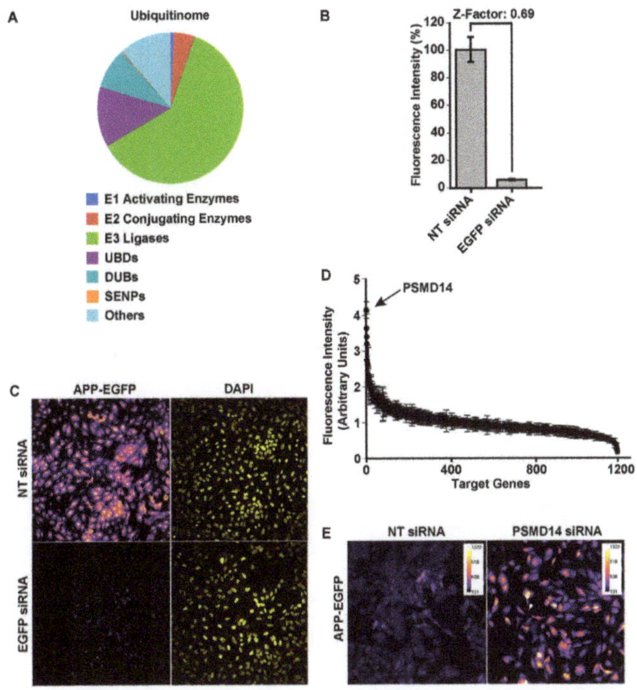

Figure 1. High-content siRNA screening assay revealed PSMD14 as a novel regulator of amyloid precursor protein (APP) levels. (**A**) Graphical distribution of the targets evaluated in the primary high content siRNA screening using the siRNA "ubiquitinome" library in H4 cells. (**B**) Quantification of the total fluorescence intensity of reporter APP-EGFP cells transfected for 72 h with NT siRNA and EGFP siRNA. Bars represent the mean ± SD with a statistical Z factor = 0.69. (**C**) High content images (20×) captured in reporter APP-EGFP cells transfected for 72 h with non-target (NT) siRNA and EGFP siRNA. (**D**) Graphical representation of total fluorescence intensity of all 1187 genes analyzed in primary siRNA screening with the reporter APP-EGFP cells. PSMD14 appears indicated as the top hit. (**E**) High content images (20×) in pseudo color of reporter APP-EGFP cells transfected for 72 h with siRNA SMARTpool targeted against PSMD14 (PSMD14 siRNA) in comparison to cells transfected with NT siRNA. The fluorescence intensity in these images was reduced to avoid saturation with the PSMD14 siRNA. Scale Bar of the images indicates the scale of fluorescence intensity.

Further, we validated these results by analyzing the effect of PSMD14 KD on the levels of full-length endogenous APP in parental H4 cells by western blot. The effect of all four PSMD14 siRNAs in KD cells tested on endogenous APP levels are shown in Figure 2A (first panel, lanes 3-6) and compared to un-transfected cells (Mock) or cells transfected with NT siRNA (Figure 2A, first panel, lanes 1 and 2, respectively). In addition, we found that silencing PSMD14 caused a strong increase in high molecular weight Ub conjugates, consistent with the role of PSMD14 as a proteasomal subunit with DUB activity (Figure 2A, second panel, lanes 3–6 compared to lanes 1 and 2). For quantification analysis, we used the most effective siRNA for the reduction in PSMD14 expression that corresponded to siRNA#1 according to quantitative reverse transcription PCR (RT-qPCR) (Figure 2D). We found a 3.13-fold increase in full-length endogenous APP levels in PSMD14 KD cells, compared to Mock cells (Figure 2B).

In addition, we observed a 5.22-fold decrease in PSMD14 levels (Figure 2C). Moreover, efficiency KD of PSMD14 was determined by RT-qPCR (Figure 2D) and confirmed by western blot (Figure 2A, third panel, lanes 3–6 compared to lanes 1 and 2). Finally, to evaluate whether the increase in APP protein levels in PSMD14 KD cells could be the result of an up-regulation of APP transcription, APP mRNA levels were evaluated by RT-qPCR observing no significant changes (Figure 2E). Altogether, these findings indicate that the silencing of PSDM14 caused a robust increase in APP protein levels, a finding that could suggest impairment in its turnover due to trafficking alterations.

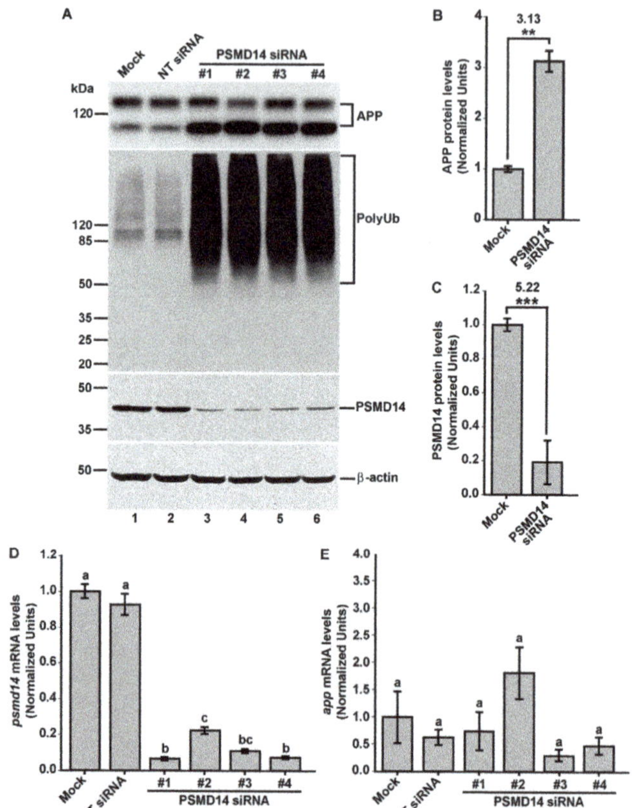

Figure 2. PSMD14 is validated as a regulator of the endogenous APP levels. (**A**) Protein extracts of parental H4 cells either untransfected (Mock), transfected with NT siRNA, or transfected with four different PSMD14 siRNA sequences for 72 h were analyzed by western blot. Polyclonal antibodies to endogenous APP (CT695) and to Ub (that recognizes all types of Ub conjugates), and monoclonal antibodies to PSMD14 (clone D18C7) and to β-actin (clone BA3R), were tested. The position of molecular mass markers are indicated on the left. Densitometric quantification of the levels of endogenous APP (**B**) and PSMD14 (**C**) in H4 cells transfected with PSMD14 siRNA#1, compared to untransfected cells (Mock). Statistical significance was determined by Student's t-test. Bars represent the mean ± SD of biological replicates (APP n =5; PSMD14 n = 4). **$p < 0.01$ and ***$p < 0.001$. (**D**) mRNA levels of *psmd14* and (**E**) mRNA levels of *app* were measured using RT-qPCR from parental H4 cells transfected for 72 h. All data were normalized for TATA binding protein expression in either untransfected cells (Mock), cells transfected with NT siRNA or cells transfected with four different PSMD14 siRNAs duplexes. Statistical significance was determined by One-Way ANOVA, followed by Tukey's test. Bars represent the mean ± SD of biological replicates (*psmd14* n = 3; *app* n = 3). Different letters above the mean bars apply to significant differences between groups $p < 0.01$.

3.2. Acute Inhibition of the Deubiquitinating Enzyme PSMD14 of the 19S RP Accumulates APP in a Swollen Golgi Apparatus

To determine whether PSMD14 is involved in the control of endogenous APP levels, we investigated the effect of incubating cells with Capzimin (CZM), a potent and specific inhibitor of the DUB activity of PSMD14 [42], a subunit of the 19S RP of the proteasome (Figure 3A). We observed that CZM led to a significant increment in full-length endogenous APP levels in a dose-dependent manner, compared to untreated cells (Figure 3B,C). Interestingly, compared to untreated cells, and in contrast to CZM, treatment with a standard concentration of 10 μM MG132, a reversible inhibitor of the β1, β 2 and β 5 subunits of the 20S catalytic core of the proteasome [43,44] (Figure 3A), caused a significant decrease in full-length endogenous APP levels (Figure 3B,C). To confirm the inhibition of PSMD14 by CZM, the effect of this inhibitor on the levels of high molecular weight Ub conjugates was tested, comparing it with MG132. In agreement with the effect of PSMD14 KD, it was observed that CZM caused a robust increase in high molecular weight Ub conjugates in a dose-dependent manner, finding a maximum effect with 10 μM CZM (Figure 3D). However, we noticed that 10 μM MG132 caused a more powerful increase in Ub conjugates (Figure 3D). To confirm the accumulation of Ub conjugates by these treatments, we performed immunofluorescence analysis. We found that 10 μM CZM (Figure 3E, middle panel) and 10 μM MG132 (Figure 3E, right panel) showed the appearance of Ub conjugates puncta, compared to untreated cells (Figure 3E, left panel), which showed the expected cytosolic distribution of Ub. Together, our results confirm that acute inhibition of PSMD14 by CZM replicates the phenotype obtained by PSMD14 KD regarding the impact on the endogenous APP levels. Thus, CZM offers a pharmacological tool addressing whether the activity of PSDM14 is required for APP membrane trafficking.

To unveil this possibility, immunofluorescence analysis of endogenous APP in parental H4 cells was performed. We observed that the treatment with 10 μM CZM resulted in a perinuclear redistribution of APP, which is highly indicative of Golgi apparatus localization (Figure 4D, compared to 4A). Moreover, similar to the PSMD14 KD, it was observed that the CZM treatment caused a significant 2.50-fold increase in APP total fluorescence intensity compared to untreated cells (Figure 4G). We also observed a significant 1.92-fold increase in the amount of APP in the area positive to the Golgi matrix protein GM130 (GM130), compared to the total area (Figure 4F, compared to 4C and Figure 4H). Moreover, we observed that the treatment with CZM caused a significant 1.30-fold increase in the total cell area (Figure 4I). In addition, we found that CZM caused the swelling of the Golgi apparatus (Figure 4E compared to Figure 4B), a phenotype that is shown in a zoom in Figure 4J. To confirm this phenotype, measurements of GM130 images were performed from confocal 3D reconstructions from the Z-stacks of cells treated with CZM (Figure 4K). We observed a 1.53-fold increase in the Golgi apparatus volume in CZM treated cells, compared to control cells (Figure 4L). Similar results were observed in quantitative 2D image analysis, observing a significant 2.53-fold increase in the Golgi apparatus area upon CZM treatment (Figure S3). Altogether, our results show that acute inhibition of PSMD14 by CZM accumulates APP in a swollen Golgi apparatus. These findings strongly suggest that CZM impairs the Golgi apparatus due to perturbations of the trafficking through this organelle.

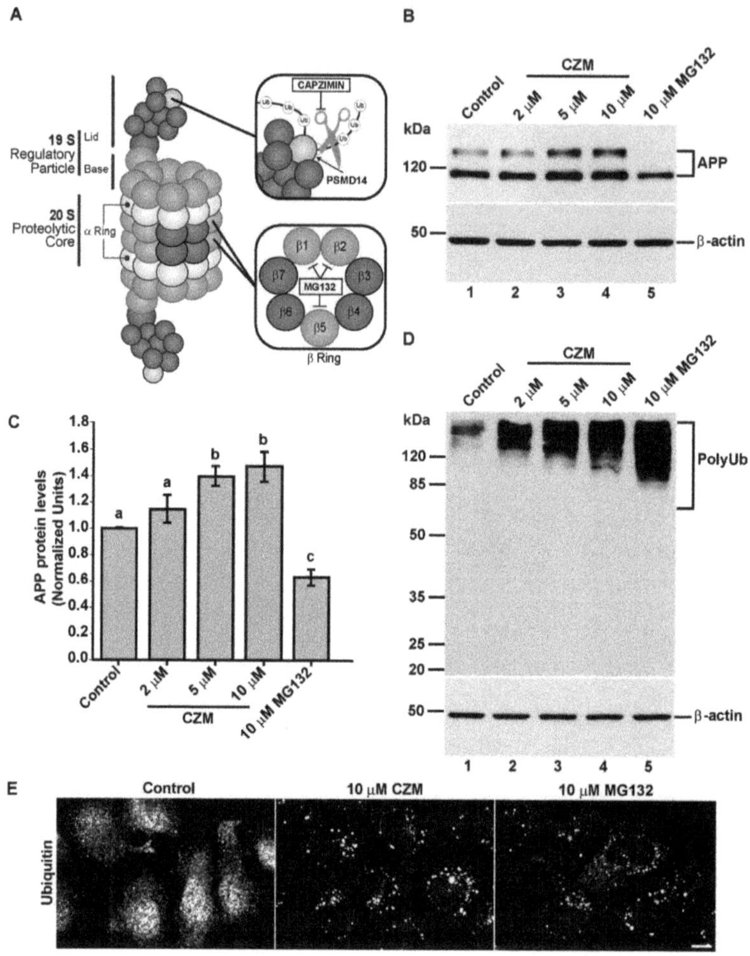

Figure 3. Acute inhibition of PSMD14 by Capzimin CZM shows a similar phenotype as that of PSMD14 KD on the levels of APP and high molecular weight Ub conjugates. (**A**) Schematic diagram of the molecular targets of Capzimin and MG132 in the 19S RP and 20S catalytic core of the proteasome, respectively. (**B**) Parental H4 cells were treated either with vehicle (DMSO; Control), or increasing doses of CZM for 4 h, or MG132 for 6 h. Protein extracts were analyzed by western blot with a polyclonal antibody to endogenous APP. Monoclonal antibody to β-actin (clone BA3R) was used as a loading control. The position of molecular mass markers is indicated on the left. (**C**) Densitometric quantification of APP protein levels as shown in (**D**). Statistical significance was determined by one-way ANOVA, followed by Tukey's test. Bars represent the mean ± SD of biological replicates (n = 4). Different letters above the mean bars apply to significant differences between groups $p < 0.05$. (**D**) Parental H4 cells were treated as in (**B**), and the protein extracts were analyzed by western blot with a polyclonal antibody to Ub that recognizes all types of Ub conjugate. Monoclonal antibody to β-actin (clone BA3R) was used as a loading control. The position of molecular mass markers is indicated on the left. (**E**) Immunofluorescence microscopy images of the cellular localization of Ub in parental H4 cells treated with either the vehicle (DMSO; Control), CZM for 4 h or MG132 for 6 h. Cells were fixed, permeabilized and stained with a mouse monoclonal antibody to Ub (clone P4D1) followed by Alexa-488-conjugated donkey anti-mouse IgG. Scale bar, 10 µm. (n = 3).

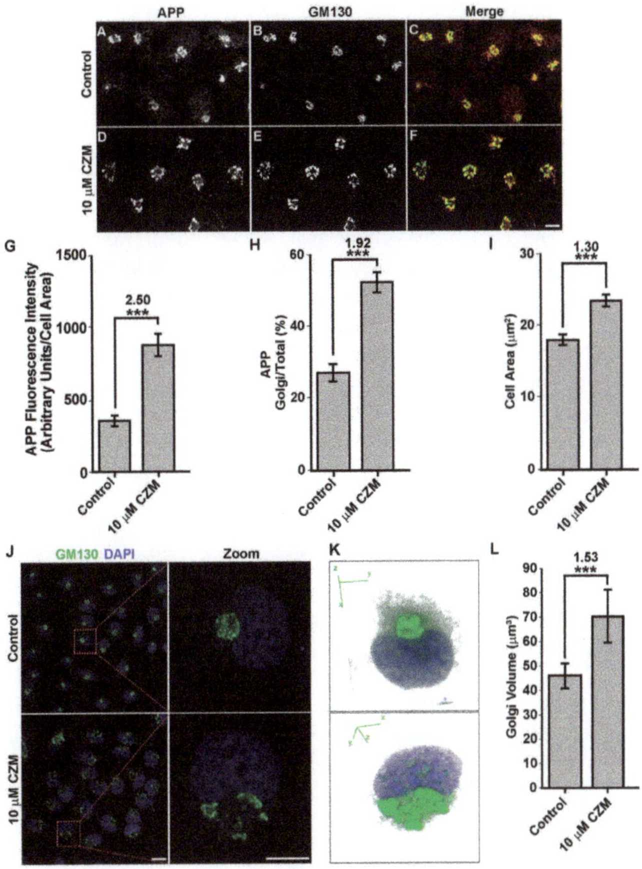

Figure 4. Acute inhibition of PSMD14 by CZM triggers the accumulation of APP in a swollen Golgi apparatus. Immunofluorescence analysis of endogenous APP in H4 parental cells treated either with the vehicle (DMSO; Control) (**A–C**) or CZM (**D–F**) for 4 h. Cells were fixed, permeabilized, and double stained with a rabbit polyclonal antibody to APP (CT695) (**A**,**D**) and a mouse monoclonal antibody to GM130 (clone35/GM130) (**B**,**E**), followed by Alexa-594-conjugated donkey anti-Rabbit IgG and Alexa-488-conjugated donkey anti-Mouse IgG. Merging of the images generated the third picture (**C**,**F**). Scale bar, 10 mm. (**G**) Quantitative analysis of the mean of total fluorescence intensity of APP upon treatment with CZM, in comparison to control cells. The statistical significance was determined by Student's t-test. Bars represent the mean ± SD of the fluorescent signal per cell area (n = 43 cells). ***$p < 0.001$. (**H**) Quantitative analysis of the fraction of APP colocalizing with GM130 under CZM treatment and compared to control cells. Statistical significance was determined by Student's t-test. Bars represent the mean ± SD of the fluorescent signal per cell area (n = 43 cells). ***$p < 0.001$. (**I**) Quantitative analysis of the cell area. Statistical significance was determined by Student's t-test. Bars represent the mean ± SD of the cell area (n = 43 cells) **$p < 0.001$. (**J**) Immunofluorescence microscopy analysis of GM130 in parental H4 cells treated either with the vehicle (DMSO; Control) or CZM for 4 h. Cells were fixed, permeabilized and stained with mouse monoclonal antibody to GM130 (clone 35/GM130) followed by Alexa-488-conjugated donkey anti-mouse IgG, and nuclei were stained with DAPI. Scale bar, 10 µm. (**K**) 3D reconstructions of the Golgi apparatus using GM130 as Golgi marker were generated from Z-stacks (250 nm). (**L**) Golgi Volume was measured from 3D reconstructions as shown in (**K**). Statistical significance was determined by Student's t-test. Bars represent the means ± SEM (n = 20 cells). *** $p < 0.001$.

3.3. Acute Inhibition of the Deubiquitinating Enzyme PSMD14 Perturbs Golgi-to-ER Retrograde Transport

The Golgi apparatus is a highly dynamic organelle that requires fine regulation of trafficking pathways in order to maintain its size, shape and composition. In particular, it has been shown that Golgi-to-ER retrograde transport plays a crucial role in the maintenance of Golgi morphology. Indeed, inhibition of this specific trafficking pathway results in a significant swelling of this organelle [32,45–49]. To investigate whether the swelling of the Golgi by the inhibition of PSMD14 with CZM could be the result of the inhibition of the Golgi-to-ER retrograde transport, we first validated our findings in HeLa cells. We measured Golgi apparatus volume in 3D reconstructions from the Z-stacks of untreated HeLa cells or treated with CZM and MG132 by using Giantin as a Golgi apparatus reporter. Similar to our findings in H4 cells, we found that the inhibition of PSMD14 DUB activity by CZM causes a 2.14-fold increase in Golgi apparatus volume (Figure 5A, middle panel compared to the left panel and 5B). Interestingly, in contrast to CZM, we observed that MG132 caused no effect on Golgi apparatus volume (Figure 5A, right panel compared to the left panel and 5B), strongly suggesting that Golgi homeostasis is regulated by the PSMD14 DUB activity. Then, we investigated whether this phenotype was the result of Golgi-to-ER retrograde transport inhibition. We assessed the distribution of stably overexpressed KDEL (Lys-Asp-Glu-Leu) Receptor 1 (KDELR1) fused to GFP (KDELR1-GFP) in HeLa cells. As was previously reported [32,50], overexpressed KDELR1-GFP is mostly distributed to the ER due to its efficient transport from the Golgi apparatus to the ER (Figure 5C, left panel and 5D). Thus, impairment of Golgi-to-ER retrograde transport causes the accumulation of KDELR1-GFP in the Golgi apparatus, working as an assay to identify novel regulators of this pathway [32]. Interestingly, we found that CZM caused a rapid time-dependent accumulation of KDELR1-GFP at the Golgi apparatus, observing a maximal effect after 90 min of treatment (Figure S4). At this time, we observed a significant 1.50-fold increase in the amount of KDELR1-GFP within the Giantin-positive Golgi apparatus, compared to the total area (Figure 5C, middle panel compared to the left panel, and 5D). As before, we tested the effect of MG132, observing no effect on the distribution of KDELR1-GFP (Figure 5C, right panel compared to left panel, and 5D), which strongly supports that MG132 has no effect on Golgi-to-ER retrograde transport. Next, to confirm the blockage of Golgi-to-ER retrograde transport by the acute inhibition of the PSMD14, we tracked the redistribution of the transiently overexpressed thermo-sensitive KDELR1-VSVG-YFP chimera (vesicular stomatitis virus G protein fused to KDEL receptor 1 and YFP) in HeLa cells upon treatment with CZM by live cell imaging at different temperatures. Briefly, at a permissive temperature of 32 °C, KDELR1-VSVG-YFP cycles between the Golgi apparatus and the ER, showing distribution mainly at the Golgi apparatus. Upon shifting to a restrictive temperature of 40°C KDELR1-VSVG-YFP is progressively accumulated at the ER due to its specific retention in this compartment at this temperature. In addition, as the Golgi-to-ER retrograde transport is not affected at 40°C, the rapid decay of KDELR1-VSVG-YFP fluorescence at the Golgi apparatus is a measurement used to evaluate inhibition of Golgi-to-ER retrograde transport pathway [51]. We found that in control cells, nearly 50% of the fluorescence of KDELR1-VSVG-YFP decayed after 15 min of shifting the temperature to 40 °C (Figure 5E, upper panels and 5F) indicating that retrograde transport is working normally [51]. In contrast, when cells were treated with CZM, the decay of KDELR1-VSVG-YFP fluorescence was not apparent, confirming that acute inhibition of the PSMD14 blocked Golgi-to-ER retrograde transport (Figure 5E, lower panels and 5F). In contrast, MG132-treated cells showed similar results to controls (data not shown), confirming that the 20S proteasome has no effect on retrograde transport. Importantly, we found that CZM is able to block retrograde transport for short times having no effect on the catalytic activity of the 20S proteasome (Figure S5). In contrast, a rapid and robust inhibition of the catalytic activity of the 20S proteasome was observed with MG132 (Figure S5). These findings strongly indicate that acute inhibition of PSMD14 DUB activity by CZM acts as a powerful blocker of Golgi-to-ER retrograde transport, explaining the swelling of the Golgi apparatus and the accumulation of protein cargoes such as APP at this location.

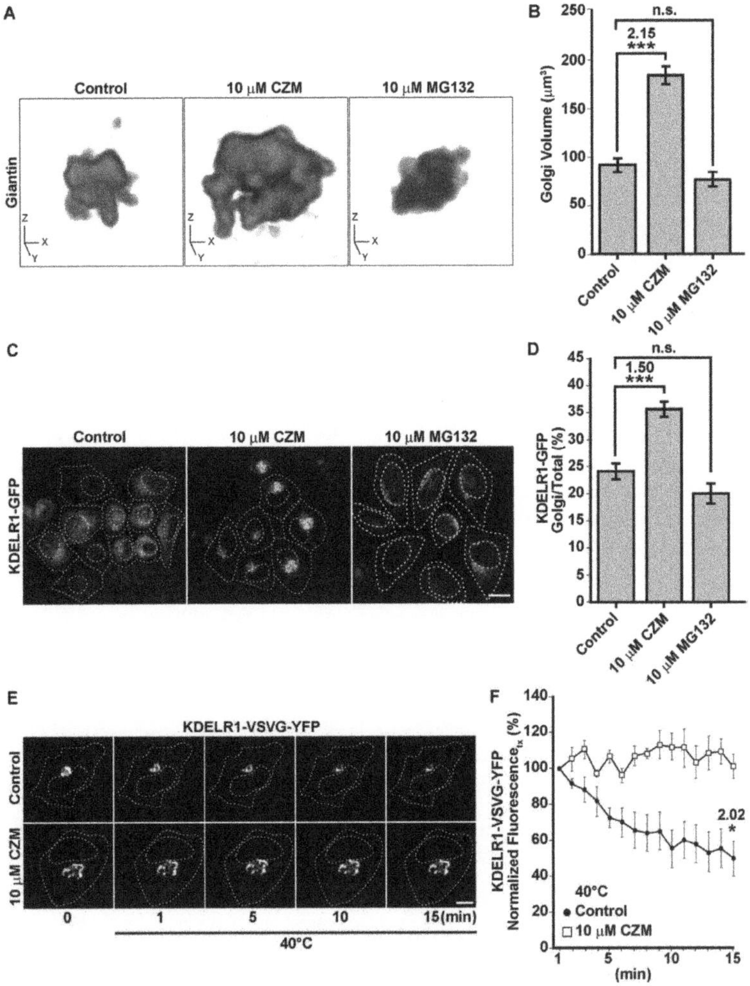

Figure 5. The PSMD14 DUB inhibitor CZM impairs Golgi-to-ER retrograde transport. (**A**) Three-dimensional reconstructions of the Golgi apparatus using giantin as a Golgi marker were generated from Z-stacks (250 nm) obtained from HeLa cells stably expressing KDELR1-GFP treated for 90 min either with vehicle (DMSO; Control), CZM or MG132. (**B**) Golgi volume was measured from 3D reconstructions as shown in (**A**). Statistical significance was determined by Student's t-test. Bars represent the means ± SEM (n = 30 cells). *** $p < 0.001$. (**C**) HeLa cells stably expressing KDELR1-GFP were treated for 90 min either with vehicle (DMSO; Control), CZM or MG132. Cells were fixed, and representative confocal images were acquired. (**D**) Measurement of giantin and total KDELR1-GFP total fluorescent intensity. Statistical significance was determined by Student's t-test. Bars represent the means ± SEM (n = 34 cells). *** $p < 0.001$. (**E**) H4 cells were transiently transfected to express the thermo-sensitive retrograde transport reporter KDELR-VSVG-YFP. Cells were kept at 32 °C to allow KDELR-VSVG-YFP localization at the Golgi. Cells were then shifted to 40 °C (restrictive temperature) and images acquired at 1 min interval for 15 min. (**F**) Quantitative image analysis was performed to measure the integrated fluorescence of KDELR-VSVG-YFP at the Golgi at 1 min interval for 15 min. Statistical significance was determined by Student's t-test. Bars represent the mean ± SEM (n = 3 cells). *$p < 0.05$.

3.4. Inhibition of Golgi-to-ER Retrograde Transport by CZM Has a Negative Impact on Macroautophagy

Several lines of evidence have shown that Golgi-to-ER retrograde transport plays a relevant role in autophagosome biogenesis at the level of the ER [52–55]. In addition, it has been shown that PSMD14 participates in the activation of the aggresome clearance by cleaving K63 Ub chains of aggregate proteins [56,57]. Thus, we investigated the effect of acute inhibition of PSMD14 DUB activity on the levels of the classical marker microtubule-associated protein 1 light chain 3B (LC3B) [58], compared to the treatment with MG132 (Figure 6A). We observed that CZM alone did not increase the levels of LC3B-II (Figure 6A and Figure S6B). As expected, and in contrast to CZM, higher levels of LC3B-II were found with MG132 treatment (Figure 6A). In agreement with these findings, we observed that CZM alone did not increase the number of autophagosomes (Figure 6E compared to 6B), in contrast to the effect of MG132 (Figure S6A). In addition, parental H4 cells were treated under nutrient starvation with Earle's balanced salt solution (EBSS), an established culture medium used for the activation of autophagosomal formation [59]. As expected, we found that starvation strongly increased the number of autophagosomes (Figure 6C), compared to cells treated under normal nutrients conditions (Figure 6B). In contrast, it was observed that treatment with CZM abolished the appearance of autophagosomes upon EBSS treatment (Figure 6D compared to 6C). We also tested the effect of CZM in cells under normal nutrients conditions, but in the absence or presence of Torin-1, a potent and selective inhibitor of mammalian target of rapamycin complexes (mTORC1/2), a trigger of autophagosomal formation [60]. Similar to the results with EBSS, we observed that CZM prevented the appearance of autophagosomes promoted by Torin-1 (Figure 6G compared to 6F). To confirm these results, we quantified the number of LC3-positive structures under all conditions tested, observing a significant decrease in the number of autophagosomes when cells, treated with EBSS solution or Torin-1, were also treated with CZM (Figure 6H). In addition, we biochemically validated these results performing western blot analysis of endogenous LC3B (Figure 6I). We found that the levels of LC3B-II were strongly increased with EBSS or Torin-1 treatment (Figure 6I, lanes 2 and 4). In contrast, when cells were treated with EBSS or Torin-1 in the presence of CZM, LC3B-II levels did not change (Figure 6I, lanes 3 and 5) compared to control cells (Figure 6I, lane 1), results that are quantified and depicted in Figure 6K. Interestingly, this biochemical analysis also showed that CZM increased LC3B-I levels under all conditions tested (Figure 6I, lanes 3, 5 and 6, and Figure 6J) compared to control cells (Figure 6I, lane 1) suggesting that the reduction of the autophagosomal structures caused by CZM was not due to a reduction in LC3B total levels. Altogether, these results confirm that acute inhibition of PSMD14 DUB activity acts as a potent blocker of autophagosome biogenesis induced by EBSS or Torin-1. In addition, these findings suggest that blockage of autophagosomal biogenesis could be a consequence of Golgi-to-ER retrograde transport inhibition.

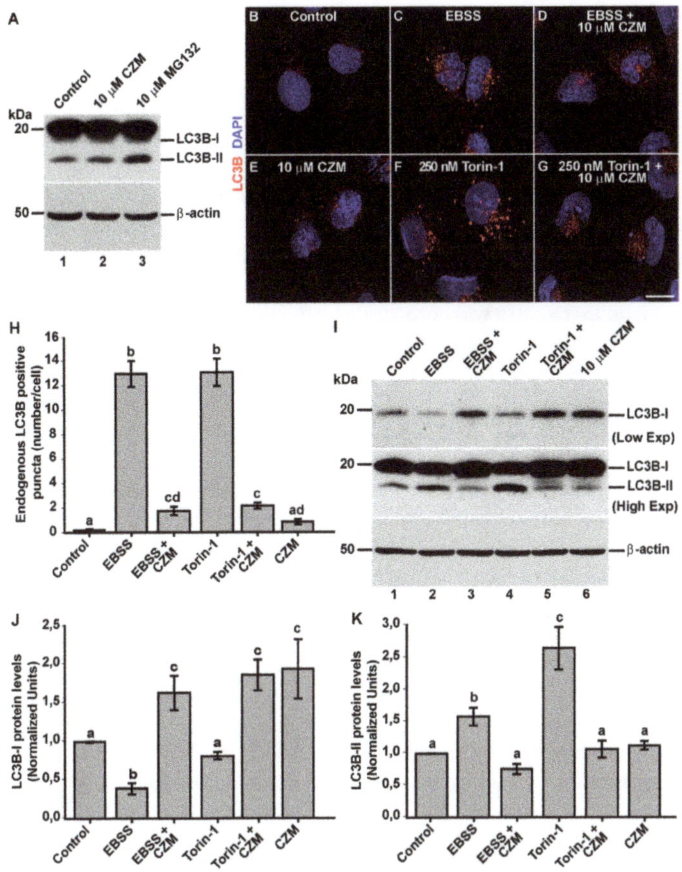

Figure 6. Inhibition of autophagosome formation by CZM. (**A**) Parental H4 cells were treated either with the vehicle (DMSO; Control), CZM for 4 h or MG132 for 6 h and protein extracts were analyzed by western blot with a polyclonal antibody to LC3B. Monoclonal antibody to β-actin (clone BA3R) was used as a loading control. The positions of the molecular mass markers are indicated on the left. Immunofluorescence microscopy analysis of the subcellular localization of endogenous LC3B in parental H4 cells treated with either the vehicle (DMSO; Control) (**B**), EBSS for 4 h (**C**), CZM for 6 h (**E**) or Torin-1 for 4 h (**F**). EBSS (**D**) and Torin-1 (**G**) were tested using a 2-h pretreatment with CZM followed by the treatment with EBSS or Torin-1 for 4 h in the presence of CZM. Cells were fixed, permeabilized and stained with a rabbit polyclonal antibody to LC3B followed by Alexa-594-conjugated donkey anti-Rabbit IgG, and nuclei were stained with DAPI. Scale bar 10 μm. (**H**) Quantification of the puncta positive to LC3B. Statistical significance was determined by one-way ANOVA, followed by Tukey's test. Bars represent the mean ± SEM (n = 50 cells). Different letters above the mean bars indicate the significant differences between groups $p < 0.05$. (**I**) Protein extracts from parental H4 cells treated as in (**B–G**) were analyzed by western blot with a rabbit polyclonal antibody to LC3B. Monoclonal antibody to β-actin (clone BA3R) was used as a loading control. The position of molecular mass markers is indicated on the left. (**J**) Densitometric quantification of LC3B-I levels and (**K**) LC3B-II levels. Statistical significance was determined by One-Way ANOVA, followed by Tukey's test. Bars represent the mean ± SEM of biological replicates (LC3B-I n = 3; LC3B-II n = 3). Different letters above the mean bars indicate the significant differences between groups $p < 0.05$.

3.5. Inhibition of Golgi-to-ER Retrograde Transport by CZM Accumulates RAB1A and ATG9A at the Golgi Apparatus

To evaluate this hypothesis, we tested the effect of acute inhibition of the PSMD14 DUB activity on the distribution of proteins implicated in the initial steps of autophagosome formation, which traffics in early compartments of the secretory pathway. We first tested RAB1A, a small GTPase with an essential role in the initiation of autophagy, facilitating the recruitment of the unc-51-like kinase 1 (ULK1) complex to subdomains of the ER, a crucial early step during autophagosome formation [61,62]. We found that CZM treatment caused a redistribution of RAB1A to the perinuclear zone (Figure 7A). Measuring these images confirmed a significant increase in RAB1A in this area (Figure 7B), a result that was accompanied by a decrease in RAB1A in the cell periphery defined as radial zone (Figure 7B). Importantly, we found that RAB1B was not affected by the same condition (data not shown). Moreover, we observed that RAB1A was distributed in the swollen Golgi apparatus, similar to GM130, upon CZM treatment (Figure S7). With the same aim, we studied the distribution of ATG9A, an essential transmembrane protein involved in macroautophagy, which plays a crucial role in the early steps of autophagosome formation [63,64]. ATG9A-containing vesicles are formed from the Golgi apparatus translocating to the ER to form the initiation site at the ER for autophagosome formation [65]. Similar to RAB1A, we found that acute inhibition of PSMD14 caused a redistribution of ATG9A to the perinuclear zone, together with a decrease in its distribution in the radial zone (Figure 7C). The quantification analysis of these images is shown in Figure 7C. Moreover, and similar to RAB1A, we observed that ATG9A is distributed to the swollen GM130-Golgi apparatus area upon CZM treatment (Figure S8). These results strongly indicate that blockage of Golgi-to-ER retrograde transport by acute inhibition of PSMD14 DUB activity causes the retention at the Golgi apparatus of key proteins implicated in early steps of autophagosome formation. The PSDM14-dependent inhibition of autophagy explains the accumulation of APP at the Golgi apparatus since macroautophagy has been recently demonstrated as a positive regulator of protein secretion from the Golgi apparatus [66]. Collectively, these results show the strong interplay between membrane transport and autophagy through a novel mechanism involving the proteasome complex through the deubiquitinating activity of PSMD14.

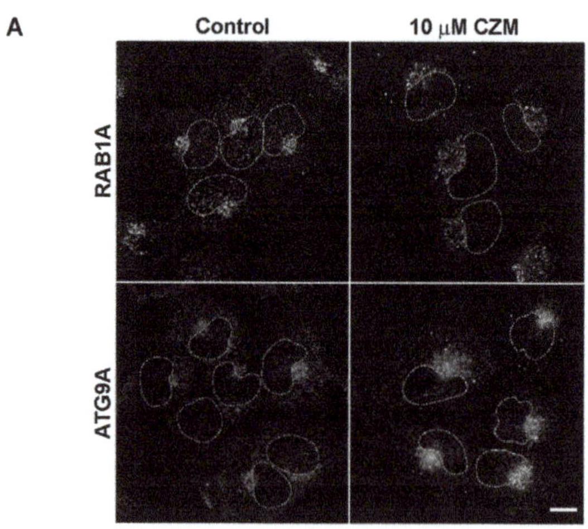

Figure 7. *Cont.*

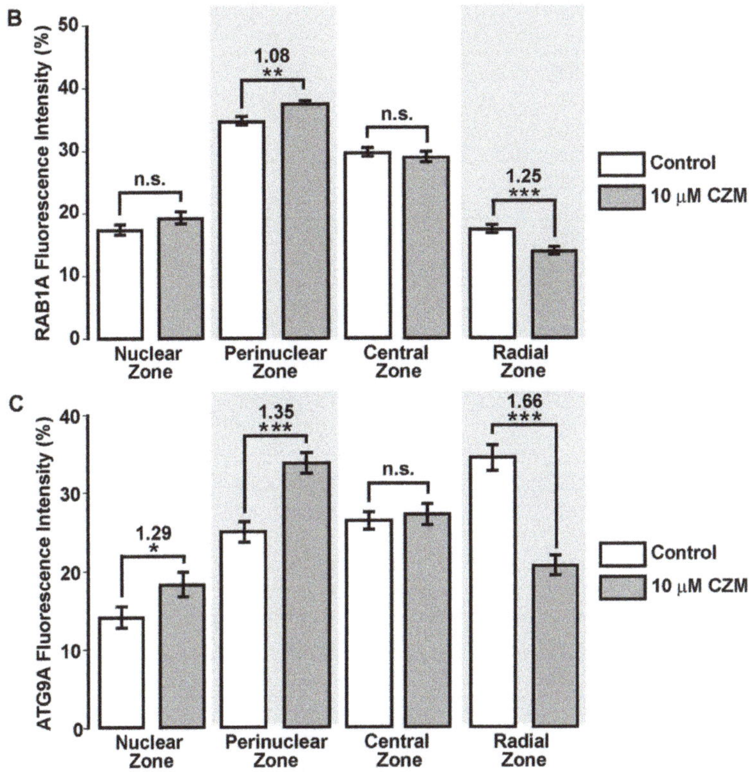

Figure 7. Redistribution of RAB1A and ATG9A to the Golgi apparatus with CZM. (**A**) Immunofluorescence analysis of endogenous RAB1A and ATG9A in H4 parental cells treated for 4 h either with the vehicle (DMSO; Control) (left panel) or CZM (right panel). Cells were fixed, permeabilized, and stained with a rabbit monoclonal antibody to RAB1A (clone D3X9S) (upper panel) and a rabbit monoclonal antibody to ATG9A (clone EPR2450(2)) (lower panel), followed by Alexa-594-conjugated donkey anti-Rabbit IgG. Scale Bar, 10 μm. (**B**) Quantitative analysis of the fluorescence intensity of RAB1A upon treatment with CZM, in comparison to control cells. Statistical significance was determined by Student's t-test. Bars represent the mean ± SEM of the fluorescent signal per cell area (n = 227 cells). ** $p < 0.01$; *** $p < 0.001$; n.s., not significant. (**C**) Quantitative analysis of the fluorescence intensity of ATG9A upon treatment with CZM, in comparison to control cells. Statistical significance was determined by Student's t-test. Bars represent the mean ± SEM of the fluorescent signal per cell area (n = 95 cells). * $p < 0.05$; *** $p < 0.001$; n.s., not significant.

4. Discussion

We report here that PSMD14 DUB activity, a subunit of the 19S RP of the proteasome, functions as a novel regulator of autophagosome formation. To our knowledge, this is the first report demonstrating that impairment of the proteasome can have a negative impact on the initiation of macroautophagy. In general, several studies have shown that the inhibition of the 20S catalytic core with the use of blockers of the β-subunits triggers the enhancement of the biogenesis of LC3B-positive autophagosomes [67–72]. In this regard, we showed that MG132 caused an increment of the LC3II/LC3I ratio and a significant decrease in full-length endogenous APP levels, strongly indicating a reduction in APP by macroautophagy [38,73–76]. Here, we show that blockage of the PSMD14 DUB activity, a component of the 19S RP, plays a negative role in the biogenesis of LC3B-positive autophagosomes, given new insights about the control of macroautophagy.

In addition, and because only CZM but not MG132 blocked Golgi-to-ER retrograde transport, a pathway implicated in the initiation of autophagosomes [52–55], we postulated that PSMD14 DUB activity controls macroautophagy by a process independent of bulk proteasomal degradation but dependent on K63-Ub chains. In agreement with this possibility, K63-Ub chains have been directly involved in the control of protein membrane trafficking in C. *elegans* [77] and as a regulator of mTORC and macroautophagy [78,79]. Likewise, free unanchored K63-Ub chains released in a PSMD14-dependent manner have shown to be crucial in coordinating the elimination of protein aggregates by macroautophagy [56], relocating the aggregates to the aggresome for final autophagic clearance by a mechanism related with deacetylase HDAC6 activity [80]. Here, we unveil the first insight into the regulation of Golgi-to-ER retrograde transport by K63-Ub chains, highlighting the deubiquitinating enzyme PSMD14 as a key regulator in the control of early events of the secretory pathway. In this regard, and in agreement with the role of the deubiquitination in membrane protein trafficking [81,82], we hypothesize that Golgi-to-ER retrograde transport must be controlled by the deubiquitination of cytosolic proteins modified with K63-Ub chains.

In this regard, key regulatory proteins involved in Golgi-to-ER transport are regulated by the state of ubiquitination/deubiquitination. In yeast, the deletion of the DUB Ube3p and its co-factor Bre5p accumulates ubiquitinated β´-COP facilitating its rapid degradation by the proteasome [81,83]. As β´-COP is a subunit of the COP-I coatomer complex, a key machinery implicated in Golgi-to-ER retrograde transport, reduction in β´-COP levels perturbs this trafficking pathway [81]. Interestingly, it has been previously proposed that full-length APP can traffic retrogradely between the Golgi apparatus and the ER mediated by the COP-I complex. In fact, the silencing of COP-I subunits by siRNAs induces APP accumulation at the Golgi apparatus, which reduces APP proteolytic processing [84].

Two other proteins that participates in this retrograde trafficking pathway, such as PKA and UVRAG [32,85], have been shown to be regulated by ubiquitination. Blockers of PKA signaling cause the inhibition of this trafficking pathway, a process that is accompanied by the swelling of the Golgi apparatus [32,49]. It has been shown that the catalytic PKA subunit (PKAc) is ubiquitinated by the CHIP E3 ligase, resulting in proteasomal degradation of PKAc and signaling shutdown [86]. UVRAG mediates the interaction of β´-COP with ER tethers and COP-I coatomer for efficient fusion of retrograde vesicles to the ER, a crucial step during Golgi-to-ER retrograde transport [53]. UVRAG is ubiquitinated by SMURF1 [87] and its silencing causes the swelling of the Golgi apparatus and the inhibition of the Golgi-to-ER retrograde transport [53]. Whether these proteins are regulated by K63-Ub chains is unknown and should be further addressed.

In addition, we propose that Golgi-to ER retrograde transport inhibition by PSMD14 dysfunction might result in the accumulation of structural and/or autophagy regulatory elements. In fact, the inhibition of PSMD14 caused the accumulation of ATG9A at the Golgi. ATG9A is a multispanning membrane protein essential for autophagy [63,64]. ATG9A is actively transported through Golgi-to-ER retrograde transport [65] and participates as a key player in the biogenesis of autophagosomes [63,88]. Moreover, along with the accumulation of ATG9A at the Golgi, we found increased levels of RAB1A at the Golgi apparatus. RAB1A is an essential small GTPase that participates in the recruitment of the ULK1 complex to subdomains of the ER for autophagy initiation [61,62]. Interestingly, and similar to PSMD14 inhibition, ATG9A deficiency causes an increase in LC3B-I levels, accompanied by a reduction in the number of autophagosome structures [63,88,89]. Together, our results strongly support that PSMD14 inhibition perturbs autophagosome biogenesis due to the sequestration of key proteins of this process at the Golgi apparatus. In addition, it supports the hypothesis of a closed intersection between Golgi-to-ER retrograde and autophagy pathway [53,85]. whereas PSMD14 DUB activity emerged as a new regulatory element of this intersection.

The PSDM14-dependent inhibition of macroautophagy might affect protein secretion from the Golgi apparatus since autophagy has been recently demonstrated as a positive regulator of this process [66], which is in agreement with the strong accumulation of APP at the Golgi apparatus. The crosstalk between macroautophagy and protein secretion was recently discovered, observing that the arrival of cargo to the Golgi apparatus induces lysosome repositioning to the perinuclear region along with the activation of the autophagy flux by a mechanism dependent on KDELR signaling [66]. Importantly, the abolishment of either lysosomal repositioning or autophagosome biogenesis strongly reduced cargo exit from the Golgi apparatus, demonstrating that protein transport from the Golgi apparatus to the plasma membrane requires macroautophagy [66]. Collectively, the results demonstrate the strong functional interplay between membrane transport and macroautophagy mediated by a novel mechanism involving the proteasome complex through the deubiquitinating activity of PSMD14 (Model in Figure 8). We propose that inhibition of PSDM14 DUB activity by CZM is a new strategy for causing inhibition of the Golgi-to-ER retrograde pathway. Thus, CZM should now be considered as a new pharmacological tool to study the impact of Golgi-to-ER retrograde transport inhibition in other cell biology processes, such as autophagy. Moreover, because CZM was discovered as a new drug for cancer treatment [42], it would now be interesting to investigate whether part of its anti-cancer effects could be the result of inhibiting the Golgi-to-ER retrograde pathway. Interestingly, recent studies have shown that PSMD14 is upregulated (mRNA and protein levels) in different tumoral cell types [90,91]. Whether cancer cells are more dependent on the Golgi-to-ER retrograde pathway than normal cells, as it happens regarding the mechanisms of protein quality control [92], is still unclear. CZM could offer an interesting tool to evaluate this hypothesis, positioning PSMD14 as a promising target for therapeutic intervention.

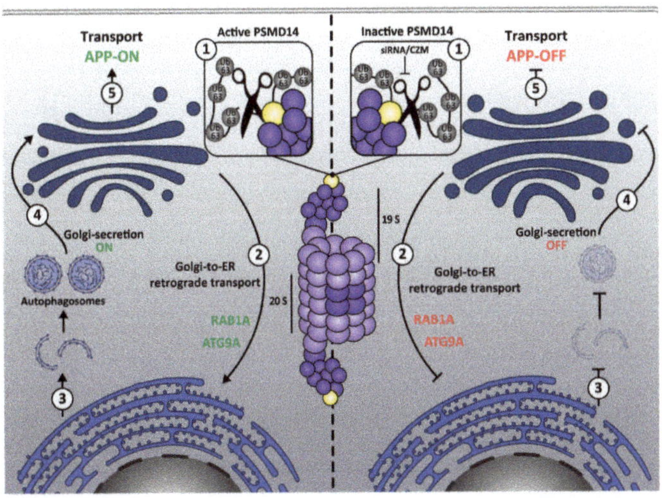

Figure 8. Model of the mechanism underlying the regulation of protein membrane trafficking and macroautophagy by the proteasome 19S RP PSMD14 DUB activity. The model depicts the closed interplay between membrane transport and macroautophagy by a novel mechanism involving the proteasome complex through the deubiquitinating activity of PSMD14. We propose that active PSMD14 and the K63-Ub chains (**1**, left panel) positively regulate Golgi-to-ER retrograde transport (**2**, left panel), a pathway implicated in the retrieval of key proteins for autophagosome biogenesis and macroautophagy (**3**, left panel). Reduction of free K63-Ub chains by inactive PSMD14 (**1**, right panel) results on the blockage of Golgi-to-ER retrograde transport (**2**, right panel) causing the accumulation of ATG9A and RAB1A at the Golgi apparatus. Thus, blockage of Golgi-to-ER retrograde transport inhibits the biogenesis of autophagosomes and macroautophagy (**3**, right panel). Macroautophagy has been recently demonstrated to act as a potent positive regulator of protein transport from the Golgi apparatus to the cell surface (Golgi-secretion ON; **4**, left panel). Thus, inhibition of macroautophagy upon inactive PSMD14 (siRNA/CZM) blocks protein transport from the Golgi apparatus to the cell surface (Golgi-secretion OFF; **4** right panel), explaining the effect on APP transport (**5** left and **5** right panels).

Supplementary Materials: The following are available online http://www.mdpi.com/2073-4409/9/3/777/s1. Figure S1: siRNA sequences directed against human PSMD14 used for Validation Stage; Figure S2: Primer pairs sequences used for RT-qPCR; Figure S3: The PSMD14 DUB inhibitor CZM increases the Golgi apparatus area; Figure S4: CZM causes the accumulation of KDELR1-GFP at the Golgi apparatus; Figure S5: Effect of CZM on proteasome activity; Figure S6: Effect of CZM and MG132 on basal macroautophagy; Figure S7: Distribution of RAB1A upon CZM treatment; Figure S8: ATG9A is distributed in the swollen Golgi apparatus upon CZM treatment.

Author Contributions: Conceptualization, H.A.B., A.R.-F. and P.V.B.; Data curation, H.A.B., K.C, A.E.G, G.E.V., Y.C, S.H., E.A.-M., C.C.-T., A.R.-F. and P.V.B.; Formal analysis, H.A.B., K.C., A.E.G., G.E.V., Y.C., S.H., E.A.-M., C.C.-T., B.K., A.R.-F. and P.V.B; Funding acquisition, R.T.H, A.R.-F. and P.V.B.; Investigation, H.A.B., G.K., A.E.G., G.E.V., E.A.-M., C.C.-T. and S.B.; Methodology, H.A.B, J.C., A.R.F. and P.V.B.; Resources, G.E.V., S.B., A.S., G.K., B.K., G.A.M., J.C., R.T.H., A.R.-F. and P.V.B; Supervision, A.R.-F. and P.V.B.; Writing—original draft, H.A.B., A.R.-F. and P.V.B; Writing—review and editing, H.A.B, S.B., A.S., B.K., G.A.M., J.C., A.R.-F., R.T.H and P.V.B. All authors have read and agreed to the published version of the manuscript.

Funding: This research was funded by Fondo Nacional de Desarrollo Científico y Tecnológico of Chile (FONDECYT; http://www.conicyt.cl/fondecyt) No. 1171649 to P.V.B. & No. 11150532 to A.R.F.; Associative Investigation Program (PIA; https://www.conicyt.cl/pia) including No. ACT-172066 to P.V.B. & No. AFB-170005 to P.V.B.; Academy Insertion Program (PAI; http://www.conicyt.cl/pai) No. 79150075 to A.R.F.; Fondo de Equipamiento Científico y Tecnológico of Chile (FONDEQUIP; http://www.conicyt.cl/fondequip) No. EQM150118 to P.V.B.; Cooperation International Programme (CONICYT-RCUK; https://www.conicyt.cl/pci) No. DPI20140068 to P.V.B.; H.A.B., A.E.G., and G.E.V. were supported by National Ph.D Fellowships No. 21130315, 201110746 and 21130511, respectively, (CONICYT; https://www.conicyt.cl/becasconicyt); Programa de Mejoramiento de la Calidad y la Equidad de la Educación superior, MECESUP AUS1203 and Vicerrectoría de Investigación de la Universidad Austral de Chile No. D #2015-02 to H.A.B., D #2013-07 to A.E.G. and D #2015-05 to G.E.V.

Acknowledgments: We thank Gonzalo Astroza (Universidad Austral de Chile) and Ellis Jaffray (GRE, University of Dundee) for technical assistance. Raymond Deshaies (Division of Biology & Biological Engineering, California Institute of Technology) and Yuyong Ma (Department of Chemistry and Biochemistry, University of California) for providing Capzimin for this study.

Conflicts of Interest: The authors declare no conflict of interest. The funders had no role in the design of the study; in the collection, analyses, or interpretation of data; in the writing of the manuscript, or in the decision to publish the results.

References

1. Shinde, S.R.; Maddika, S. Post translational modifications of Rab GTPases. *Small GTPases* **2017**, *9*, 49–56. [CrossRef] [PubMed]
2. Luo, P.M.; Boyce, M. Directing Traffic: Regulation of COPI Transport by Post-translational Modifications. *Front. Cell Dev. Boil.* **2019**, *7*, 190. [CrossRef] [PubMed]
3. Li, M.; Brooks, C.L.; Wu-Baer, F.; Chen, D.; Baer, R.; Gu, W. Mono- Versus Polyubiquitination: Differential Control of p53 Fate by Mdm2. *Science* **2003**, *302*, 1972–1975. [CrossRef] [PubMed]
4. Komander, D.; Rape, M. The Ubiquitin Code. *Annu. Rev. Biochem.* **2012**, *81*, 203–229. [CrossRef] [PubMed]
5. Foot, N.; Henshall, T.; Kumar, S. Ubiquitination and the Regulation of Membrane Proteins. *Physiol. Rev.* **2017**, *97*, 253–281. [CrossRef]
6. Pelham, H.R. Membrane Traffic: GGAs Sort Ubiquitin. *Curr. Boil.* **2004**, *14*, R357–R359. [CrossRef]
7. Yang, B.; Kumar, S. Nedd4 and Nedd4-2: Closely related ubiquitin-protein ligases with distinct physiological functions. *Cell Death Differ.* **2010**, *17*, 68–77. [CrossRef]
8. Tan, J.; Evin, G. β-Site APP-cleaving enzyme 1 trafficking and Alzheimer's disease pathogenesis. *J. Neurochem.* **2012**, *120*, 869–880. [CrossRef]
9. Raiborg, C.; Stenmark, H. The ESCRT machinery in endosomal sorting of ubiquitylated membrane proteins. *Nature* **2009**, *458*, 445–452. [CrossRef]
10. Ren, X.; Hurley, J.H. VHS domains of ESCRT-0 cooperate in high-avidity binding to polyubiquitinated cargo. *EMBO J.* **2010**, *29*, 1045–1054. [CrossRef]
11. Clague, M.J.; Liu, H.; Urbe, S. Governance of Endocytic Trafficking and Signaling by Reversible Ubiquitylation. *Dev. Cell* **2012**, *23*, 457–467. [CrossRef] [PubMed]
12. Haglund, K. Distinct monoubiquitin signals in receptor endocytosis. *Trends Biochem. Sci.* **2003**, *28*, 598–604. [CrossRef] [PubMed]
13. Clague, M.J.; Urbe, S. Ubiquitin: Same Molecule, Different Degradation Pathways. *Cell* **2010**, *143*, 682–685. [CrossRef] [PubMed]

14. Kwon, Y.T.; Ciechanover, A.J. The Ubiquitin Code in the Ubiquitin-Proteasome System and Autophagy. *Trends Biochem. Sci.* **2017**, *42*, 873–886. [CrossRef] [PubMed]
15. Ciechanover, A.; Kwon, Y.T. Degradation of misfolded proteins in neurodegenerative diseases: Therapeutic targets and strategies. *Exp. Mol. Med.* **2015**, *47*, e147. [CrossRef] [PubMed]
16. Clague, M.J.; Urbe, S. Endocytosis: The DUB version. *Trends Cell Boil.* **2006**, *16*, 551–559. [CrossRef] [PubMed]
17. Komander, D.; Clague, M.J.; Urbe, S. Breaking the chains: Structure and function of the deubiquitinases. *Nat. Rev. Mol. Cell Boil.* **2009**, *10*, 550–563. [CrossRef]
18. Mevissen, T.E.; Komander, D. Mechanisms of Deubiquitinase Specificity and Regulation. *Annu. Rev. Biochem.* **2017**, *86*, 159–192. [CrossRef]
19. Bienko, M.; Green, C.; Crosetto, N.; Rudolf, F.; Zapart, G.; Coull, B.; Kannouche, P.; Wider, G.; Peter, M.; Lehmann, A.R.; et al. Ubiquitin-Binding Domains in Y-Family Polymerases Regulate Translesion Synthesis. *Science* **2005**, *310*, 1821–1824. [CrossRef]
20. Bienko, M.; Green, C.; Sabbioneda, S.; Crosetto, N.; Matic, I.; Hibbert, R.G.; Begovic, T.; Niimi, A.; Mann, M.; Lehmann, A.R.; et al. Regulation of Translesion Synthesis DNA Polymerase η by Monoubiquitination. *Mol. Cell* **2010**, *37*, 396–407. [CrossRef]
21. Husnjak, K.; Dikic, I. Ubiquitin-Binding Proteins: Decoders of Ubiquitin-Mediated Cellular Functions. *Annu. Rev. Biochem.* **2012**, *81*, 291–322. [CrossRef] [PubMed]
22. Lai, A.; Sisodia, S.S.; Trowbridge, I.S. Characterization of sorting signals in the beta-amyloid precursor protein cytoplasmic domain. *J. Boil. Chem.* **1995**, *270*, 3565–3573. [CrossRef]
23. Perez, R.G.; Soriano, S.; Hayes, J.D.; Ostaszewski, B.; Xia, W.; Selkoe, D.J.; Chen, X.; Stokin, G.B.; Koo, E.H. Mutagenesis identifies new signals for beta-amyloid precursor protein endocytosis, turnover, and the generation of secreted fragments, including Abeta42. *J. Boil. Chem.* **1999**, *274*, 18851–18856. [CrossRef] [PubMed]
24. Burgos, P.V.; Mardones, G.A.; Rojas, A.L.; DaSilva, L.L.; Prabhu, Y.; Hurley, J.H.; Bonifacino, J.S. Sorting of the Alzheimer's Disease Amyloid Precursor Protein Mediated by the AP-4 Complex. *Dev. Cell* **2010**, *18*, 425–436. [CrossRef] [PubMed]
25. Watanabe, T.; Hikichi, Y.; Willuweit, A.; Shintani, Y.; Horiguchi, T. FBL2 Regulates Amyloid Precursor Protein (APP) Metabolism by Promoting Ubiquitination-Dependent APP Degradation and Inhibition of APP Endocytosis. *J. Neurosci.* **2012**, *32*, 3352–3365. [CrossRef] [PubMed]
26. El Ayadi, A.; Stieren, E.S.; Barral, J.M.; Boehning, D. Ubiquilin-1 regulates amyloid precursor protein maturation and degradation by stimulating K63-linked polyubiquitination of lysine Proc. *Natl. Acad. Sci. USA* **2012**, *109*, 13416–13421. [CrossRef]
27. Morel, E.; Chamoun, Z.; Lasiecka, Z.M.; Chan, R.B.; Williamson, R.L.; Vetanovetz, C.; Dall'Armi, C.; Simoes, S.; Du Jour, K.S.P.; McCabe, B.D.; et al. Phosphatidylinositol-3-phosphate regulates sorting and processing of amyloid precursor protein through the endosomal system. *Nat. Commun.* **2013**, *4*, 2250. [CrossRef]
28. Williamson, R.L.; Laulagnier, K.; Miranda, A.M.; Fernandez, M.A.; Wolfe, M.S.; Sadoul, R.; Di Paolo, G. Disruption of amyloid precursor protein ubiquitination selectively increases amyloid β (Aβ) 40 levels via presenilin 2-mediated cleavage. *J. Boil. Chem.* **2017**, *292*, 19873–19889. [CrossRef]
29. Cooper, E.M.; Cutcliffe, C.; Kristiansen, T.Z.; Pandey, A.; Pickart, C.M.; Cohen, R.E. K63-specific deubiquitination by two JAMM/MPN+ complexes: BRISC-associated Brcc36 and proteasomal Poh1. *EMBO J.* **2009**, *28*, 621–631. [CrossRef]
30. Prabhu, Y.; Burgos, P.V.; Schindler, C.; Farías, G.G.; Magadár, J.G.; Bonifacino, J.S. Adaptor protein 2–mediated endocytosis of the β-secretase BACE1 is dispensable for amyloid precursor protein processing. *Mol. Boil. Cell* **2012**, *23*, 2339–2351. [CrossRef]
31. Bustamante, H.A.; Rivera-Dictter, A.; Cavieres, V.A.; Muñoz, V.C.; González, A.; Lin, Y.; Mardones, G.A.; Burgos, P.V. Turnover of C99 is Controlled by a Crosstalk between ERAD and Ubiquitin-Independent Lysosomal Degradation in Human Neuroglioma Cells. *PLoS ONE* **2013**, *8*, e83096. [CrossRef] [PubMed]
32. Cancino, J.; Capalbo, A.; Di Campli, A.; Giannotta, M.; Rizzo, R.; Jung, J.E.; Di Martino, R.; Persico, M.; Heinklein, P.; Sallese, M.; et al. Control Systems of Membrane Transport at the Interface between the Endoplasmic Reticulum and the Golgi. *Dev. Cell* **2014**, *30*, 280–294. [CrossRef] [PubMed]
33. Bett, J.S.; Ibrahim, A.F.M.; Garg, A.K.; Rocha, S.; Hay, R. siRNA Screening to Identify Ubiquitin and Ubiquitin-like System Regulators of Biological Pathways in Cultured Mammalian Cells. *J. Vis. Exp.* **2014**. [CrossRef] [PubMed]

34. Mackay, C.; Carroll, E.; Ibrahim, A.F.; Garg, A.; Inman, G.; Hay, R.; Alpi, A. E3 ubiquitin ligase HOIP attenuates apoptotic cell death induced by cisplatin. *Cancer Res.* **2014**, *74*, 2246–2257. [CrossRef] [PubMed]
35. Livak, K.J.; Schmittgen, T.D. Analysis of relative gene expression data using real-time quantitative PCR and the 2(-Delta Delta C(T)) method. *Methods* **2001**, *25*, 402–408. [CrossRef] [PubMed]
36. Pfaffl, M.W. A new mathematical model for relative quantification in real-time RT-PCR. *Nucleic Acids Res.* **2001**, *29*, 45. [CrossRef]
37. Mlynarczuk-Bialy, I.; Doeppner, T.R.; Golab, J.; Nowis, D.; Wilczyński, G.; Parobczak, K.; Wigand, M.E.; Hajdamowicz, M.; Bialy, L.; Aniolek, O.; et al. Biodistribution and Efficacy Studies of the Proteasome Inhibitor BSc2118 in a Mouse Melanoma Model. *Transl. Oncol.* **2014**, *7*, 570–579. [CrossRef]
38. González, A.E.; Muñoz, V.C.; Cavieres, V.A.; Bustamante, H.A.; Cornejo, V.-H.; Januário, Y.C.; González, I.; Hetz, C.; DaSilva, L.L.; Rojas-Fernández, A.; et al. Autophagosomes cooperate in the degradation of intracellular C-terminal fragments of the amyloid precursor protein via the MVB/lysosomal pathway. *FASEB J.* **2017**, *31*, 2446–2459. [CrossRef]
39. Moffat, J.; Reiling, J.H.; Sabatini, D.M. Off-target effects associated with long dsRNAs in Drosophila RNAi screens. *Trends Pharmacol. Sci.* **2007**, *28*, 149–151. [CrossRef]
40. Verma, R.; Aravind, L.; Oania, R.; McDonald, W.H.; Yates, J.R.; Koonin, E.V.; Deshaies, R. Role of Rpn11 Metalloprotease in Deubiquitination and Degradation by the 26S Proteasome. *Science* **2002**, *298*, 611–615. [CrossRef]
41. Yao, T.; Cohen, R.E. A cryptic protease couples deubiquitination and degradation by the proteasome. *Nature* **2002**, *419*, 403–407. [CrossRef] [PubMed]
42. Li, J.; Yakushi, T.; Parlati, F.; Mackinnon, A.L.; Perez, C.; Ma, Y.; Carter, K.P.; Colayco, S.; Magnuson, G.; Brown, B.; et al. Capzimin is a potent and specific inhibitor of proteasome isopeptidase RpnNat. *Chem. Biol.* **2017**, *13*, 486–493.
43. Tsubuki, S.; Saito, Y.; Tomioka, M.; Ito, H.; Kawashima, S. Differential Inhibition of Calpain and Proteasome Activities by Peptidyl Aldehydes of Di-Leucine and Tri-Leucine. *J. Biochem.* **1996**, *119*, 572–576. [CrossRef] [PubMed]
44. Kisselev, A.F.; Goldberg, A.L. Proteasome inhibitors: From research tools to drug candidates. *Chem. Boil.* **2001**, *8*, 739–758. [CrossRef]
45. Wieland, F.T.; Gleason, M.L.; Serafini, T.A.; Rothman, J.E. The rate of bulk flow from the endoplasmic reticulum to the cell surface. *Cell* **1987**, *50*, 289–300. [CrossRef]
46. Martínez-Menárguez, J.A.; Geuze, H.J.; Slot, J.W.; Klumperman, J. Vesicular Tubular Clusters between the ER and Golgi Mediate Concentration of Soluble Secretory Proteins by Exclusion from COPI-Coated Vesicles. *Cell* **1999**, *98*, 81–90. [CrossRef]
47. Klumperman, J. Transport between ER and Golgi. *Curr. Opin. Cell Boil.* **2000**, *12*, 445–449. [CrossRef]
48. Thor, F.; Gautschi, M.; Geiger, R.; Helenius, A. Bulk Flow Revisited: Transport of a Soluble Protein in the Secretory Pathway. *Traffic* **2009**, *10*, 1819–1830. [CrossRef]
49. Tenorio, M.J.; Luchsinger, C.; Mardones, G.A. Protein Kinase A Activity Is Necessary for Fission and Fusion of Golgi to Endoplasmic Reticulum Retrograde Tubules. *PLoS ONE* **2015**, *10*, e0135260. [CrossRef]
50. Hsu, V.W.; Shah, N.; Klausner, R.D. A brefeldin A-like phenotype is induced by the overexpression of a human ERD-2-like protein, ELP-1. *Cell* **1992**, *69*, 625–635. [CrossRef]
51. Cole, N.B.; Ellenberg, J.; Song, J.; DiEuliis, D.; Lippincott-Schwartz, J. Retrograde Transport of Golgi-localized Proteins to the ER. *J. Cell Boil.* **1998**, *140*, 1–15. [CrossRef] [PubMed]
52. Lewis, M.J.; Rayner, J.C.; Pelham, H.R.B. A novel SNARE complex implicated in vesicle fusion with the endoplasmic reticulum. *EMBO J.* **1997**, *16*, 3017–3024. [CrossRef] [PubMed]
53. He, S.; Ni, D.; Ma, B.; Lee, J.-H.; Zhang, T.; Ghozalli, I.; Pirooz, S.D.; Zhao, Z.; Bharatham, N.; Li, B.; et al. PtdIns(3)P-bound UVRAG coordinates Golgi–ER retrograde and Atg9 transport by differential interactions with the ER tether and the beclin 1 complex. *Nature* **2013**, *15*, 1206–1219. [CrossRef] [PubMed]
54. Lemus, L.; Ribas, J.L.; Sikorska, N.; Goder, V. An ER-Localized SNARE Protein Is Exported in Specific COPII Vesicles for Autophagosome Biogenesis. *Cell Rep.* **2016**, *14*, 1710–1722. [CrossRef] [PubMed]
55. Chen, Q.; Xiao, Y.; Chai, P.; Zheng, P.; Teng, J.; Chen, J. ATL3 Is a Tubular ER-Phagy Receptor for GABARAP-Mediated Selective Autophagy. *Curr. Boil.* **2019**, *29*, 846–855.e6. [CrossRef]
56. Hao, R.; Nanduri, P.; Rao, Y.; Panichelli, R.S.; Ito, A.; Yoshida, M.; Yao, T.-P. Proteasomes activate aggresome disassembly and clearance by producing unanchored ubiquitin chains. *Mol. Cell* **2013**, *51*, 819–828. [CrossRef]

57. Nanduri, P.; Hao, R.; Fitzpatrick, T.; Yao, T.-P. Chaperone-mediated 26S Proteasome Remodeling Facilitates Free K63 Ubiquitin Chain Production and Aggresome Clearance*. *J. Boil. Chem.* **2015**, *290*, 9455–9464. [CrossRef]
58. Kabeya, Y.; Mizushima, N.; Ueno, T.; Yamamoto, A.; Kirisako, T.; Noda, T.; Kominami, E.; Ohsumi, Y.; Yoshimori, T. LC3, a mammalian homologue of yeast Apg8p, is localized in autophagosome membranes after processing. *EMBO J.* **2000**, *19*, 5720–5728. [CrossRef]
59. Munafó, D.B.; Colombo, M.I. A novel assay to study autophagy: Regulation of autophagosome vacuole size by amino acid deprivation. *J. Cell Sci.* **2001**, *114*, 3619–3629.
60. Thoreen, C.C.; Kang, S.A.; Chang, J.W.; Liu, Q.; Zhang, J.; Gao, Y.; Reichling, L.J.; Sim, T.; Sabatini, D.M.; Gray, N.S. An ATP-competitive mammalian target of rapamycin inhibitor reveals rapamycin-resistant functions of mTORCJ. *Biol. Chem.* **2009**, *284*, 8023–8032. [CrossRef]
61. Winslow, A.R.; Chen, C.-W.; Corrochano, S.; Acevedo-Arozena, A.; Gordon, D.E.; A Peden, A.; Lichtenberg, M.; Menzies, F.M.; Ravikumar, B.; Imarisio, S.; et al. α-Synuclein impairs macroautophagy: Implications for Parkinson's disease. *J. Cell Boil.* **2010**, *190*, 1023–1037. [CrossRef] [PubMed]
62. Webster, C.P.; Smith, E.F.; Bauer, C.S.; Moller, A.; Hautbergue, G.; Ferraiuolo, L.; Myszczynska, M.; Higginbottom, A.; Walsh, M.J.; Whitworth, A.J.; et al. The C9orf72 protein interacts with Rab1a and the ULK 1 complex to regulate initiation of autophagy. *EMBO J.* **2016**, *35*, 1656–1676. [CrossRef] [PubMed]
63. Orsi, A.; Razi, M.; Dooley, H.C.; Robinson, D.; Weston, A.; Collinson, L.M.; Tooze, S.A. Dynamic and transient interactions of Atg9 with autophagosomes, but not membrane integration, are required for autophagy. *Mol. Boil. Cell* **2012**, *23*, 1860–1873. [CrossRef] [PubMed]
64. Kishi-Itakura, C.; Koyama-Honda, I.; Itakura, E.; Mizushima, N. Ultrastructural analysis of autophagosome organization using mammalian autophagy-deficient cells. *J. Cell Sci.* **2014**, *127*, 4984. [CrossRef]
65. Karanasios, E.; Walker, S.; Okkenhaug, H.; Manifava, M.; Hummel, E.; Zimmermann, H.; Ahmed, Q.; Domart, M.-C.; Collinson, L.M.; Ktistakis, N.T. Autophagy initiation by ULK complex assembly on ER tubulovesicular regions marked by ATG9 vesicles. *Nat. Commun.* **2016**, *7*, 12420. [CrossRef]
66. Tapia, D.; Jiménez, T.; Zamora, C.; Espinoza, J.; Rizzo, R.; González-Cárdenas, A.; Fuentes, D.; Hernández, S.; Cavieres, V.A.; Soza, A.; et al. KDEL receptor regulates secretion by lysosome relocation- and autophagy-dependent modulation of lipid-droplet turnover. *Nat. Commun.* **2019**, *10*, 735. [CrossRef]
67. Iwata, A.; Riley, B.E.; Johnston, J.; Kopito, R.R. HDAC6 and Microtubules Are Required for Autophagic Degradation of Aggregated Huntingtin. *J. Boil. Chem.* **2005**, *280*, 40282–40292. [CrossRef]
68. Ding, W.-X.; Ni, H.-M.; Gao, W.; Yoshimori, T.; Stolz, N.B.; Ron, D.; Yin, X.-M. Linking of Autophagy to Ubiquitin-Proteasome System Is Important for the Regulation of Endoplasmic Reticulum Stress and Cell Viability. *Am. J. Pathol.* **2007**, *171*, 513–524. [CrossRef]
69. Lan, D.; Wang, W.; Zhuang, J.; Zhao, Z. Proteasome inhibitor-induced autophagy in PC12 cells overexpressing A53T mutant α-synuclein. *Mol. Med. Rep.* **2014**, *11*, 1655–1660. [CrossRef]
70. Peng, H.; Yang, J.; Li, G.; You, Q.; Han, W.; Li, T.; Gao, D.; Xie, X.; Lee, B.-H.; Du, J.; et al. Ubiquitylation of p62/sequestosome1 activates its autophagy receptor function and controls selective autophagy upon ubiquitin stress. *Cell Res.* **2017**, *27*, 657–674. [CrossRef]
71. Milani, M.; Rzymski, T.; Mellor, H.R.; Pike, L.; Bottini, A.; Generali, D.; Harris, A.L. The Role of ATF4 Stabilization and Autophagy in Resistance of Breast Cancer Cells Treated with Bortezomib. *Cancer Res.* **2009**, *69*, 4415–4423. [CrossRef]
72. Albornoz, N.; Bustamante, H.; Soza, A.; Burgos, P. Cellular Responses to Proteasome Inhibition: Molecular Mechanisms and Beyond. *Int. J. Mol. Sci.* **2019**, *20*, 3379. [CrossRef]
73. Zhou, F.; Van Laar, T.; Huang, H.; Zhang, L. APP and APLP1 are degraded through autophagy in response to proteasome inhibition in neuronal cells. *Protein Cell* **2011**, *2*, 377–383. [CrossRef]
74. Swaminathan, G.; Zhu, W.; Plowey, E.D. BECN1/Beclin 1 sorts cell-surface APP/amyloid β precursor protein for lysosomal degradation. *Autophagy* **2016**, *12*, 2404–2419. [CrossRef] [PubMed]
75. Wang, B.-J.; Her, G.M.; Hu, M.-K.; Chen, Y.-W.; Tung, Y.-T.; Wu, P.-Y.; Hsu, W.-M.; Lee, H.; Jin, L.-W.; Hwang, S.-P.L.; et al. ErbB2 regulates autophagic flux to modulate the proteostasis of APP-CTFs in Alzheimer's disease. *Proc. Natl. Acad. Sci. USA* **2017**, *114*, E3129–E3138. [CrossRef]
76. Yang, C.; Cai, C.-Z.; Song, J.; Tan, J.-Q.; Durairajan, S.S.K.; Iyaswamy, A.; Wu, M.-Y.; Chen, L.-L.; Yue, Z.; Li, M.; et al. NRBF2 is involved in the autophagic degradation process of APP-CTFs in Alzheimer disease models. *Autophagy* **2017**, *13*, 2028–2040. [CrossRef] [PubMed]

77. Zhang, J.; Liu, J.; Norris, A.; Grant, B.D.; Wang, X. A novel requirement for ubiquitin-conjugating enzyme UBC-13 in retrograde recycling of MIG-14/Wntless and Wnt signaling. *Mol. Biol. Cell* **2018**, *29*, 2098–2112. [CrossRef] [PubMed]
78. Deng, L.; Jiang, C.; Chen, L.; Jin, J.; Wei, J.; Zhao, L.; Chen, M.; Pan, W.; Xu, Y.; Chu, H.; et al. The Ubiquitination of RagA GTPase by RNF152 Negatively Regulates mTORC1 Activation. *Mol. Cell* **2015**, *58*, 804–818. [CrossRef] [PubMed]
79. Ji, C.H.; Kim, H.Y.; Heo, A.J.; Lee, S.H.; Lee, M.J.; Bin Kim, S.; SrinivasRao, G.; Mun, S.R.; Cha-Molstad, H.; Ciechanover, A.; et al. The N-Degron Pathway Mediates ER-phagy. *Mol. Cell* **2019**, *75*, 1058–1072.e9. [CrossRef]
80. Kawaguchi, Y.; Kovacs, J.J.; McLaurin, A.; Vance, J.; Ito, A.; Yao, T.-P. The Deacetylase HDAC6 Regulates Aggresome Formation and Cell Viability in Response to Misfolded Protein Stress. *Cell* **2003**, *115*, 727–738. [CrossRef]
81. Cohen, M.M.; Stutz, F.; Dargemont, C. Deubiquitination, a New Player in Golgi to Endoplasmic Reticulum Retrograde Transport. *J. Boil. Chem.* **2003**, *278*, 51989–51992. [CrossRef] [PubMed]
82. Millard, S.; Wood, S. Riding the DUBway: Regulation of protein trafficking by deubiquitylating enzymes. *J. Cell Boil.* **2006**, *173*, 463–468. [CrossRef]
83. Cohen, M.M.; Stutz, F.; Belgareh, N.; Haguenauer-Tsapis, R.; Dargemont, C. Ubp3 requires a cofactor, Bre5, to specifically de-ubiquitinate the COPII protein, Sec23. *Nature* **2003**, *5*, 661–667. [CrossRef] [PubMed]
84. Bettayeb, K.; Chang, J.C.; Luo, W.; Aryal, S.; Varotsis, D.; Randolph, L.; Netzer, W.J.; Greengard, P.; Flajolet, M. δ-COP modulates Aβ peptide formation via retrograde trafficking of APP. *Proc. Natl. Acad. Sci. USA* **2016**, *113*, 5412–5417. [CrossRef] [PubMed]
85. He, S.; O'Connell, D.; Zhang, X.; Yang, Y.; Liang, C. The intersection of Golgi-ER retrograde and autophagic trafficking. *Autophagy* **2013**, *10*, 180–181. [CrossRef]
86. Rinaldi, L.; Donne, R.D.; Catalanotti, B.; Torres-Quesada, O.; Enzler, F.; Moraca, F.; Nisticò, R.; Chiuso, F.; Piccinin, S.; Bachmann, V.; et al. Feedback inhibition of cAMP effector signaling by a chaperone-assisted ubiquitin system. *Nat. Commun.* **2019**, *10*, 2572. [CrossRef]
87. Feng, X.; Jia, Y.; Zhang, Y.; Ma, F.; Zhu, Y.; Hong, X.; Zhou, Q.; He, R.; Zhang, H.; Jin, J.; et al. Ubiquitination of UVRAG by SMURF1 promotes autophagosome maturation and inhibits hepatocellular carcinoma growth. *Autophagy* **2019**, *15*, 1130–1149. [CrossRef]
88. Saitoh, T.; Fujita, N.; Hayashi, T.; Takahara, K.; Satoh, T.; Lee, H.; Matsunaga, K.; Kageyama, S.; Omori, H.; Noda, T.; et al. Atg9a controls dsDNA-driven dynamic translocation of STING and the innate immune response. *Proc. Natl. Acad. Sci. USA* **2009**, *106*, 20842–20846. [CrossRef]
89. Mattera, R.; Park, S.Y.; De Pace, R.; Guardia, C.M.; Bonifacino, J.S. AP-4 mediates export of ATG9A from the trans-Golgi network to promote autophagosome formation. *Proc. Natl. Acad. Sci. USA* **2017**, *114*, E10697–E10706. [CrossRef]
90. Song, Y.; Li, S.; Ray, A.; Das, D.S.; Qi, J.; Samur, M.K.; Tai, Y.-T.; Munshi, N.; Carrasco, R.D.; Chauhan, D.; et al. Blockade of deubiquitylating enzyme Rpn11 triggers apoptosis in multiple myeloma cells and overcomes bortezomib resistance. *Oncogene* **2017**, *36*, 5631–5638. [CrossRef]
91. Wang, C.-H.; Lu, S.-X.; Liu, L.-L.; Li, Y.; Yang, X.; He, Y.-F.; Chen, S.-L.; Cai, S.; Wang, H.; Yun, J. POH1 Knockdown Induces Cancer Cell Apoptosis via p53 and Bim. *Neoplasia* **2018**, *20*, 411–424. [CrossRef] [PubMed]
92. Deshaies, R. Proteotoxic crisis, the ubiquitin-proteasome system, and cancer therapy. *BMC Boil.* **2014**, *12*, 94. [CrossRef] [PubMed]

cells

MDPI

Article

Tissue-Specific Impact of Autophagy Genes on the Ubiquitin–Proteasome System in *C. elegans*

Sweta Jha and Carina I. Holmberg *

Medicum, Department of Biochemistry and Developmental Biology, Faculty of Medicine, University of Helsinki, Haartmaninkatu 8, 00290 Helsinki, Finland; sweta.jha@helsinki.fi

* Correspondence: carina.holmberg@helsinki.fi; Tel.: +358-5-0448-6449

Received: 3 July 2020; Accepted: 6 August 2020; Published: 8 August 2020

check for updates

Abstract: The ubiquitin–proteasome system (UPS) and the autophagy–lysosomal pathway (ALP) are the two main eukaryotic intracellular proteolytic systems involved in maintaining proteostasis. Several studies have reported on the interplay between the UPS and ALP, however it remains largely unknown how compromised autophagy affects UPS function in vivo. Here, we have studied the crosstalk between the UPS and ALP by investigating the tissue-specific effect of autophagy genes on the UPS at an organismal level. Using transgenic *Caenorhabditis elegans* expressing fluorescent UPS reporters, we show that the downregulation of the autophagy genes *lgg-1* and *lgg-2 (ATG8/LC3/GABARAP)*, *bec-1 (BECLIN1)*, *atg-7 (ATG7)* and *epg-5* (mEPG5) by RNAi decreases proteasomal degradation, concomitant with the accumulation of polyubiquitinated proteasomal substrates in a tissue-specific manner. For some of these genes, the changes in proteasomal degradation occur without a detectable alteration in proteasome tissue expression levels. In addition, the *lgg-1* RNAi-induced reduction in proteasome activity in intestinal cells is not dependent on *sqst-1*/p62 accumulation. Our results illustrate that compromised autophagy can affect UPS in a tissue-specific manner, and demonstrate that UPS does not function as a direct compensatory mechanism in an animal. Further, a more profound understanding of the multilayered crosstalk between UPS and ALP can facilitate the development of therapeutic options for various disorders linked to dysfunction in proteostasis.

Keywords: autophagy; ubiquitin–proteasome system; crosstalk; tissue specificity; *C. elegans*

1. Introduction

Protein homeostasis (proteostasis) is a dynamic balance between protein biogenesis and degradation, and is essential for cell survival and growth. The pool of intracellular and various extracellular proteins is constantly being replaced by newly synthesized proteins, and therefore protein degradation must be selective and tightly regulated. In eukaryotes, the ubiquitin–proteasome system (UPS) and autophagy–lysosome pathway (ALP) are the two major intracellular proteolytic systems that mediate protein turnover.

UPS is the main pathway responsible for the degradation of soluble and short-lived misfolded proteins, both in the cytosol and the nucleus. Proteasomal substrates are first polyubiquitinated via the action of three classes of enzymes: ubiquitin-activating enzymes (E1), ubiquitin-conjugating enzymes (E2) and ubiquitin ligases (E3), reviewed in [1,2]. The substrates are then degraded by the evolutionarily conserved 26S proteasome, which consists of a central barrel-shaped core particle (CP or 20S proteasome) enclosing the peptide hydrolysis activity of the complex. The 20S comprises four stacked heteroheptameric rings: two rings of α subunits at each end and two middle β subunit rings harboring the proteolytic activities. The 20S core particle is capped by one or two 19S regulatory particles (RP), required for substrate recognition, the removal of the attached ubiquitin chains, the ATP-dependent unfolding of the substrate, and transfer into the core for proteolysis, as reviewed in [3–5].

Long-lived misfolded proteins and defective cellular organelles are in turn degraded by the ALP. There are three main types of autophagy reviewed in [6], including macroautophagy, hereafter referred to as autophagy. The initiation of autophagy starts by the formation of an isolation membrane, the phagophore, which elongates to engulf the substrate(s), thus forming the double-layered autophagosome. This then fuses with late endosomes and lysosomes, leading to the formation of autolysosomes, where the substrate(s) is degraded by lysosomal hydrolases [2,7,8]. Autophagy is mediated through the conserved action of several proteins, from yeast to mammalian cells. A class-III phosphatidylinositol 3-kinase (PI3K) complex, together with lipid kinase VPS34 and the regulatory protein Beclin1, initiate the autophagic process by mediating vesicle nucleation. Two ubiquitin-like conjugation systems, encoded by the autophagy-related genes (atg), then regulate the elongation of the isolation membrane. The first conjugation system comprises ATG-12 and ATG-5, conjugated via a covalent bond mediated by the E1-like ATG-7 and E2-like ATG-10. The second conjugation system involves ATG-8, which conjugates with lipid phosphatidylethanolamine in a process regulated by ATG-7 and the E3-like ATG-3. The ATG-5-ATG-12 conjugate is detached after autophagosome formation, whereas ATG-8 remains in the autophagosome in its conjugated form. Currently, yeast ATG-8 and its mammalian homolog LC3 are the most commonly used markers for autophagic activity [9–12]. Another set of proteins functions at the last fusion step. RAB-7, a member of the Rab GTPase family, is present in late endosomes, and is involved in regulating the fusion to late lysosomes [12]. Additionally, mEPG-5 is required for the maturation of autophagosomes and their fusion to lysosomes [13,14]. The activation of autophagy can be induced by several stimuli, such as amino-acid deprivation, starvation or various stress conditions [15,16].

The UPS and ALP were previously believed to have exceedingly distinct substrates, and they were regarded as independent proteolytic systems, but accumulating evidence has established extensive crosstalk between the two, reviewed in [17]. Several studies using human cell lines and mice have reported that pharmacological and genetic inhibition of the UPS leads to the activation of autophagy [18–21]. However, the blocking of autophagy has reportedly more complex and contrasting results, both inhibiting and activating UPS [22–25].

Both the autophagy pathway and the UPS are highly conserved in the nematode *Caenorhabditis elegans*, reviewed in [26–28]. It has been reported that the rate of autophagic flux can vary in different *C. elegans* tissues during aging and stress [29,30]. Similarly, we have previously shown that UPS activity varies in a tissue-specific manner in *C. elegans* [31–33]. Here, we have investigated the crosstalk between UPS and ALP at an organismal level using *C. elegans*. We reveal that the downregulation of certain autophagy genes that function at different ALP steps elicits distinct tissue-specific effects on UPS in *C. elegans*, and that the observed changes in proteasomal degradation can occur either concomitant with, or without, a detectable change in proteasome tissue expression levels.

2. Material and Methods

2.1. *C. elegans* and Growth Conditions

C. elegans strains were grown and maintained under standard conditions at 20 °C as described previously [34]. N2 (Bristol) and MAH215 strains were obtained from the Caenorhabditis Genetics Center (CGC). NL2099[*rrf-3(pk1426)II*] mutant strain was a kind gift from Dr. G. Wong (University of Eastern Finland). YD116[*rrf-3(pk1426);xzIs2[unc-54p::UIM2::ZsProSensor]*] reporter strain was generated by crossing NL2099[*rrf-3(pk1426)II*] strain with YD114[*xzIs2[unc-54p::UIM2::ZsProSensor]*] [35]. Information on the cloning of the plasmid has been described previously in Matilainen et al., 2013, as a precursor for the YD90[*xzIs1[vha-6p::UIM2::ZsProSensor]*] strain [32]. All strains used in this study are listed in Table S1.

2.2. *C. elegans* RNA Interference (RNAi)

RNAi was performed using the previously described feeding protocol [36]. The HT115 bacterial strain carrying the empty *pL4440* expression vector was used as a control in all experiments. Double stranded RNA expression was induced by adding 0.4 mM of isopropyl-β-D-thiogalactopyranoside (I6758, Sigma, St. Louis, MO, USA) during peak culture growth and its concentration was further increased to 0.8 mM just prior to the seeding of the feeding plates. Unless otherwise indicated, age-synchronized L1 larvae (day 1) were placed on control and RNAi-seeded plates, targeting either *lgg-1, lgg-2, bec-1, atg-7, rab-7, epg-5* or *sqst-1* (*C32D5.9, ZK593.6, T19E7.3, M7.5, W03C9.3, C56C10.12* or *T12G3.1*, respectively, Source BioScience, J. Ahringer library). Double RNAi of *lgg-1* and *sqst-1* was performed by mixing *lgg-1* and *sqst-1* RNAi cultures in 1:1 ratio according to their optical density prior to seeding.

2.3. Microscopy and Image Analysis

Age-synchronized animals were imaged at first day of adulthood (day 4). Groups of animals were mounted on 3% agarose pad on glass slides and immobilized using 0.5 mM levamisole diluted in M9 buffer. The polyubiquitin reporter animals were imaged with a Zeiss Axio Imager upright epifluorescence microscope using 10× 0.3 NA EC Plan Neofluar objective. Images were quantified using the Fiji ImageJ software.

For the UbG76V-Dendra2 and Dendra2 strains, images of a group of animals were taken before and immediately after proteins were converted from green to red fluorescence using 405 nm UV light (considered 0 h after conversion). Animals were recovered on corresponding feeding plates and reimaged after 6 h (intestinal reporter strains) or 24 h (body-wall muscle reporter strains). A Zeiss Axio Imager Z2 upright epifluorescence microscope was used for photoconversion and images were acquired with a 10 × 0.3 NA EC Plan Neofluar objective. Fluorescent intensities were quantified using the Fiji ImageJ software. Relative intensities were calculated by setting the absolute values of fluorescence intensity before and after photoconversion as 100%, for green and red fluorescence signals, respectively. For both the polyubiquitin reporter and the UPS activity reporter, images were exported into tiff-format and quantified using the original black and white version of the images without modifications. The background was subtracted using the corresponding command in Fiji software. The threshold was selected from the brightest image and this same threshold was applied to all images from the same experiment. The average of mean intensity was analyzed.

The autophagy dual marker reporter strain was imaged using Zeiss LSM880 confocal microscope (Motorized Zeiss Axio Observer Z1 inverted microscope). Z-stack images were acquired at 0.8 μm slice intervals with a 63× 1.4 NA plan-Apochromat objective. The z-stack images were converted to maximum intensity projection format using ZEN 2.1 (black version) and converted to tiff-format using Zen 2 lite (blue version). The number of puncta was calculated manually.

All images were processed with Adobe Photoshop CC 2018 software. When the brightness of an image was increased to make the fluorescent signal clearly visible, all corresponding images from the same experiment were modified in the same way. The number of animals imaged for each treatment and their *p*-values relative to the control are listed in Tables S2, S3 and S5 for the polyubiquitin reporter, UPS activity reporter and autophagy dual marker reporter, respectively.

2.4. Quantitative Real-Time PCR

Age-synchronized RNAi-treated animals were collected in M9 buffer at first day of adulthood (day 4) and stored at −80 °C. Total RNA was extracted using NucleoSpin RNA kit (Macherey-Nagel, Düren, Germany) and RNA concentration was measured with Nanodrop spectrophotometer at 260 nm. RT-PCR was performed using Maxima First Strand cDNA Synthesis Kit for RT-qPCR (Thermo Scientific, Waltham, MA, USA). The quantitative real-time PCR was done using Maxima SYBR Green/ROX qPCR Master Mix (2X) (Thermo Scientific) and LightCycler 480 (Roche, Basel, Switzerland) quantitative PCR machine. The data from qPCR were normalized to the geometric mean of mRNA concentration of three reference genes (*act-1*, *cdc-42* and *pmp-3*) [37]. The qPCR oligos used in this study are listed in Table S6.

2.5. In-Gel Proteasome Activity Assay and Western Blotting

RNAi-treated age-synchronized young adult (day 4) *rrf-3(pk1426)* animals were collected in M9 buffer prior to freezing at −80 °C for both the in-gel activity assay and Western blot samples. For in-gel activity assay, the animals were lysed in native gel lysis buffer using Dounce homogenizer as previously described [38]. The in-gel assay was performed as described earlier, but with slight modifications [32,38]. The native 3.5% acrylamide gels were run for 30 min at 20 mA and then for 2 h at 40 mA on an ice bath in a cold room (+4 °C). The gels were developed using developing buffer including 80 µM of the fluorogenic proteasome substrate suc-LLVY-AMC (I-1395, Bachem, Bubendorf, Switzerland). The gels were imaged with MultiImage Light Cabinet using FluorChem 8900 software (Alpha Innotech Corporation). After imaging, Coomassie staining of the gels was performed using Colloidal Blue Staining Kit (Invitrogen) to assess sample loading. The fluorescent signal was modified similarly for all images from the same experiment with Adobe Photoshop CC 2018 software, and signal intensities were quantified using Fiji ImageJ software.

For Western blotting, the animals were lysed either according to the native gel protocol (detection of proteasome 20S alpha subunits) or by sonication using Western blot lysis buffer (50 mM Hepes (pH 7.4), 150 mM NaCl, 5 mM EDTA, 20 mM NEM, 10 µM MG-132 and protease inhibitor cocktail (Roche); detection of polyubiquitinated proteins). Samples were run on SDS-PAGE gel and immunoblotted onto a nitrocellulose membrane using Trans-Blot Turbo transfer system (Bio-Rad). Anti-20S alpha antibody (for proteasome 20S α-subunits 1–3 and 5–7, BML-PW8195, Enzo Life Sciences, 1:1000 dilution), FK-1 antibody (for polyubiquitinated proteins, BML-PW8805, Enzo Life Sciences, 1:500 dilution) and anti-α-tubulin antibody (for α-tubulin, T5168, Sigma, 1:10 000 dilution) were used for immunoblotting. The secondary antibodies for the proteasome 20S α-subunits and α-tubulin were anti-mouse IgG-HRP conjugates (W4021, Promega, 1:10,000 dilution), and anti-mouse IgM-HRP conjugates (401225, Calbiochem, 1:10,000 dilution) were used for polyubiquitinated proteins. Image Studio software (Licor) was used for imaging and quantifying the signals. A summary of the number of experiments and *p*-values relative to the respective controls is included in Table S4.

2.6. Immunohistochemical Analysis

Age-synchronized young adult (day 4) animals were collected in M9 buffer and fixed with 10% (*v*/*v*) phosphate buffered formalin as described earlier [33]. The fixed animals were embedded in 2% agar and the paraffin-embedded agar blocks were cut into 4 µm sections. Immunohistochemical staining was performed using Dako REAL™ EnVision™ Detection System, Peroxidase/DAB+, Rabbit/mouse kit (Dako) and the anti-20S alpha antibody (BML-PW8195, Enzo Life Sciences, 1:1000 dilution) [33]. The staining specificity of the anti-20S alpha antibody was previously established in *C. elegans* by omitting the primary antibody or by pre-absorption with purified 20S proteasome [33]. The stained slides were imaged with a Zeiss Axio Imager Z2 upright epifluorescence microscope using 10× 0.3 NA EC Plan Neofluar objective or 63× 1.4 NA Plan-Apochromat objective. Images were converted to tiff-format using Zen 2 lite (blue version). Immunostaining evaluation was performed independently and blindly by three investigators without prior knowledge of original treatment conditions. Staining intensity was scored as 0 for negative, 1 for mild, 2 for moderate and 3 for strong positive immunoreactivity. Information on the number of experiments and *p*-values relative to controls is presented in Table S4.

2.7. Statistical Analysis

Statistical significance was determined using the student's *t*-test (two-tailed) in all quantifications and the treated group was compared to either control or to another treatment, as indicated in the figures.

3. Results

3.1. Downregulation of Autophagy Genes Affects the Accumulation of Polyubiquitinated Proteins in a Tissue-Specific Manner In Vivo

To investigate the interconnection between the UPS and autophagy at an organismal level, we downregulated the autophagy genes influencing different stages of ALP (Figure S1A) and analyzed the subsequent effects on the UPS in *C. elegans.* The RNAi-sensitive *rrf-3(pk1426)* strain was exposed to various RNAi treatments, targeting the individual autophagy genes *lgg-1*, *lgg-2* (homologs of mammalian LC3/GABARAP), *bec-1* (BECLIN 1), *atg-7* (ATG7), *rab-7* or *epg-5* (mEPG-5) (Figure S1B), resulting in a clear downregulation of their corresponding mRNA levels (Figure S1C). To monitor the outcome from the compromised autophagy gene expression, we used the dual fluorescence mCherry::GFP::LGG-1 reporter strain [29] to quantify the number of autophagosomes (APs) present in the live animals. As expected, RNAi against *lgg-1* or *bec-1* decreased the number of puncta that were positive for autophagosomes, whereas knockdown of *lgg-2*, *atg-7*, *rab-7* or *epg-5* significantly increased the formation of APs (Figure S2A, Table S5), validating our experimental setup.

Next, we investigated the accumulation of polyubiquitinated proteins in live animals using our previously developed tissue-specific fluorescent polyubiquitin reporter, which binds to endogenous Lys-48-linked polyubiquitinated proteasomal substrates [32,39]. An increase in the fluorescence of the reporter corresponded with an accumulation of proteasomal substrates upon impaired proteasome function. Interestingly, animals expressing the polyubiquitin reporter in intestinal cells displayed increased accumulation of polyubiquitinated proteins upon introduction of RNAi against *lgg-1*, *lgg-2*, *bec-1* or *atg-7* (Figure 1A,C, Table S2). In contrast, the downregulation of *rab-7* or *epg-5* did not affect the reporter fluorescence in intestinal cells (Figure 1A,C). Additionally, we examined the effect of the knockdown of autophagy genes in animals expressing the polyubiquitin reporter in body-wall muscle cells.

Our results revealed that *lgg-2*, *bec-1* or *epg-5* RNAi increased the amounts of polyubiquitinated proteins in these cells, whereas *lgg-1*, *atg-7* or *rab-7* RNAi did not change the intensity of reporter fluorescence (Figure 1B,D, Table S2). When we investigated the total amounts of polyubiquitinated proteins present in whole animal lysates by Western blotting, we did not detect any differences between the control and RNAi-treated animals (Figure S3A,B). Taken together, our data indicate the tissue-specific effect of certain autophagy genes on the accumulation of polyubiquitinated proteasomal substrates.

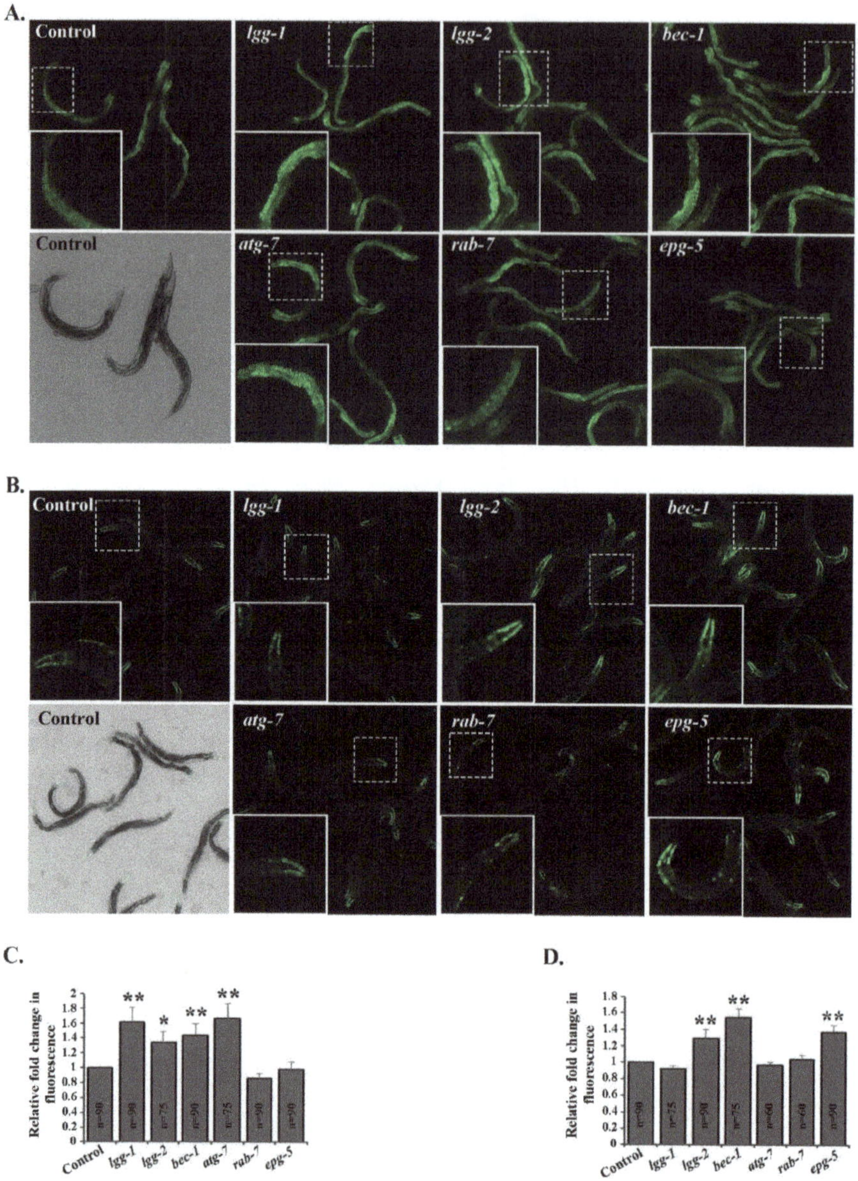

Figure 1. Downregulation of autophagy genes affects accumulation of fluorescent polyubiquitin reporters in a tissue-specific manner in vivo. Representative fluorescence micrographs of control, *lgg-1*, *lgg-2*, *bec-1*, *atg-7*, *rab-7* or *epg-5* RNAi-treated animals expressing the polyubiquitin reporter in the intestinal (**A**) and body-wall muscle (**B**) cells, respectively. Quantification of the fluorescent signal of the polyubiquitin reporter in intestinal (**C**) and body-wall muscle (**D**) cells, respectively. Graphs show average fold change in the amount of fluorescence signal compared to control RNAi (set as 1). Results are the mean of quantifications from 6 independent experiments (n = number of animals) (See Table S2). Error bars, SEM, * $p < 0.05$ and ** $p < 0.01$ compared to the control.

3.2. Tissue-Specific Differences in UPS Activity In Vivo Upon Knockdown of Autophagy Genes

Next, we wanted to investigate whether the observed tissue-specific accumulation of polyubiquitinated proteins upon downregulation of autophagy genes could be a result of changes in proteasome activity. For this purpose, we took advantage of our previously generated photoconvertible UPS reporter system, which is based on the photoswitchable green-to-red fluorescent protein Dendra2 fused to the non-hydrolyzable ubiquitin moiety UbG76V [32,40]. The UbG76V-Dendra2 is targeted by the UPS, and degraded by the proteasome. After an UV light-induced photoconversion, the decrease in red fluorescence intensity of UbG76V-Dendra2 over time reflects UPS-mediated protein degradation, and in a manner independent of the translation of new reporter proteins. Here, we examined whether the modulation of autophagy gene expression directly affects the degradation of UbG76V-Dendra2 in different tissues.

Animals expressing UbG76V-Dendra2 in intestinal cells were fed either control, *lgg-1*, *lgg-2*, *bec-1*, *atg-7*, *rab-7* or *epg-5* RNAi bacteria, and analyzed for potential changes in reporter degradation. Six hours after photoconversion, approximately 10–12% of UbG76V-Dendra2 had been degraded in the *lgg-1*, *lgg-2*, *bec-1* or *atg-7* RNAi-treated animals, compared with the roughly 30% degradation rate observed in control animals (Figure 2A,B, Table S3). This slower degradation correlates with the increased accumulation of polyubiquitinated proteins (Figure 1A,C). In agreement with these results, RNAi against *rab-7* or *epg-5* did not affect UPS activity in intestinal cells (Figure 2A,B, Table S3). In contrast, the degradation rate of the UbG76V-Dendra2 reporter in body-wall muscle cells changed solely upon *epg-5* RNAi, resulting in approximately 15% difference compared with the control (Figure 3A,B, Table S3). Surprisingly, we did not detect a change in reporter degradation upon *lgg-2* and *bec-1* RNAi, although these treatments caused the accumulation of polyubiquitin reporter in the muscle cells. Further, the fluorescence in animals expressing the control Dendra2 in intestinal or body-wall muscle cells remained stable after RNAi against autophagy genes over the measured time period (Figures S4 and S5, Table S3). Altogether, our data demonstrate that the knockdown of specific autophagy genes induces distinct effects on UPS activity in intestinal and body-wall muscle cells in vivo.

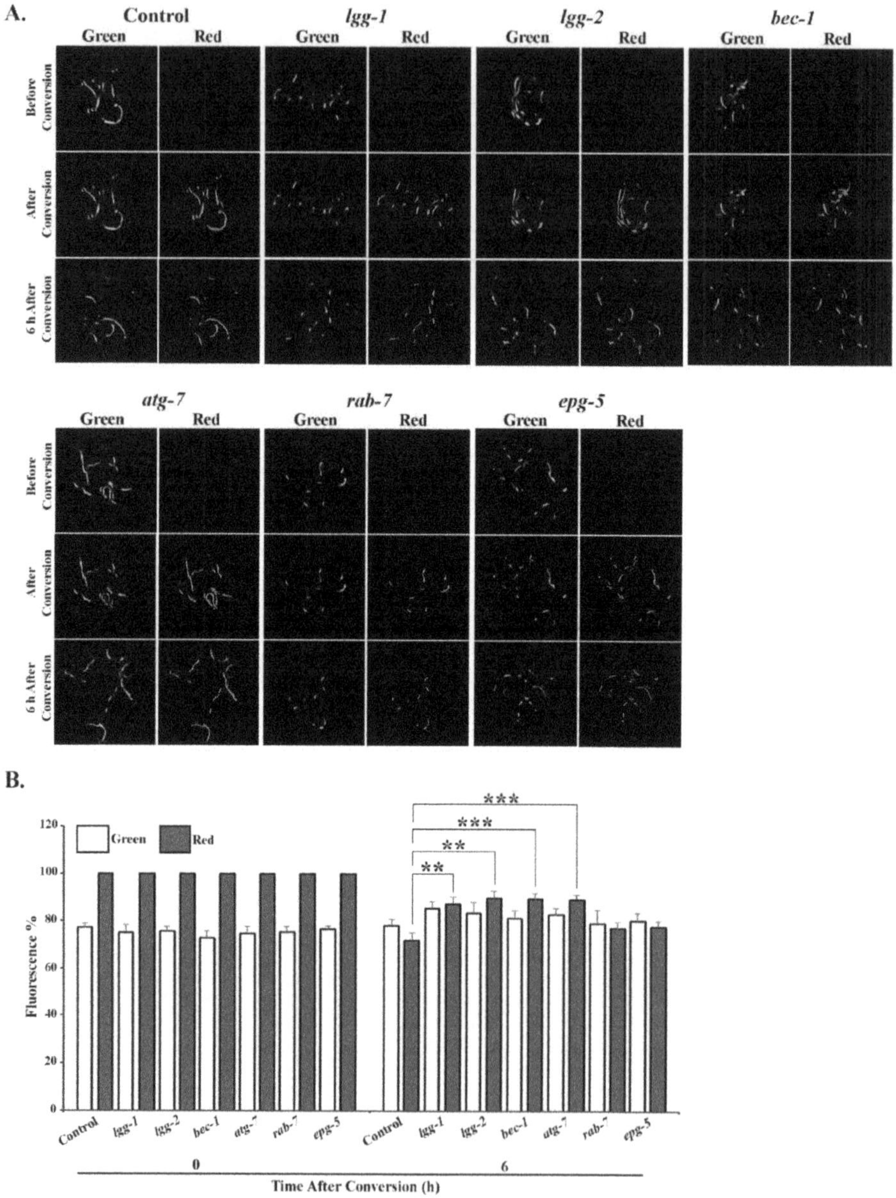

Figure 2. RNAi of *lgg-1*, *lgg-2*, *bec-1* or *atg-7* decreases UPS activity in intestinal cells in vivo. (**A**) The images represent fluorescent micrographs of control, *lgg-1*, *lgg-2*, *bec-1*, *atg-7*, *rab-7* or *epg-5* RNAi-treated N2 animals expressing UbG76V-Dendra2 reporter in intestinal cells before, immediately after and 6 h after photoconversion. (**B**) Quantification of UbG76V-Dendra2 degradation rate in intestinal cells. Graph shows the average percentage of green or red fluorescence relative to initial fluorescence intensity (set as 100%), or intensity at the point of photoconversion (set as 100%), respectively. The decrease in red fluorescence of UbG76V-Dendra2 reflects the UPS activity in the tissue. Results are the mean of quantification of 5 independent experiments (number of animals (n) = 105) (See Table S3). Error bars, SEM, ** $p < 0.01$ and *** $p < 0.001$ compared to the control.

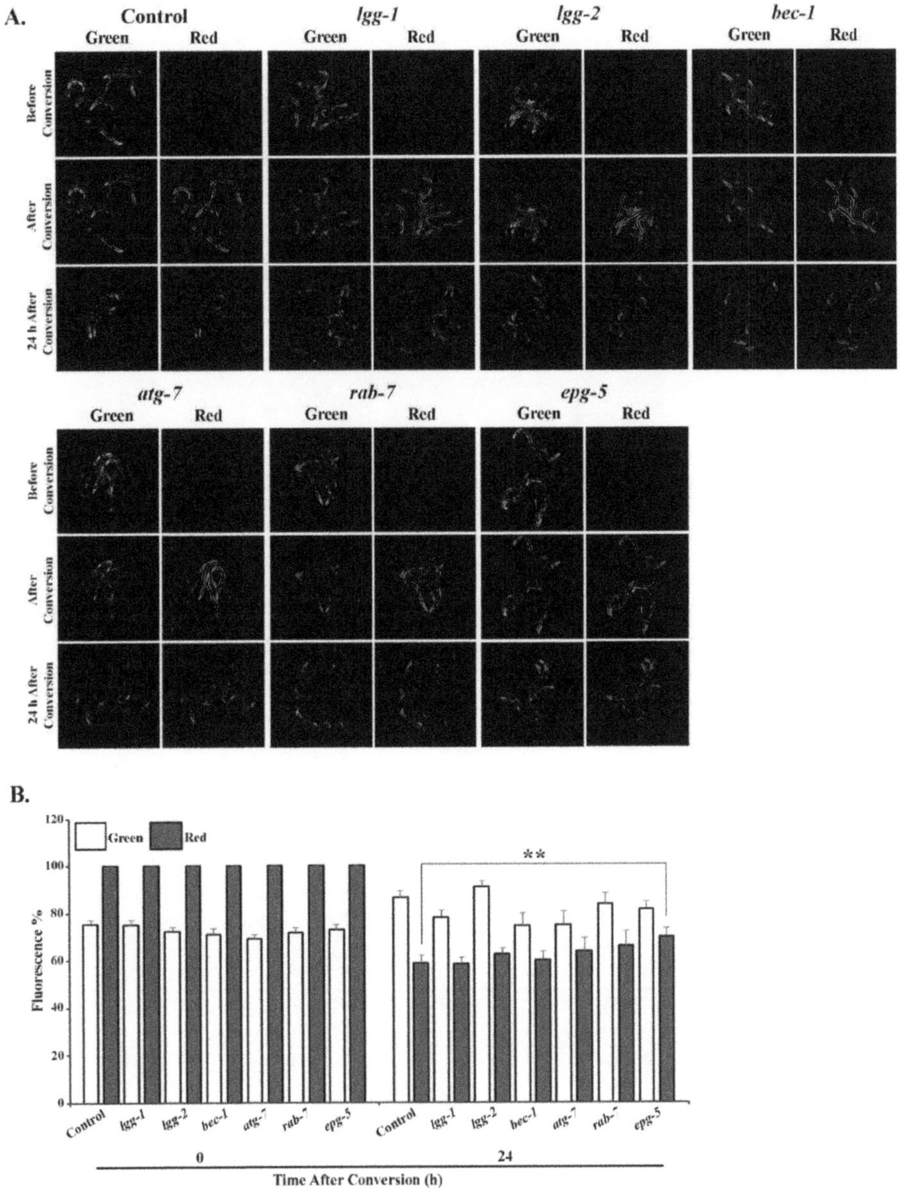

Figure 3. RNAi of *epg-5* decreases UPS activity in body-wall muscle cells in vivo. (**A**) The images represent fluorescent micrographs of control, *lgg-1*, *lgg-2*, *bec-1*, *atg-7*, *rab-7* or *epg-5* RNAi-treated wild-type animals expressing the UbG76V-Dendra2 reporter in body-wall muscle cells before, immediately after and 24 h after photoconversion. (**B**) Quantification of UbG76V-Dendra2 degradation rate in body-wall muscle cells. Graph shows the average percentage of green or red fluorescence relative to initial fluorescence intensity (set as 100%), or intensity at the point of photoconversion (set as 100%), respectively. The decrease in red fluorescence of UbG76V-Dendra2 reflects the UPS activity in the tissue. Results are the mean of quantification of 5 independent experiments (number of animals (n) = 100) (See Table S3). Error bars, SEM, ** $p < 0.01$ compared to the control.

3.3. Downregulation of lgg-1 Affects UPS Function Independently of p62/SQST-1 Accumulation

Previous studies have reported that polyubiquitinated proteins accumulate in mammalian cells upon autophagy inhibition via a process mediated by increased levels of the autophagosome cargo protein p62 [23,41]. We performed RNAi against *sqst-1*, the *C. elegans* homolog of mammalian p62, in combination with *lgg-1* RNAi to prevent p62 accumulation. Simultaneous downregulation of these two genes did not counteract the *lgg-1* RNAi-induced accumulation of polyubiquitinated proteins, nor the reduction in proteasomal substrate degradation in intestinal cells (Figure 4A,B). This suggests that the decreased UPS activity observed upon *lgg-1* RNAi is independent of *sqst-1* upregulation, as *sqst-1* mRNA levels were efficiently decreased upon both single or double gene RNAi treatment (Figure S1D). *sqst-1* RNAi alone did not elicit an effect on proteasome activity in intestinal cells (Figure 4A,B). Our data imply that modulating autophagy via decreased *lgg-1* expression leads to impaired proteasome activity, in a SQST-1-independent manner, in *C. elegans* intestinal cells.

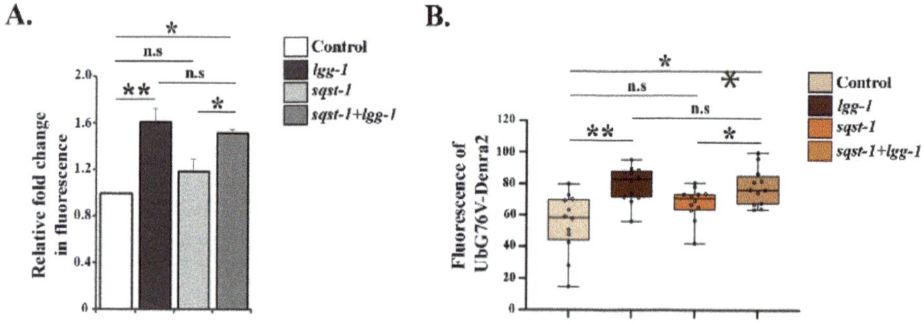

Figure 4. *lgg-1* RNAi affects UPS function independently of SQST-1. (**A**) Quantification of polyubiquitin reporter fluorescence in intestinal cells upon control, *lgg-1*, *sqst-1* or combined *sqst-1* and *lgg-1* RNAi. Graphs show average fold change in the amount of fluorescent signal compared to control RNAi treatment (set as 1). Results are the mean of quantifications of 4 independent experiments (number of animals (n) = 90). (**B**) Quantification of UbG76V-Dendra2 degradation rate in intestinal cells. Graph shows the average percentage of red fluorescence remaining 6 h after photoconversion upon respective RNAi treatments. Results are the mean of quantification of 3 independent experiments (number of animals (n) = 65) (See Tables S2 and S3). Error bars, SEM, * $p < 0.05$, ** $p < 0.01$, and n.s. = non-significant compared to control or indicated treatment.

3.4. Depletion of Autophagy Genes Affects Proteasome Activity or Expression

In addition to the UPS in vivo experiments, we investigated proteasome activity in whole animal lysates by performing an in-gel activity assay whereby the lysates were resolved on a native acrylamide gel, and the in-gel activity was measured with the fluorogenic suc-LLVY-AMC proteasome substrate [38]. Knockdown of *lgg-1* or *epg-5* by RNAi decreased proteasome activity by approximately 0.2-fold, compared to the control (Figure 5A,B, Table S4), which is in agreement with our in vivo data (Figures 2A,B and 3A,B, Table S3). Interestingly, we did not detect any effect on proteasome activity after *lgg-2*, *bec-1* or *atg-7* RNAi treatments (Figure 5A,B, Table S4), suggesting that the in vivo tissue-specific proteasome activity responses (Figures 2 and 3, Table S3) are masked in whole animal lysates.

We next investigated total proteasome amounts in whole worm lysates with Western blot, using an antibody against proteasome 20S alpha subunits. The downregulation of *lgg-2* or *bec-1* resulted in approximately a 0.2-fold decrease in total proteasome abundance, whereas no change could be detected with *lgg-1*, *atg-7*, *rab-7* or *epg-5* RNAi (Figure 5C,D, Table S4).

Taken together, our in vivo and in vitro results reveal that the depletion of autophagy genes causes different proteasomal responses, i.e., *lgg-1* and *epg-5* RNAi modulate proteasome activity without detectably changing its level, whereas *lgg-2* and *bec-1* RNAi in turn decrease total proteasome amount in whole animal lysates without causing changes in proteasome activity.

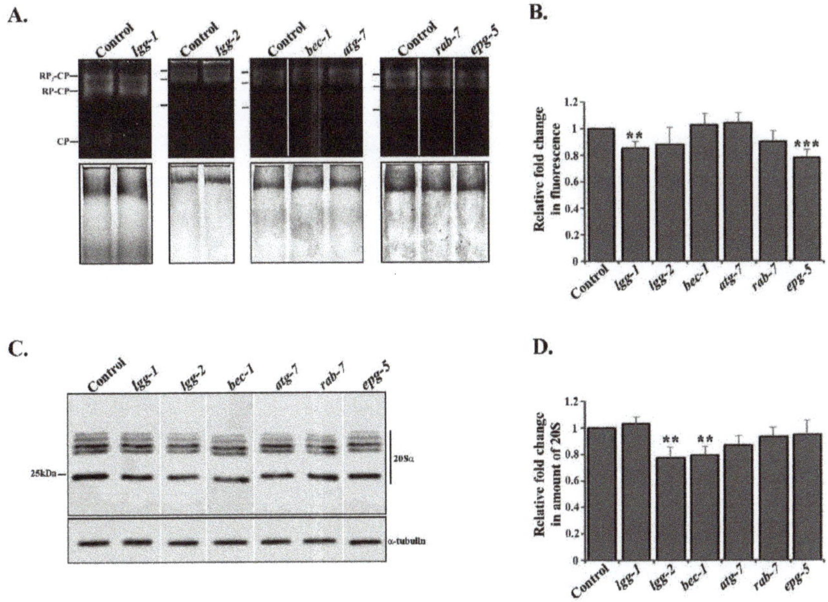

Figure 5. Autophagy genes affect proteasome activity or expression. (**A**) Proteasome activity investigated by in-gel activity assay with fluorogenic suc-LLVY-AMC proteasome substrate from whole animal lysates of *rrf-3(pk1426)* animals treated with control, *lgg-1*, *lgg-2*, *bec-1*, *atg-7*, *rab-7* or *epg-5* RNAi (upper panel), and Coomassie staining of total protein of the same gel (lower panel). The marker lines represent RP_2-CP, RP-CP and CP, respectively. (**B**) Quantification of in-gel proteasome activity. The bands corresponding to RP_2-CP, RP-CP and CP were used for quantification. Graph shows average fold change in fluorescent signal compared to control RNAi (set as 1). Results are the mean of quantifications from 10 independent experiments (See Table S4). Error bars, SEM, ** $p < 0.01$ and *** $p < 0.001$ compared to the control. (**C**) Lysates of *rrf-3(pk1426)* animals treated with control, *lgg-1*, *lgg-2*, *bec-1*, *atg-7*, *rab-7* or *epg-5* RNAi separated on SDS-PAGE and immunoblotted against proteasome 20S alpha subunits (upper panel) and α-tubulin (lower panel). (**D**) Quantification of proteasome immunoblots. Graph shows average fold change in levels of 20S alpha subunits normalized against α-tubulin. Results are the mean of quantifications from 7 independent experiments (See Table S4). Error bars, SEM, ** $p < 0.01$ compared to the control.

3.5. RNAi of *lgg-2* or *bec-1* Causes Distinct Responses in Proteasome Tissue Expression

To further investigate proteasome expression at the tissue and cell resolution level in *C. elegans*, we used an immunohistochemical approach for the detection of the 20S proteasome [33]. In support of our Western blot data from the whole animal lysates, we detected decreased proteasome amounts after the downregulation of *lgg-2* or *bec-1* by RNAi (Figure 6). Importantly, the decrease in proteasome expression was not uniform, but displayed distinct tissue-specificity. *bec-1* RNAi decreased proteasome amounts in both intestinal and body-wall muscle cells (approximately by 0.25-fold compared to control RNAi, Figure 6A–D, Table S4), whereas *lgg-2* RNAi caused a decrease in proteasome expression only in the body-wall muscle cells (approximately by 0.2-fold, Figure 6C,D, Table S4). In agreement with our Western blot analysis, no changes were detected with *epg-5*, *lgg-1*, *atg-7* or *rab-7* RNAi (Figure 6 and Figure S6, Table S4). Altogether, our studies reveal that the downregulation of various autophagy genes leads to distinct responses in proteasome tissue expression levels.

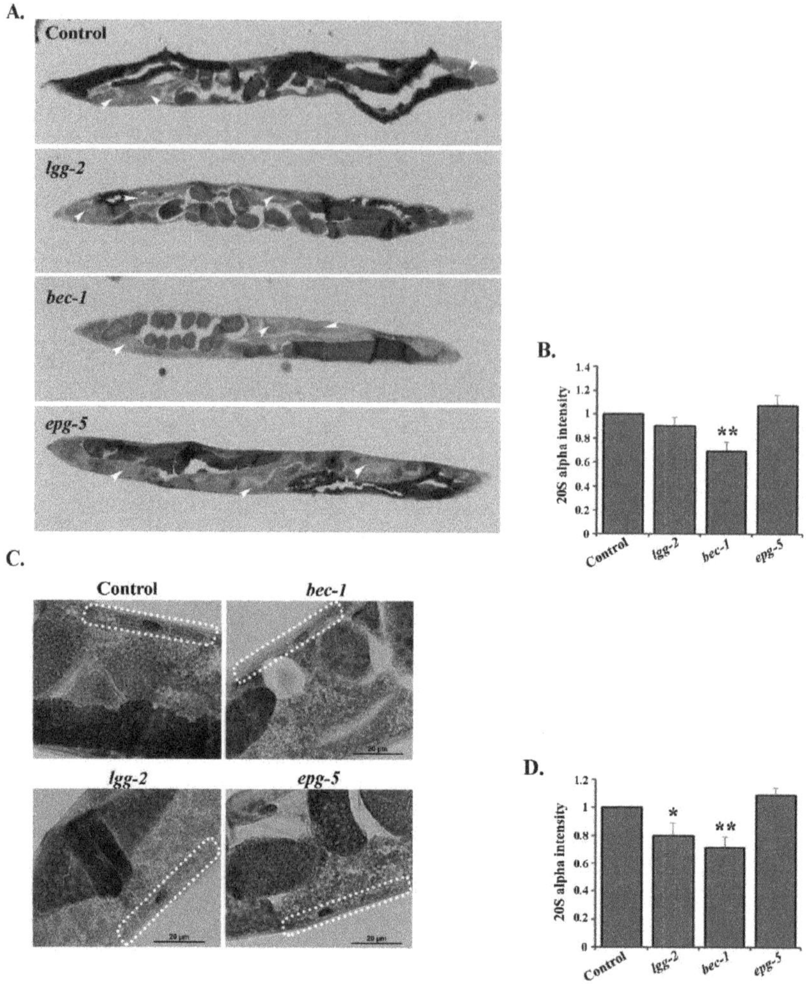

Figure 6. Knockdown of *lgg-2* and *bec-1* results in tissue-specific proteasome expression responses. (**A**) Images representing proteasome immunoreactivity of *rrf-3(pk1426)* animals treated with control, *lgg-2*, *bec-1* or *epg-5* RNAi in the intestinal cells (indicated by white arrowheads). (**B**) Quantification of immunoreactivity in intestinal cells. Graph shows the average fold change in staining intensity of 20S alpha antibody compared to control treatment (set as 1). Results are the mean of quantifications from 3 independent experiments (number of intestinal cells (n) = 80) (See Table S4). Error bars, SEM, ** $p < 0.01$ compared to the control. (**C**) Higher magnification images of *rrf-3(pk1426)* animals treated with control, *lgg-2*, *bec-1* or *epg-5* RNAi in the body-wall muscle cells (outlined with white dash line). (**D**) Quantification of immunoreactivity in the body-wall muscle cells. Graph shows the average fold change in staining intensity against 20S alpha antibody. Results are the mean of quantifications from 3 independent experiments (number of muscle cells (n) = 45) (See Table S4). Error bars, SEM, * $p < 0.05$, ** $p < 0.01$ compared to control.

4. Discussion

In this study, we use *C. elegans* as an organismal approach to evaluate for the first time how compromising autophagy at different steps affects UPS in different tissues. We show that individual depletions of several autophagy genes cause tissue-specific effects on UPS activity and proteasome abundance. Though several previous studies using cell lines and mice have shown that the impairment of proteasome function results in the induction of autophagy [18–21,42,43], studies on the inhibition of autophagy have reported more complex results, including both UPS activation and inhibition. For example, the pharmacological inhibition of autophagy and the knockdown of *Atg-5* or *Atg-7* in human colon cancer cell lines have been shown to increase proteasomal activity [24]. Similarly, Kim et al. reported that *Atg-5* deficient mouse embryonic fibroblast exhibited elevated proteasome activity [22]. On the other hand, *Atg-7* deficient mouse neurons display accumulations of ubiquitinated substrates without observed changes in proteasomal function [44]. Additionally, Korolchuk et al. reported that arresting autophagy by siRNA knockdown of *Atg-7* and *Atg-12*, or with autophagy inhibitors, causes impaired degradation of at least two proteasomal substrates, UbG76V-GFP and p53, in HeLa cells, although without affecting total proteasome activity in the cell lysates [23]. Furthermore, autophagy blockage due to lysosomal enzymatic deficiency causes decreased proteasomal activity in neuroblastoma cells [25]. Here, we show that downregulation of *bec-1*, *atg-7*, *lgg-1*, *lgg-2* or *epg-5* impairs UPS function in *C. elegans*. Importantly, our results reveal that, depending on which autophagy gene is depleted, the effect on UPS varies. For example, downregulation of *lgg-1* or *epg-5* by RNAi affects the UPS and proteasomal activity in in vivo and in vitro assays, without a concomitant change in proteasome expression, whereas *lgg-2* or *bec-1* RNAi decreases UPS activity in vivo, as well as the amount of the proteasome (Figures 2, 3, 5 and 6). The results of *lgg-1* and *epg-5* RNAi suggest that their effect could be mediated through a regulator of the proteasome. In line with this suggestion, we have previously shown that the proteasome-associated deubiquitinase UBH-4 regulates proteasome activity in *C. elegans* in an insulin/IGF-1 signaling-mediated manner [32], and that animals with mutations in this pathway display altered proteasome activity without a concomitant change in proteasome tissue expression [32,33]. Further studies are needed to uncover the underlying mechanisms of *lgg-1* and *epg-5* knockdown in proteasome activity.

Our previous research has demonstrated that UPS activity varies in a tissue-specific manner in live *C. elegans*, and is affected by aging and stress, such as proteasomal impairment and elevated growth temperature [31,32,35]. The rate of autophagic flux has also been reported to vary in different *C. elegans* tissues [29,30]. Chang et al. showed that autophagy activity declines in several tissues in wild-type animals, whereas long-lived mutants display spatiotemporal regulation of autophagy [29]. Additionally, Chapin et al. reported that both basal autophagy as well as stress-induced autophagy can vary in different tissues [30]. Recently, it was shown that affecting proteostasis through the downregulation of chaperones involved in disaggregation results in increased autophagy and decreased UPS in *C. elegans* and human HEK293 cells [45]. Interestingly, the same study revealed that both autophagy and UPS are impaired in tissues expressing the human disease-linked amyloid protein $A\beta_{3-42}$ and Q_{40} in *C. elegans* [45]. Here, we reveal that a widespread downregulation of distinct autophagy genes generates tissue-specific UPS responses. We show that the downregulation of *bec-1*, *lgg-1*, *lgg-2* and *atg-7* decreases UPS activity in intestinal cells, whereas *epg-5* RNAi affects UPS activity in muscle cells (Figures 2 and 3). Similarly, we detect tissue-specific decreases in proteasome expression upon autophagy impairment, as *bec-1* RNAi decreases the amount of proteasome in both intestinal and muscle cells, but *lgg-2* RNAi affects proteasome expression only in muscle cells (Figure 6). This tissue specificity is not due to a difference in RNAi efficiency, as gene-specific RNAi phenotypes are observed for both cell types. It is noteworthy that our data suggest that a particular proteasomal substrate can be degraded while some substrates accumulate upon compromised autophagy.

Our experimental approaches for studying UPS activity and proteasome expression enable us to perform intra-tissue analysis at the organismal level. The *bec-1* RNAi-induced accumulations of the UPS substrate UbG76V-Dendra2, as well as those of the endogenous pool of K-48-linked polyubiquitinated proteins in the intestine, coincided with the decreased expression of proteasome in the same tissue, thus displaying a mechanism by which BEC-1 affects intestinal UPS function. The instances wherein we did not observe a correlation between changes in proteasome level and UPS activity (i.e., downregulation of *lgg-1*, *lgg-2*, *atg-7* and *epg-5*) could be due to, for example, changes in substrate delivery to the proteasome, posttranslational modifications, or proteasome interaction with regulators. Korolchuk et al. has demonstrated that, in HeLa cells, the accumulation of distinct proteasomal substrates upon compromised autophagy is dependent on increased p62 levels without a change in proteasome activity, suggesting difficulties in substrate delivery to the proteasome [23]. Here, we show that the accumulation of endogenous K-48-linked polyubiquitinated proteasomal substrates, as well as the decreased degradation of the proteasomal substrate UbG76V-Dendra2 upon impaired autophagy by *lgg-1* RNAi, is independent of increased levels of SQST-1 in *C. elegans* intestinal cells. As p62/SQST-1 has preference for binding to K-63-linked polyubiquitinated proteins [46], this may explain why the pool of endogenous K-48-linked proteasomal substrates is not affected. Moreover, it was recently shown that overexpression of *sqst-1* in *C. elegans* leads to a modest increase in proteasome activity, and a reduced amount of ubiquitinated proteins, as analyzed in whole animal lysates [47].

Our data demonstrate that UPS does not function as a direct compensatory mechanism upon ALP impairment in a multicellular organism. Importantly, our results emphasize the existence of tissue-specific responses in UPS activity and proteasome expression to disturbed autophagy. Whilst previous research has largely focused on how UPS and autophagy are individually regulated, our study provides new information that could be implemented for the simultaneous modulation of UPS and ALP. This is of particular interest, as dysfunctions in these proteolytic systems are associated with various, often aging-related, disorders, such as neurodegenerative diseases (e.g., Parkinson's disease and Alzheimer's disease), autoimmune diseases and different types of cancer. In particular, the tissue-specific effects of autophagy gene depletions on UPS revealed here could provide insight into designing therapies for altering proteasome activity in a specific tissue. Our results create further interest in discovering the mechanisms by which the depletion of different autophagy genes elicits distinct UPS responses. More detailed understanding of UPS and ALP, and their multifaceted crosstalk in varying physiological conditions, will increase our understanding of proteostasis regulation.

Supplementary Materials: The following are available online at http://www.mdpi.com/2073-4409/9/8/1858/s1. Figure S1: Efficient downregulation of autophagy genes following RNAi. Figure S2: Expression of *mCherry::GFP::LGG-1* reporter in *C. elegans* following RNAi of autophagy genes. Figure S3: Downregulation of autophagy genes does not affect the amount of polyubiquitinated proteins in whole animal lysates. Figure S4: Knockdown of autophagy genes does not affect the stability of Dendra2 reporter in intestinal cells. Figure S5: Knockdown of autophagy genes does not affect the stability of Dendra2 reporter in body-wall muscle cells. Figure S6: RNAi of *lgg-1*, *atg-7* and *rab-7* does not affect proteasome tissue expression. Table S1: List of *C. elegans* strains used in this study. Table S2: Numbers of imaged polyubiquitin reporter animals. Table S3: Numbers of imaged UbG76V-Dendra2 and Dendra2 reporter animals. Table S4: Significance values of in vitro assays. Table S5: Numbers of imaged *mCherry::GFP::LGG1* reporter animals. Table S6: List of qPCR oligonucleotides used in this study.

Author Contributions: Conceptualization, S.J. and C.I.H.; Data curation, S.J.; Formal analysis, S.J.; Funding acquisition, C.I.H.; Investigation, S.J.; Methodology, S.J.; Project administration, C.I.H.; Resources, C.I.H.; Supervision, C.I.H.; Validation, S.J. and C.I.H.; Writing—original draft, S.J.; Writing—review and editing, C.I.H. All authors have read and agreed to the published version of the manuscript.

Funding: This research was funded by the Academy of Finland (297776) and the Sigrid Jusélius Foundation to CIH. S.J. was funded by the Doctoral Program in Biomedicine, Doctoral School in Health Science, University of Helsinki. Open access funding provided by University of Helsinki.

Acknowledgments: We thank the Biomedicum Imaging Unit (BIU), Faculty of Medicine, University of Helsinki, for their help with microscopy and image analysis, Holmberg lab members and Professor Dan Lindholm for their advice and valuable comments. Some strains were provided by the CGC, which is funded by NIH Office of Research Infrastructure Programs (P40 OD010440).

Conflicts of Interest: The authors declare no conflict of interest.

References

1. Collins, G.A.; Goldberg, A.L. The Logic of the 26S Proteasome. *Cell* **2017**, *169*, 792–806. [CrossRef]
2. Kocaturk, N.M.; Gozuacik, D. Crosstalk Between Mammalian Autophagy and the Ubiquitin-Proteasome System. *Front. Cell Dev. Biol.* **2018**, *6*, 128. [CrossRef]

3. Finley, D. Recognition and processing of ubiquitin-protein conjugates by the proteasome. *Annu. Rev. Biochem.* **2009**, *78*, 477–513. [CrossRef] [PubMed]
4. Schmidt, M.; Finley, D. Regulation of proteasome activity in health and disease. *Biochim. Biophys. Acta* **2014**, *1843*, 13–25. [CrossRef] [PubMed]
5. Coll-Martínez, B.; Crosas, B. How the 26S Proteasome Degrades Ubiquitinated Proteins in the Cell. *Biomolecules* **2019**, *9*, 395. [CrossRef] [PubMed]
6. Wen, X.; Klionsky, D.J. At a glance: A history of autophagy and cancer. *Semin. Cancer Biol.* **2019**. [CrossRef] [PubMed]
7. Mizushima, N. A brief history of autophagy from cell biology to physiology and disease. *Nat. Cell Biol.* **2018**, *20*, 521–527. [CrossRef] [PubMed]
8. Nam, T.; Han, J.H.; Devkota, S.; Lee, H.W. Emerging Paradigm of Crosstalk between Autophagy and the Ubiquitin-Proteasome System. *Mol. Cells* **2017**, *40*, 897–905.
9. Nakatogawa, H.; Ichimura, Y.; Ohsumi, Y. Atg8, a ubiquitin-like protein required for autophagosome formation, mediates membrane tethering and hemifusion. *Cell* **2007**, *130*, 165–178. [CrossRef]
10. Nakatogawa, H. Two ubiquitin-like conjugation systems that mediate membrane formation during autophagy. *Essays Biochem.* **2013**, *55*, 39–50.
11. Meng, Y.C.; Lou, X.L.; Yang, L.Y.; Li, D.; Hou, Y.Q. Role of the autophagy-related marker LC3 expression in hepatocellular carcinoma: A meta-analysis. *J. Cancer Res. Clin. Oncol.* **2020**, *146*, 1103–1113. [CrossRef] [PubMed]
12. Gutierrez, M.G.; Munafo, D.B.; Beron, W.; Colombo, M.I. Rab7 is required for the normal progression of the autophagic pathway in mammalian cells. *J. Cell Sci.* **2004**, *117*, 2687–2697. [CrossRef] [PubMed]
13. Tian, Y.; Li, Z.; Hu, W.; Ren, H.; Tian, E.; Zhao, Y.; Lu, Q.; Huang, X.; Yang, P.; Li, X.; et al. *C. elegans* screen identifies autophagy genes specific to multicellular organisms. *Cell* **2010**, *141*, 1042–1055. [CrossRef] [PubMed]
14. Zhao, H.; Zhao, Y.G.; Wang, X.; Xu, L.; Miao, L.; Feng, D.; Chen, Q.; Kovacs, A.L.; Fan, D.; Zhang, H. Mice deficient in Epg5 exhibit selective neuronal vulnerability to degeneration. *J. Cell Biol.* **2013**, *200*, 731–741. [CrossRef] [PubMed]
15. Mizushima, N.; Yoshimori, T.; Levine, B. Methods in mammalian autophagy research. *Cell* **2010**, *140*, 313–326. [CrossRef] [PubMed]
16. Ohsumi, Y. Historical landmarks of autophagy research. *Cell Res.* **2014**, *24*, 9–23. [CrossRef]
17. Pohl, C.; Dikic, I. Cellular quality control by the ubiquitin-proteasome system and autophagy. *Science* **2019**, *366*, 818–822. [CrossRef]
18. Fan, T.; Huang, Z.; Chen, L.; Wang, W.; Zhang, B.; Xu, Y.; Pan, S.; Mao, Z.; Hu, H.; Geng, Q. Associations between autophagy, the ubiquitin-proteasome system and endoplasmic reticulum stress in hypoxia-deoxygenation or ischemia-reperfusion. *Eur. J. Pharmacol.* **2016**, *791*, 157–167. [CrossRef]
19. Shen, Y.F.; Tang, Y.; Zhang, X.J.; Huang, K.X.; Le, W.D. Adaptive changes in autophagy after UPS impairment in Parkinson's disease. *Acta Pharmacol. Sin.* **2013**, *34*, 667–673. [CrossRef]
20. Zheng, Q.; Su, H.; Tian, Z.; Wang, X. Proteasome malfunction activates macroautophagy in the heart. *Am. J. Cardiovasc. Dis.* **2011**, *1*, 214–226.
21. Li, C.; Wang, X.; Li, X.; Qiu, K.; Jiao, F.; Liu, Y.; Kong, Q.; Liu, Y.; Wu, Y. Proteasome Inhibition Activates Autophagy-Lysosome Pathway Associated with TFEB Dephosphorylation and Nuclear Translocation. *Front. Cell Dev. Biol.* **2019**, *7*, 170. [CrossRef] [PubMed]
22. Kim, E.; Park, S.; Lee, J.H.; Mun, J.Y.; Choi, W.H.; Yun, Y.; Lee, J.; Kim, J.H.; Kang, M.J.; Lee, M.J. Dual Function of USP14 Deubiquitinase in Cellular Proteasomal Activity and Autophagic Flux. *Cell Rep.* **2018**, *24*, 732–743. [CrossRef] [PubMed]
23. Korolchuk, V.I.; Mansilla, A.; Menzies, F.M.; Rubinsztein, D.C. Autophagy inhibition compromises degradation of ubiquitin-proteasome pathway substrates. *Mol. Cell* **2009**, *33*, 517–527. [CrossRef] [PubMed]
24. Wang, X.J.; Yu, J.; Wong, S.H.; Cheng, A.S.; Chan, F.K.; Ng, S.S.; Cho, C.H.; Sung, J.J.; Wu, W.K. A novel crosstalk between two major protein degradation systems: Regulation of proteasomal activity by autophagy. *Autophagy* **2013**, *9*, 1500–1508. [CrossRef] [PubMed]
25. Qiao, L.; Zhang, J. Inhibition of lysosomal functions reduces proteasomal activity. *Neurosci. Lett.* **2009**, *456*, 15–19. [CrossRef] [PubMed]
26. Chen, Y.; Scarcelli, V.; Legouis, R. Approaches for Studying Autophagy in Caenorhabditis elegans. *Cells* **2017**, *6*, 27. [CrossRef]

27. Palmisano, N.J.; Meléndez, A. Autophagy in *C. elegans* development. *Dev. Biol.* **2019**, *447*, 103–125. [CrossRef]
28. Papaevgeniou, N.; Chondrogianni, N. The ubiquitin proteasome system in Caenorhabditis elegans and its regulation. *Redox Biol.* **2014**, *2*, 333–347. [CrossRef]
29. Chang, J.T.; Kumsta, C.; Hellman, A.B.; Adams, L.M.; Hansen, M. Spatiotemporal regulation of autophagy during Caenorhabditis elegans aging. *eLife* **2017**, *6*. [CrossRef]
30. Chapin, H.C.; Okada, M.; Merz, A.J.; Miller, D.L. Tissue-specific autophagy responses to aging and stress in *C. elegans*. *Aging (Albany NY)* **2015**, *7*, 419–434. [CrossRef]
31. Hamer, G.; Matilainen, O.; Holmberg, C.I. A photoconvertible reporter of the ubiquitin-proteasome system in vivo. *Nat. Methods* **2010**, *7*, 473–478. [CrossRef] [PubMed]
32. Matilainen, O.; Arpalahti, L.; Rantanen, V.; Hautaniemi, S.; Holmberg, C.I. Insulin/IGF-1 signaling regulates proteasome activity through the deubiquitinating enzyme UBH-4. *Cell Rep.* **2013**, *3*, 1980–1995. [CrossRef] [PubMed]
33. Mikkonen, E.; Haglund, C.; Holmberg, C.I. Immunohistochemical analysis reveals variations in proteasome tissue expression in *C. elegans*. *PLoS ONE* **2017**, *12*, e0183403. [CrossRef] [PubMed]
34. Brenner, S. The genetics of Caenorhabditis elegans. *Genetics* **1974**, *77*, 71–94. [PubMed]
35. Pispa, J.; Matilainen, O.; Holmberg, C.I. Tissue-specific effects of temperature on proteasome function. *Cell Stress Chaperones* **2020**, *25*, 563–572. [CrossRef]
36. Timmons, L.; Court, D.L.; Fire, A. Ingestion of bacterially expressed dsRNAs can produce specific and potent genetic interference in Caenorhabditis elegans. *Gene* **2001**, *263*, 103–112. [CrossRef]
37. Vandesompele, J.; De Preter, K.; Pattyn, F.; Poppe, B.; Van Roy, N.; De Paepe, A.; Speleman, F. Accurate normalization of real-time quantitative RT-PCR data by geometric averaging of multiple internal control genes. *Genome Biol.* **2002**, *3*, RESEARCH0034. [CrossRef]
38. Elsasser, S.; Schmidt, M.; Finley, D. Characterization of the proteasome using native gel electrophoresis. *Methods Enzymol.* **2005**, *398*, 353–363.
39. Matilainen, O.; Jha, S.; Holmberg, C.I. Fluorescent Tools for In Vivo Studies on the Ubiquitin-Proteasome System. *Methods Mol. Biol.* **2016**, *1449*, 215–222.
40. Li, X.; Matilainen, O.; Jin, C.; Glover-Cutter, K.M.; Holmberg, C.I.; Blackwell, T.K. Specific SKN-1/Nrf stress responses to perturbations in translation elongation and proteasome activity. *PLoS Genet.* **2011**, *7*, e1002119. [CrossRef]
41. Komatsu, M.; Waguri, S.; Koike, M.; Sou, Y.S.; Ueno, T.; Hara, T.; Mizushima, N.; Iwata, J.; Ezaki, J.; Murata, S.; et al. Homeostatic levels of p62 control cytoplasmic inclusion body formation in autophagy-deficient mice. *Cell* **2007**, *131*, 1149–1163. [CrossRef] [PubMed]
42. Ding, W.X.; Ni, H.M.; Gao, W.; Yoshimori, T.; Stolz, D.B.; Ron, D.; Yin, X.M. Linking of autophagy to ubiquitin-proteasome system is important for the regulation of endoplasmic reticulum stress and cell viability. *Am. J. Pathol.* **2007**, *171*, 513–524. [CrossRef] [PubMed]
43. Rideout, H.J.; Lang-Rollin, I.; Stefanis, L. Involvement of macroautophagy in the dissolution of neuronal inclusions. *Int. J. Biochem. Cell Biol.* **2004**, *36*, 2551–2562. [CrossRef] [PubMed]
44. Komatsu, M.; Waguri, S.; Chiba, T.; Murata, S.; Iwata, J.; Tanida, I.; Ueno, T.; Koike, M.; Uchiyama, Y.; Kominami, E.; et al. Loss of autophagy in the central nervous system causes neurodegeneration in mice. *Nature* **2006**, *441*, 880–884. [CrossRef] [PubMed]
45. Feleciano, D.R.; Juenemann, K.; Iburg, M.; Brás, I.C.; Holmberg, C.I.; Kirstein, J. Crosstalk Between Chaperone-Mediated Protein Disaggregation and Proteolytic Pathways in Aging and Disease. *Front. Aging Neurosci.* **2019**, *11*, 9. [CrossRef]
46. Seibenhener, M.L.; Babu, J.R.; Geetha, T.; Wong, H.C.; Krishna, N.R.; Wooten, M.W. Sequestosome 1/p62 is a polyubiquitin chain binding protein involved in ubiquitin proteasome degradation. *Mol. Cell. Biol.* **2004**, *24*, 8055–8068. [CrossRef]
47. Kumsta, C.; Chang, J.T.; Lee, R.; Tan, E.P.; Yang, Y.; Loureiro, R.; Choy, E.H.; Lim, S.H.Y.; Saez, I.; Springhorn, A.; et al. The autophagy receptor p62/SQST-1 promotes proteostasis and longevity in *C. elegans* by inducing autophagy. *Nat. Commun.* **2019**, *10*, 5648. [CrossRef]

cells

MDPI

Article

Functional Characterisation of the Autophagy ATG12~5/16 Complex in *Dictyostelium discoideum*

Malte Karow [1], Sarah Fischer [1], Susanne Meßling [1], Roman Konertz [1], Jana Riehl [1], Qiuhong Xiong [2], Ramesh Rijal [3], Prerana Wagle [4], Christoph S. Clemen [5,6,7] and Ludwig Eichinger [1,*]

1 Centre for Biochemistry, Institute of Biochemistry I, Medical Faculty, University of Cologne, 50931 Cologne, Germany; mail@maltekarow.com (M.K.); sarah.fischer@uni-koeln.de (S.F.); susanne_messling@web.de (S.M.); roman.konertz@uni-koeln.de (R.K.); jriehl@uni-koeln.de (J.R.)

2 Institute of Biomedical Sciences, Shanxi University, No. 92 Wucheng Road, Taiyuan 030006, China; xiongqiuhong67@hotmail.com

3 Department of Biology, Texas A&M University, College Station, TX 77843-3474, USA; rrijal@bio.tamu.edu

4 Bioinformatics Core Facility, CECAD Research Center, University of Cologne, 50931 Cologne, Germany; ake37@uni-koeln.de

5 Institute of Aerospace Medicine, German Aerospace Center (DLR), 51147 Cologne, Germany; Christoph.Clemen@uni-koeln.de

6 Center for Physiology and Pathophysiology, Institute of Vegetative Physiology, Medical Faculty, University of Cologne, 50931 Cologne, Germany

7 Institute of Neuropathology, University Hospital Erlangen, Friedrich-Alexander University Erlangen-Nürnberg, 91054 Erlangen, Germany

* Correspondence: ludwig.eichinger@uni-koeln.de; Tel.: +49-221-478-6928; Fax: +49-221-478-97524

Received: 19 March 2020; Accepted: 6 May 2020; Published: 9 May 2020

check for updates

Abstract: Macroautophagy, a highly conserved and complex intracellular degradative pathway, involves more than 20 core autophagy (ATG) proteins, among them the hexameric ATG12~5/16 complex, which is part of the essential ubiquitin-like conjugation systems in autophagy. *Dictyostelium discoideum atg5* single, *atg5/12* double, and *atg5/12/16* triple gene knock-out mutant strains displayed similar defects in the conjugation of ATG8 to phosphatidylethanolamine, development, and cell viability upon nitrogen starvation. This implies that ATG5, 12 and 16 act as a functional unit in canonical autophagy. Macropinocytosis of TRITC dextran and phagocytosis of yeast were significantly decreased in ATG5$^-$ and ATG5$^-$/12$^-$ and even further in ATG5$^-$/12$^-$/16$^-$ cells. In contrast, plaque growth on *Klebsiella aerogenes* was about twice as fast for ATG5$^-$ and ATG5$^-$/12$^-$/16$^-$ cells in comparison to AX2, but strongly decreased for ATG5$^-$/12$^-$ cells. Along this line, phagocytic uptake of *Escherichia coli* was significantly reduced in ATG5$^-$/12$^-$ cells, while no difference in uptake, but a strong increase in membrane association of *E. coli*, was seen for ATG5$^-$ and ATG5$^-$/12$^-$/16$^-$ cells. Proteasomal activity was also disturbed in a complex fashion, consistent with an inhibitory activity of ATG16 in the absence of ATG5 and/or ATG12. Our results confirm the essential function of the ATG12~5/16 complex in canonical autophagy, and furthermore are consistent with autophagy-independent functions of the complex and its individual components. They also strongly support the placement of autophagy upstream of the ubiquitin-proteasome system (UPS), as a fully functional UPS depends on autophagy.

Keywords: ATG12~5/16 complex; autophagy; *Dictyostelium*; ubiquitin-like protein; phagocytosis; pinocytosis; proteasome; UPS

1. Introduction

Macroautophagy (hereafter autophagy for simplicity) is the major route for the lysosomal degradation of long-lived proteins, protein aggregates, and mal-functioning organelles in all eukaryotic cells [1]. Most likely, it evolved in unicellular organisms as a survival mechanism in response to starvation [2,3]. Autophagy is active at basal levels in most cell types and is involved in the regulation of a wide range of cellular functions. Due to its crucial roles, impaired autophagy contributes to the etiology of multiple human diseases, such as cancer, neurodegeneration, muscular dystrophy, and lipid-storage disorders [4–6].

Autophagy is characterised by the de novo formation of a double-membrane autophagosome, which matures into the autolysosome through fusion of its outer membrane with the lysosome. This complex cellular process, which serves to selectively or non-selectively sequester cargo, can be subdivided into initiation, maturation, and lysosomal degradation phases. It engages around 20 core autophagy (ATG) proteins that act in a sequential way and are organised in different molecular complexes [7–9]. In the initiation phase, an isolation membrane (phagophore in yeast) is generated at the endoplasmic reticulum (ER), which is expanded in the maturation phase by a membrane delivery and two ubiquitin-like conjugation systems. In the first ubiquitin-like conjugation system, the ubiquitin-like protein ATG12 is activated through the formation of a thioester bond of its C-terminal glycine with a cysteine of the E1-like enzyme ATG7 [10]. Subsequently, ATG12 is transferred to the cysteine of the E2-like enzyme ATG10 [11]. In the last step, the C-terminus of ATG12 is linked via an isopeptide bond to the ε-amino residue of an absolutely conserved lysine of ATG5 [12]. Two ATG12~5 conjugates interact non-covalently via ATG5 with the N-terminal region of an ATG16 dimer to the heterohexameric ATG12~5/16 complex ("~" indicates the covalent bond between ATG12 and ATG5 and "/" the non-covalent interaction between ATG5 and ATG16) [13]. In the final step of the second ubiquitin-like system, the ubiquitin-like protein ATG8 (LC3 in mammals) is covalently attached through the E3-like enzymatic activity of the ATG12~5/16 complex to the lipid phosphatidylethanolamine (PE) on the expanding autophagosomal membrane [10,14–17] (Figure 1).

In in vitro assays, it was shown that the yeast ATG12~5/16 complex directly binds to membranes and that this binding is mediated by ATG5, inhibited by ATG12, and activated by ATG16 [18]. The interaction of the ATG12~5/16 complex with the isolation membrane is in vivo further mediated by two distinct and specific mechanisms. (i) ATG12 interacts with the Ulk1 kinase complex (ATG1 kinase complex in yeast) that is crucial for the generation of the isolation membrane [19,20]. (ii) ATG16 binds to the PROPPIN family protein WIPI2b (ATG21 in yeast), which interacts with PtIns3P, generated by the PtIns3P-kinase complex at the isolation membrane [21,22]. Together, these mechanisms provide the required affinity and specificity for the correct targeting of the complex to the isolation membrane. Within the ATG12~5/16 complex, the ATG12~5 conjugate possesses the E3 ligase activity that promotes the conjugation of ATG8 to PE of the autophagosomal membrane [17,23]. Mice lacking either ATG5, ATG12, or ATG16L1, the mammalian ortholog of *D. discoideum* ATG16, survive the embryonic phase, but die one day after birth, corroborating the importance of an intact ATG12~5/16 complex for postnatal survival [24].

The social amoeba *Dictyostelium discoideum* has a unique life cycle, with motile unicellular and multicellular stages, and the organism serves as a model for a number of biological problems that are relevant to human health [25–27]. In the presence of sufficient food supply, the amoebae grow as separate, independent cells, which divide by binary fission and take up bacteria via phagocytosis. Upon starvation, up to 100,000 solitary amoebae aggregate by chemotaxis towards cAMP. The aggregate transforms via distinct morphological states into a mature fruiting body, composed of a ball of spores supported by a thin, long stalk made of vacuolised dead cells [28]. Since development takes place in the absence of external nutrients, *D. discoideum* cells must mobilise a large fraction of the required energy for biosynthetic needs and morphogenesis by autophagy and glycogenolysis [3]. Consequently, *D. discoideum* is well-established for the investigation of the autophagic process and powerful methods to monitor and quantify autophagy in this organism have been developed [29,30].

The *D. discoideum* autophagy machinery is more similar to higher eukaryotes than to yeast [3,9] and ectopic expression of tagged proteins, as well as the generation of multiple gene knock-out strains, is generally straightforward [31,32]. Furthermore, novel conserved autophagy genes have been discovered in *D. discoideum* and the analysis of single or double knock-out mutants of core autophagy genes revealed informative phenotypes [33–41].

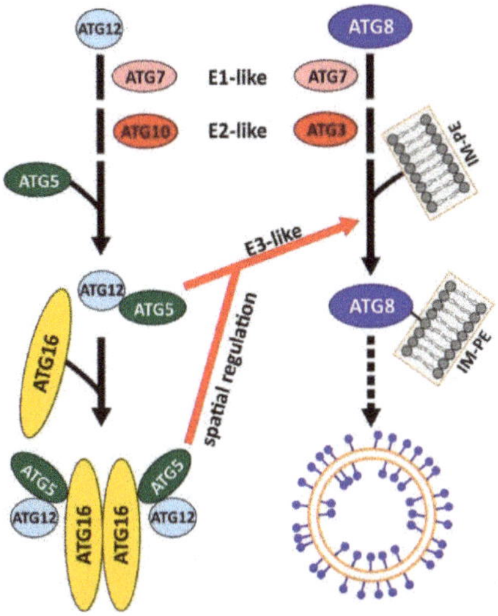

Figure 1. Schematic depiction of the components and their interrelations of the two ubiquitin-like conjugation systems in autophagy. The ATG12 (left) and the ATG8 (LC3 in mammals) (right) conjugation systems are represented. Similar to protein ubiquitination, the ubiquitin-like protein ATG12 is activated by the E1 enzyme ATG7 and transferred to the E2 enzyme ATG10. Finally, ATG12 is covalently attached to its target protein ATG5 and two ATG12~5 conjugates in turn associate non-covalently with an ATG16 dimer and form a heterohexameric complex. Likewise, the ubiquitin-like protein ATG8 is also activated by ATG7, transferred to the E2 enzyme ATG3, and finally conjugated to PE via an amide bond. The ATG12~5/16 complex catalyses via its E3-like activity the conjugation of ATG8 to PE at the isolation membrane. The different components of the two conjugation systems are not drawn to scale. IM-PE, isolation membrane containing phosphatidylethanolamine. Modified from [39]. See text for further details.

We here generated ATG5⁻, ATG5⁻/12⁻, and ATG5⁻/12⁻/16⁻ cells and analysed their phenotypes in development, cell viability, growth, phagocytosis, macropinocytosis, and protein homeostasis. This is, to our knowledge, the first report of the analysis of an ATG5/12 double and ATG5/12/16 triple mutant. We observed complex, and for some cellular processes, opposite phenotypes of varying severity in the generated knock-out strains. We find similar phenotypes for the single, double, and triple knock-out mutants in cellular processes known to depend on canonical autophagy. This implies that deletion of any of the components in the ATG12~5/16 complex destroys its function in these processes. Our results further support autophagy-independent functions of the complex and its individual components, as well as the placement of autophagy upstream of the ubiquitin–proteasome system (UPS).

2. Materials and Methods

2.1. Dictyostelium Strains, Growth and Development

D. discoideum AX2 was used as a wild-type strain. The ATG5⁻, ATG5⁻/12⁻, and ATG5⁻/12⁻/16⁻ strains were generated by the replacement of the *atg5* gene with the knock-out construct in AX2, ATG12⁻ and

ATG12⁻/16⁻ cells, respectively [39]. We have isolated one ATG5⁻/12⁻/16⁻ mutant, two independent ATG5⁻/12⁻, and three independent ATG5⁻ mutants. We observed no phenotypic difference in the analysis of the independent knock-out mutants. The gene replacement construct is depicted in Figure S1A and transformation was carried out as described [39]. The strains used in this study are listed in Table 1. All *D. discoideum* strains were grown at 22 °C in AX2 liquid nutrient medium [42] on plates (10 cm diameter), in Erlenmeyer flasks with shaking at 160 rpm [43] or on *Klebsiella aerogenes*-overlaid SM agar plates [44,45]. For antibiotic-resistant strains, the AX2 medium was supplemented with 5 µg/mL Blasticidin S (ICN Biomedicals GmbH, Eschwege, Germany). Logarithmic growth phase cells (2–4 × 10^6 cells/mL) were used for all experiments.

Table 1. *D. discoideum* strains used in this study.

Strains	Summary	References
AX2	Axenically growing derivate of wild isolate NC-4	[46]
ATG5⁻	ATG5 null mutant	This work
ATG5⁻/12⁻	ATG5/12 double null mutant	This work
ATG5⁻/12⁻/16⁻	ATG5/12/16 triple null mutant	This work
ATG12⁻	ATG12 null mutant	[39]
ATG16⁻	ATG16 null mutant	[40]
ATG12⁻/16⁻	ATG12/16 double null mutant	[39]

For growth experiments of wild-type AX2 as well as ATG5⁻, ATG5⁻/12⁻, and ATG5⁻/12⁻/16⁻ mutants, cells were inoculated in 25 mL of AX2 medium at a density of 2 × 10^4 cells/mL. For each strain, the cell titres of three separate flasks for each independent experiment were determined every 24 h for up to 10 days, and the generation time was calculated from three timepoints within the logarithmic growth phase. Each point in the graph in Figure 5A represents the average generation time of one experiment. Three (ATG5⁻/12⁻/16⁻), four (AX2 and ATG5⁻/12⁻), and six (ATG5⁻) independent experiments were performed. Growth on *K. aerogenes* was quantitated by measuring plaque size on a bacterial lawn every 24 h for 7 days using a stereomicroscope (M205 C, Leica, Wetzlar, Germany) and the accompanying Leica LAS X software (v.3.3.0, Leica, Wetzlar, Germany). We have performed four independent experiments for AX2, three for ATG5⁻/12⁻ and ATG5⁻/12⁻/16⁻, and seven for ATG5⁻ cells. In each independent experiment, we have measured between 10 and 23 plaques for each mutant. The average increase in the plaque size per 24 h was calculated. Analysis of cell survival upon nitrogen starvation and developmental experiments were carried out as described [38,40].

2.2. Primary Antibodies

For the generation of ATG5-specific polyclonal antibodies, the coding sequence for amino acids 150 to 280 of ATG5 was amplified by PCR and cloned into the pGEX-6P-1 expression vector. The GST fusion protein was expressed in *E. coli* ArcticExpress RIL (Stratagene GmbH, Heidelberg, Germany), purified using Glutathione Sepharose 4B beads (GE Healthcare GmbH, Solingen, Germany) and released through cleavage with PreScission protease (GE Healthcare GmbH, Solingen, Germany). The purified ATG5 polypeptide was used for immunisation of rabbits (BioGenes GmbH, Berlin, Germany). Subsequently, the generated ATG5 anti-serum (rabbit polyclonal antibody #28612) was affinity-purified using a purified, recombinant ATG5$^{150-280}$ polypeptide.

For immunoblotting, the affinity-purified ATG5 antibody was used at a 1:2500 dilution. ATG12 was detected with mouse monoclonal antibody K89-141-1 at a 1:10 dilution [39], ATG16 with rabbit polyclonal antibody #21105 at 1:1000 dilution [40], actin with mouse monoclonal antibody Act1-7 at a 1:100 dilution [47], ubiquitin with mouse monoclonal antibody P4D1 at 1:1000 dilution (Cell Signaling Technology, Frankfurt, Germany), and the proteasomal subunit psmA7 with mouse monoclonal antibody 171-337-2 at a 1:50 dilution [48].

For immunofluorescence analysis, the ubiquitin antibody P4D1, as well as a rabbit polyclonal antibody directed against ATG8a (courtesy of Jason King, University of Sheffield, Sheffield, UK), were used at a 1:1000 dilution in PBG buffer (1× PBS containing 0.5% BSA and 0.045% fish gelatin).

2.3. SDS-PAGE and Western Blotting

SDS-PAGE and Western blotting of total cell lysates were performed with 4×10^5 cells per lane as described [49–51]. The secondary antibodies used were anti-mouse and anti-rabbit IgG conjugated to horseradish peroxidase (Sigma-Aldrich, Darmstadt, Germany) at a 1:10,000 dilution, followed by chemiluminescence detection. Images were recorded using an Intas ECL Chemostar documentation system. Band intensity and protein molecular mass were determined using LabImage 1D L-340 software (Intas Science Imaging Instruments GmbH, Göttingen, Germany).

2.4. Macropinocytosis and Phagocytosis Assays

Quantitative macropinocytosis of TRITC-labelled dextran and quantitative phagocytosis of TRITC-labelled heat-killed yeast cells were performed as described [40]. For phagocytosis of *E. coli* bioparticles, *D. discoideum* cells were prepared as described for fluorescence microscopy without fixation [39]. *E. coli* (K-12 strain) BioParticles labelled with Alexa-Fluor 594 (Thermo Fisher Scientific, Schwerte, Germany) were added to the cells at a 50-fold excess relative to *D. discoideum* cells. After 30 min of incubation, cells were washed four times with Soerensen's phosphate buffer (2.0 mM Na2HPO4, 14.6 mM KH2PO4, pH 6.0). Cells were fixed, nuclei were stained with DAPI and immunofluorescence microscopy was performed as described [39].

2.5. Random Cell Motility Analysis

Cell motility measurements of the different *D. discoideum* strains were conducted using the ibidi 2D μ-slides for chemotaxis according to the manufacturer's instructions with minor changes. *D. discoideum* log phase cells were harvested, washed with Soerensen phosphate buffer and starved for 4 h in Soerensen phosphate buffer. A total of 6 μL of a cell suspension with 3×10^6 cells/mL was injected into the central channel of the ibidi μ-slide. The central channel connects two large reservoirs and also represents the observation area. Cells were allowed to attach for 10 min; non-adherent cells were removed by two injections of 10 μL Soerensen phosphate buffer and the right and the left reservoir were filled with 65 μL Soerensen phosphate buffer. The random movement of *D. discoideum* cells in the central channel was documented in a TIRF microscope, and the accompanying Leica LAS X software (v.3.3.0) over a period of 30 min by taking images every 30 s. After the experiment, the images were transferred to Image J (https://imagej.nih.gov/ij/), and for each strain cell tracking and determination of the velocity of 20 cells were carried out with the "manual tracking" plugin.

2.6. Proteasomal Activity Analysis

Proteasomal activity measurements were performed using the established protocol from skeletal muscle tissue [52] with minor changes as described [53]. The specific proteasomal activity was determined by using a linear regression of the relative luminescence in the linear range (30 to 60 min) followed by normalisation to the amount of psmA7. Five independent experiments were performed to calculate the specific proteasomal activity.

2.7. Fluorescence Microscopy

Immunofluorescence microscopy was essentially done as described [39]. Fixed *D. discoideum* cells were incubated with the indicated primary antibodies; secondary antibodies were Alexa-Fluor 488 conjugated goat anti-mouse and Alexa-Fluor 568 conjugated goat anti-rabbit IgG at a 1:10,000 dilution (Invitrogen GmbH, Darmstadt, Germany). Nuclei were stained with 1 μg/mL 4′,6-diamidino-2-phenylindole (DAPI, Sigma-Aldrich, Darmstadt, Germany). Images of fixed cells were taken with an inverted Leica TCS SP5 confocal laser scanning microscope (Leica, Wetzlar, Germany) with a 100 × HC Pl APO 1.4 oil immersion objective. Excitation of DAPI was at 405 nm, emission at 460–480; excitation of Alexa-Fluor 488 was at 488 nm, emission at 515–535 nm; excitation of TRITC and of Alexa-Fluor 569 and 594 was at 561 nm and emission at 570–580 nm, 590–610 nm, and 605–625 nm, respectively. Images were processed using the Leica MMAF Acquisition- and Analysis software, Adobe Photoshop CS, and Corel Draw 2017.

2.8. Bioinformatics and RNA$_{seq}$

Multiple sequence alignments were performed with Clustal Omega version 1.2.4 [54], and sequence conservation was scored by Jensen–Shannon divergence [55]. Asparagine-rich and low-complexity regions in *D. discoideum* ATG5 were identified using the SMART webtool [56]. To analyse the ATG5 domain structure, InterPro version 72.0 [57] was used after excluding the identified asparagine-rich regions. The homology model of *D. discoideum* ATG5 was created with SWISS-MODEL ProMod3 version 1.3.0 [58] and visualised with DeepView version 4.1 [59]. The structure of the human ATG12~5/16N complex, which contains amino acids 11–43 of ATG16L1, was used as a template (PDB ID: 4gdk) [60]. Isolation and quality control of total RNA from vegetative *D. discoideum* cells for RNA$_{seq}$ experiments were performed as described [39,61]. Six biological replicates of each strain were analysed. Obtained sequences were filtered and preprocessed as described [62], aligned to the AX4 reference genome [63], and evaluated using QuickNGS version 1.26 [62]. RNA$_{seq}$ analysis was done as described by using the DESeq2 package [39,64]. The full RNA$_{seq}$ results will be published separately.

2.9. Statistics and Reproducibility

Unless otherwise indicated, all data shown are derived from at least three independent experiments. The analysis of the statistical significance of experimentally detected differences was carried out with the software environment R, version 3.5.2 [65] and the packages lawstat, version 3.2 [66], dunn.test, version 1.3.5 [67], car [68], and SuperExactTest [69]. The Shapiro–Wilk test served as a test of normality and Levene's test assessed the equality of variances for data. If the null hypothesis was accepted for both tests, one-way ANOVA was used as a parametric test, followed by Tukey's test as an appropriate post hoc analysis. For data which were visualised relative to a control (wild-type AX2), a two-way ANOVA with the non-normalised data and the experimental run as a second factor was used [70]. A rejected null-hypothesis of Shapiro-Wilk test and/or Levene's test led to the implementation of the non-parametric Kruskal-Wallis test in conjunction with the Dunn-Bonferroni test [71] as post hoc analysis. Three levels of significance were defined as follows: *p*-value: <0.05 = significant *; <0.01 = very significant **; <0.001 = highly significant ***. Details are given in the figure legends. For data with parametric statistics (ANOVA), mean values and the standard error of the mean were used for data visualisation. For nonparametric statistics (Kruskal–Wallis test), median and the interquartile range were used.

3. Results

3.1. D. discoideum ATG5 Is highly Conserved

The core autophagy protein ATG5 was originally described in *Saccharomyces cerevisiae* [72]. It is conserved in all eukaryotes and is composed of two ubiquitin-like domains (UblDs), which flank a helix-rich domain (HRD) [7,60,73,74]. In *D. discoideum* ATG5, the ubiquitin-like domains are interrupted by asparagine-rich (N-rich), low-complexity regions (Figure 2A). As no structural model of *D. discoideum* ATG5 existed, we created a homology model with the human ATG12~5/16N complex (PDB ID: 4gdk) as template [60]. The model suggests that the three identified domains of *D. discoideum* ATG5 (in grey, orange, and blue) adopt a functional fold despite the interruption of the two UblDs by low-complexity regions (in red). The UblDs mediate the interaction with the N-terminal part of ATG16 (in rose) containing the ATG5 interaction motif (AFIM), and the HRD is responsible for the interaction with

ATG12 (in yellow) (Figure 2B). The *D. discoideum* ATG5 protein sequence, exempt from the N-rich low complexity regions, is 41% identical with *Drosophila melanogaster*, 42% with *Caenorhabditis elegans*, 43% with *Homo sapiens*, and 49% with *Danio rerio* ATG5. Multiple sequence alignments revealed a number of highly conserved regions in ATG5, which were mostly located in parts with α-helices and β-sheets (Figure S2). The lysine residue, which forms a covalent bond with the C-terminal glycine of ATG12, is absolutely conserved and located in the first α-helix of the HRD (Figure 2B,C).

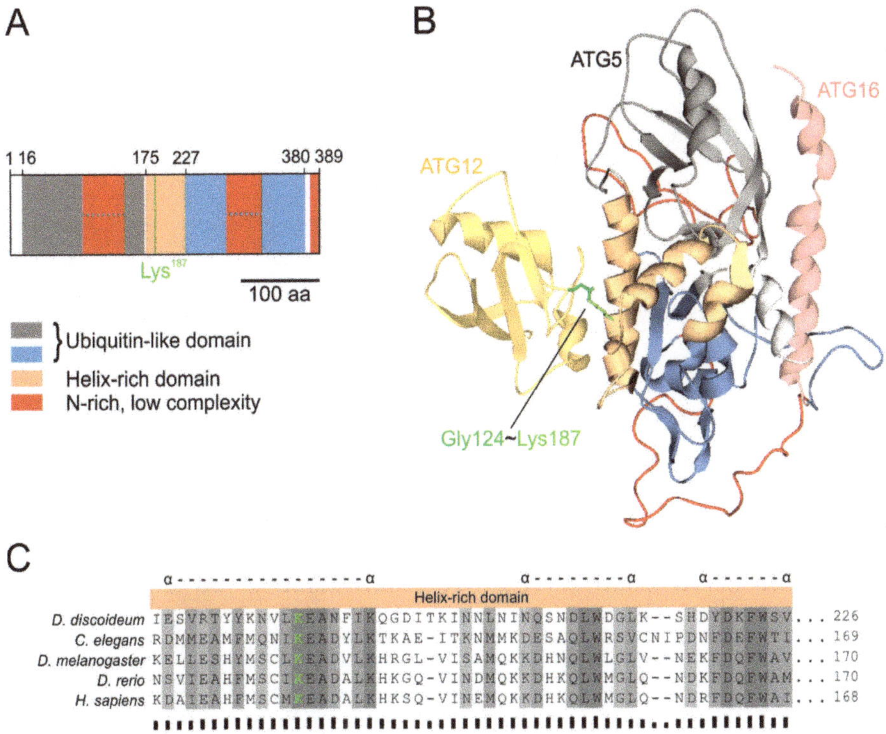

Figure 2. ATG5 domain composition, 3D structure and multiple sequence alignment of the helix-rich domain (HRD). (**A**) Schematic representation of the predicted ATG5 domains. Two ubiquitin-like domains (UblDs) flank the HRD, which contains Lys187 (green line), that forms the covalent bond with the C-terminal glycine of ATG12. Both UblDs are interrupted by N-rich, low-complexity regions. The domains and regions were predicted by InterPro [57] and SMART [56], respectively. (**B**) Predicted 3D structure of the *D. discoideum* ATG12~5/16N complex as ribbon diagram by homology modelling [58] using the human complex as template (PDB ID: 4gdk) [60]. ATG12 is shown in yellow, the ATG16N in rose, and for ATG5 the same colors as in (**A**) were used, to emphasise the domains and regions. The amino acid side chain of ATG5 Lys187 is shown in green. (**C**) Multiple sequence alignment of the conserved HRD of ATG5 orthologues from different organisms. The multiple sequence alignment was performed with Clustal Omega [54]. Amino acid residues are numbered on the right and sequence similarity is indicated by shading. Dark grey represents identical amino acid residues, medium grey highlights amino acids with very similar properties (roughly equivalent to > 0.5 scoring in the Gonnet PAM 250 matrix) and light grey amino acids with slightly similar properties (roughly equivalent to scoring ≤ 0.5 and > 0 in the Gonnet PAM 250 matrix). Sequence conservation is indicated below the alignment by bar sizes [55]. Alpha helices as predicted in the homology model for *D. discoideum* ATG5 in (**B**) are marked with α above the alignment. The complete sequence alignment of ATG5 orthologs is shown in Figure S2.

To analyse the cellular functions of the ATG12~5/16 complex and its individual proteins, we generated knock-out mutants of *atg5* in the AX2 wild-type, ATG12⁻ and ATG12⁻/16⁻ background. In addition, we included in some assays the previously generated ATG12⁻, ATG16⁻ and ATG12⁻/16⁻ cells [39]. The identity of the different gene replacement mutants (Table 1) was confirmed by PCR analysis of genomic DNA (Figure S1B,C) and by immunoblotting with specific antibodies (Figure 3A). In cells which expressed ATG5 and ATG12, we detected with ATG5- and ATG12-specific antibodies a single band of around 68 kDa, that corresponds in size to the ATG12~5 conjugate. This conjugate was absent in ATG5⁻, ATG12⁻, ATG5⁻/12⁻, ATG12⁻/16⁻, and ATG5⁻/12⁻/16⁻ cells, but we detected unconjugated ATG5 of about 46 kDa in ATG12⁻, and ATG12⁻/16⁻ cells (Figure 3A). As unconjugated ATG5 was not detectable in AX2 and ATG16⁻ cells, conjugation of ATG12 to ATG5 appears to be very efficient in vivo. Furthermore, ATG12 was not detectable in cells lacking ATG5, indicating that ATG12 is rapidly degraded in the absence of ATG5 (Figure 3A).

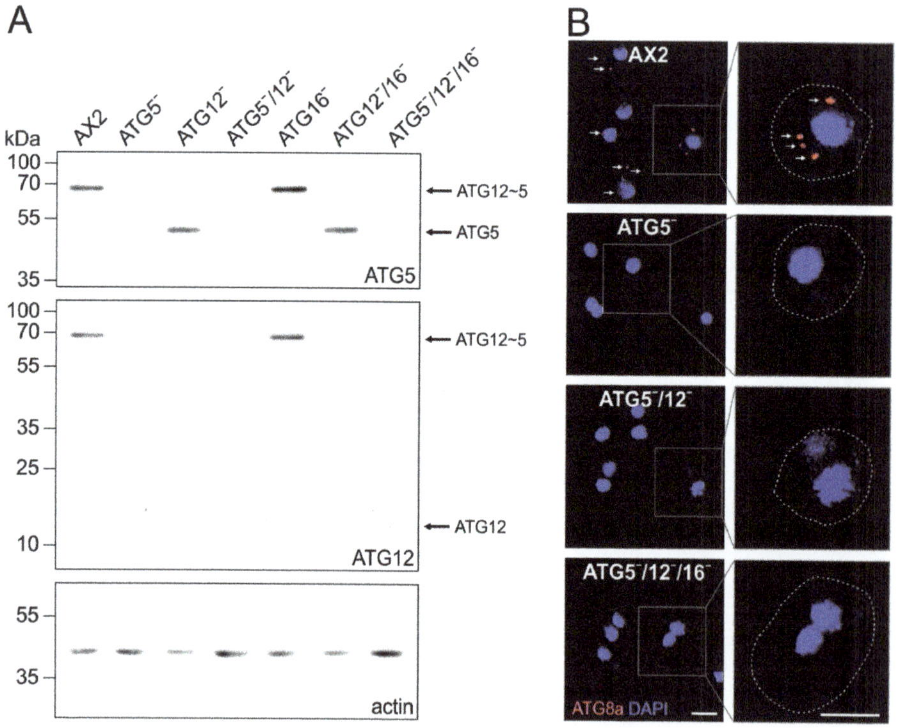

Figure 3. Verification of the different mutant strains by immunoblotting and immunofluorescence analysis of ATG8a-positive autophagosomes. (**A**) Immunoblotting of total cell lysates of wild-type AX2, ATG5⁻, ATG12⁻, ATG5⁻/12⁻, ATG16⁻, ATG12⁻/16⁻, and ATG5⁻/12⁻/16⁻ cells. The ATG12~5 conjugate was detected at about 68 kDa in AX2 and ATG16⁻ cell lysates but not in *atg5* and *atg12* knock-out strains. Unconjugated ATG5 of about 46 kDa was detected in ATG12⁻ and ATG12⁻/16⁻ cells. No unconjugated ATG12 of about 14 kDa was detectable. Actin was used as a loading control. ATG5, ATG12 and actin were visualised on the same membrane. Top row, ATG5 pAb; middle row, ATG12 mAb; bottom row, actin mAb. (**B**) Immunofluorescence microscopy of AX2, ATG5⁻, ATG5⁻/12⁻ and ATG5⁻/12⁻/16⁻ cells. Cells were fixed with cold methanol and stained with the ATG8a pAb. Puncta representing ATG8a-positive autophagosomes (arrows) were only detected in AX2 wild-type cells. Cell boundaries in the enlarged insets are indicated by dotted lines. Nuclei were visualised by DAPI staining. Scale bar, 5 μm.

The E3-like activity of the ATG12~5/16 complex is indispensable for the conjugation of ATG8 proteins (ATG8a and b in *D. discoideum*; LC3s and GABARAPs in mammals) to PE of the autophagosomal membrane. Therefore, we next investigated the presence of ATG8a-positive autophagosomes in AX2, ATG5⁻, ATG5⁻/12⁻ and ATG5⁻/12⁻/16⁻ cells by immunofluorescence analysis. We found that cells lacking ATG5 contained no ATG8a-positive autophagosomes, while these were clearly detectable in AX2 wild-type cells (Figure 3B).

3.2. Cellular Processes Dependent on Canonical Autophagy Are Similarly Impaired in the Different atg5 Knock-Out Strains

Starvation triggers *D. discoideum* cells to enter development, where, in response to secreted cAMP, up to 100,000 cells aggregate and ultimately form, via a number of defined morphological states, a mature fruiting body. This process depends on canonical autophagy as it requires recycled simple molecular building blocks and energy for the synthesis of new macromolecules. As a consequence, autophagy-deficient *D. discoideum* strains generally display developmental problems of varying severity [3]. Similar to our results for ATG9-, ATG12-, and ATG16-deficient cells [38–40], we found severe developmental problems on phosphate agar plates for ATG5⁻, ATG5⁻/12⁻, and ATG5⁻/12⁻/16⁻ cells. In comparison to AX2 wild-type cells, fruiting body formation of all mutant strains took approximately twice as long, with morphological aberrations along the developmental process. For example, in comparison to AX2, mutant cells usually generated more mounds and their fruiting bodies were extremely tiny and had a crippled shape with thickened stalks. Of note, there were no obvious differences in the developmental phenotypes of all three mutants, the single ATG5⁻, the double ATG5⁻/12⁻ and the triple ATG5⁻/12⁻/16⁻ strains (Figure 4A).

Cell survival upon nitrogen starvation also strongly depends on autophagy. We monitored cell viability of AX2 wild-type and ATG5⁻, ATG5⁻/12⁻ and ATG5⁻/12⁻/16⁻ cells in amino-acid-free medium for five days by determining the number of colony-forming units (CFUs) on a lawn of *K. aerogenes* every 24 h. After five days of nitrogen starvation, approximately 35% of AX2 cells were still viable. By contrast, less than 20% of the ATG5⁻, ATG5⁻/12⁻, and ATG5⁻/12⁻/16⁻ mutants had retained viability (Figure 4B). Again, the same as for development, the single ATG5⁻, the double ATG5⁻/12⁻, and the triple ATG5⁻/12⁻/16⁻ strains had similar defects. Similar results for these processes were obtained in previous studies for ATG12⁻, ATG16⁻ and ATG12⁻/16⁻ strains [39,40]. We conclude that cellular processes that depend on canonical autophagy are dependent on an intact ATG12~5/16 complex. Depletion of any of the proteins of the ATG12~5/16 complex results in similarly severe problems in canonical autophagy (Figures 3B and 4).

3.3. Cell Growth, Macropinocytosis and Phagocytosis Are Severely Affected in Strains Lacking ATG5

Axenic *D. discoideum* strains are capable of consuming liquid nutrients by macropinocytosis [75]. To analyse a possible role of ATG5 and the ATG12~5/16 complex in macropinocytosis, we first determined cell growth in complex medium. We found that ATG5⁻, ATG5⁻/12⁻ and ATG5⁻/12⁻/16⁻ cells grow significantly slower than AX2, and generation times were increased by about 36%, 65%, and 108%, respectively (Figure 5A). Thus, in contrast to development and cell viability upon nitrogen starvation, the cell growth phenotype increased in severity from the single to the double and triple mutant. A similar result, i.e., an increase in severity from single to double mutant, was also found for ATG12⁻ and ATG12⁻/16⁻ cells [39]. The growth defect could be caused by less efficient macropinocytosis and/or a deficiency in the intracellular utilisation of nutrients. Therefore, we next analysed the macropinocytic uptake of TRITC-labelled dextran, which cannot be metabolised by *D. discoideum* cells. All three mutant strains showed significantly reduced macropinocytic activity in comparison to AX2 wild-type cells. After 2 h, the relative fluorescence of ATG5⁻, ATG5⁻/12⁻ and ATG5⁻/12⁻/16⁻ cells was 31%, 27% and 52%, respectively, lower than for AX2 (Figure 5B). The decrease in ATG5⁻ and ATG5⁻/12⁻ cells was similar, while the deletion of ATG16 in the double mutant resulted in a further decrease in macropinocytic activity. This suggests that ATG16 has a supportive function in macropinocytosis also in the absence of the ATG12~5 conjugate. In summary, the stepwise increase in generation times from ATG5⁻ to ATG5⁻/12⁻ and ATG5⁻/12⁻/16⁻ cells appear to be caused by a decrease in both macropinocytosis and the intracellular utilisation of nutrients.

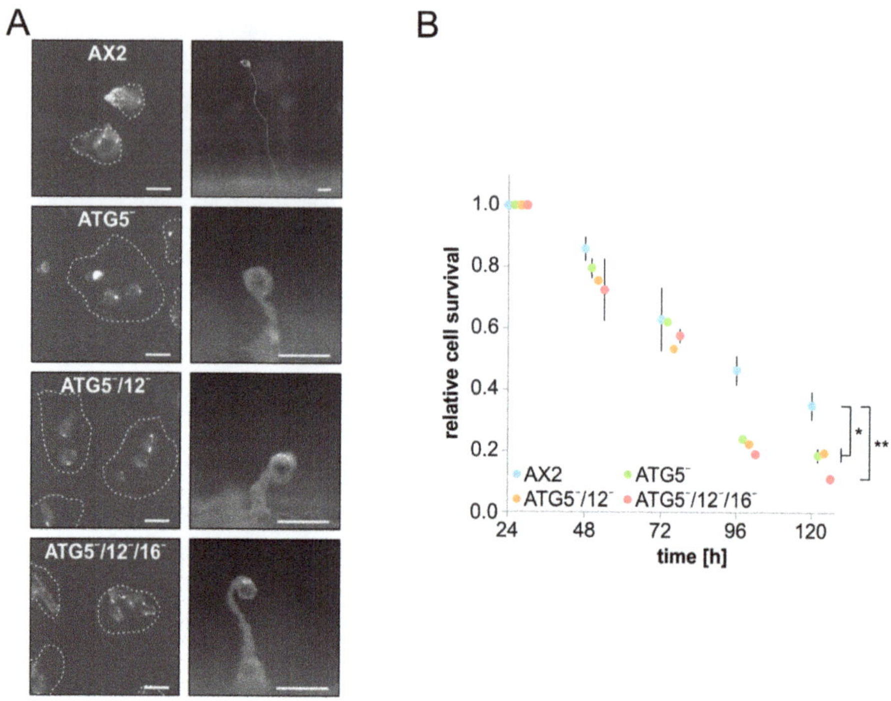

Figure 4. Development and cell survival upon nitrogen starvation of AX2 and mutant strains. (**A**) Development of AX2, ATG5⁻, ATG5⁻/12⁻ and ATG5⁻/12⁻/16⁻ cells on phosphate agar plates. 5×10^7 cells of each strain were plated homogeneously. On the left, the tipped mound stage for AX2 after 15 h, and for mutant cells after 30 h (top view), and on the right an exemplary fruiting body of AX2 after 24 h, and of mutant cells after 48 h (side view), is shown. The white dashed line encircles tipped mounds. Scale bars, 100 µm. Note the different scale for mutant fruiting bodies. (**B**) Cell survival of AX2, ATG5⁻, ATG5⁻/12⁻ and ATG5⁻/12⁻/16⁻ cells upon nitrogen starvation. AX2 and mutant strains were grown in SIH medium without amino acids, and cell survival was determined every 24 h for 5 days. Relative cell survival after 24 h was set to 1 for each strain. Cell survival of mutant cells was significantly reduced in comparison to AX2 after 96 and 120 h. Mean values and standard errors of the mean of three independent experiments are shown. For statistical analysis, one-way ANOVA and Tukey's test as post hoc analysis were used. *, significant (p-value < 0.05); **, very significant (p-value < 0.01).

Because *D. discoideum* cells can also feed on yeast and bacteria, we next analysed phagocytosis. TRITC-labelled yeast was co-incubated with the different strains and the resulting fluorescence was measured for 3 hours. The overall kinetics, with an almost linear increase in fluorescence throughout the assay, was similar for all strains. However, in comparison to AX2 the final fluorescence values of ATG5⁻, ATG5⁻/12⁻ and ATG5⁻/12⁻/16⁻ cells were with 41%, 30%, and 68%, respectively, which was significantly lower. The decrease in phagocytic activity was nearly identical for ATG5⁻ and ATG5⁻/12⁻ cells, while it was more pronounced in the *atg5/12/16* triple knock-out strain (Figure 6A). It appears, therefore, that ATG16 also has a role in phagocytosis of yeast, which is independent of the ATG12~5 conjugate, or that in the absence of ATG5 or ATG5 and 12, ATG16 may have some residual function in the phagocytosis of yeast. The severity of the phenotypes of the three mutant strains was similar to the ones observed for macropinocytosis (see Figure 5B).

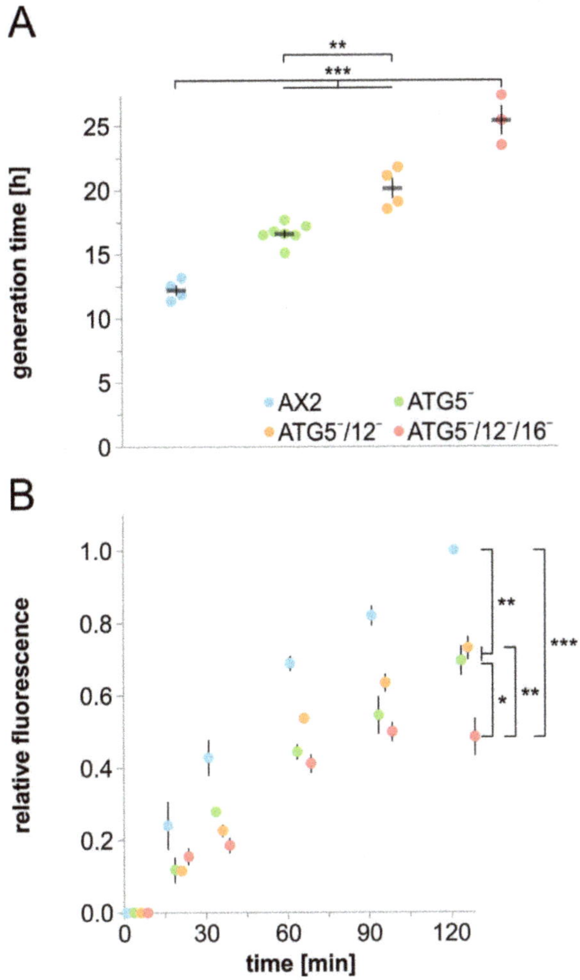

Figure 5. Analysis of cell growth and macropinocytosis in AX2 and mutant strains. (**A**) Generation time of AX2, ATG5⁻, ATG5⁻/12⁻ and ATG5⁻/12⁻/16⁻ cells in shaking culture. Cell titres of three parallel shaking cultures were determined every 24 h. Mean values and SEM of three (ATG5⁻/12⁻/16⁻), four (AX2 and ATG5⁻/12⁻) and seven (ATG5⁻) independent experiments are shown. For statistical analysis, one-way ANOVA and Tukey's test as post hoc analysis were used. (**B**) Macropinocytosis of TRITC-labelled dextran. AX2, ATG5⁻, ATG5⁻/12⁻ and ATG5⁻/12⁻/16⁻ cells were adjusted to 6×10^6 cells/mL and co-incubated with TRITC-labelled dextran. Intracellular fluorescence was determined at t0 and after 15, 30, 60, 90, and 120 min. The final fluorescence of AX2 was set to 1. Color assignment of strains is as in panel A. Mean values and SEM of three independent experiments are shown. For statistical analysis, two-way ANOVA and Tukey's test as post hoc analysis were used. *, significant (p-value < 0.05); **, very significant (p-value < 0.01); ***, highly significant (p-value < 0.001).

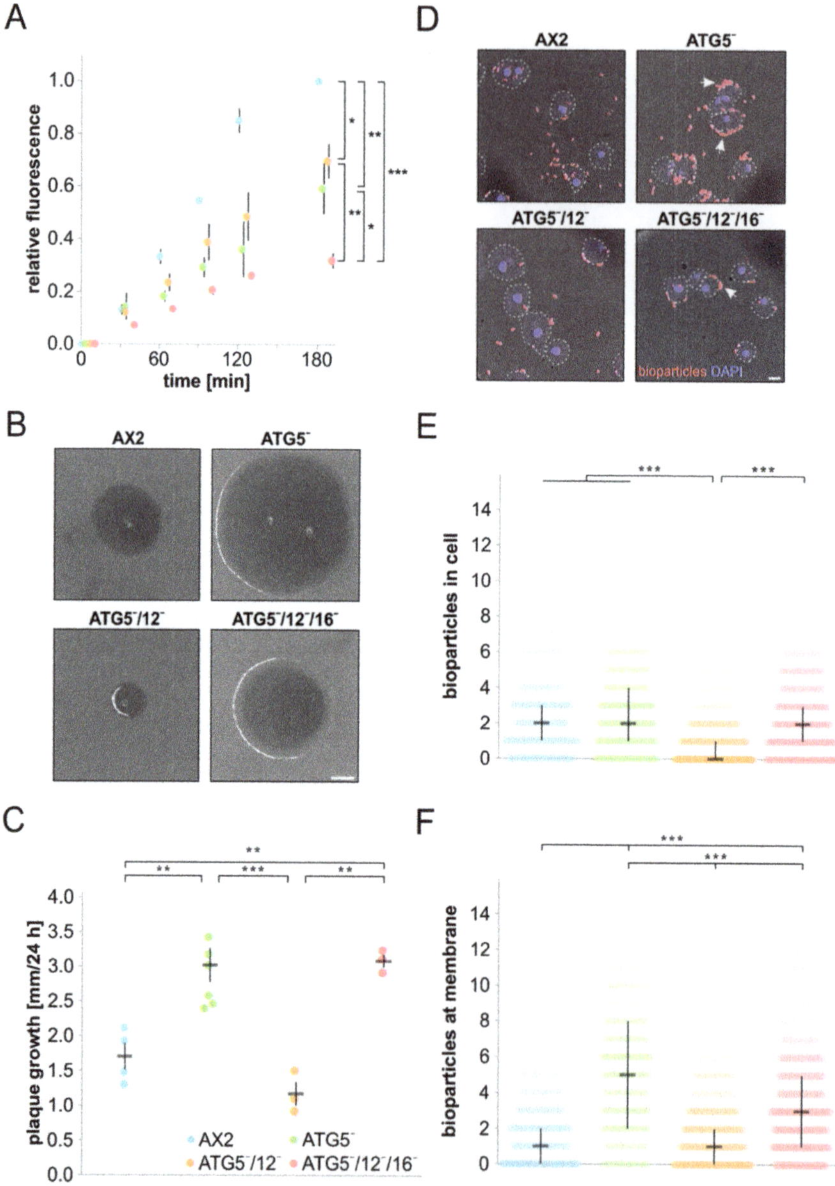

Figure 6. Phagocytosis of yeast and bacteria by AX2 and mutant strains. (**A**) Phagocytosis of TRITC-labelled yeast. AX2, ATG5⁻, ATG5⁻/12⁻ and ATG5⁻/12⁻/16⁻ cells were adjusted to 6×10^6 cells/mL and co-incubated with a six-fold excess of TRITC-labelled yeast for 180 min. Color assignment of strains is as in panel C. Analysis was done as for macropinocytosis (see Figure 5B). (**B**) Growth on *K. aerogenes*. Representative images of plaques formed by AX2, ATG5⁻, ATG5⁻/12⁻ and ATG5⁻/12⁻/16⁻ cells after 96 h growth on a lawn of *K. aerogenes*. Scale bar is 1 mm. (**C**) Quantitation of plaque growth for AX2, ATG5⁻, ATG5⁻/12⁻ and ATG5⁻/12⁻/16⁻ cells. Three days after plating the cells, plaque diameters were determined every 24 h for five days and the average growth per 24 h was calculated. Mean values

and SEM of three (ATG5⁻/12⁻ and ATG5⁻/12⁻/16⁻), four (AX2) and seven (ATG5⁻) independent experiments are shown. For statistical analysis, one-way ANOVA and Tukey's test as post hoc analysis were used. (**D**) Representative images of phagocytosis of *E. coli* BioParticles by AX2, ATG5⁻, ATG5⁻/12⁻ and ATG5⁻/12⁻/16⁻ cells. Cells were co-incubated with fluorescently labelled *E. coli* for 30 min. After fixation of cells, fluorescence and phase contrast microscopy was performed. Some of the membrane bound *E. coli* are marked by arrows. Nuclei were stained with DAPI and cell outlines are indicated by white dashed lines. Scale bar, 5 μm. (**E**) and (**F**) Quantitation of phagocytosed and membrane bound *E. coli*. The amount of *E. coli* BioParticles within the cells (**E**) and at the cell membrane (**F**) was determined for 100 cells of each strain in each of three independent experiments. Every data point represents a single cell. Additionally, the median and interquartile range are shown. Color assignment of strains is as in panel C. For statistical analysis, Kruskal–Wallis test and the Dunn–Bonferroni test were used as post hoc analysis. *, significant (p-value < 0.05); **, very significant (p-value < 0.01); ***, highly significant (p-value < 0.001).

We next investigated the clearance of *K. aerogenes* and phagocytosis of fluorescent *E. coli* bioparticles by AX2 and the three mutant strains. In the *K. aerogenes* clearing assay, a low number of *D. discoideum* cells are spread on a bacterial lawn, where they start clearing the bacteria by phagocytosis. After around 72 h, tiny plaques, which were each initiated by a single *D. discoideum* cell, become visible on the bacterial lawn. The plaques increased steadily in size over the next few days, and we found after 96 h in comparison to AX2 increased plaque sizes for ATG5⁻ and ATG5⁻/12⁻/16⁻ cells, respectively, while ATG5⁻/12⁻ plaques were much smaller (Figure 6B). The differences in plaque sizes were caused by faster and slower plaque growth over time for the respective mutant cells (Figure 6C). This result suggests that ATG16 somehow inhibits growth on *K. aerogenes* in the absence of the ATG12~5 conjugate (ATG5⁻/12⁻ cells), while this enigmatic activity is not seen with ATG5⁻ cells. On the contrary, this strain and the ATG5⁻/12⁻/16⁻ strain grew much faster on a lawn of *K. aerogenes*.

Differences in plaque size between the different strains could also be caused by differences in cell motility, and we next investigated random cell motility of AX2 wild-type and the different knock-out cells. Our analysis did not reveal significant differences between AX2 and ATG5⁻, ATG5⁻/12⁻, and ATG5⁻/12⁻/16⁻ cells, respectively (Table S1). To analyse phagocytosis more directly, we used *E. coli* BioParticles labelled with Alexa-Fluor 594 and quantitated phagocytosed and plasma membrane adherent bacteria by fluorescence microscopy (Figure 6D). With respect to phagocytosis, we found that wild-type AX2, ATG5⁻ and ATG5⁻/12⁻/16⁻ cells had ingested a comparable number of *E. coli* on average, while their uptake was significantly reduced in ATG5⁻/12⁻ cells (Figure 6E). In contrast, the number of plasma-membrane-bound *E. coli* was comparable for AX2 and ATG5⁻/12⁻ cells, while their number was significantly increased for ATG5⁻ and ATG5⁻/12⁻/16⁻ cells (Figure 6F). Therefore, the smaller plaque size on *K. aerogenes* for ATG5⁻/12⁻ cells is likely caused by a decrease in phagocytosis, while the larger plaque sizes of ATG5⁻ and ATG5⁻/12⁻/16⁻ cells might be a consequence of many particles at the plasma membrane, which are not ingested, i.e., these cells appear to be inefficient "eaters". Of note, the phenotypes of the ATG5⁻, ATG5⁻/12⁻ and ATG5⁻/12⁻/16⁻ strains differed with respect to the phagocytosis of bacteria and yeast.

3.4. Protein Homeostasis Is Severely Disturbed in Mutant Strains

Autophagy and the UPS are the two major cellular systems for the degradation of non-functional or superfluous proteins and organelles. Proteins destined for degradation are marked by ubiquitin. We therefore examined global protein ubiquitination in total cell lysates of wild-type and mutant strains with the P4D1 monoclonal antibody which detects ubiquitin, polyubiquitin and ubiquitinated proteins (Figure S3A). The quantification of five independent experiments revealed an approximately 3- to 4-fold increase in the amount of ubiquitinated proteins for ATG5⁻ and ATG5⁻/12⁻ cells and a 2.5-fold increase for ATG5⁻/12⁻/16⁻ cells (Figure 7A). We next investigated by immunofluorescence microscopy with the P4D1 antibody whether this increase in global protein ubiquitination would cause protein aggregates in the mutants. Indeed, in contrast to AX2 cells we detected many large ubiquitin-positive protein aggregates in all three mutant strains, indicating a severe imbalance in protein homeostasis (Figure 7B). To determine a possible contribution of the UPS, we performed proteasomal activity assays of AX2, ATG5$^-$, ATG5$^-$/12$^-$ and ATG5$^-$/12$^-$/16$^-$ cells. ATG16$^-$ cells, for which we had previously determined an approximately 50% decrease in proteasomal activity, were used as an additional control [40]. Similar to ATG16$^-$ cells, we found an approximately 50% reduction in proteasomal activity for ATG5$^-$/12$^-$/16$^-$ cells, while ATG5$^-$ and ATG5$^-$/12$^-$ cells displayed a very high reduction of 90% (Figure 7C). The observed decrease in measured proteasomal activity could be caused by a decrease in the number or activity of the proteasomes. RNA$_{seq}$ data (complete analysis will be published separately) revealed no significant changes in the differential regulation of proteasomal genes (Table S2). In addition, proteasome number, based on the protein level of the psmA7 core subunit as a representative of the 20S proteasome, was not significantly changed in the mutant strains (Figure 7D and Figure S3B). We also excluded the possibility that proteasomes of mutant strains, but not of wild-type cells, could have gotten lost in the process of preparing the lysates for the proteasomal activity assay by Western blot analysis (Figure S3C). We conclude that not the number of proteasomes, but indeed their overall activity, is strongly decreased in the mutants. This decrease was highly significant in ATG5$^-$/12$^-$/16$^-$, cells and much stronger in the ATG5$^-$ and ATG5$^-$/12$^-$ cells.

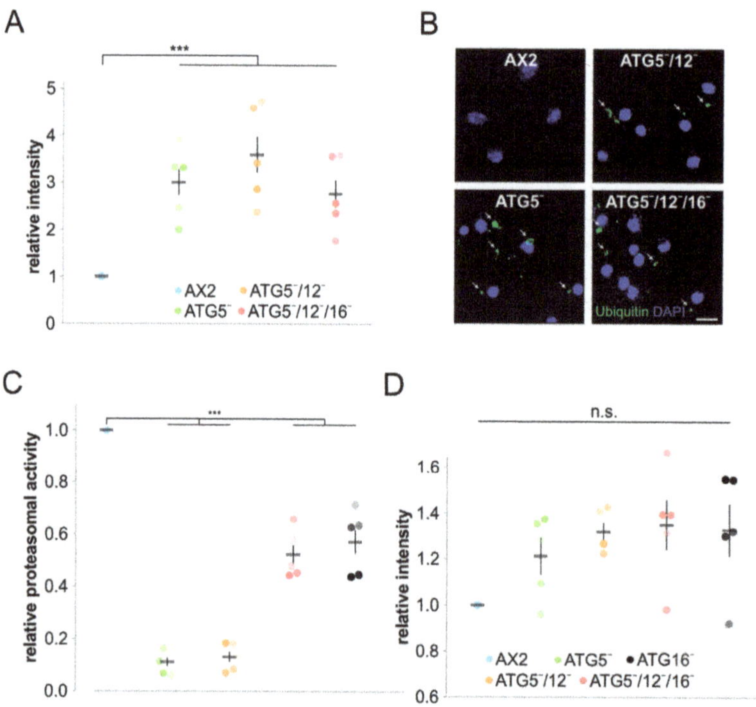

Figure 7. Protein homeostasis is disturbed. (**A**) Quantification of global protein ubiquitination of AX2, ATG5$^-$, ATG5$^-$/12$^-$, and ATG5$^-$/12$^-$/16$^-$ cells. Western blots of total cell lysates were stained with mAb P4D1 and signal intensities of ubiquitinated proteins were quantitated and normalised with the actin signal. The signal intensity of AX2 was set to 1. Mean values and SEM of five independent experiments are shown. (**B**) Immunofluorescence microscopy of AX2, ATG5$^-$, ATG5$^-$/12$^-$, and ATG5$^-$/12$^-$/16$^-$ cells. Cells were fixed with cold methanol and stained with the mAb P4D1. Ubiquitin-positive protein aggregates in mutant strains are marked by arrows. Nuclei were visualised by DAPI staining. Scale bar,

5 μm. (**C**) Proteasomal activity of AX2, ATG5$^-$, ATG5$^-$/12$^-$, ATG5$^-$/12$^-$/16$^-$, and ATG16$^-$ cells. The assay was performed as described [39]. The chymotrypsin-like activity of AX2 was set to 1. Color assignment of strains is as in panel D. Mean values and SEM of five independent experiments are shown. (**D**) Quantification of the amount of the proteasomal subunit A7 (psmA7) in AX2, ATG5$^-$, ATG5$^-$/12$^-$, ATG5$^-$/12$^-$/16$^-$, and ATG16$^-$ cells. The psmA7 signal intensities were quantitated after Western blotting and normalised with the actin signal. The signal intensity of AX2 was set to 1. Mean values and SEM of five independent experiments are shown. For all statistical analyses, two-way ANOVA and Tukey's test were used as post hoc analysis. n.s., not significant; ***, highly significant (p-value < 0.001).

4. Discussion

The ATG12~5/16 complex acts as an E3 ligase for the covalent attachment of ATG8 family proteins (LC3s, GABARAPs and ATG8s) to the headgroup of the membrane lipid PE and is essential for canonical autophagy. In this complex, ATG16 is responsible for binding to the correct membrane via protein–protein and protein–lipid interactions, ATG12 mediates binding of the ATG3~8 conjugate, and ATG5 holds the E3 ligase activity [19,76,77]. All activities of the complex together ensure the correct conjugation of ATG8 family proteins to PE of target membranes. In canonical autophagy, ATG8 family proteins are involved in membrane elongation and fusion processes and act as receptors for cargo to be ultimately degraded in autolysosomes [76]. To further unravel the functions of the ATG12~5/16 complex and the possible roles of its individual proteins in addition to canonical autophagy, we have generated and analysed *atg5* single, *atg5/12* double, and *atg5/12/16* triple knock-out mutants. In previous work, we have shown that the *atg12* single and *atg12/16* double knock-out mutants have severe and complex phenotypes, which implied additional functions of ATG12 and ATG16 in autophagy-independent processes [39].

4.1. ATG5 Is Required for Conjugation of ATG8 to the Autophagosomal Membrane and for ATG12 Stability

The crystal structure of the human ATG12~5 conjugate in complex with the N-terminal part of ATG16 (ATG16N) confirmed the ubiquitin-like fold for ATG12 and revealed for ATG5 two ubiquitin-like domains (UblDs) that flank a helix-rich domain (HRD), and an additional N-terminal α-helix [60]. Modelling of the corresponding *D. discoideum* protein sequences suggests a highly similar structure, despite the interruption of each UblD with a low-complexity, asparagin-rich sequence stretch. In the predicted structure, the ubiquitin-like protein ATG12 and the N-terminal α-helix of ATG16 are located on opposite sides of ATG5 and there is no common interface of ATG12 and ATG16N (Figure 2B). Hence, no interaction between ATG12 and ATG16 is expected in ATG5-deficient cells. The interaction of ATG5 with ATG16N is mediated by the two UblDs, while its HRD is responsible for the interaction with ATG12. The HRD contains the absolutely conserved lysine residue, that is required for the formation of the ATG12~5 conjugate (Figure 2B,C). In mouse embryonic fibroblasts (MEFs) deficient for ATG5 or upon expression of the conjugation-defective ATG5^{K130R} mutant in the *atg5* knock-out background, the PE-conjugated form of LC3, also known as LC3-II, could not be detected. This demonstrates the absolute requirement of ATG5 and its conserved lysine for the E3 ligase activity of the ATG12~5/16 complex in mouse [60,78]. Similarly, we could not detect ATG8a-positive autophagosomes in immunofluorescence microscopy in ATG5-deficient *D. discoideum* cells (Figure 3B). These results do not disprove the formation of autophagosomes and we have previously shown that the average number of autophagosomes is only slightly reduced in ATG12$^-$, ATG16$^-$ and ATG12$^-$/16$^-$ cells; however, fusion with lysosomes and autophagic flux was strongly impaired [39,40]. The formation of autophagosomes has also been observed in the absence of ATG5, ATG3 or ATG7 in MEFs and of all six ATG8 family members in HeLa cells [79–83]. Thus, ATG5, ATG12, ATG16 and the other components of the two ubiquitin-like conjugation systems are less important for autophagosome formation but are critical for efficient autophagosome lysosome fusion and degradation of the inner autophagosomal membrane and the cargo [9,39,82,84].

Our Western blot analyses of single, double and triple knock-out mutants of components of the ATG12~5/16 complex revealed very efficient conjugation of ATG12 to ATG5 in vivo, because we could only detect the ATG12~5 conjugate of around 68 kDa, but not monomeric ATG12 or ATG5 in wild-type AX2 and ATG16-deficient cells. In addition, the presence of ATG5 appears to be required for the stability of ATG12, because unconjugated ATG12 was not detectable in ATG5-deficient cells (Figure 3A). In contrast, unconjugated ATG5 of about 46 kDa appears to be stable and was readily detected in ATG12⁻ and ATG12⁻/16⁻ strains (Figure 3A) [39]. It was reported that free ATG12 has an estimated half-life of 30 min and is rapidly degraded in a proteasome-dependent manner in MEFs. Furthermore, the level of free ATG12 was unchanged in ATG3-, ATG5-, and ATG7-deficient MEFs, despite an up-regulation of the *atg12* mRNA [85]. A possible reason for the substantial difference in stability between the free and conjugated forms is, that ATG5 masks a destabilising region of ATG12 in the ATG12~5 conjugate, which is exposed in free ATG12 [60,85]. Of note, although we could neither detect free ATG12 nor ATG12~7 nor ATG12~10 conjugates in ATG5⁻ cells, we find significant differences in the phenotypes of ATG5⁻ and ATG5⁻/12⁻ cells (Figure 5A, Figure 6B–F, Table 2). Therefore, residual amounts of these proteins must be present and appear to be responsible for the observed differences in the ATG5⁻ and ATG5⁻/12⁻ phenotypes.

Table 2. Classification of phenotypes for different cellular processes for ATG5⁻, ATG5⁻/12⁻ and ATG5⁻/12⁻/16⁻ cells. "+" and "−" indicate an increase and a decrease, respectively, in the corresponding activity in the mutants in comparison to AX2 cells. Multiple "+" and "−" indicate the severity of the phenotype.

Type	Cellular Process	Strains		
		ATG5⁻	**ATG5⁻/12⁻**	**ATG5⁻/12⁻/16⁻**
(i)	Development	−	−	−
	Cell viability	−	−	−
	Conjugation of ATG8 to PE	−	−	−
(ii)	Generation time	−	−−	−−−
	Macropinocytosis	−	−	−−
	Phagocytosis of yeast	−	−	−−
(iii)	Growth on *K. aerogenes*	++	−	+
	Phagocytosis/membrane association of bacteria	++	−	+
(iv)	Proteasomal activity	−−	−−	−

4.2. The Dictyostelium ATG5⁻, ATG5⁻/12⁻ and ATG5⁻/12⁻/16⁻ Mutants Have Complex and Distinct Phenotypes

ATG5, ATG12 and ATG16 act in canonical autophagy together as a heterohexameric complex, which provides the E3-like ligase activity for conjugation of ATG8 to PE [17]. The single, double, and triple mutants have identical phenotypes in cellular processes, dependent on canonical autophagy, e.g., the conjugation of ATG8 to PE, development, and survival under nitrogen deprivation (Table 2, type (i), Figures 3B and 4) [39]. Hence, the ATG12~5/16 complex acts as a functional unit in canonical autophagy, and each of its components is absolutely required. However, we also found distinct and sometimes opposite phenotypes in ATG5⁻, ATG5⁻/12⁻, and ATG5⁻/12⁻/16⁻ cells (Figure 5A, Figure 6B–F, Figure 7C). One possible explanation is that the loss of individual proteins in the complex autophagy system leads to changes in other proteins or complexes that may take on new roles such as, e.g., inhibiting membrane flux. Alternatively, these complex phenotypes could be caused by differences in the importance of the respective proteins for either non-canonical autophagy or autophagy-independent functions. Indeed, in recent years more and more autophagy-independent roles have been described for core autophagy proteins besides their function in canonical autophagy [86–89]. This is also the case for the components of the ATG12~5/16 complex. For example, the ATG12~5 conjugate acts as a suppressor of innate antiviral immune signalling in mouse embryonic fibroblasts [90]. Furthermore, the C-terminal β-propeller domain, composed of seven WD40 repeats, of ATG16, is required for LC3-associated phagocytosis (LAP), but is dispensable for canonical autophagy [91–94].

We could distinguish three additional main classes of phenotypes. A basically similarly severe phenotype of the single and double mutant, and an increase in the defect in the triple mutant (Table 2, type (ii)). We noticed this phenotype for macropinocytosis of TRITC dextran and phagocytosis of yeast (Figures 5B and 6A). We propose that the further decrease in the triple mutant is due to an independent, albeit reduced, activity of ATG16 in this process in the absence of the ATG12~5 conjugate. Furthermore, we observed a similar phenotype of ATG5⁻ and ATG5⁻/12⁻/16⁻ cells and the adverse effect in ATG5⁻/12⁻ cells in the uptake of bacteria (Table 2, type (iii)). At present, we do not have a good explanation for this phenotype. We think that it could be related to the function of the ATG12~5/16 complex in LAP [95]. Possibly, the ATG12~5/16 complex plays an inhibitory role in phagocytosis of bacteria in the AX2 wild-type situation, which would explain the difference between AX2 and ATG5⁻/12⁻/16⁻ cells. The inhibitory function of the complex in LAP would be mediated by ATG16, and therefore becomes fully visible in ATG5⁻/12⁻ cells. In contrast, in the ATG5⁻ cells the inhibitory activity of ATG16 would be somehow overcompensated by ATG12 (Figure 6B–F). Of note, as already reported for ATG12⁻ and ATG12⁻/16⁻ cells [39], we observed clear differences in phagocytosis of yeast and *K. aerogenes* for ATG5⁻, ATG5⁻/12⁻, and ATG5⁻/12⁻/16⁻ cells (Figure 6). These could be caused by differences in receptors and/or intracellular killing, as *Dictyostelium* is apparently able to discriminate between different types of bacteria and other microorganisms [96,97]. Finally, we noted a very severe defect in proteasomal activity in ATG5⁻ and ATG5⁻/12⁻ cells, which was significantly less severe in the triple mutant (Table 2, type (iv); see below).

4.3. Protein Homeostasis Is Severely Impaired in ATG5 Deficient Strains

Protein homeostasis is crucial for cell welfare and is accomplished by a delicate balance between protein synthesis and degradation. Autophagy and the UPS are of utmost importance for the clearance and recycling of proteins and other cellular material, and defects in either system cause serious cellular problems. Although the two degradation machineries have different substrate preferences and separate molecular mechanisms, it is clear that the two pathways are interrelated and share molecular determinants and substrates [89,98–100]. Several reports have shown that there is compensatory upregulation of autophagy upon the inhibition of proteasomal activity [101–104]. On the other hand, the effect of autophagy deficiency on the activity of the UPS is still controversial, and the compensatory activation of the UPS, no change, and also a decrease in proteasomal activity have been reported [105–108]. The reason for these variable results is not yet known, however, differences in the used experimental systems might be responsible. In *D. discoideum*, we and others found an increase in ubiquitinated proteins and the appearance of ubiquitin-positive protein aggregates in all studied autophagy-compromised strains [33,36,38–40,53]. Furthermore, we showed significantly reduced proteasomal activity of varying severity in ATG8a⁻, ATG8a⁻/b⁻, ATG9⁻, ATG9⁻/16⁻, ATG12⁻, and ATG12⁻/16⁻ cells [36,39,40]. Our results demonstrate that ATG5⁻, ATG5⁻/12⁻, and ATG5⁻/12⁻/16⁻ cells also suffer from a strong imbalance in protein homeostasis. In comparison to AX2 wild-type cells, we see in mutant cells an increase in ubiquitinated proteins and the appearance of ubiquitin-positive protein aggregates (Figure 7A,B and Figure S3A). Furthermore, we found an approximately 50% decrease in proteasomal activity in ATG5⁻/12⁻/16⁻ and ATG16⁻ cells and a significant further decrease to approximately 10% in ATG5⁻ and ATG5⁻/12⁻ cells in comparison to AX2 (Figure 7C). Proteasome number, as evidenced by RNA_{seq}, and quantitation of the 20S proteasomal subunit psmA7, was unchanged in the mutant strains (Figure 7D, Figure S3B and Table S2). The marked further decrease in proteasomal activity in cells deficient for ATG5 suggests that ATG16 on its own somehow has a negative or inhibitory effect, be it via direct or indirect interaction(s). Possibly, ATG16 acts via PSMD1 and PSMD2 (two subunits of the 19S regulatory particle) by mediating their degradation via autophagy [100], however, further work is needed to unravel the responsible pathway. In this respect, in mammals the half-life and cellular concentration of ATG12 and ATG16 themselves appear to be regulated by the UPS [85,109].

We conclude that in *D. discoideum* the UPS is dependent on intact autophagy for full activity. The exact molecular principle is not clear; however, an attractive possibility is that proteasomal subunits experience post-translational modifications, e.g., ubiquitination, during their lifetime, which cause a more or less pronounced decrease in the proteolytic activity of the proteasome. These less active or inactive proteasomes are sensed and delivered for degradation by proteaphagy, a recently described novel type of selective autophagy [110–112]. We propose that less functional or non-functional proteasomes are less efficiently degraded in autophagy-constrained strains and their proportion in the total proteasome pool, which does not change, increases. This would then lead to the observed general decline in proteasomal activity.

Our results disclose novel, diverse, and complex phenotypes for different cellular processes in ATG5⁻, ATG5⁻/12⁻, and ATG5⁻/12⁻/16⁻ cells. They are consistent with autophagy-independent functions of the ATG12~5/16 complex and its components, in addition to its function in canonical autophagy. Further work is needed to fully unravel all independent activities of ATG5, ATG12, and ATG16, as well as possible additional roles of the ATG12~5, ATG12~7, and ATG12~10 conjugates.

Supplementary Materials: The following supplementary information is available online http://www.mdpi.com/2073-4409/9/5/1179/s1. Figure S1: Generation and verification of ATG5⁻, ATG5⁻/12⁻ and ATG5⁻/12⁻/16⁻ cells. Figure S2: Multiple sequence alignment of complete sequences of ATG5 orthologs. Figure S3: Western blot analysis of global protein ubiquitination and proteasomal subunit psmA7 (SU7) expression in AX2 and mutant strains. Table S1: Random cell motility. Table S2: List of transcriptional regulation of proteasomal genes.

Author Contributions: M.K. carried out the molecular lab work, performed data analysis, and drafted the manuscript; S.F., S.M., J.R., Q.X., and R.R. performed experiments and contributed reagents; C.S.C. contributed to analysis; P.W., and R.K. contributed to analysis of RNA_{seq} datasets; L.E. conceived of the study, designed and coordinated it, and finalised the manuscript. All authors have read and agreed to the published version of the manuscript.

Funding: This research was funded by the German Research Foundation (Deutsche Forschungsgemeinschaft: CRC670 TP01) and supported by Köln Fortune.

Acknowledgments: We thank Jason King (University of Sheffield) for sharing the ATG8a pAb, Rolf Müller for purification of recombinant ATG5, and Maria Stumpf for technical assistance.

Conflicts of Interest: The authors declare no conflict of interest. The funders had no role in the design of the study; in the collection, analyses, or interpretation of data; in the writing of the manuscript; or in the decision to publish the results.

References

1. Stanley, R.E.; Ragusa, M.J.; Hurley, J.H. The beginning of the end: How scaffolds nucleate autophagosome biogenesis. *Trends Cell Biol.* **2014**, *24*, 73–81. [CrossRef]
2. Mizushima, N.; Klionsky, D.J. Protein turnover via autophagy: Implications for metabolism. *Annu. Rev. Nutr.* **2007**, *27*, 19–40. [CrossRef]
3. Mesquita, A.; Cardenal-Munoz, E.; Dominguez, E.; Munoz-Braceras, S.; Nunez-Corcuera, B.; Phillips, B.A.; Tabara, L.C.; Xiong, Q.; Coria, R.; Eichinger, L.; et al. Autophagy in dictyostelium: Mechanisms, regulation and disease in a simple biomedical model. *Autophagy* **2017**, *13*, 24–40. [CrossRef]
4. Mizushima, N.; Levine, B.; Cuervo, A.M.; Klionsky, D.J. Autophagy fights disease through cellular self-digestion. *Nature* **2008**, *451*, 1069–1075. [CrossRef]
5. Schneider, J.L.; Cuervo, A.M. Autophagy and human disease: Emerging themes. *Curr. Opin. Genet. Dev.* **2014**, *26*, 16–23. [CrossRef]
6. Kiriyama, Y.; Nochi, H. The function of autophagy in neurodegenerative diseases. *Int. J. Mol. Sci.* **2015**, *16*, 26797–26812. [CrossRef]
7. Noda, N.N.; Inagaki, F. Mechanisms of autophagy. *Annu. Rev. Biophys.* **2015**, *44*, 101–122. [CrossRef]
8. Boya, P.; Reggiori, F.; Codogno, P. Emerging regulation and functions of autophagy. *Nat. Cell Biol.* **2013**, *15*, 713–720. [CrossRef]
9. Fischer, S.; Eichinger, L. Dictyostelium discoideum and autophagy—A perfect pair. *Int. J. Dev. Biol.* **2019**, *63*, 485–495. [CrossRef]

10. Tanida, I.; Mizushima, N.; Kiyooka, M.; Ohsumi, M.; Ueno, T.; Ohsumi, Y.; Kominami, E. Apg7p/cvt2p: A novel protein-activating enzyme essential for autophagy. *Mol. Biol. Cell* **1999**, *10*, 1367–1379. [CrossRef]
11. Shintani, T.; Mizushima, N.; Ogawa, Y.; Matsuura, A.; Noda, T.; Ohsumi, Y. Apg10p, a novel protein-conjugating enzyme essential for autophagy in yeast. *EMBO J.* **1999**, *18*, 5234–5241. [CrossRef] [PubMed]
12. Mizushima, N.; Noda, T.; Yoshimori, T.; Tanaka, Y.; Ishii, T.; George, M.D.; Klionsky, D.J.; Ohsumi, M.; Ohsumi, Y. A protein conjugation system essential for autophagy. *Nature* **1998**, *395*, 395–398. [CrossRef] [PubMed]
13. Mizushima, N.; Noda, T.; Ohsumi, Y. Apg16p is required for the function of the apg12p–apg5p conjugate in the yeast autophagy pathway. *EMBO J.* **1999**, *18*, 3888–3896. [CrossRef]
14. Kabeya, Y.; Kamada, Y.; Baba, M.; Takikawa, H.; Sasaki, M.; Ohsumi, Y. Atg17 functions in cooperation with atg1 and atg13 in yeast autophagy. *Mol. Biol. Cell* **2005**, *16*, 2544–2553. [CrossRef]
15. Kirisako, T.; Baba, M.; Ishihara, N.; Miyazawa, K.; Ohsumi, M.; Yoshimori, T.; Noda, T.; Ohsumi, Y. Formation process of autophagosome is traced with apg8/aut7p in yeast. *J. Cell Biol.* **1999**, *147*, 435–446. [CrossRef]
16. Kirisako, T.; Ichimura, Y.; Okada, H.; Kabeya, Y.; Mizushima, N.; Yoshimori, T.; Ohsumi, M.; Takao, T.; Noda, T.; Ohsumi, Y. The reversible modification regulates the membrane-binding state of apg8/aut7 essential for autophagy and the cytoplasm to vacuole targeting pathway. *J. Cell Biol.* **2000**, *151*, 263–276. [CrossRef]
17. Hanada, T.; Noda, N.N.; Satomi, Y.; Ichimura, Y.; Fujioka, Y.; Takao, T.; Inagaki, F.; Ohsumi, Y. The atg12-atg5 conjugate has a novel e3-like activity for protein lipidation in autophagy. *J. Biol. Chem.* **2007**, *282*, 37298–37302. [CrossRef]
18. Romanov, J.; Walczak, M.; Ibiricu, I.; Schuchner, S.; Ogris, E.; Kraft, C.; Martens, S. Mechanism and functions of membrane binding by the atg5-atg12/atg16 complex during autophagosome formation. *EMBO J.* **2012**, *31*, 4304–4317. [CrossRef]
19. Harada, K.; Kotani, T.; Kirisako, H.; Sakoh-Nakatogawa, M.; Oikawa, Y.; Kimura, Y.; Hirano, H.; Yamamoto, H.; Ohsumi, Y.; Nakatogawa, H. Two distinct mechanisms target the autophagy-related e3 complex to the pre-autophagosomal structure. *eLife* **2019**, *8*, e43088. [CrossRef]
20. Nishimura, T.; Kaizuka, T.; Cadwell, K.; Sahani, M.H.; Saitoh, T.; Akira, S.; Virgin, H.W.; Mizushima, N. Fip200 regulates targeting of atg16l1 to the isolation membrane. *EMBO Rep.* **2013**, *14*, 284–291. [CrossRef]
21. Dooley, H.C.; Razi, M.; Polson, H.E.; Girardin, S.E.; Wilson, M.I.; Tooze, S.A. Wipi2 links lc3 conjugation with pi3p, autophagosome formation, and pathogen clearance by recruiting atg12-5-16l1. *Mol. Cell* **2014**, *55*, 238–252. [CrossRef] [PubMed]
22. Juris, L.; Montino, M.; Rube, P.; Schlotterhose, P.; Thumm, M.; Krick, R. Pi3p binding by atg21 organises atg8 lipidation. *EMBO J.* **2015**, *34*, 955–973. [CrossRef]
23. Fujita, N.; Itoh, T.; Omori, H.; Fukuda, M.; Noda, T.; Yoshimori, T. The atg16l complex specifies the site of lc3 lipidation for membrane biogenesis in autophagy. *Mol. Biol. Cell* **2008**, *19*, 2092–2100. [CrossRef] [PubMed]
24. Kuma, A.; Hatano, M.; Matsui, M.; Yamamoto, A.; Nakaya, H.; Yoshimori, T.; Ohsumi, Y.; Tokuhisa, T.; Mizushima, N. The role of autophagy during the early neonatal starvation period. *Nature* **2004**, *432*, 1032–1036. [CrossRef]
25. Muller-Taubenberger, A.; Kortholt, A.; Eichinger, L. Simple system–substantial share: The use of dictyostelium in cell biology and molecular medicine. *Eur. J. Cell Biol.* **2013**, *92*, 45–53. [CrossRef]
26. Bozzaro, S.; Eichinger, L. The professional phagocyte dictyostelium discoideum as a model host for bacterial pathogens. *Curr. Drug Targets* **2011**, *12*, 942–954. [CrossRef]
27. Williams, R.S.; Boeckeler, K.; Graf, R.; Muller-Taubenberger, A.; Li, Z.; Isberg, R.R.; Wessels, D.; Soll, D.R.; Alexander, H.; Alexander, S. Towards a molecular understanding of human diseases using dictyostelium discoideum. *Trends Mol. Med.* **2006**, *12*, 415–424. [CrossRef]
28. Annesley, S.J.; Fisher, P.R. Dictyostelium discoideum—A model for many reasons. *Mol. Cell. Biochem.* **2009**, *329*, 73–91. [CrossRef]
29. Dominguez-Martin, E.; Cardenal-Munoz, E.; King, J.S.; Soldati, T.; Coria, R.; Escalante, R. Methods to monitor and quantify autophagy in the social amoeba dictyostelium discoideum. *Cells* **2017**, *6*, 18. [CrossRef]
30. Klionsky, D.J.; Abdelmohsen, K.; Abe, A.; Abedin, M.J.; Abeliovich, H.; Acevedo Arozena, A.; Adachi, H.; Adams, C.M.; Adams, P.D.; Adeli, K.; et al. Guidelines for the use and interpretation of assays for monitoring autophagy (3rd edition). *Autophagy* **2016**, *12*, 1–222. [CrossRef]

31. Faix, J.; Linkner, J.; Nordholz, B.; Platt, J.L.; Liao, X.H.; Kimmel, A.R. The application of the cre-loxp system for generating multiple knock-out and knock-in targeted loci. *Methods Mol. Biol.* **2013**, *983*, 249–267.
32. Fey, P.; Dodson, R.J.; Basu, S.; Chisholm, R.L. One stop shop for everything dictyostelium: Dictybase and the dicty stock center in 2012. *Methods Mol. Biol.* **2013**, *983*, 59–92.
33. Calvo-Garrido, J.; Escalante, R. Autophagy dysfunction and ubiquitin-positive protein aggregates in dictyostelium cells lacking vmp1. *Autophagy* **2010**, *6*, 100–109. [CrossRef]
34. Calvo-Garrido, J.; King, J.S.; Munoz-Braceras, S.; Escalante, R. Vmp1 regulates ptdins3p signaling during autophagosome formation in dictyostelium discoideum. *Traffic* **2014**, *15*, 1235–1246. [CrossRef]
35. Munoz-Braceras, S.; Calvo, R.; Escalante, R. Tipc and the chorea-acanthocytosis protein vps13a regulate autophagy in dictyostelium and human hela cells. *Autophagy* **2015**, *11*, 918–927. [CrossRef]
36. Messling, S.; Matthias, J.; Xiong, Q.; Fischer, S.; Eichinger, L. The two *Dictyostelium discoideum* autophagy 8 proteins have distinct autophagic functions. *Eur. J. Cell Biol.* **2017**, *96*, 312–324. [CrossRef]
37. Otto, G.P.; Wu, M.Y.; Kazgan, N.; Anderson, O.R.; Kessin, R.H. Dictyostelium macroautophagy mutants vary in the severity of their developmental defects. *J. Biol. Chem.* **2004**, *279*, 15621–15629. [CrossRef]
38. Tung, S.M.; Unal, C.; Ley, A.; Pena, C.; Tunggal, B.; Noegel, A.A.; Krut, O.; Steinert, M.; Eichinger, L. Loss of dictyostelium atg9 results in a pleiotropic phenotype affecting growth, development, phagocytosis and clearance and replication of legionella pneumophila. *Cell. Microbiol.* **2010**, *12*, 765–780. [CrossRef]
39. Fischer, S.; Rijal, R.; Frommolt, P.; Wagle, P.; Konertz, R.; Faix, J.; Messling, S.; Eichinger, L. Functional characterization of ubiquitin-like core autophagy protein atg12 in dictyostelium discoideum. *Cells* **2019**, *8*, 72. [CrossRef]
40. Xiong, Q.; Unal, C.; Matthias, J.; Steinert, M.; Eichinger, L. The phenotypes of atg9, atg16 and atg9/16 knock-out mutants imply autophagy-dependent and -independent functions. *Open Biol.* **2015**, *5*, 150008. [CrossRef]
41. Yamada, Y.; Schaap, P. The proppin bcas3 and its interactor kinkya localize to the early phagophore and regulate autophagy. *Autophagy* **2020**, 1–16. [CrossRef]
42. Watts, D.J.; Ashworth, J.M. Growth of myxamoebae of the cellular slime mould *Dictyostelium discoideum* in axenic culture. *Biochem. J.* **1970**, *119*, 171–174. [CrossRef]
43. Brink, M.; Gerisch, G.; Isenberg, G.; Noegel, A.A.; Segall, J.E.; Wallraff, E.; Schleicher, M. A dictyostelium mutant lacking an f-actin cross-linking protein, the 120-kd gelation factor. *J. Cell Biol.* **1990**, *111*, 1477–1489. [CrossRef]
44. Sussman, M. Biochemical and genetic methods in the study of cellular slime mold development. *Methods Cell Physiol.* **1966**, *2*, 397–410.
45. Williams, K.L.; Newell, P.C. A genetic study of aggregation in the cellular slime mould dictyostelium discoideum using complementation analysis. *Genetics* **1976**, *82*, 287–307.
46. Raper, K.B. Dictyostelium discoideum, a new species of slime mold from decaying forest leaves. *J. Agric. Res.* **1935**, *50*, 135–147.
47. Simpson, P.A.; Spudich, J.A.; Parham, P. Monoclonal antibodies prepared against dictyostelium actin: Characterization and interactions with actin. *J. Cell Biol.* **1984**, *99*, 287–295. [CrossRef]
48. Schauer, T.M.; Nesper, M.; Kehl, M.; Lottspeich, F.; Muller-Taubenberger, A.; Gerisch, G.; Baumeister, W. Proteasomes from dictyostelium discoideum: Characterization of structure and function. *J. Struct. Biol.* **1993**, *111*, 135–147. [CrossRef]
49. Laemmli, U.K. Cleavage of structural proteins during the assembly of the head of bacteriophage t4. *Nature* **1970**, *227*, 680–685. [CrossRef]
50. Matthias, J.; Messling, S.; Eichinger, L. The two dictyostelium autophagy eight proteins, atg8a and atg8b, associate with the autophagosome in succession. *Eur. J. Cell Biol.* **2016**, *95*, 15–25. [CrossRef]
51. Towbin, H.; Staehelin, T.; Gordon, J. Electrophoretic transfer of proteins from polyacrylamide gels to nitrocellulose sheets: Procedure and some applications. *Proc. Natl. Acad. Sci. USA* **1979**, *76*, 4350–4354. [CrossRef]
52. Strucksberg, K.H.; Tangavelou, K.; Schroder, R.; Clemen, C.S. Proteasomal activity in skeletal muscle: A matter of assay design, muscle type, and age. *Anal. Biochem.* **2010**, *399*, 225–229. [CrossRef]
53. Arhzaouy, K.; Strucksberg, K.H.; Tung, S.M.; Tangavelou, K.; Stumpf, M.; Faix, J.; Schroder, R.; Clemen, C.S.; Eichinger, L. Heteromeric p97/p97r155c complexes induce dominant negative changes in wild-type and autophagy 9-deficient dictyostelium strains. *PLoS ONE* **2012**, *7*, e46879. [CrossRef]

54. Chojnacki, S.; Cowley, A.; Lee, J.; Foix, A.; Lopez, R. Programmatic access to bioinformatics tools from embl-ebi update: 2017. *Nucleic Acids Res.* **2017**, *45*, W550–W553. [CrossRef]
55. Capra, J.A.; Singh, M. Predicting functionally important residues from sequence conservation. *Bioinformatics* **2007**, *23*, 1875–1882. [CrossRef]
56. Letunic, I.; Bork, P. 20 years of the smart protein domain annotation resource. *Nucleic Acids Res.* **2018**, *46*, D493–D496. [CrossRef]
57. Mitchell, A.L.; Attwood, T.K.; Babbitt, P.C.; Blum, M.; Bork, P.; Bridge, A.; Brown, S.D.; Chang, H.Y.; El-Gebali, S.; Fraser, M.I.; et al. Interpro in 2019: Improving coverage, classification and access to protein sequence annotations. *Nucleic Acids Res.* **2019**, *47*, D351–D360. [CrossRef]
58. Waterhouse, A.; Bertoni, M.; Bienert, S.; Studer, G.; Tauriello, G.; Gumienny, R.; Heer, F.T.; de Beer, T.A.P.; Rempfer, C.; Bordoli, L.; et al. Swiss-model: Homology modelling of protein structures and complexes. *Nucleic Acids Res.* **2018**, *46*, W296–W303. [CrossRef]
59. Guex, N.; Peitsch, M.C. Swiss-model and the swiss-pdbviewer: An environment for comparative protein modeling. *Electrophoresis* **1997**, *18*, 2714–2723. [CrossRef]
60. Otomo, C.; Metlagel, Z.; Takaesu, G.; Otomo, T. Structure of the human atg12~atg5 conjugate required for lc3 lipidation in autophagy. *Nat. Struct. Mol. Biol.* **2013**, *20*, 59–66. [CrossRef]
61. Farbrother, P.; Wagner, C.; Na, J.; Tunggal, B.; Morio, T.; Urushihara, H.; Tanaka, Y.; Schleicher, M.; Steinert, M.; Eichinger, L. Dictyostelium transcriptional host cell response upon infection with legionella. *Cell. Microbiol.* **2006**, *8*, 438–456. [CrossRef] [PubMed]
62. Wagle, P.; Nikolic, M.; Frommolt, P. Quickngs elevates next-generation sequencing data analysis to a new level of automation. *BMC Genom.* **2015**, *16*, 487. [CrossRef] [PubMed]
63. Eichinger, L.; Pachebat, J.A.; Glockner, G.; Rajandream, M.A.; Sucgang, R.; Berriman, M.; Song, J.; Olsen, R.; Szafranski, K.; Xu, Q.; et al. The genome of the social amoeba Dictyostelium discoideum. *Nature* **2005**, *435*, 43–57. [CrossRef] [PubMed]
64. Love, M.I.; Huber, W.; Anders, S. Moderated estimation of fold change and dispersion for rna-seq data with deseq2. *Genome Biol.* **2014**, *15*, 550. [CrossRef]
65. RCoreTeam. *R: A Language and Environment for Statistical Computing*; R Foundation for Statistical Computing: Vienna, Austria, 2018.
66. Gastwirth, J.L.; Gel, Y.R.; Hui, W.L.W.; Lyubchich, V.; Miao, W.; Noguchi, K. Lawstat: Tools for Biostatistics, Public Policy, and Law. R package version 3.3. 2017.
67. Dinno, A. Dunn.Test: Dunn's Test of Multiple Comparisons using Rank Sums. R package version. 2017.
68. Fox, J.; Weisberg, S. (Eds.) *An R Companion to Applied Regression*, 3rd ed.; Sage Publications: Thousand Oaks, CA, USA, 2018; p. 577.
69. Wang, M.; Zhao, Y.; Zhang, B. Efficient test and visualization of multi-set intersections. *Sci. Rep.* **2015**, *5*, 16923. [CrossRef]
70. Lew, M. Good statistical practice in pharmacology. Problem 2. *Br. J. Pharmacol.* **2007**, *152*, 299–303. [CrossRef]
71. Dunn, O.J. Multiple comparisons using rank sums. *Technometrics* **1964**, *6*, 241–252. [CrossRef]
72. Tsukada, M.; Ohsumi, Y. Isolation and characterization of autophagy-defective mutants of *Saccharomyces cerevisiae*. *FEBS Lett.* **1993**, *333*, 169–174. [CrossRef]
73. Matsushita, M.; Suzuki, N.N.; Obara, K.; Fujioka, Y.; Ohsumi, Y.; Inagaki, F. Structure of atg5.Atg16, a complex essential for autophagy. *J. Biol. Chem.* **2007**, *282*, 6763–6772. [CrossRef]
74. Yamaguchi, M.; Noda, N.N.; Yamamoto, H.; Shima, T.; Kumeta, H.; Kobashigawa, Y.; Akada, R.; Ohsumi, Y.; Inagaki, F. Structural insights into atg10-mediated formation of the autophagy-essential atg12-atg5 conjugate. *Structure* **2012**, *20*, 1244–1254. [CrossRef]
75. Soll, D.R.; Yarger, J.; Mirick, M. Stationary phase and the cell cycle of *Dictyostelium discoideum* in liquid nutrient medium. *J. Cell Sci.* **1976**, *20*, 513–523.
76. Lystad, A.H.; Carlsson, S.R.; Simonsen, A. Toward the function of mammalian atg12-atg5-atg16l1 complex in autophagy and related processes. *Autophagy* **2019**, *15*, 1485–1486. [CrossRef]
77. Mizushima, N. The atg conjugation systems in autophagy. *Curr. Opin. Cell Biol.* **2019**, *63*, 1–10. [CrossRef]
78. Mizushima, N.; Yamamoto, A.; Hatano, M.; Kobayashi, Y.; Kabeya, Y.; Suzuki, K.; Tokuhisa, T.; Ohsumi, Y.; Yoshimori, T. Dissection of autophagosome formation using apg5-deficient mouse embryonic stem cells. *J. Cell Biol.* **2001**, *152*, 657–668. [CrossRef]

79. Kishi-Itakura, C.; Koyama-Honda, I.; Itakura, E.; Mizushima, N. Ultrastructural analysis of autophagosome organization using mammalian autophagy-deficient cells. *J. Cell Sci.* **2014**, *127*, 4089–4102. [CrossRef]
80. Nguyen, T.N.; Padman, B.S.; Usher, J.; Oorschot, V.; Ramm, G.; Lazarou, M. Atg8 family lc3/gabarap proteins are crucial for autophagosome-lysosome fusion but not autophagosome formation during pink1/parkin mitophagy and starvation. *J. Cell Biol.* **2016**, *215*, 857–874. [CrossRef]
81. Sou, Y.S.; Waguri, S.; Iwata, J.; Ueno, T.; Fujimura, T.; Hara, T.; Sawada, N.; Yamada, A.; Mizushima, N.; Uchiyama, Y.; et al. The atg8 conjugation system is indispensable for proper development of autophagic isolation membranes in mice. *Mol. Biol. Cell* **2008**, *19*, 4762–4775. [CrossRef]
82. Tsuboyama, K.; Koyama-Honda, I.; Sakamaki, Y.; Koike, M.; Morishita, H.; Mizushima, N. The atg conjugation systems are important for degradation of the inner autophagosomal membrane. *Science* **2016**, *354*, 1036–1041. [CrossRef]
83. Uemura, T.; Yamamoto, M.; Kametaka, A.; Sou, Y.S.; Yabashi, A.; Yamada, A.; Annoh, H.; Kametaka, S.; Komatsu, M.; Waguri, S. A cluster of thin tubular structures mediates transformation of the endoplasmic reticulum to autophagic isolation membrane. *Mol. Cell. Biol.* **2014**, *34*, 1695–1706. [CrossRef]
84. Renna, M.; Rubinsztein, D.C. Macroautophagy without lc3 conjugation? *Cell Res.* **2017**, *27*, 5–6. [CrossRef]
85. Haller, M.; Hock, A.K.; Giampazolias, E.; Oberst, A.; Green, D.R.; Debnath, J.; Ryan, K.M.; Vousden, K.H.; Tait, S.W. Ubiquitination and proteasomal degradation of atg12 regulates its proapoptotic activity. *Autophagy* **2014**, *10*, 2269–2278. [CrossRef] [PubMed]
86. Galluzzi, L.; Green, D.R. Autophagy-independent functions of the autophagy machinery. *Cell* **2019**, *177*, 1682–1699. [CrossRef] [PubMed]
87. Malhotra, R.; Warne, J.P.; Salas, E.; Xu, A.W.; Debnath, J. Loss of atg12, but not atg5, in pro-opiomelanocortin neurons exacerbates diet-induced obesity. *Autophagy* **2015**, *11*, 145–154.
88. Mauthe, M.; Langereis, M.; Jung, J.; Zhou, X.; Jones, A.; Omta, W.; Tooze, S.A.; Stork, B.; Paludan, S.R.; Ahola, T.; et al. An sirna screen for atg protein depletion reveals the extent of the unconventional functions of the autophagy proteome in virus replication. *J. Cell Biol.* **2016**, *214*, 619–635. [CrossRef]
89. Nam, T.; Han, J.H.; Devkota, S.; Lee, H.W. Emerging paradigm of crosstalk between autophagy and the ubiquitin-proteasome system. *Mol. Cells* **2017**, *40*, 897–905.
90. Jounai, N.; Takeshita, F.; Kobiyama, K.; Sawano, A.; Miyawaki, A.; Xin, K.Q.; Ishii, K.J.; Kawai, T.; Akira, S.; Suzuki, K.; et al. The atg5 atg12 conjugate associates with innate antiviral immune responses. *Proc. Natl. Acad. Sci. USA* **2007**, *104*, 14050–14055. [CrossRef]
91. Fletcher, K.; Ulferts, R.; Jacquin, E.; Veith, T.; Gammoh, N.; Arasteh, J.M.; Mayer, U.; Carding, S.R.; Wileman, T.; Beale, R.; et al. The wd40 domain of atg16l1 is required for its non-canonical role in lipidation of lc3 at single membranes. *EMBO J.* **2018**, *37*, e97840. [CrossRef]
92. Florey, O.; Kim, S.E.; Sandoval, C.P.; Haynes, C.M.; Overholtzer, M. Autophagy machinery mediates macroendocytic processing and entotic cell death by targeting single membranes. *Nat. Cell Biol.* **2011**, *13*, 1335–1343. [CrossRef]
93. Lai, S.C.; Devenish, R.J. Lc3-associated phagocytosis (lap): Connections with host autophagy. *Cells* **2012**, *1*, 396–408. [CrossRef]
94. Rai, S.; Arasteh, M.; Jefferson, M.; Pearson, T.; Wang, Y.; Zhang, W.; Bicsak, B.; Divekar, D.; Powell, P.P.; Naumann, R.; et al. The atg5-binding and coiled coil domains of atg16l1 maintain autophagy and tissue homeostasis in mice independently of the wd domain required for lc3-associated phagocytosis. *Autophagy* **2019**, *15*, 599–612. [CrossRef]
95. Heckmann, B.L.; Green, D.R. Lc3-associated phagocytosis at a glance. *J. Cell Sci.* **2019**, *132*. [CrossRef]
96. Lima, W.C.; Balestrino, D.; Forestier, C.; Cosson, P. Two distinct sensing pathways allow recognition of klebsiella pneumoniae by dictyostelium amoebae. *Cell. Microbiol.* **2014**, *16*, 311–323. [CrossRef]
97. Nasser, W.; Santhanam, B.; Miranda, E.R.; Parikh, A.; Juneja, K.; Rot, G.; Dinh, C.; Chen, R.; Zupan, B.; Shaulsky, G.; et al. Bacterial discrimination by dictyostelid amoebae reveals the complexity of ancient interspecies interactions. *Curr. Biol.* **2013**, *23*, 862–872. [CrossRef]
98. Gao, Z.; Gammoh, N.; Wong, P.-M.; Erdjument-Bromage, H.; Tempst, P.; Jiang, X. Processing of autophagic protein lc3 by the 20s proteasome. *Autophagy* **2010**, *6*, 126–137. [CrossRef]
99. Pohl, C.; Dikic, I. Cellular quality control by the ubiquitin-proteasome system and autophagy. *Science* **2019**, *366*, 818–822. [CrossRef]

100. Xiong, Q.; Fischer, S.; Karow, M.; Muller, R.; Messling, S.; Eichinger, L. Atg16 mediates the autophagic degradation of the 19s proteasomal subunits psmd1 and psmd2. *Eur. J. Cell Biol.* **2018**, *97*, 523–532. [CrossRef]
101. Ding, W.X.; Ni, H.M.; Gao, W.; Yoshimori, T.; Stolz, D.B.; Ron, D.; Yin, X.M. Linking of autophagy to ubiquitin-proteasome system is important for the regulation of endoplasmic reticulum stress and cell viability. *Am. J. Pathol.* **2007**, *171*, 513–524. [CrossRef]
102. Ji, C.H.; Kwon, Y.T. Crosstalk and interplay between the ubiquitin-proteasome system and autophagy. *Mol. Cells* **2017**, *40*, 441–449.
103. Korolchuk, V.I.; Menzies, F.M.; Rubinsztein, D.C. Mechanisms of cross-talk between the ubiquitin-proteasome and autophagy-lysosome systems. *FEBS Lett.* **2010**, *584*, 1393–1398. [CrossRef]
104. Pandey, U.B.; Nie, Z.; Batlevi, Y.; McCray, B.A.; Ritson, G.P.; Nedelsky, N.B.; Schwartz, S.L.; DiProspero, N.A.; Knight, M.A.; Schuldiner, O.; et al. Hdac6 rescues neurodegeneration and provides an essential link between autophagy and the ups. *Nature* **2007**, *447*, 859–863. [CrossRef]
105. Komatsu, M.; Waguri, S.; Chiba, T.; Murata, S.; Iwata, J.; Tanida, I.; Ueno, T.; Koike, M.; Uchiyama, Y.; Kominami, E.; et al. Loss of autophagy in the central nervous system causes neurodegeneration in mice. *Nature* **2006**, *441*, 880–884. [CrossRef]
106. Korolchuk, V.I.; Menzies, F.M.; Rubinsztein, D.C. A novel link between autophagy and the ubiquitin-proteasome system. *Autophagy* **2009**, *5*, 862–863. [CrossRef]
107. Qiao, L.; Zhang, J. Inhibition of lysosomal functions reduces proteasomal activity. *Neurosci. Lett.* **2009**, *456*, 15–19. [CrossRef]
108. Wang, X.; Terpstra, E.J. Ubiquitin receptors and protein quality control. *J. Mol. Cell. Cardiol.* **2013**, *55*, 73–84. [CrossRef]
109. Fujita, N.; Saitoh, T.; Kageyama, S.; Akira, S.; Noda, T.; Yoshimori, T. Differential involvement of atg16l1 in crohn disease and canonical autophagy: Analysis of the organization of the atg16l1 complex in fibroblasts. *J. Biol. Chem.* **2009**, *284*, 32602–32609. [CrossRef]
110. Marshall, R.S.; Li, F.; Gemperline, D.C.; Book, A.J.; Vierstra, R.D. Autophagic degradation of the 26s proteasome is mediated by the dual atg8/ubiquitin receptor rpn10 in arabidopsis. *Mol. Cell* **2015**, *58*, 1053–1066. [CrossRef]
111. Marshall, R.S.; McLoughlin, F.; Vierstra, R.D. Autophagic turnover of inactive 26s proteasomes in yeast is directed by the ubiquitin receptor cue5 and the hsp42 chaperone. *Cell Rep.* **2016**, *16*, 1717–1732. [CrossRef]
112. Waite, K.A.; De-La Mota-Peynado, A.; Vontz, G.; Roelofs, J. Starvation induces proteasome autophagy with different pathways for core and regulatory particles. *J. Biol. Chem.* **2016**, *291*, 3239–3253. [CrossRef]

cells

MDPI

Article

Impairment of Proteasome and Autophagy Underlying the Pathogenesis of Leukodystrophy

Dar-Shong Lin [1,2,*], Che-Sheng Ho [3], Yu-Wen Huang [4], Tsu-Yen Wu [4], Tsung-Han Lee [4], Zo-Darr Huang [4], Tuan-Jen Wang [5], Shun-Jie Yang [4] and Ming-Fu Chiang [6,7,8,*]

1 Department of Pediatrics, Mackay Memorial Hospital, Taipei 10449, Taiwan
2 Department of Medicine and Institute of Biomedical Sciences, Mackay Medical College, New Taipei 25245, Taiwan
3 Department of Pediatric Neurology, Mackay Memorial Hospital, Taipei 10449, Taiwan; pedcsho@mmh.org.tw
4 Department of Medical Research, Mackay Memorial Hospital, Taipei 10449, Taiwan; wendyhuang1219@gmail.com (Y.-W.H.); linws@mmh.org.tw (T.-Y.W.); randy.b746mmh@gmail.com (T.-H.L.); darr.9475@mmh.org.tw (Z.-D.H.); jes53832.f631@mmh.org.tw (S.-J.Y.)
5 Department of Laboratory Medicine, Mackay Memorial Hospital, Taipei 10449, Taiwan; dj.wang@mmh.org.tw
6 Department of Neurosurgery, Mackay Memorial Hospital, Taipei 10449, Taiwan
7 Mackay Medicine, Nursing and Management College, Taipei 11260, Taiwan
8 Graduate Institute of Injury Prevention and Control, Taipei Medical University, Taipei 11031, Taiwan
* Correspondence: dslin@mmh.org.tw (D.-S.L.); chiang66@gmail.com (M.-F.C.); Tel.: +886-2-2809-4661 (D.-S.L.)

Received: 5 March 2020; Accepted: 28 April 2020; Published: 1 May 2020

Abstract: Impairment of the ubiquitin-proteasome-system (UPS) and autophagy causing cytoplasmic aggregation of ubiquitin andp62 have been implicated in the pathogenesis of most neurodegenerative disorders, yet, they have not been fully elucidated in leukodystrophies. The relationship among impairment of UPS, autophagy, and globoid cell leukodystrophy (GLD), one of the most common demyelinating leukodystrophies, is clarified in this study. We examined the ubiquitin and autophagy markers in the brains of twitcher mice, a murine model of infantile GLD, and in human oligodendrocytes incubated with psychosine. Immunohistochemical examinations showed spatiotemporal accumulation of ubiquitin- and p62-aggregates mainly in the white matter of brain and spinal cord at disease progression. Western blot analysis demonstrated a significant accumulation of ubiquitin, p62, and LC3-II in insoluble fraction in parallel with progressive demyelination and neuroinflammation in twitcher brains. In vitro study validated a dose- and time-dependent cytotoxicity of psychosine upon autophagy and UPS machinery. Inhibition of autophagy and UPS exacerbated the accumulation of insoluble ubiquitin, p62, and LC3-II proteins mediated by psychosine cytotoxicity as well as increased cytoplasmic deposition of ubiquitin- and p62-aggregates, and accumulation of autophagosomes and autolysosomes. Further, the subsequent accumulation of reactive oxygen species and reduction of mitochondrial respiration led to cell death. Our studies validate the impairment of proteasome and autophagy underlying the pathogenesis of GLD. These findings provide a novel insight into pathogenesis of GLD and suggest a specific pathomechanism as an ideal target for therapeutic approaches.

Keywords: leukodystrophies; globoid cell leukodystrophy; psychosine; autophagy; p62; ubiquitin; proteasome

1. Introduction

Impairment of the ubiquitin proteasome system (UPS) and autophagy have been implicated in the pathogenesis of several neurodegenerative diseases, yet, their role in the pathomechanism of leukodystrophies has not been fully elucidated. Leukodystrophies is a group of devastating neurodegenerative diseases resulting from inherited defects in myelin sheath formation and/or maintenance within the nervous system. The X-linked adrenoleukodystrophy, metachromatic leukodystrophy, and globoid cell leukodystrophy (GLD) caused by lysosomal or peroxisomal enzymes deficiency are the most common demyelinating leukodystrophies; complete rescue therapy is still a challenge and the pathomechanism remains elusive [1–3]. Among the leukodystrophies, GLD is the most distinctive type in that greater than 90% of affected individuals are infants [4,5]. Affected infants manifest with developmental delay, hypotonia, and spasticity as early as three months of age, progression involves atrophy, ataxia, epilepsy, cognitive deficits, and eventually death before 2 years of age [4,5]. Currently, GLD is included in newborn screening programs in New York and Kentucky, USA, for rapid diagnosis and early initiation of human stem cell transplantation, though an optimal therapy is still needed [6]. Mechanistic insights into the pathogenesis may prompt a novel therapeutic approach for this disease and other leukodystrophies.

GLD is a lysosomal storage disease caused by an autosomal recessive mutation in the galactosylcerebrosidase (GALC) gene and is characterized by progressive demyelination and astrogliosis in the nervous system [2,4]. It has been shown that enzymatic deficiency of GALC impairs the efficiency of catabolism of galactosylceramide and galactosylsphingosine (psychosine), resulting in progressive accumulation of psychosine in oligodendrocytes, Schwann cells, and neurons. Of note, galactosylceramide does not accumulate in the central nervous system of GLD patients as it can also be degraded by G_{M1} ganglioside β-galactosidase [7], while psychosine can only be hydrolyzed by GLAC and thus accumulates excessively in brain of GLD individuals [4]. The psychosine hypothesis as introduced to explain the pathogenetic mechanism of GLD and has been validated in both cellular and animal models of GLD [8,9]. Cytotoxicity of psychosine causes broad demyelination, reduced remyelination, as well as degeneration of axons and Purkinje cells in the nervous system [4,10–12]. To date, many of the studies on pathogenesis and development of therapeutic strategies in GLD have been performed on the naturally occurring twitcher mouse, which is an authentic murine model for infantile GLD [13,14]. Twitcher mice appear normal at birth but start twitching at postnatal day 21, have tremors and steady weight loss around day 25, show apparent limb weakness with kyphotic posture at day 30, have severe hind limb paralysis and intension tremors at day 35, show rapid deterioration after day 40, and early death is usually around day 42 [14,15]. Correlated with the clinical phenotype of twitcher mice, infiltration of macrophages and astrogliosis are first recognized in the cerebellum arbor vita, brain stem, and spinal cord after postnatal day 20, progress to cerebral white matter after day 25, and are evident in cerebral and cerebellar gray matter after day 30. Consistently, demyelination is recognized in cerebellar arbor vita, brain stem, and spinal cord after postnatal day 20, progresses in a caudal-to-rostral order, and is profound in cerebral and cerebellar white matter, brain stem, and spinal cord after day 30 [14,16]. Several therapeutic strategies, including substrate reduction, bone marrow transplantation, gene therapy, and a combination of both cell and transgene delivery, have been applied in twitcher mice and showed variable efficacy, while the complete rescue of phenotype and pathogenesis in GLD is still a challenge [11,17–20].

Accumulation of toxic metabolite psychosine contributes to the pathogenesis of GLD, though the molecular mechanism is still far from clear. Based on the psychosine hypothesis, several studies have been completed in cells supplemented with exogenous psychosine to unveil the pathological role of psychosine in oligodendrocytes [9,21–31]. It has been shown that psychosine accumulation in membrane microdomains results in the disruption of lipid rafts, alteration of membrane integrity, and inhibition of protein kinase C to the plasma membrane in neurons and oligodendrocytes [22,26]. The preferential accumulation of psychosine in cell membranes causes lipid raft clustering, impairment of tyrosine kinase receptor A-mediated signaling cascades and endocytosis, decreased microtubule stability, and defective axonal transport in neural cells [21,31]. Recently, it has been identified that psychosine accelerates fibrillization of α-synuclein and alters α-synuclein conformation [23]. Moreover, brain samples from twitcher mice and patients affected with the infantile and late-onset forms of GLD have demonstrated neuronal inclusions of α-synuclein, which distributes abundantly in brain areas with the highest levels of psychosine accumulation [23]. The psychosine-mediated alteration of protein conformation and aggregation of misfolded proteins suggest impairment of degradation on misfolded and damaged proteins contributing to the pathogenesis of GLD.

Synthesis of new proteins and degradation of damaged proteins are coordinately regulated and balanced to maintain the cellular protein homeostasis (proteostasis). UPS and macroautophagy (hereafter autophagy) are the two major pathways for eukaryotic intracellular protein catabolism. The UPS is the major protein degradation system and consists of concerted actions of enzymes and ubiquitin to tag damaged proteins for degradation. Enzymatic components E1 and ubiquitin-conjugating E2 enzymes prepare ubiquitin for conjugation, and E3 ligases link the activated ubiquitin from E2s onto the aberrant and/or misfolded soluble proteins, leading to an increasing polyubiquitination chain [32]. The polyubiquitinated proteins are recognized by the 26S proteasome for degradation into small peptides and amino acids. The UPS predominantly degrades short-lived cytosolic and nuclear proteins, including regulator and misfolded proteins as well as damaged proteins. While large misfolded proteins and damaged organelles, which cannot pass through the narrow chamber of the proteasome for degradation, are delivered to and degraded in lysosomes via the process of the autophagy pathway. When autophagy is induced, an isolation membrane emerging from the cellular organelle membrane network engulfs the cargo to form a double-membrane autophagosome, which fuses with the lysosome to form an autolysosome for subsequent lysosomal degradation. The autophagy process is coordinately controlled by autophagy-related (ATG) proteins, among which microtubule-associated protein 1 light chain 3 (LC3), a mammalian Atg8 homolog, undergoes a unique ubiquitin-like conjugation to phosphatidylethanolamine on the autophagic membrane to formLC3-phosphatidylethanolamine conjugate (LC3-II), mediating subsequent autophagosome formation [33]. The autophagy adaptorp62 interacts directly with LC3 to control traffic of ubiquitinated cargo into autophagosomes [34]. During autophagic flux, both LC3 and p62 proteins as well as ubiquitinated cargoes are subject to degradation in autolysosomes. The autophagy essentially degrades long-lived proteins, large protein complexes, aggregated proteins, and organelles.

Dysfunction of UPS or/and autophagy cause aberrant accumulation of misfolded proteins, which have been linked to pathogenesis in several neurodegenerative diseases, such as Lewy bodies and Lewy neurites in Parkinson's disease and dementia with Lewy bodies or oligodendroglial cytoplasmic inclusions in multiple system atrophy [35–37]. Additionally, impairment of autophagy has also been observed in several lysosomal storage diseases (LSDs), where various stages of the autophagic flux are disrupted in different LSDs [38]. Accumulation of autophagosomes, autolysosomes, and autophagic substrates as well as an impaired lysosomal activity have been demonstrated in a murine model of Niemann–Pick type C1 disease and disease-relevant cells [39], while brain or neurons and astrocytes cultured from Gaucher disease mice have manifested with an accumulation of cytoplasmic aggregates containing ubiquitinated protein, insoluble α-synuclein, and p62 [38]. In GLD, activation of autophagy has been documented in the disease-cell model and cytoplasmic aggregates of p62 in neurons have been observed in the twitcher brain at an early symptomatic stage [24,29,30]. Nonetheless, the impact of psychosine upon autophagy and UPS as well as the role of autophagy and UPS in the pathogenesis of GLD are largely elusive. Herein we show defects in both autophagy and UPS pathways in a murine model of GLD, leading to a progressive accumulation of insoluble ubiquitin and p62 proteins concomitant with cytoplasmic aggregates of p62 and ubiquitin in the brain at disease progression. Furthermore, an in vitro study using human oligodendrocyte MO3.13 cells validates a time- and dose-dependent cytotoxicity of psychosine upon UPS and autophagy, and recapitulates the pathogenesis observed in the GLD murine model. Our work, therefore, suggests that psychosine-mediated impairment of UPS and autophagy underlies the pathogenesis of GLD.

2. Materials and Methods

2.1. Animals

The heterozygous (*twi*/+) twitcher mice with a congenic C57Bl/6 background were obtained from the Jackson Laboratory (Bar Harbor, ME, USA) and maintained under standard housing conditions in the animal research facility of our institution. Wild-type (+/+) and homozygous (*twi*/*twi*) twitcher mice were generated by strict inbred mating of heterozygous twitcher mice. Genomic DNA was isolated from tissue sample of 2-day-old pups. The genotypes were determined in genomic DNA by a PCR amplification with a sense primer 5′-ATGAGACTGAAATTGGTAGACAGC-3′ and an anti-sense primer 5′-ATGGCCCACTGTCTTCAGGTGATA-3′ specific for the mutant allele, followed by EcoRV digestion as previously described [40,41]. All animal procedures were approved and performed according to the guidelines established by the Animal Care and Use Committee of our institution. Mice were euthanized under anesthesia with pentobarbital. Brains and spinal cords were removed immediately, snapped frozen with liquid nitrogen, and sorted at −80 °C before biochemical analysis. Alternatively, the hemisphere and spinal cords were post-fixed in 4% paraformaldehyde (PFA) overnight at 4 °C, cryoprotected in 30% sucrose, and then quickly frozen in OCT compound (TissueTek, Sakura Finetek, Torrance, CA, USA). Cryosectioned slices of 8-μm thickness were obtained and stored at −80 °C before immunohistochemistry.

2.2. Cell Culture and Treatment

Human oligodendrocytes MO3.13 cells were maintained in Dulbecco's modified eagle's medium (DMEM; Invitrogen, Eugene, OR, USA),supplemented with 10% (*v*/*v*) fetal bovine serum (FBS; Gibco, Rockford, IL, USA) and 1%penicillin G/streptomycin sulfate at 37 °C in a humidified atmosphere of 5% (*v*/*v*) CO_2.When indicated, theMO3.13 cells were plated in standard 6-well cell culture plates overnight. The next day, MO3.13 cells were exposed to different concentrations of psychosine (Sigma-Aldrich, St. Louis, MO, USA)/ DMSO (Sigma-Aldrich) for 24 h. For the selective experiment, MO3.13 cells were treated with or without 20 μM psychosine for 72 h in the presence of inhibitors 20 μM chloroquine (Sigma-Aldrich) and/or 0.25 μM MG132 (Sigma-Aldrich) during the last 24 h.

2.3. Immunofluorescence Staining

Brain cryosections were fixed with 4% PFA for 20 min, microwave heated in Tris-ETDA pH9 buffer for 3 min, quenched with 3% H_2O_2 in methanol for 5 min, and permeabilized with 0.5% Triton X-100 for 10 min. The sections were then blocked in 5% horse serum/0.25% Triton X-100 in phosphate-buffered saline (PBS) for 30 min at room temperature and incubated with primary antibodies (Table 1) over-night at room temperature. After extensive washing with PBS, the sections were visualized with secondary antibodies conjugated with Alex Flour 488 or Alex Flour 594 (Molecular Probes, Eugene, OR, USA) and counterstained with DAPI.

Cultured cells were fixed with 4% paraformaldehyde in PBS for 10 min at room temperature, washed extensively with PBS, then permeabilized with cold methanol for 10 min, microwave heated in Tris-ETDA pH9 buffer for 3 min, and washed with PBS. Cells were blocked with 10% goat serum/0.25% Triton X-100/PBS for 30 min at room temperature, incubated with primary antibodies (Table 1) for 2h at room temperature, washed 3 times with PBS, then incubated with secondary antibodies conjugated with Alex Flour 488 or Alex Flour 594 (Molecular Probes) for 1 h at room temperature. The cells were washed 3 times with PBS, counterstained with DAPI, and visualized under a fluorescence microscope.

Table 1. List of primary antibodies for immunohistochemistry (IHC), immunocytochemistry (ICC), and Western blot (WB) analyses.

Antibodies	Host	Supplier	IHC Dilution	ICC Dilution	WB Dilution
Anti-LC3B	Rabbit	Sigma-Aldrich (St. Louis, MO, USA)			1:1000
Anti-P62	Mouse	Abcam (Cambridge, UK)	1:300		1:2000
Anti-P62	Rabbit	Proteintech (Chicago, IL, USA)	1:300	1:300	
Anti-Ubiquitin	Mouse	Abcam (Cambridge, UK)	1:300	1:300	
Anti-Ubiquitin	Rabbit	Cell Signaling (Danvers, MA, USA)	1:200		1:1000
Anti-Nrf2	Rabbit	Abcam (Cambridge, UK)			1:1000
Anti-Keap1	Rabbit	Proteintech (Chicago, IL, USA)			1:1000
Anti-NQO1	Rabbit	Genetex (Irvine, CA)			1:1000
Anti-Histone H3	Mouse	Cell Signaling (Danvers, MA, USA)			1:1000
Anti-β-Actin	Mouse	Sigma-Aldrich (St. Louis, MO, USA)			1:5000
Anti-NeuN	Mouse	Millipore (Burlington, MA, USA)	1:100		
Anti-Calbindin	Mouse	Sigma-Aldrich (St. Louis, MO, USA)	1:100		
Anti-GFAP	Rabbit	Dako (Santa Clara, CA)	1:200		
Anti-GFAP	Mouse	Invitrogen (Eugene, OR, USA)			1:2000
Anti-MBP	Mouse	Abcam (Cambridge, UK)	1:100		
Anti-MBP	Mouse	Millipore (Burlington, MA, USA)			1:1000
Anti-PLP	Rabbit	Abcam (Cambridge, UK)	1:100		
Anti-PLP	Mouse	Abcam (Cambridge, UK)	1:100		
Anti-Iba1	Rabbit	Biocare (Pacheco, CA, USA)	1:100		
Anti-LAMP1	Rabbit	Genetex (Irvine, CA)	1:100		

2.4. Preparation of Protein Samples

Proteins were extracted from deep-frozen tissues and cells, respectively, lysed with T-PER™ Tissue Protein Extraction Reagent (ThermoFisher Scientific) supplemented with Halt™ protease inhibitor cocktail (Thermo Fisher Scientific). The tissue was then homogenized in Eppendorf Scientific tubes with a pellet pestle at 4 °C. The lysate was centrifuged at 17,000× *g* for 20 min at 4 °C and the supernatant was collected as a detergent soluble fraction according to methods described previously [42]. The insoluble pellet was dissolved in T-PER™ Tissue Protein Extraction Reagent supplemented with 2% sodium dodecyl sulfate (SDS; Merk, Darmstadt, Germany), sonicated on ice, and centrifuged at 17,000× *g* for 20 min at 4 °C. The supernatant was collected as an insoluble fraction [42]. The concentration of each protein fraction was measured using the BCA protein assay (Thermo Fisher Scientific, Rockford, IL, USA) and subjected to Western blotting analysis.

2.5. Western Blot Analysis and Immunoprecipitation

Each protein sample was subjected to sodium dodecyl sulfate-polyacrylamide gel electrophoresis (SDS-PAGE) using10% or 12% polyacrylamide gels, and transferred onto a Poly-vinylidenedifluoride (PVDF) membrane (Millipore). The membranes were blocked with 0.5% skimmed milk in TBST (20 mM Tris-HCl, pH 7.5, 150 mM NaCl, 0.1% Tween20) for 1 h at room temperature, washed with TBST, and immunoblotted with the primary antibody (Table 1) in TBST overnight at 4 °C. The PVDF membrane was then washed, incubated with a secondary antibody conjugated to horseradish peroxidase for 2 h at room temperature, washed again, and then visualized by Immobilon Western Chemiluminescent HP Substrate (Millipore), which provides high sensitivity over a broad detection range, and detected by ChemiDoc-It 815 Imaging System (Analytik Jena, Upland, CA, USA), which provides a broad dynamic range of imaging for linear ranges of detection. To avoid oversaturation of signal intensity, multiple images with different exposure times were captured sequentially by ChemiDoc-It 815 Imaging System and the image with the best results was selected for quantitative analysis by VisionWorks software (Analytik Jena). Furthermore, if a protein of interest was expressed at relatively low or high levels, the amount of the loaded sample was increased or diluted, accordingly, to fit within the linear range of detection.

To verify the interaction between insoluble p62 and ubiquitin, immunoprecipitation (IP) was performed using Pierce protein A/G magnetic beads (Thermo Fisher Scientific) and magnetic bead-based separation according to the manufacturer's protocol. Briefly, the anti-p62/anti-ubiquitin antibody was added to prepared protein A/G magnetic beads and then mixed and incubated on a rotating platform for 15 min at room temperature. The magnetic beads were collected and washed with Modified Coupling Buffer three times, then incubated with disuccinimidyl suberate for 30 min. The antibody-crosslinked magnetic beads were washed three times with Elute Buffer followed by two washes with Modified Coupling Buffer. The lysates were incubated with antibody-crosslinked magnetic beads overnight at 4 °C. The beads were washed twice with Modified Coupling Buffer and once with ultrapure water and then incubated with Elute Buffer for 5 min at room temperature. The eluate containing the target antigen was collected by magnetic separation with beads, and neutralized with neutralization buffer. The IP products were detected by Western blot analysis as described above.

2.6. Detection of Autolysosomes

Briefly, MO3.13 cells treated with or without psychosine in the presence of autophagy and/or proteasome inhibitors were stained with Cyto-ID Green detection reagent (Enzo Life Science, Taipei, Taiwan) and LysoTrakcer Red (Thermo Fisher Scientific) and counterstained with Hoechst 33342, according to the manufacturer's protocol. Images were obtained and recorded using the ImageXpress Micro 4 system (Molecular Devices) at 40× magnification in 18 fields of view per well and analyzed by the Multi-Wavelength Cell Scoring Application Module. The cells were selected based on both Cyto-ID and LysoTracker Red fluorescence by setting the maximal and minimal diameters and the minimal fluorescence intensity relative to the background from both channels. The integrated intensity/cell which represented the fluorescence of each cell was used to measure Cyto-ID and LysoTracker Red co-expression in different groups. Colocalization of both Cyto-ID and LysoTracker Red fluorescence corresponded to the puncta of autolysosomes.

2.7. Reactive Oxygen Species (ROS) Detection

Levels of reactive oxygen species (ROS) were measured using the fluorescent probe MitoSOX™ Red according to the manufacture's protocol (Thermo Fisher Scientific). Briefly, theMO3.13 cells were cultured in a 6 cm dish and treated without or with psychosine/Chloroquine/MG132 in a humidified atmosphere of 5% (*v*/*v*) CO_2 at 37 °C. Cells were harvested by trypsinization, incubated with MitoSOX™ Red for 10 min at 37 °C in the dark, and centrifuged. Cells were resuspended in pre-warmed Hank's Balanced Salt Solution (HBSS), centrifuged, washed with pre-warmed HBSS, centrifuged again, and resuspended in pre-warmed HBSS. Cells were then analyzed on a flow cytometer for ROS detection, using the 510 nm laser for excitation and detected at 580 nm.

2.8. Mitochondrial Respiration

Bioenergetic profiles were determined using theXF24 extracellular flux analyzer (Seahorse Biosciences, Lexington, MA, USA). MO3.13 cells were seeded on XF24 cell culture microplates and maintained in a humidified atmosphere of 5% (*v*/*v*) CO_2 at 37 °C. The MO3.13 cells were treated without or with psychosine/ Chloroquine/MG132 as previously described. Before assay, MO3.13 cells were equilibrated in unbuffered DMEM medium (Gibco) supplemented with 25 mM glucose, 1 mM sodium pyruvate, and 2 mM L-Glutamine and transferred to a non-CO2 incubator for 1 h before measurement. Mitochondrial oxygen consumption rate (OCR) was measured in real-time for 20 min. Each plotted value of real-time assessment of mitochondrial respiration was normalized with cell number and results were presented as mean ± S.D.

2.9. Cell Viability Assay

The MO3.13 cells were seeded on a 96-well plate and treated as previously described in a humidified atmosphere of 5% (*v*/*v*) CO_2 at 37 °C. Cell numbers were quantified by the CyQUANT cell proliferation assay kit according to the manufacture's protocol (Thermo Fisher Scientific). Briefly, cells were incubated with CyQUANT dye at 37 °C for 60 min and fluorescence intensity was measured on a plate reader at OD = 530nM.

2.10. Statistical Analysis

All data were obtained from at least three independent experiments and results were expressed as the mean ± S.D., unless stated otherwise. Student's *t*-test was used for two-group comparison and ANOVA test was used for multiple comparisons; *p*-values less than 0.05 were considered significant.

3. Results

3.1. Cytoplasmic Aggregates Containing Ubiquitin and p62 in the Twitcher Brain

Dysfunction of the UPS and autophagy impairs protein homeostasis leading to accumulation of toxic and misfolded proteins and ubiquitin-positive aggregates in neurons, which is a pathologic hallmark of many neurodegenerative diseases [35–38]. To investigate whether dysfunction of both UPS and autophagy contribute to the pathogenesis of GLD, we first examined the expression patterns of ubiquitin and p62, molecular indicators of UPS and autophagy, respectively, in the brain of twitcher mice. Antibodies directed against ubiquitin and p62 were used in the analysis of immunohistochemistry. For the first time, spatiotemporal accumulation and distribution of ubiquitin- and p62-aggregates during the progression of the disease were documented in the twitcher brain (Figures 1 and 2). At postnatal day 21, ubiquitin aggregates appeared sparsely in the spinal cord, brain stem, and cerebellum arbor vita, and aggregates were observed scarcely in the corpus callosum (Figure 1). At postnatal day 28, the accumulation of ubiquitin aggregates was more evident in the brain stem, spinal cord, and cerebellum arbor vitae. Ubiquitin aggregates were detected moderately in the corpus callosum, septum, thalamus, and hypothalamus, and scarcely in the cortical layer VI along the corpus callosum and in granular and molecular layers of the cerebellum (Figure 1). At postnatal day 35, variable puncta of ubiquitin aggregates distributed diffusely in the spinal cord, brain stem, cerebellum arbor vitae, and corpus callosum. The thalamus, hypothalamus, and septum demonstrated more evident distribution of ubiquitin aggregates (Figure 1). At postnatal day 42, ubiquitin aggregates were abundant in the spinal cord, brain stem, cerebellum arbor vitae, cerebellar peduncle, fornix, internal capsule, and corpus callosum. The cortex and granular and molecular layers of the cerebellum showed moderate ubiquitin aggregates. Ubiquitin aggregates were observed scarcely in the hippocampus (Figure 1).

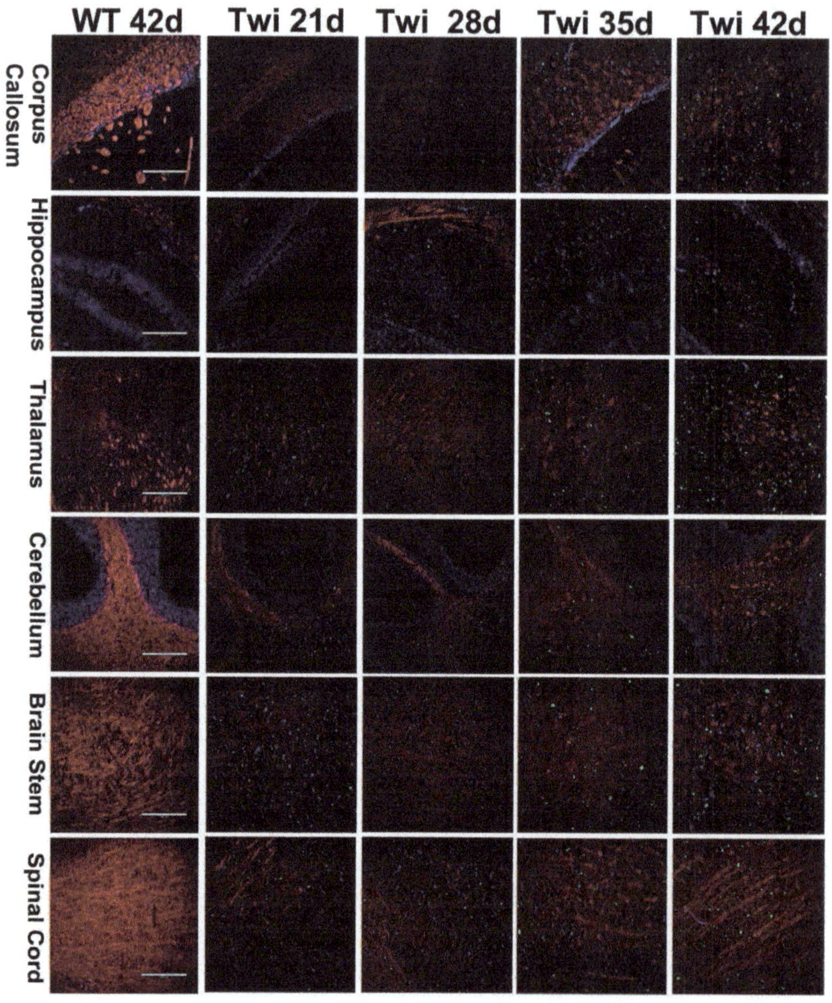

Figure 1. Spatial and temporal distribution of ubiquitin aggregates in the brain of twitcher mice. Distribution of ubiquitin aggregates in different regions of the twitcher (Twi) brain at ages 21 to 42 days. The twitcher brain and wild-type (WT) brain were immuno-stained with anti-ubiquitin (in green) and anti-PLP (in red), respectively, and nuclei were counterstained with DAPI (in blue). Scale bars: 200 µm.

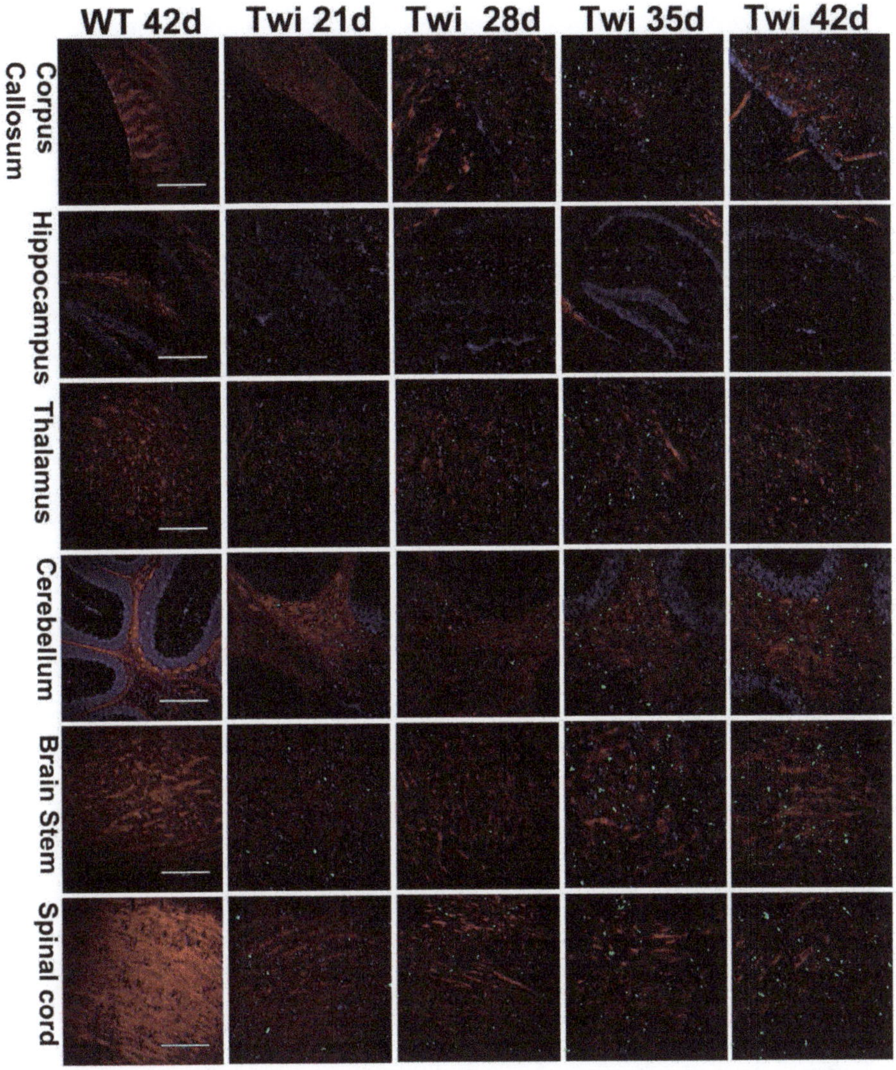

Figure 2. Progressive accumulation of p62 aggregates in the brain of twitcher mice. Distribution of p62 aggregates in different regions of the twitcher (Twi) brain at ages 21 to 42 days. The twitcher brain and wild-type (WT) brain were immuno-stained with anti-p62 (in green) and anti-PLP (in red), respectively, and nuclei were counterstained with DAPI (in blue). Scale bars: 200 µm.

Previous studies have observed the presence of p62 aggregates at postnatal day 20 and day 30 in the brain of twitcher mice, and suggested the dysregulation of autophagy as a molecular pathogenesis of GLD [29]. However, the anatomical and temporal distribution of p62 aggregates in the twitcher brain has not been elucidated. Using immunohistochemistry staining of the twitcher brain at different postnatal days, we were able to determine the regional and temporal distribution of p62 aggregates (Figure 2). Intriguingly, the distribution of p62 aggregates was in line with that of ubiquitin aggregates in the twitcher brain. P62 aggregates were detected sparsely in the brain stem, spinal cord, and cerebellum arbor vitae at postnatal day 21, then accumulated progressively and were abundant in these regions as well as in the fornix, internal capsule, and cerebellar peduncle at postnatal day 42.

While in regions of the thalamus, hypothalamus, and septum, p62 aggregates were sparse at postnatal day 28 and were more evident at postnatal day 42. Consistent with the aggregation of ubiquitin in twitcher brains, p62 aggregates were rare in the cerebral cortex and granular and molecular layers of cerebellum at postnatal day 28, and were moderate at postnatal day 42. While the hippocampus showed scarce p62 aggregates at postnatal day 42. Overall, the spatiotemporal distribution of p62 aggregates was identical to the progressive aggregation of ubiquitin in twitcher brains.

3.2. p62 Colocalized with Ubiquitin in Aggregates

Ubiquitin colocalized with p62 in cytoplasmic inclusions has often been documented in several neurodegenerative disease pathologies [43]. Thus, to further investigate whether the ubiquitin colocalized with p62 in cytoplasmic aggregates as observed in our study, the sections of twitcher brain were double immunolabeled for p62 and ubiquitin with specific antibodies, respectively (Figure 3). Intriguingly, our results revealed that ubiquitin-positive signals were almost exclusively colocalized with p62 in cytoplasmic aggregates at all time points from postnatal day 21 to day 42. Further, p62- and ubiquitin-aggregates accumulated progressively and widely distributed in the brain of twitcher mice in a caudal-to-rostral axis as the disease progressed in line with the deterioration of demyelination and phenotype, suggesting the correlation between disturbed proteostasis and pathogenesis in GLD. However, p62- and ubiquitin-aggregates were not colocalized with lysosome marker LAMP1 (Figure 3), suggesting the deposition of cytoplasmic aggregates as a result of impaired proteostasis.

3.3. Colocalization of Aggregates in Degenerative Neurons and White Matter

To investigate the localization of ubiquitin and p62 aggregates, neuronal cells and brain slices from wild-type and twitcher mice were double immunolabeled with anti-ubiquitin or anti-p62 together with antibodies specific for different brain cell types to visualize astrocytes, microglia, oligodendrocytes, neurons, and Purkinje cells (Figure 3). A previous study by Del Grosso and colleagues revealed the colocalization of p62 aggregates in neurons at the early symptom stage (postnatal day 20) of twitcher mice, albeit the extent and affected regions were not indicated [29]. In the present study, we further documented that ubiquitin- and p62-aggregates are seen conjointly with the neurons in the brain stem, spinal cord, and granular layer of cerebellum and were sparsely found at postnatal day 28 and moderately at day 42 (Figure 3). Moreover, neurons with either ubiquitin or p62 aggregates showed attenuated signals of both nuclei and neuronal protein marker NeuN, indicating the progression of neuronal degeneration.

However, ubiquitin- and p62-aggregates were not detected in either astrocytes or microglia (Figure 3). Further, in the present study, Purkinje cells were devoid of cytoplasmic aggregates (Figure 3), and this supports our previous study showing atrophy of Purkinje cells and neural degeneration in the twitcher brain [11]. It has been assumed that the degeneration of granule cells leads to lower secretion of neurotrophic factors, which are required for modulating the dendritic differentiation of Purkinje cells and cerebellar plasticity, and the absence of which would lead to impaired dendritogenesis, synaptogenesis of Purkinje cells, and subsequently cell death [11,44,45].

Intriguingly, ubiquitin- and p62-aggregates were almost exclusively localized in the white matter of the nervous system (Figures 1–3), including the corpus callosum, fornix, internal capsule, cerebellum arbor vitae, cerebellar peduncle, fiber plexus, bundles in the brain stem and spinal cord, as well as fiber arrays in the cortex, midbrain, thalamus, and hypothalamus. More specifically, small puncta of aggregates were localized in the perinuclear region of oligodendrocytes (Figure 3). While large pleomorphic aggregates were embedded among corrupted fiber bundles of white matter, they were more discernible as demyelination progressed (Figure 3). The progressive accumulation of ubiquitin- and p62-aggregates in the white matter is paralleled by the progression of demyelination and deterioration of phenotype, implicating cytotoxicity of inclusions upon oligodendrocyte viability contributes to the progressive demyelination in GLD.

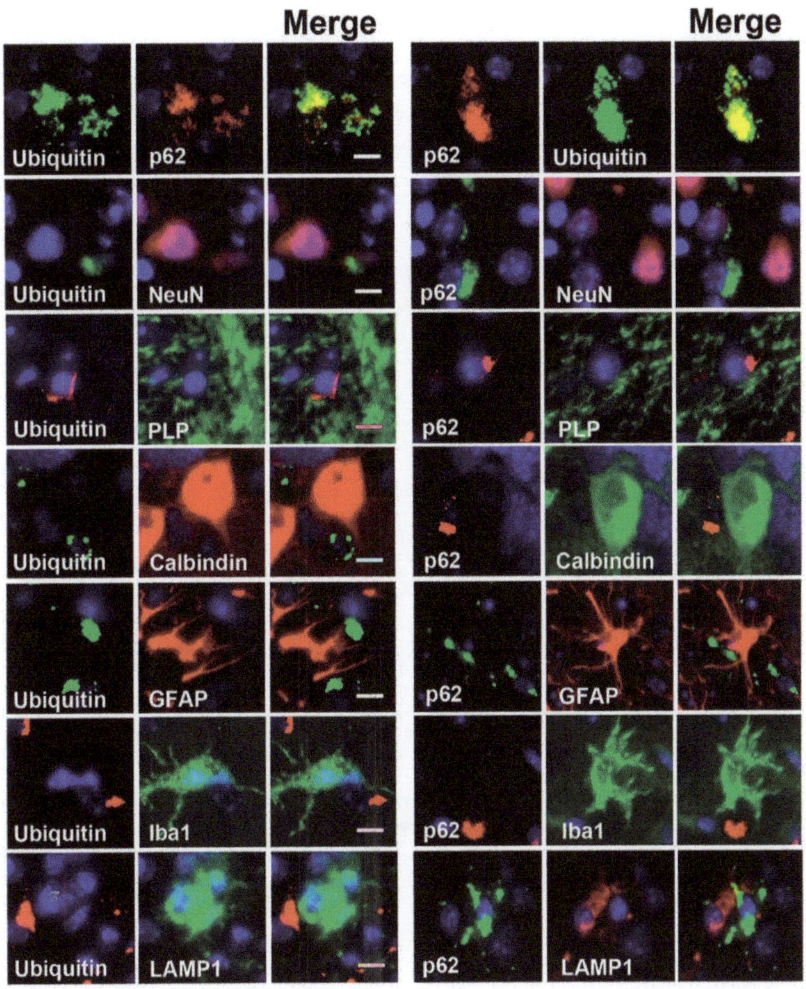

Figure 3. Accumulation of ubiquitin/p62 aggregates in oligodendrocytes and neurons. Sections of twitcher brain were double labeled with anti-ubiquitin/anti-p62 and anti-NeuN (neuron)/anti-PLP (oligodendrocytes)/anti-calbindin (Purkinje cells)/anti-GFAP (astrocyte)/anti-Iba1 (microglia)/anti-LAMP1 (lysosome), respectively, and nuclei were counter-stained with DAPI (blue). Scale bars: 10 μm.

3.4. Dysfunction of Autophagy and the Proteasome-Ubiquitin System in the Twitcher Brain

To biochemically investigate the molecular markers of autophagy and UPS, we extracted detergent soluble and insoluble proteins from the brain stem of the twitcher mice and wild-type (WT) mice at postnatal days 21, 28, 35, and 42, and quantified the expression of LC3, p62, and ubiquitin after Western blot analysis (Figures 4a and 5a).

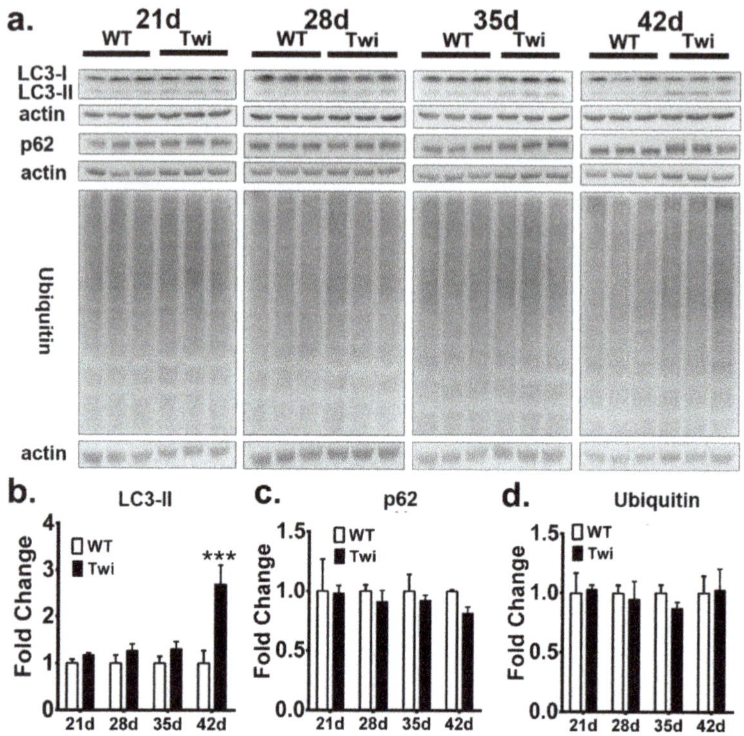

Figure 4. Increased autophagic flux in twitcher brain. **a** Western blot analysis of the total LC3-II, p62, and ubiquitin content in brain stems from wild-type (WT) and twitcher (Twi) mice at ages 21 days, 28 days, 35 days, and 42 days. Soluble fraction was analyzed and β-actin (actin) was used for the loading control. **b**, **c**, **d** Quantification of soluble LC3-II (**b**), p62 (**c**), and ubiquitin (**d**). Values are expressed as mean ± S.D. ($n = 3$). Statistical significance $^{***}p < 0.001$.

The quantitative analysis of molecular markers of autophagy in fractions of soluble proteins showed that LC3 II was significantly increased in the twitcher brain at postnatal day 42 when compared to that of the WT brain (Figure 4b), while the levels of p62 in the twitcher brain at all time point were compatible with that of the WT brain (Figure 4c), indicating the activation of autophagy. Similarly, the level of ubiquitin between WT and twitcher brains was identical at all time points (Figure 4d). Intriguingly, the quantitative analysis of the insoluble protein fractions revealed a prominent accumulation of LC3 II, p62, and ubiquitin in twitcher brains compared to that of WT brains, suggesting dysfunction of autophagy and UPS (Figure 5b–d). Levels of insoluble LC3 II, p62, and ubiquitin were significantly increased in the twitcher brain at postnatal day 35 and day 42 (Figure 5b–d), when the twitcher mice showed a rapid deterioration of clinical phenotypes. Further, the binding of insoluble p62 and ubiquitin was confirmed by the immunoprecipitation analysis (Figure 5e). Taken together with our immunofluorescent analysis, it is concluded that ubiquitin and p62 inclusions are insoluble, cytosolic protein aggregates. The accumulation of insoluble ubiquitin proteins and cytoplasmic deposition of ubiquitin-positive aggregates suggests accumulation of oligomerized protein aggregates and indicates defects of UPS commonly associated with many neurodegenerative proteinopathies. While the accumulation of LC3 II and p62 in insoluble protein fractions together with cytoplasmic p62 aggregates indicates interruption of autophagic flux, the accumulation of insoluble ubiquitin and broad deposition of ubiquitin aggregates implicates the dysfunction of UPS.

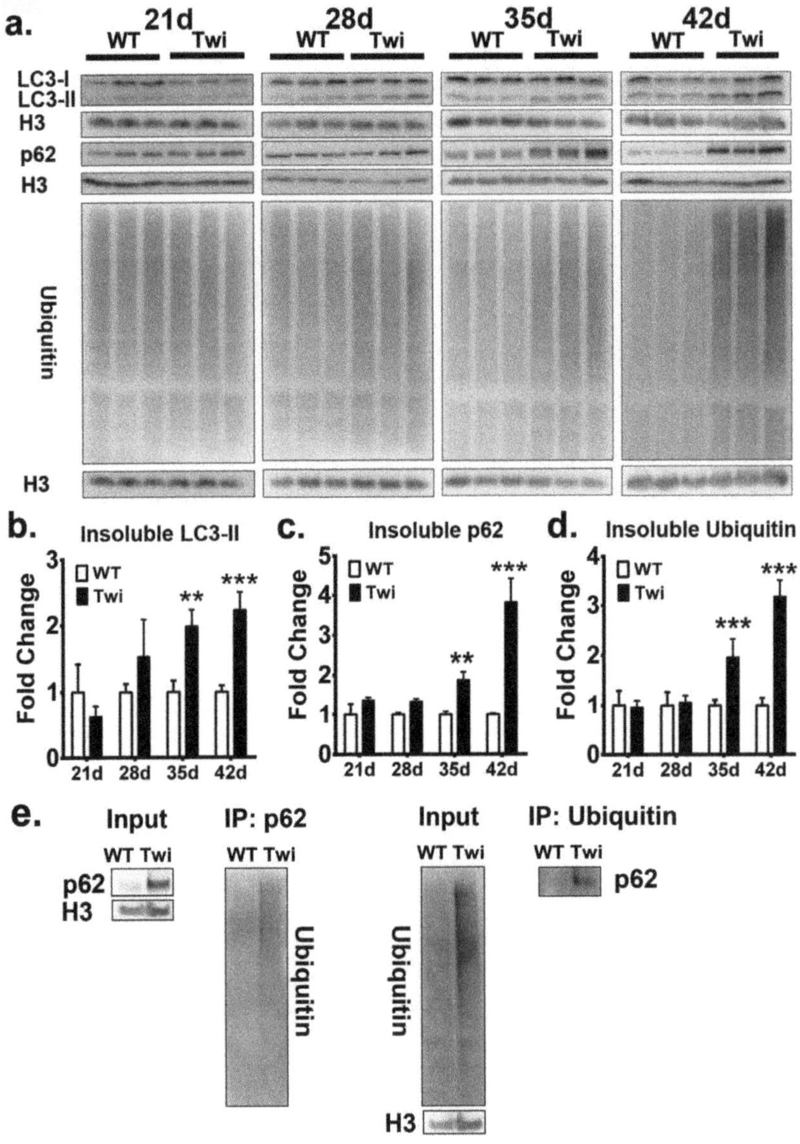

Figure 5. Defective autophagy and ubiquitin-proteasome system in twitcher brains. **a:** Western blot analysis of the total LC3-II, p62, and ubiquitin content in brain stems from wild-type (WT) and twitcher (Twi) mice at ages 21 days, 28 days, 35 days, and 42 days. Insoluble fraction was analyzed and histon-H3 (H3) was used for the loading control. b, c, d Quantification of insoluble LC3-II (**b**), p62, (**c**) and ubiquitin (**d**). Interaction between insoluble p62 and ubiquitin was verified by immunoprecipitation analysis (**e**). Values are expressed as mean ± S.D. ($n = 3$). Statistical significance, $^{**}p < 0.01$, $^{***}p < 0.001$.

Additionally, the expressions of GFAP and MBP proteins, molecular markers of astrocytes and myelin, respectively, were also determined in both twitcher and WT mice (Figure 6). In concordance with the progression of neuroinflammation and demyelination in the pathogenesis of GLD, an increasing level of GFAP and decreasing expression of MBP were demonstrated in the twitcher brain when compared to those of a WT brain. Accordingly, accumulation of insoluble p62 and insoluble ubiquitin concomitant with the formation of p62- and ubiquitin-aggregates in the twitcher brain were in line with the progression of neuroinflammation, demyelination, and the deterioration of the disease.

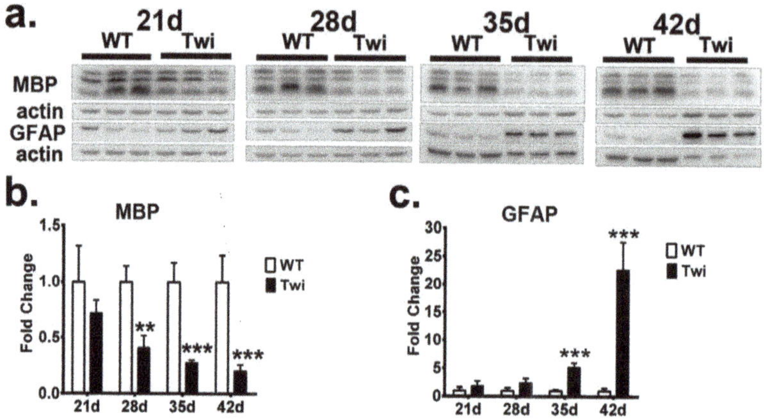

Figure 6. Progressive astrocytosis and demyelination in twitcher brain. (**a**) Western blot analysis of the total MBP and GFAP content in brain stems from wild-type (WT) and twitcher (Twi) mice at ages 21 days, 28 days, 35 days, and 42 days of age. **b**, **c** Quantification of MBP (**b**) and GFAP (**c**). β-actin (actin) was used for the loading control. Values are expressed as mean ± S.D. ($n = 3$). Statistical significance, ** $p< 0.01$, *** $p < 0.001$.

3.5. Psychosine-Induced Autophagy and Cell Death in Time-and Dose-Dependent Toxicity

To elucidate the impact of psychosine upon autophagy and UPS, the human oligodendrocyte cell line MO3.13 treated with psychosine was used as a cellular model to recapitulate the pathomechanism of GLD. In contrast to a previous study showing synergic effects of psychosine and starvation on the activation of autophagy in MO3.13 cells [30], our study used full medium to investigate the sole impact of psychosine on autophagic flux without the starvation-induced autophagy. MO3.13 cells were cultured in the full medium for 24 h in the presence of different concentrations of psychosine, and the expression levels of autophagic markers LC3 and p62 proteins were determined (Figure 7a). The Western blot analysis demonstrated that LC3 II protein expression was elevated progressively in the presence of increasing concentrations of psychosine compared with that of the untreated control (Figure 7b).In the presence of psychosine, the levels of LC3 II increased significantly to 1.8-, 4.8-, 7-, and 8.2-fold of those of the control at concentrations of 5, 10, 15, and 20 μM, respectively, and remarkably, to 21.2-fold of that of the control at concentration of30μM, suggesting the activation of autophagy in line with the concentration of psychosine. Consistent with the expression of LC3 II proteins, the levels of p62 also increased progressively in the presence of psychosine at concentrations of 5–15 μM, though these increases were not significant statistically, and accumulated significantly at concentrations of 20μM and 30 μM (Figure 7c), suggesting an impairment of autophagy upon degradation of autophagic substrates at high doses. This observation validates a dose-dependent cytotoxicity of psychosine upon the activation and function of autophagy.

Further, the cytotoxicity of psychosine upon cell viability was measured in MO3.13 cells treated with different concentrations of psychosine for 24 h (Figure 7d). At 5 and 10μM, psychosine did not cause cell death after the incubation for 24 h, while psychosine at 15 μM for 24 h caused 85% cell survival, albeit not significantly, and psychosine at 20 and 30 μM caused a significant decrease of cell viability to 74.2% and 56.7% cell survival at 24 h, respectively. Thus, psychosine at high concentrations (30 μM) was lethal to oligodendrocytes.

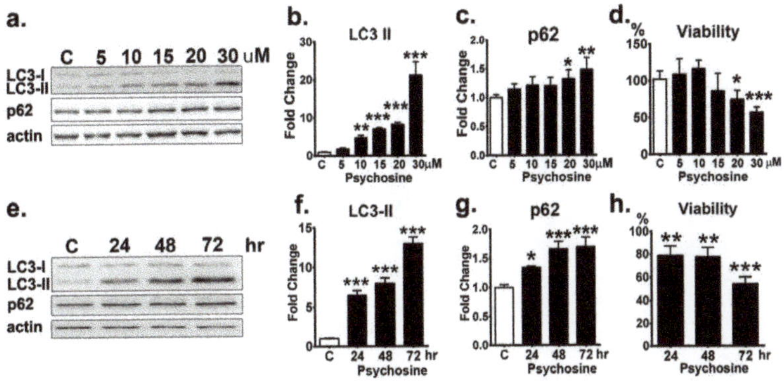

Figure 7. Dose- and time-dependent cytotoxicity of psychosine upon MO3.13 cells. (**a**) Western blot analysis of total LC3-II and p62 in MO313 cells treated with different concentration of psychosine (psy) for 24 h. β-actin (actin) was used for the loading control. (**b**,**c**,**d**) Quantification of total LC3-II (**b**) and p62 (**c**) and cell viability (**d**). e Western blot analysis of total LC3-II and p62 in MO313 cells treated with 20 μM psychosine for 72 h. (**f**,**g**,**h**) Quantification of total LC3-II (f) and p62 (**g**) and cell viability (**h**). Values are expressed as mean ± S.D. (Western blot, $n = 3$; Viability, $n = 6$). Statistical significance $*p < 0.05$, $**p < 0.01$, $***p < 0.001$. MO313 cells without treatment are labeled as a control group (C).

Finally, the time-dependent impact of psychosine at 20 μM for 72 h upon autophagy of oligodendrocytes was determined (Figure 7e). Western blot analysis demonstrated that LC3-II protein expression was increased significantly and progressively from 24 h to 72 h with exposure to psychosine (Figure 7f). In line with the increase of LC3-II protein expression, levels of p62 protein were also increased significantly and gradually; levels were indistinguishable between 48 h and 72 h (Figure 7g). Of note, psychosine at 20 μM caused a prominent and significant decrease of cell viability to 54.3% survival under the incubation for 72 h, indicating time-dependent cytotoxicity. (Figure 7h). Collectively, our results indicate a dose- and time-dependent cytotoxicity of psychosine upon the capacity of autophagy and cell viability of oligodendrocytes.

3.6. Autophagy Inhibition Exacerbates the Accumulation of p62 Aggregates

To elucidate further the role of impaired autophagy in the pathogenesis of GLD, autophagic flux and UPS were measured in the presence and absence of an autophagy inhibitor chloroquine in MO3.13 cells incubated with 20μM psychosine for 72 h (Figure 8a). Protein expression and solubility of LC3-II, p62, and ubiquitin were analyzed to measure the autophagy flux and UPS. As expected, chloroquine significantly increased LC3-II turnover and levels of LC3-II in both MO3.13 cells treated with and without psychosine (Figure 8b), where the increase of LC3-II expression and turnover was prominent in psychosine-treated cells with concomitant incubation of chloroquine (Figure 8b). Of note, inhibition of autophagy induced accumulation of insoluble LC3-II in MO3.13 control cells and further exacerbated the level of insoluble LC3-IIin psychosine-treated cells (Figure 8e). The level of p62 protein was also increased significantly after the treatment of chloroquine in control cells, while chloroquine did not further increase the already elevated levels of p62 protein expression in psychosine-treated cells (Figure 8c). However, inhibition of the autophagic flux by chloroquine significantly increased the insoluble p62 protein accumulation, which was more remarkable in MO3.13 cells treated with psychosine, while the level of insoluble p62 protein was nearly undetectable in untreated control cells (Figure 8f).

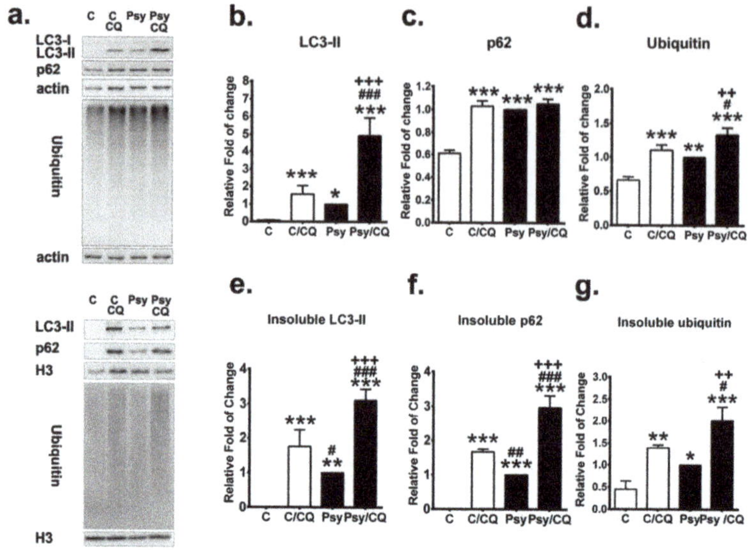

Figure 8. Autophagy inhibition increases accumulation of insoluble p62 and ubiquitin. (**a**) Western blot analysis of the total LC3-II, p62, and ubiquitin in MO3.13 cells treated without (control: C)/with psychosine (Psy) for 72 h concomitant with/without chloroquine (CQ) for the last 24 h. Soluble and insoluble fractions were analyzed. β-actin (actin) and histon-H3 (H3) were used for the loading control in soluble and insoluble fractions, respectively. (**b**–**d**) Quantification of soluble LC3-II (**b**), p62 (**c**), and ubiquitin (**d**). (**e**–**g**) Quantification of insoluble LC3-II (**e**), p62 (**f**), and ubiquitin (**g**). Values are expressed as mean ± S.D. (n = 3). Statistical significance * $p < 0.05$, ** $p < 0.01$, *** $p < 0.001$ in comparison to control (C). C/CQ vs. Psy, or Psy/CQ, # $p < 0.05$, ## $p < 0.01$, ### $p < 0.001$. Psy vs. Psy/CQ, ++ $p < 0.01$, +++ $p < 0.001$.

The level of ubiquitin was a useful parameter for measuring of the UPS process and was determined in cells treated with and without chloroquine in the presence and absence of psychosine. The significant increase of ubiquitin expression induced by psychosine was further augmented significantly after the addition of chloroquine, which also increased the expression of ubiquitin significantly in control cells (Figure 8d). Intriguingly, levels of insoluble ubiquitin were increased significantly in cells treated with psychosine, indicating a dysfunction of the UPS process, and were further accumulated significantly after the concomitant addition of chloroquine (Figure 8g). Taken together, these findings indicate that inhibition of autophagic flux further exacerbates the accumulation of LC3-II, p62, and ubiquitin in insoluble protein fractions mediated by psychosine cytotoxicity.

During the process of autophagy, LC3-II and p62 are incorporated into the autophagosomal membrane, which fuses with lysosomes to form autolysosomes, degrading the misfolded and/or damaged proteins [34]. Therefore, an increase in LC3-IIexpression alone does not provide a complete picture of autophagy in the cells, while observations and measurement of the concurrent formation of the autolysosome depict the autophagic flux [46]. To quantify the autophagy induced by psychosine in cells, the formation of autolysosomes was determined by staining MO3.13 cells with both Cyto-ID and LysoTracker Red for autophagosomes and lysosomes, respectively. It has to be noted that the use of fluorescent probes Cyto-ID and LysoTracker Red have been broadly applied to visualize the autophagy dynamics in live cells in many studies [47,48]. Recent studies further indicated that Cyto-ID labels autophagic compartments with minimal and negligible staining of lysosomes [47], while the LysoTracker Red labels the lysosomes specifically [49]. There is compelling evidence validating the identification of autophagosomes by CytoID, lysosomes by LysoTracker Red, and autolysosomes by colocalization of CytoID and LysoTracker Red [48,50,51]. The process of autophagic flux was thus evaluated by direct imaging and quantification of autolysosomes in cells (Figure 9a,b). The puncta of autolysosomes in cells were significantly increased from 1.8-fold to 2.4-fold compared to that of the untreated control after addition of chloroquine in psychosine-treated cells, and a 2-fold increase compared to that of the control cells treated with chloroquine (Figure 9b). Images of autolysosomes in cells showed numerous enlarged puncta accumulated in the cytoplasm after the addition of chloroquine, especially in those cells with concomitant incubation of psychosine (Figure 9a).

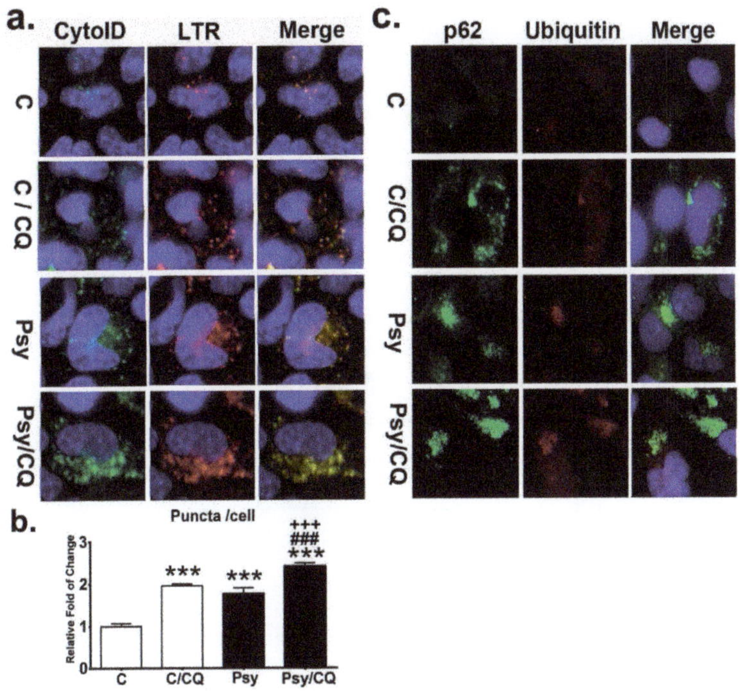

Figure 9. Accumulation of autolysosomes and p62 aggregates by impaired autophagic flux. (**a**) Autolysosome was determined by the colocalization of the Cyto-ID fluorescence and LysoTracker Red (LTR) in MO3.13 cells treated with/without psychosine (Psy) concomitant with/without chloroquine (CQ) and counter-stained with Hoechst. (**b**) Quantification of autolysosome puncta in cells. (**c**) Cytoplasmic deposition of p62 and ubiquitin by chloroquine. Cells were double labeled with anti-p62 and anti-ubiquitin and counterstained with DAPI. Values are expressed as mean ± S.D. ($n = 3$). Statistical significance *** $p< 0.001$ in comparison to control (C). C/CQ vs. Psy, or Psy/CQ, $^{\#\#\#}p < 0.001$. Psy vs. Psy/CQ, $^{+++}p < 0.001$.

To evaluate whether impairment of autophagy leads to aggregation of p62 and ubiquitin as demonstrated in the twitcher brain, immunefluorescent staining for both p62 and ubiquitin was performed in MO3.13 cells with and without treatment of chloroquine in the presence and absence of psychosine (Figure 9c). It was shown that psychosine induced cytoplasmic aggregates with p62, which was more prominent with concomitant treatment with chloroquine. The control cells showed a few small puncta of p62, which was evident after the addition of chloroquine, in accordance with the inhibition of autophagy. Furthermore, small puncta of ubiquitin-positive aggregates were observed in psychosine treated MO3.13 cells and were increased moderately after the addition of chloroquine. Whereas small ubiquitin-positive puncta were scarcely observed in control cells after treatment with chloroquine.

3.7. Inhibition of Both Proteasome and Autophagy Augments Aggregation with Ubiquitin and p62

The impact of UPS dysfunction upon induction of p62- and ubiquitin-positive aggregates as observed in the twitcher brain was further investigated in the MO3.13 cellular model, which was exposed to the proteasome inhibitor MG132 at a subtoxic concentration (Figure 10a). It has been shown that MG132potently inhibits the chymotrypsin-like activity of the proteasome, leading to exacerbation of the burden of misfolded and damaged proteins and concomitant cellular stress [52]. At a subtoxic concentration, MG132 did not increase the levels of LC3-II and p62 protein expression in the soluble fraction of control cells but increased the level of insoluble p62 protein (Figure 10b, 10c, 10e, 10f). While in the psychosine-treated cells, MG132 significantly increased the expression of LC3-II and p62 proteins, which were further accumulated significantly in the insoluble phase (Figure 10b,c,e,f). Consistently, the level of ubiquitin in control cells, in either the soluble or insoluble phase, was not altered by a subtoxic concentration of MG132 (Figure 10d,g). Of note, MG132 increased the accumulation of insoluble ubiquitin significantly from an initial 2-fold of control to 4-fold of control in psychosine-treated cells (Figure 10g).

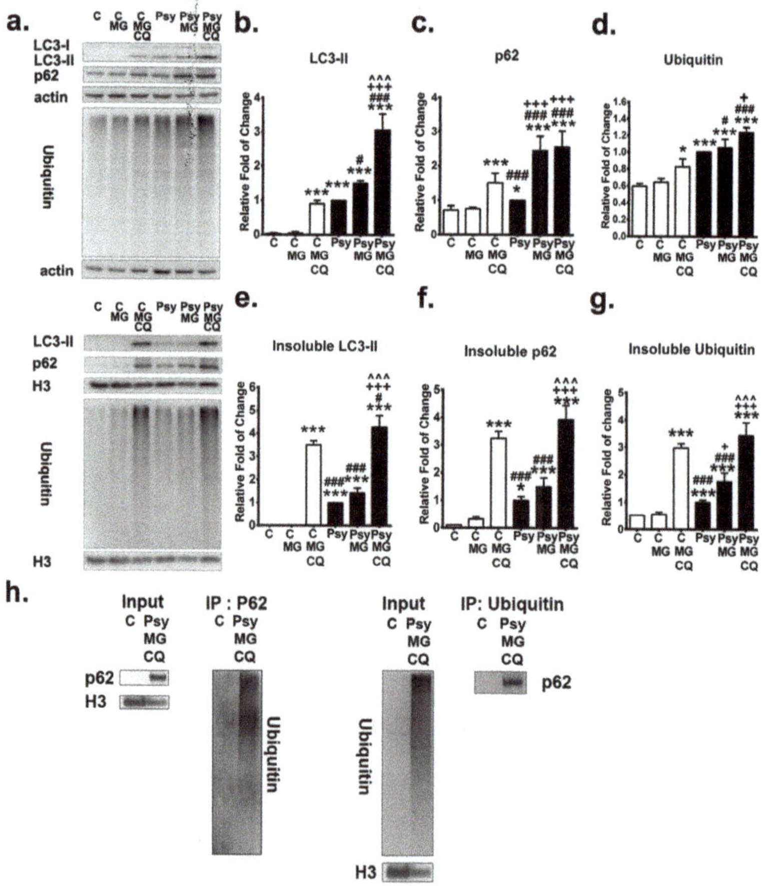

Figure 10. Inhibition of autophagy and proteasome exacerbates accumulation of p62 and ubiquitin. (**a**) Western blot analysis of the total LC3-II, p62, and ubiquitin in MO3.13 cells treated without (control: C)/with psychosine (Psy) for 72 h concomitant with/without MG132 (MG) with/without chloroquine (CQ) for the last 24 h. Soluble and insoluble fractions were analyzed. β-actin (actin) and histon-H3 (H3) were used for the loading control in soluble and insoluble fractions, respectively. (**b**–**d**) Quantification

of soluble LC3-II (**b**), p62 (c)and ubiquitin (**d**). (**e–g**) Quantification of insoluble LC3-II (**e**), p62 (**f**) and ubiquitin (**g**). (h) Interaction between insoluble p62 and ubiquitin was verified by immunoprecipitation analysis (**h**).Values were expressed as mean ± S.D. ($n = 3$). Statistical significance $^{*}p< 0.05$, $^{***}p < 0.001$ in comparison to control (C) or C/MG. C/MG/CQ vs. Psy, or Psy/MG, or Psy/MG/CQ, $^{\#}p< 0.05$, $^{\#\#\#}p < 0.001$. Psy vs. Psy/MG, or Psy/MG/CQ, $^{+}p < 0.05$, $^{++}p < 0.01$, $^{+++}p< 0.001$. Psy/MG vs. Psy/MG/CQ, $^{\wedge\wedge\wedge}p < 0.001$. No statistical significance between C and C/MG.

Moreover, the effect of inhibition of both autophagy and proteasomes by concomitant treatment of chloroquine and MG132 was determined (Figure 10a). Levels of LC3-II, p62, and ubiquitin proteins were significantly increased in both the soluble and insoluble phase in control cells under the simultaneous inhibition of autophagy and proteasomes (Figure 10b,g). In psychosine-treated cells, the concomitant addition of both chloroquine and MG132 induced a significant increase of LC3-II in both the soluble and insoluble phase and a robust accumulation of insoluble p62 and insoluble ubiquitin proteins (Figure 10b,g). Further, the binding of insoluble p62 and ubiquitin was confirmed by the immunoprecipitation analysis (Figure 10h).

Direct observation and quantification of autolysosomes in cells showed that MG132 alone did not further increase the accumulation of autolysosomes in psychosine-treated cells, while concomitant treatment with chloroquine and MG132 increased the accumulation of autolysosomes significantly in both control and psychosine-treated cells (Figure 11a,b). Furthermore, with immunofluorescent labeling of p62 and ubiquitin, it was demonstrated that the signal of ubiquitin-positive puncta became intense after the addition of MG132 in both control and psychosine-treated cells (Figure 11c). Of note, large pleomorphic aggregates containing ubiquitin and p62 proteins were observed in psychosine-treated cells with concomitant inhibition of autophagy and UPS, recapitulating the pathological hallmarks observed in the twitcher brain in this study.

3.8. Inhibition of Both Proteasome and Autophagy Leads to Accumulation of ROS, Reduction of Mitochondrial Respiration, and Decreased Viability

The consequence of the inhibition of proteasomes and autophagy was further explored by identifying the levels of reactive oxygen species (ROS), a function of mitochondrial respiration and cell viability. Our results showed that inhibition of autophagy and UPS induced a prominent increase of ROS levels in psychosine-treated MO3.13 cells in comparison to that of control cells (Figure 12a). Furthermore, the assay of mitochondrial respiration by direct analysis of oxygen consumption rate demonstrated a significant reduction of respiration in MO3.13 cells treated with psychosine concomitant with inhibition of autophagy and UPS (Figure 12b). Accordingly, the cell viability of psychosine treated MO3.13 cells with inhibition of autophagy and UPS was remarkably decreased to 25.4% of that of control cells (Figure 12c).

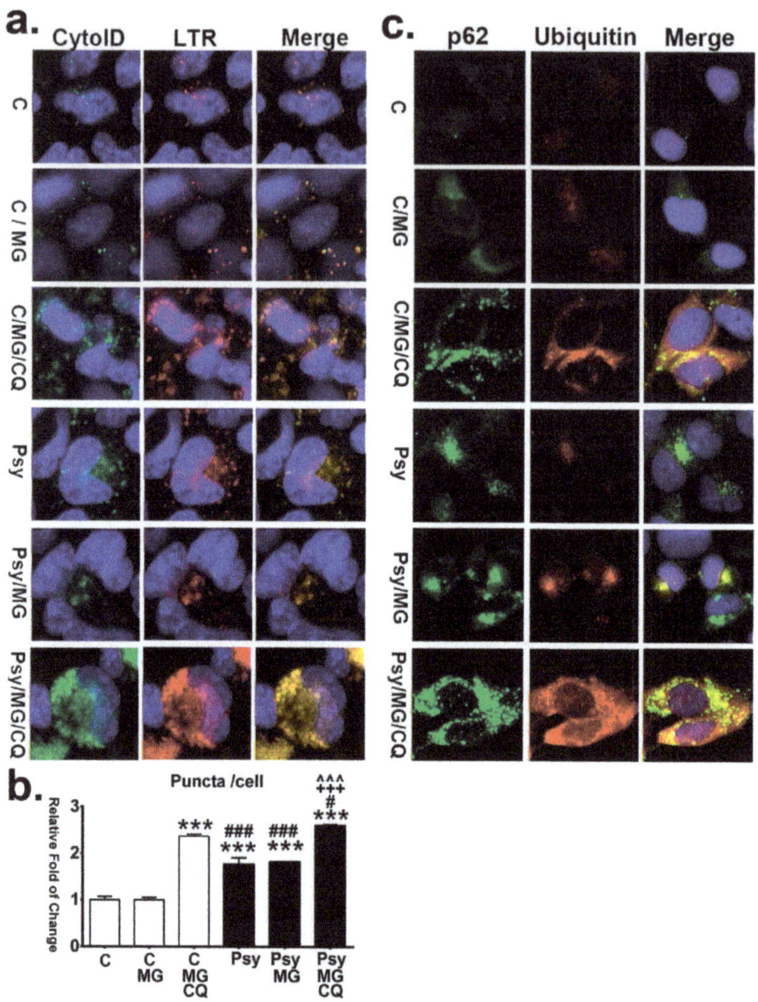

Figure 11. Impaired autophagic flux and proteasomes enhance aggressive accumulation of autolysosomes and aggregation of p62 and ubiquitin. (**a**) Autolysosome was determined by the colocalization of the Cyto-ID fluorescence and LysoTracker Red (LTR) in MO3.13 cells treated with/without psychosine (Psy) concomitant with/without MG132 (MG) with/without chloroquine (CQ), and counter-stained with Hoechst. (**b**) Quantification of autolysosome puncta in cells. (**c**) Increased p62- and ubiquitin-aggregation. Cells were double labeled with anti-p62 and anti-ubiquitin, and counterstained with DAPI. Values are expressed as mean ± S.D. ($n = 3$). Statistical significance $^{***}p<$ 0.001 in comparison to control (C) or C/MG. C/MG/CQ vs. Psy, or Psy/MG, or Psy/MG/CQ, $^{\#}p < 0.05$, $^{\#\#\#}p < 0.001$. Psy vs. Psy/MG, or Psy/MG/CQ, $^{+++}p < 0.001$. Psy/MG vs. Psy/MG/CQ, $^{\wedge\wedge\wedge}p < 0.001$. No statistical significance between C and C/MG.

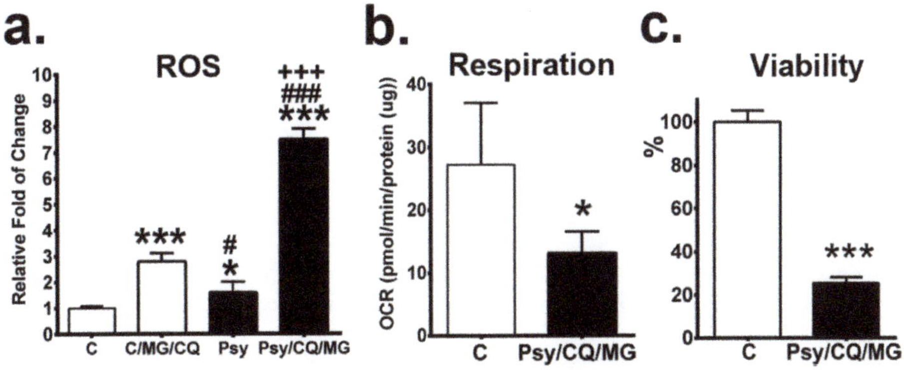

Figure 12. Inhibition of both proteasome and autophagy promotes cell death. MO3.13 cells treated without/with psychosine (Psy) for 72 hs concomitant with MG132 (MG) and chloroquine (CQ) in the last 24 h. (**a–c**) Levels of ROS (**a**), basal respiration (**b**), and cell viability (**c**) were quantified. Values areexpressed as mean ± S.D. (ROS, $n = 3$; Respiration, $n = 5$; Viability, $n = 6$). Statistical significance *$p < 0.05$, ***$p < 0.001$ in comparison to control (**C**). C/MG/CQ vs. Psy, or Psy/MG/CQ, # $p < 0.05$, ### $p < 0.001$. Psy vs. Psy/MG/CQ, +++ $p < 0.001$.

4. Discussion

Cytotoxicity of psychosine has been implicated in the pathological hallmarks of GLD, inducing demyelination, astrocytosis, and formation of multinuclear globoid-like cells in the central nervous system (CNS) [25,27,53–55]. Progressive accumulation of psychosine upregulates the production of inflammatory cytokines and generation of lysophosphatidylcholine and arachidonic acid resulting in apoptosis and cell death in oligodendrocytes [28,53,54,56]. However, the underlying mechanism of impact of psychosine on oligodendrocytes contributing to demyelination of GLD still remains largely elusive. In this study, we provide the first evidence, in vivo and in vitro, that psychosine induces impairment of UPS and autophagy, resulting in cytoplasmic accumulation of insoluble ubiquitin- and p62-aggregates, giving rise to an elevation of ROS and cell death. The spatial-temporal distribution of ubiquitin- and p62-aggregates correlates with progressive demyelination in the nervous system and provides connection to the neurological symptomatology.

Though misfolded and damaged proteins mediated by psychosine have been implicated in the neuropathy of GLD, the impact of imbalanced proteostasis upon oligodendrocytes and white matter has not been validated. Previous studies demonstrated that psychosine interacts with α-synuclein and promotes aggregation of fibrillized α-synuclein in neurons in the brains of both human and twitcher mice [23]. A recent study further demonstrated a deposition of p62 aggregates in neurons of the twitcher brain at an early symptomatic stage, albeit the locations of neuronal aggregates in regions of the twitcher brain were not indicated and presence of p62 aggregates in oligodendrocytes was not documented [29]. These findings suggest that misfolded protein aggregates are involved in the pathogenesis of GLD. In this study, we first demonstrated that the ubiquitin- and p62-aggregates, which colocalize with oligodendrocytes, accumulate ubiquitously and constitutively in white matter and parallel the progressive demyelination of disease progression. Additionally, neurons with ubiquitin- and p62-aggregates were moderate and localized in the cerebellar granular layer, brain stem, and spinal cord. Cytosolic aggregates containing ubiquitinated proteins and p62 proteins have been linked to dysfunction of UPS and autophagy, underscoring the pathogenesis of many neurodegenerative diseases and several demyelination disorders [35–37,57]. Soluble extraneous or misfolded proteins are ubiquitinated and directed to UPS for degradation, while decreased efficacy or overwhelmed proteasome systems lead to soluble ubiquitinated proteins and insoluble aggregates, which become cellular inclusions, as a hallmark of neurodegenerative proteinopathies. Damaged protein, organelles, and insoluble aggregates are engulfed in autophagosomes, which are decorated with LC3 and p62 for maturation and subsequent lysosomal degradation, whereas dysfunction of autophagy results in the formation of insoluble cytosolic aggregates associated with p62 accumulation, contributing to the pathogenesis of neurodegenerative diseases and several lysosomal storage diseases [38,58]. Of note, p62 also serves as an adaptor protein, linking UPS and autophagy by directly binding ubiquitinated proteins for degradation in autophagy [59,60]. The cytosolic accumulation of aggregates containing both ubiquitin and p62 proteins has been widely associated with compromised UPS and autophagy in many neurodegenerative diseases. Along these lines, our findings indicate lysosomal GALC deficiency contributing to the dysfunction of UPS and autophagy, which underlies the vulnerability to demyelination and neurodegeneration in GLD.

To address the role of autophagy implicated in the pathogenesis of demyelination in GLD, several human and murine cell lines with exogenous supplements of psychosine have been used as a disease-cell model [9,21–31]. In a study by Ribbens et al., murine oligodendrocyte cell lines derived from the twitcher brain showed elevated levels of LC3-II colocalized with lysosomes under exogenous treatment with 15μM psychosine for 24 h, suggesting induction of autophagy by psychosine [24]. Another study by Del Grosso et al. demonstrated an evident autophagy activation in the human oligodendrocyte cell line MO3.13 under the synergic impact of 10 μM psychosine and serum starvation for 24 h [30]. Furthermore, Del Grosso et al. showed that levels of p62 and LC3-II were elevated in twitcher fibroblasts with exposure to 100 μM, psychosine, albeit higher levels of p62 protein were also noted in psychosine-untreated twitcher fibroblasts and levels of p62 aggregates were indistinguishable between psychosine-treated and untreated twitcher cells [29]. Those previous studies all demonstrated an activation of autophagy under short-term exposure to psychosine at different concentrations in different disease-cell models. Intriguingly, in the present study using the human oligodendrocyte cell line MO3.13, we are the first to document a disease-cell model recapitulating the pathogenesis of GLD. Our results indicate that psychosine induces impairment of autophagy and UPS with a dose-and time-dependent cytotoxicity, which explains the activation of autophagy under short-term exposure of psychosine in previous studies [24,29,30]. Elevated concentrations of psychosine elicit an increase of autophagy, progressive and prolonged treatment of psychosine (20 μM for 72 h) disrupts the function of both autophagy and UPS, characterized with the accumulation of LC3-II, p62, and ubiquitin proteins in the insoluble fraction. Of note, inhibition of proteasomes and autophagy further exacerbates the accumulation of insoluble proteins p62 and ubiquitin, mediated by psychosine, and augments cytoplasmic deposition of p62- and ubiquitin-aggregates, which recapitulate the pathological findings observed in the twitcher brain in this report. Those findings implicate a time-dependent accumulation of p62- and ubiquitin-aggregates by psychosine cytotoxicity, supporting the chronological changes of pathogenesis observed in the brain of twitcher mice.

In the present study, our findings provide evidence for a link between psychosine and dysfunction in autophagy and UPS, which underlies vulnerability to progressive demyelination in GLD. Protein homeostasisis regulated by the crosstalk between UPS and autophagy, while dysfunction of autophagy and/or UPS leads to cytoplasmic deposition of p62- and ubiquitin-aggregates, underlying the molecular mechanism for neurodegenerative diseases. It is argued that the aggregates of p62 and ubiquitin in the twitcher brain are the sequential results from impairment of autophagy or UPS alone or combination. In the disease-cell model, psychosine induced a decrease of proteasome activity and accumulation of insoluble ubiquitin and cytoplasmic aggregates, indicating the impairment of UPS is caused by the psychosine toxicity. Several reports have documented mechanisms on how aggregated proteins interact with proteasomes and impair the function of UPS. It has been shown that heterogeneous aggregates of prion proteins impair UPS though reducing proteasomal gating for substrate entry into protealytic sites [61]. A recent study further validates that aggregate proteins generated from Alzheimer's disease, Parkinson's, and Huntington's disease form three-dimensional oligomers, which inhibit proteasome gates opening, thus impairing ubiquitin-dependent and ubiquitin-independent proteasome function [62]. These findings provide a partial explanation to psychosine-induced proteasome impairment, given that psychosine accelerates the formation of α-synuclein oligomers [23].

In regards to maintaining protein homeostasis, UPS impairment upregulates autophagy to compensate for the dysfunction of the proteasome [63–66]. Indeed, we observed an increase of LC3-II both in vivo and in vitro, suggesting an increment of autophagic flux, while the concomitant accumulation of p62 as an indicative marker of dysfunction of autophagy was also demonstrated in our study. The increased expression of p62can compromise the protein homeostasis of misfolded proteins, leading to the formation of ubiquitinated protein aggregates [67,68]. A previous study indicated that increased expression of p62 in expanded polyglutamine-expressing cells with inhibition of proteasome enhanced the formation of aggresomes and ubiquitinated polyglutamine inclusions [68]. Moreover, over expression of p62 accelerates the accumulation of insoluble p62 and poly-ubiquitinated proteins by hindering the function of UPS and autophagy in the degradation of misfolded proteins [67]. In the present study, the accumulation of insoluble p62 and cytoplasmic deposition of p62 aggregates in the twitcher brain and disease-cell model implicated the impairment of the autophagy pathway. Several molecular mechanisms have been linked to dysfunction of autophagy. The disruption of autophagic flux observed in the twitcher brain can be caused by an overload of autophagic substrates above the capacity of degradation or multiple lysosomal dysfunctions mediated by psychosine [69,70]. Similar to the pathogenesis of Niemann–Pick Type C, we observed an accumulation of autophagosomes, autolysosomes, and p62 in our disease-cell model, suggesting the reduction of lysosomal activity and subsequent impairment of autolysosomal cargo degradation [39]. Previous studies have indicated that psychosine disrupts lysosomal function by causing abnormalities in lipid homeostasis, endolysosomal transport, and cathepsin activity [69]. Additionally, lysosomal membrane permeability caused by the primary lysosomal deficit, which disrupts the efficacy of lysosomal degradation, promotes insoluble p62 aggregate formation into cytosolic inclusions [58]. Nonetheless, excess accumulation of p62 by impaired autophagy can hinder the trafficking of ubiquitinated proteins to the proteasome proteases for degradation, resulting in reduced degradation of UPS substrates and accumulation of aggregation-prone proteins [71]. Collectively, reciprocal regulation of autophagy and UPS is crucial for the maintenance of protein homeostasis, while an imbalance between both proteolytic pathways caused by either system alone or both systems together could result in the same pathological consequence. Accordingly, misfolded protein accumulation and aggregation cause over-production of reactive oxygen species, which contributes to cell death.

5. Conclusions

Our study, for the first time, unveils the dysfunction of UPS and autophagy contributing to the pathogenesis of demyelination and neuropathy in GLD. Here, we describe the increase of insoluble p62 and ubiquitin proteins concomitant with spatiotemporal accumulation of colocalized ubiquitin- and p62-aggregatesin alignment with a decrement of myelin protein, which underlies vulnerability to progressive demyelination in the murine model of GLD. Impairment of proteasomes and autophagy results in the accumulation of ROS, reduction of mitochondrial respiration, and eventually decreased viability. These findings provide new insights in the understanding of the molecular mechanism for the pathogenesis of GLD, and provide possible therapeutic avenues via targeting brain regions with specific pathomechanisms, modulation of proteostasis, as well as providing a sufficient antioxidant system.

Author Contributions: D.-S.L., C.-S.H., and M.-F.C. contributed to the study conception and design. Material preparation, data collection, and analysis were performed by Y.-W.H., T.-J.W., T.-Y.W., T.-H.L., Z.-D.H. and S.-J.Y. The first draft of the manuscript was written by D.-S.L. and all authors commented on previous versions of the manuscript. All authors read and approved the final manuscript.

Funding: This work was supported by the grants from Mackay Memorial Hospital [MMH-E-108-02, MMH-CT-10809] and Ministry of Science and Technology [MOST-108-2314-B-195-020].

Conflicts of Interest: The authors declare no conflict of interest.

Abbreviations

ATG	autophagy-related proteins;
GFAP	glial fibrillary acidic protein;
GLD	globoid cell leukodystrophy;
Iba-1	ionized calcium-binding adapter molecule 1;
LC3	microtubule-associated protein 1 light chain;
LC3-II	LC3-phosphatidylethanolamine conjugate;
LSD	lysosomal storage disease;
LAMP1	lysosomal-associated membrane protein 1;
MBP	myelin basic protein;
NeuN	neuronal nuclei;
PLP	proteolipd protein;
Twi	twitcher mice;
UPS	ubiquitin proteasome system;
WT	wild-type mice.

References

1. Moser, H.W.; Mahmood, A.; Raymond, G.V. X-linked adrenoleukodystrophy. *Nat. Clin. Pract. Neurol.* **2007**, *3*, 140–151. [CrossRef]
2. Wenger, D.A.; Rafi, M.A.; Luzi, P.; Datto, J.; Costantino-Ceccarini, E. Krabbe disease: Genetic aspects and progress toward therapy. *Mol. Genet. Metab.* **2000**, *70*, 1–9. [CrossRef] [PubMed]
3. Polten, A.; Fluharty, A.L.; Fluharty, C.B.; Kappler, J.; von Figura, K.; Gieselmann, V. Molecular basis of different forms of metachromatic leukodystrophy. *N. Engl. J. Med.* **1991**, *324*, 18–22. [CrossRef]
4. Suzuki, K. Twenty five years of the "psychosine hypothesis": A personal perspective of its history and present status. *Neurochem Res.* **1998**, *23*, 251–259. [CrossRef] [PubMed]
5. Beltran-Quintero, M.L.; Bascou, N.A.; Poe, M.D.; Wenger, D.A.; Saavedra-Matiz, C.A.; Nichols, M.J.; Escolar, M.L. Early progression of Krabbe disease in patients with symptom onset between 0 and 5 months. *Orphanet. J. Rare Dis.* **2019**, *14*, 46. [CrossRef] [PubMed]
6. Kwon, J.M.; Matern, D.; Kurtzberg, J.; Wrabetz, L.; Gelb, M.H.; Wenger, D.A.; Ficicioglu, C.; Waldman, A.T.; Burton, B.K.; Hopkins, P.V.; et al. Consensus guidelines for newborn screening, diagnosis and treatment of infantile Krabbe disease. *Orphanet. J. Rare Dis.* **2018**, *13*, 30. [CrossRef]
7. Kobayashi, T.; Shinnoh, N.; Goto, I.; Kuroiwa, Y. Hydrolysis of galactosylceramide is catalyzed by two genetically distinct acid beta-galactosidases. *J. Biol. Chem.* **1985**, *260*, 14982–14987.
8. Miyatake, T.; Suzuki, K. Globoid cell leukodystrophy: Additional deficiency of psychosine galactosidase. *Biochem. Biophys. Res. Commun.* **1972**, *48*, 539–543. [CrossRef]
9. Won, J.S.; Singh, A.K.; Singh, I. Biochemical, cell biological, pathological, and therapeutic aspects of Krabbe's disease. *J. Neurosci. Res.* **2016**, *94*, 990–1006. [CrossRef]
10. Graziano, A.C.; Cardile, V. History, genetic, and recent advances on Krabbe disease. *Gene* **2015**, *555*, 2–13. [CrossRef]
11. Lin, D.S.; Hsiao, C.D.; Lee, A.Y.; Ho, C.S.; Liu, H.L.; Wang, T.J.; Jian, Y.R.; Hsu, J.C.; Huang, Z.D.; Lee, T.H.; et al. Mitigation of cerebellar neuropathy in globoid cell leukodystrophy mice by AAV-mediated gene therapy. *Gene* **2015**, *571*, 81–90. [CrossRef] [PubMed]
12. Bongarzone, E.R.; Escolar, M.L.; Gray, S.J.; Kafri, T.; Vite, C.H.; Sands, M.S. Insights into the Pathogenesis and Treatment of Krabbe Disease. *Pediatr. Endocrinol. Rev.* **2016**, *13*, 689–696. [PubMed]
13. Duchen, L.W.; Eicher, E.M.; Jacobs, J.M.; Scaravilli, F.; Teixeira, F. Hereditary leucodystrophy in the mouse: The new mutant twitcher. *Brain* **1980**, *103*, 695–710. [CrossRef]
14. Taniike, M.; Suzuki, K. Spacio-temporal progression of demyelination in twitcher mouse: With clinico-pathological correlation. *Acta Neuropathol.* **1994**, *88*, 228–236. [CrossRef] [PubMed]
15. Suzuki, K.; Suzuki, K. The twitcher mouse: A model for Krabbe disease and for experimental therapies. *Brain Pathol.* **1995**, *5*, 249–258. [CrossRef]
16. Kondo, Y.; Adams, J.M.; Vanier, M.T.; Duncan, I.D. Macrophages counteract demyelination in a mouse model of globoid cell leukodystrophy. *J. Neurosci.* **2011**, *31*, 3610–3624. [CrossRef]

17. Hawkins-Salsbury, J.A.; Shea, L.; Jiang, X.; Hunter, D.A.; Guzman, A.M.; Reddy, A.S.; Qin, E.Y.; Li, Y.; Gray, S.J.; Ory, D.S.; et al. Mechanism-based combination treatment dramatically increases therapeutic efficacy in murine globoid cell leukodystrophy. *J. Neurosci.* **2015**, *35*, 6495–6505. [CrossRef]
18. Lin, D.; Donsante, A.; Macauley, S.; Levy, B.; Vogler, C.; Sands, M.S. Central nervous system-directed AAV2/5-mediated gene therapy synergizes with bone marrow transplantation in the murine model of globoid-cell leukodystrophy. *Mol. Ther.* **2007**, *15*, 44–52. [CrossRef]
19. Hu, P.; Li, Y.; Nikolaishvili-Feinberg, N.; Scesa, G.; Bi, Y.; Pan, D.; Moore, D.; Bongarzone, E.R.; Sands, M.S.; Miller, R.; et al. Hematopoietic Stem cell transplantation and lentiviral vector-based gene therapy for Krabbe's disease: Present convictions and future prospects. *J. Neurosci. Res.* **2016**, *94*, 1152–1168. [CrossRef]
20. Sands, S.A.; LeVine, S.M. Substrate reduction therapy for Krabbe's disease. *J. Neurosci. Res.* **2016**, *94*, 1261–1272. [CrossRef]
21. Teixeira, C.A.; Miranda, C.O.; Sousa, V.F.; Santos, T.E.; Malheiro, A.R.; Solomon, M.; Maegawa, G.H.; Brites, P.; Sousa, M.M. Early axonal loss accompanied by impaired endocytosis, abnormal axonal transport, and decreased microtubule stability occur in the model of Krabbe's disease. *Neurobiol. Dis.* **2014**, *66*, 92–103. [CrossRef]
22. White, A.B.; Givogri, M.I.; Lopez-Rosas, A.; Cao, H.; van Breemen, R.; Thinakaran, G.; Bongarzone, E.R. Psychosine accumulates in membrane microdomains in the brain of krabbe patients, disrupting the raft architecture. *J. Neurosci.* **2009**, *29*, 6068–6077. [CrossRef]
23. Smith, B.R.; Santos, M.B.; Marshall, M.S.; Cantuti-Castelvetri, L.; Lopez-Rosas, A.; Li, G.; van Breemen, R.; Claycomb, K.I.; Gallea, J.I.; Celej, M.S.; et al. Neuronal inclusions of alpha-synuclein contribute to the pathogenesis of Krabbe disease. *J. Pathol.* **2014**, *232*, 509–521. [CrossRef]
24. Ribbens, J.J.; Moser, A.B.; Hubbard, W.C.; Bongarzone, E.R.; Maegawa, G.H. Characterization and application of a disease-cell model for a neurodegenerative lysosomal disease. *Mol. Genet. Metab.* **2014**, *111*, 172–183. [CrossRef]
25. Haq, E.; Giri, S.; Singh, I.; Singh, A.K. Molecular mechanism of psychosine-induced cell death in human oligodendrocyte cell line. *J. Neurochem.* **2003**, *86*, 1428–1440. [CrossRef]
26. Hawkins-Salsbury, J.A.; Parameswar, A.R.; Jiang, X.; Schlesinger, P.H.; Bongarzone, E.; Ory, D.S.; Demchenko, A.V.; Sands, M.S. Psychosine, the cytotoxic sphingolipid that accumulates in globoid cell leukodystrophy, alters membrane architecture. *J. Lipid Res.* **2013**, *54*, 3303–3311. [CrossRef]
27. Giri, S.; Jatana, M.; Rattan, R.; Won, J.S.; Singh, I.; Singh, A.K. Galactosylsphingosine (psychosine)-induced expression of cytokine-mediated inducible nitric oxide synthases via AP-1 and C/EBP: Implications for Krabbe disease. *FASEB J.* **2002**, *16*, 661–672. [CrossRef]
28. Giri, S.; Khan, M.; Rattan, R.; Singh, I.; Singh, A.K. Krabbe disease: Psychosine-mediated activation of phospholipase A2 in oligodendrocyte cell death. *J. Lipid Res.* **2006**, *47*, 1478–1492. [CrossRef] [PubMed]
29. Del Grosso, A.; Angella, L.; Tonazzini, I.; Moscardini, A.; Giordano, N.; Caleo, M.; Rocchiccioli, S.; Cecchini, M. Dysregulated autophagy as a new aspect of the molecular pathogenesis of Krabbe disease. *Neurobiol. Dis.* **2019**, *129*, 195–207. [CrossRef] [PubMed]
30. Del Grosso, A.; Antonini, S.; Angella, L.; Tonazzini, I.; Signore, G.; Cecchini, M. Lithium improves cell viability in psychosine-treated MO3.13 human oligodendrocyte cell line via autophagy activation. *J. Neurosci Res.* **2016**, *94*, 1246–1260. [CrossRef] [PubMed]
31. White, A.B.; Galbiati, F.; Givogri, M.I.; Lopez Rosas, A.; Qiu, X.; van Breemen, R.; Bongarzone, E.R. Persistence of psychosine in brain lipid rafts is a limiting factor in the therapeutic recovery of a mouse model for Krabbe disease. *J. Neurosci. Res.* **2011**, *89*, 352–364. [CrossRef] [PubMed]
32. Glickman, M.H.; Ciechanover, A. The ubiquitin-proteasome proteolytic pathway: Destruction for the sake of construction. *Physiol. Rev.* **2002**, *82*, 373–428. [CrossRef]
33. Shpilka, T.; Weidberg, H.; Pietrokovski, S.; Elazar, Z. Atg8: An autophagy-related ubiquitin-like protein family. *Genome Biol.* **2011**, *12*, 226. [CrossRef] [PubMed]
34. Pankiv, S.; Clausen, T.H.; Lamark, T.; Brech, A.; Bruun, J.A.; Outzen, H.; Overvatn, A.; Bjorkoy, G.; Johansen, T. p62/SQSTM1 binds directly to Atg8/LC3 to facilitate degradation of ubiquitinated protein aggregates by autophagy. *J. Biol. Chem.* **2007**, *282*, 24131–24145. [CrossRef] [PubMed]
35. Arotcarena, M.L.; Teil, M.; Dehay, B. Autophagy in Synucleinopathy: The Overwhelmed and Defective Machinery. *Cells* **2019**, *8*, 565. [CrossRef]

36. Refolo, V.; Stefanova, N. Neuroinflammation and Glial Phenotypic Changes in Alpha-Synucleinopathies. *Front. Cell Neurosci.* **2019**, *13*, 263. [CrossRef]
37. Stefanova, N.; Kaufmann, W.A.; Humpel, C.; Poewe, W.; Wenning, G.K. Systemic proteasome inhibition triggers neurodegeneration in a transgenic mouse model expressing human alpha-synuclein under oligodendrocyte promoter: Implications for multiple system atrophy. *Acta Neuropathol.* **2012**, *124*, 51–65. [CrossRef]
38. Seranova, E.; Connolly, K.J.; Zatyka, M.; Rosenstock, T.R.; Barrett, T.; Tuxworth, R.I.; Sarkar, S. Dysregulation of autophagy as a common mechanism in lysosomal storage diseases. *Essays Biochem.* **2017**, *61*, 733–749.
39. Elrick, M.J.; Yu, T.; Chung, C.; Lieberman, A.P. Impaired proteolysis underlies autophagic dysfunction in Niemann-Pick type C disease. *Hum. Mol. Genet.* **2012**, *21*, 4876–4887. [CrossRef]
40. Sakai, N.; Inui, K.; Tatsumi, N.; Fukushima, H.; Nishigaki, T.; Taniike, M.; Nishimoto, J.; Tsukamoto, H.; Yanagihara, I.; Ozono, K.; et al. Molecular cloning and expression of cDNA for murine galactocerebrosidase and mutation analysis of the twitcher mouse, a model of Krabbe's disease. *J. Neurochem.* **1996**, *66*, 1118–1124. [CrossRef]
41. Lin, D.; Fantz, C.R.; Levy, B.; Rafi, M.A.; Vogler, C.; Wenger, D.A.; Sands, M.S. AAV2/5 vector expressing galactocerebrosidase ameliorates CNS disease in the murine model of globoid-cell leukodystrophy more efficiently than AAV2. *Mol. Ther.* **2005**, *12*, 422–430. [CrossRef] [PubMed]
42. Wirths, O. Extraction of Soluble and Insoluble Protein Fractions from Mouse Brains and Spinal Cords. *Bio-protocol* **2017**, *7*, e2422. [CrossRef]
43. Salminen, A.; Kaarniranta, K.; Haapasalo, A.; Hiltunen, M.; Soininen, H.; Alafuzoff, I. Emerging role of p62/sequestosome-1 in the pathogenesis of Alzheimer's disease. *Prog. Neurobiol.* **2012**, *96*, 87–95. [CrossRef]
44. Sadakata, T.; Kakegawa, W.; Mizoguchi, A.; Washida, M.; Katoh-Semba, R.; Shutoh, F.; Okamoto, T.; Nakashima, H.; Kimura, K.; Tanaka, M.; et al. Impaired cerebellar development and function in mice lacking CAPS2, a protein involved in neurotrophin release. *J. Neurosci.* **2007**, *27*, 2472–2482. [CrossRef] [PubMed]
45. Ohashi, R.; Sakata, S.; Naito, A.; Hirashima, N.; Tanaka, M. Dendritic differentiation of cerebellar Purkinje cells is promoted by ryanodine receptors expressed by Purkinje and granule cells. *Dev. Neurobiol.* **2014**, *74*, 467–480. [CrossRef] [PubMed]
46. Rajan, R.; Karbowniczek, M.; Pugsley, H.R.; Sabnani, M.K.; Astrinidis, A.; La-Beck, N.M. Quantifying autophagosomes and autolysosomes in cells using imaging flow cytometry. *Cytometry A* **2015**, *87*, 451–458. [CrossRef] [PubMed]
47. Guo, S.; Liang, Y.; Murphy, S.F.; Huang, A.; Shen, H.; Kelly, D.F.; Sobrado, P.; Sheng, Z. A rapid and high content assay that measures cyto-ID-stained autophagic compartments and estimates autophagy flux with potential clinical applications. *Autophagy* **2015**, *11*, 560–572. [CrossRef]
48. Park, J.T.; Lee, Y.S.; Park, S.C. Quantification of Autophagy During Senescence. *Methods Mol. Biol.* **2019**, *1896*, 149–157.
49. DeVorkin, L.; Gorski, S.M. LysoTracker staining to aid in monitoring autophagy in Drosophila. *Cold Spring Harb. Protoc.* **2014**, *2014*, 951–958. [CrossRef]
50. Lin, D.S.; Huang, Y.W.; Ho, C.S.; Hung, P.L.; Hsu, M.H.; Wang, T.J.; Wu, T.Y.; Lee, T.H.; Huang, Z.D.; Chang, P.C.; et al. Oxidative Insults and Mitochondrial DNA Mutation Promote Enhanced Autophagy and Mitophagy Compromising Cell Viability in Pluripotent Cell Model of Mitochondrial Disease. *Cells* **2019**, *8*, 65. [CrossRef]
51. Oeste, C.L.; Seco, E.; Patton, W.F.; Boya, P.; Perez-Sala, D. Interactions between autophagic and endo-lysosomal markers in endothelial cells. *Histochem. Cell Biol.* **2013**, *139*, 659–670. [CrossRef] [PubMed]
52. Lee, D.H.; Goldberg, A.L. Proteasome inhibitors: Valuable new tools for cell biologists. *Trends Cell Biol.* **1998**, *8*, 397–403. [CrossRef]
53. Jatana, M.; Giri, S.; Singh, A.K. Apoptotic positive cells in Krabbe brain and induction of apoptosis in rat C6 glial cells by psychosine. *Neurosci. Lett.* **2002**, *330*, 183–187. [CrossRef]
54. Taniike, M.; Mohri, I.; Eguchi, N.; Irikura, D.; Urade, Y.; Okada, S.; Suzuki, K. An apoptotic depletion of oligodendrocytes in the twitcher, a murine model of globoid cell leukodystrophy. *J. Neuropathol Exp. Neurol* **1999**, *58*, 644–653. [CrossRef] [PubMed]
55. Kanazawa, T.; Nakamura, S.; Momoi, M.; Yamaji, T.; Takematsu, H.; Yano, H.; Sabe, H.; Yamamoto, A.; Kawasaki, T.; Kozutsumi, Y. Inhibition of cytokinesis by a lipid metabolite, psychosine. *J. Cell Biol.* **2000**, *149*, 943–950. [CrossRef]

56. LeVine, S.M.; Brown, D.C. IL-6 and TNFalpha expression in brains of twitcher, quaking and normal mice. *J. Neuroimmunol.* **1997**, *73*, 47–56. [CrossRef]
57. Ciechanover, A.; Brundin, P. The ubiquitin proteasome system in neurodegenerative diseases: Sometimes the chicken, sometimes the egg. *Neuron* **2003**, *40*, 427–446. [CrossRef]
58. Micsenyi, M.C.; Sikora, J.; Stephney, G.; Dobrenis, K.; Walkley, S.U. Lysosomal membrane permeability stimulates protein aggregate formation in neurons of a lysosomal disease. *J. Neurosci.* **2013**, *33*, 10815–10827. [CrossRef]
59. Komatsu, M.; Waguri, S.; Koike, M.; Sou, Y.S.; Ueno, T.; Hara, T.; Mizushima, N.; Iwata, J.; Ezaki, J.; Murata, S.; et al. Homeostatic levels of p62 control cytoplasmic inclusion body formation in autophagy-deficient mice. *Cell* **2007**, *131*, 1149–1163. [CrossRef]
60. Lim, J.; Yue, Z. Neuronal aggregates: Formation, clearance, and spreading. *Dev. Cell* **2015**, *32*, 491–501. [CrossRef]
61. Deriziotis, P.; Andre, R.; Smith, D.M.; Goold, R.; Kinghorn, K.J.; Kristiansen, M.; Nathan, J.A.; Rosenzweig, R.; Krutauz, D.; Glickman, M.H.; et al. Misfolded PrP impairs the UPS by interaction with the 20S proteasome and inhibition of substrate entry. *EMBO J.* **2011**, *30*, 3065–3077. [CrossRef] [PubMed]
62. Thibaudeau, T.A.; Anderson, R.T.; Smith, D.M. A common mechanism of proteasome impairment by neurodegenerative disease-associated oligomers. *Nat. Commun.* **2018**, *9*, 1097. [CrossRef] [PubMed]
63. Demishtein, A.; Fraiberg, M.; Berko, D.; Tirosh, B.; Elazar, Z.; Navon, A. SQSTM1/p62-mediated autophagy compensates for loss of proteasome polyubiquitin recruiting capacity. *Autophagy* **2017**, *13*, 1697–1708. [CrossRef] [PubMed]
64. Pandey, U.B.; Nie, Z.; Batlevi, Y.; McCray, B.A.; Ritson, G.P.; Nedelsky, N.B.; Schwartz, S.L.; DiProspero, N.A.; Knight, M.A.; Schuldiner, O.; et al. HDAC6 rescues neurodegeneration and provides an essential link between autophagy and the UPS. *Nature* **2007**, *447*, 859–863. [CrossRef] [PubMed]
65. Kageyama, S.; Sou, Y.S.; Uemura, T.; Kametaka, S.; Saito, T.; Ishimura, R.; Kouno, T.; Bedford, L.; Mayer, R.J.; Lee, M.S.; et al. Proteasome dysfunction activates autophagy and the Keap1-Nrf2 pathway. *J. Biol Chem* **2014**, *289*, 24944–24955. [CrossRef]
66. Milan, E.; Perini, T.; Resnati, M.; Orfanelli, U.; Oliva, L.; Raimondi, A.; Cascio, P.; Bachi, A.; Marcatti, M.; Ciceri, F.; et al. A plastic SQSTM1/p62-dependent autophagic reserve maintains proteostasis and determines proteasome inhibitor susceptibility in multiple myeloma cells. *Autophagy* **2015**, *11*, 1161–1178. [CrossRef]
67. Mitsui, S.; Otomo, A.; Nozaki, M.; Ono, S.; Sato, K.; Shirakawa, R.; Adachi, H.; Aoki, M.; Sobue, G.; Shang, H.F.; et al. Systemic overexpression of SQSTM1/p62 accelerates disease onset in a SOD1(H46R)-expressing ALS mouse model. *Mol. Brain* **2018**, *11*, 30. [CrossRef]
68. Nagaoka, U.; Kim, K.; Jana, N.R.; Doi, H.; Maruyama, M.; Mitsui, K.; Oyama, F.; Nukina, N. Increased expression of p62 in expanded polyglutamine-expressing cells and its association with polyglutamine inclusions. *J. Neurochem.* **2004**, *91*, 57–68. [CrossRef]
69. Folts, C.J.; Scott-Hewitt, N.; Proschel, C.; Mayer-Proschel, M.; Noble, M. Lysosomal Re-acidification Prevents Lysosphingolipid-Induced Lysosomal Impairment and Cellular Toxicity. *PLoS Biol.* **2016**, *14*, e1002583. [CrossRef]
70. Lieberman, A.P.; Puertollano, R.; Raben, N.; Slaugenhaupt, S.; Walkley, S.U.; Ballabio, A. Autophagy in lysosomal storage disorders. *Autophagy* **2012**, *8*, 719–730. [CrossRef]
71. Korolchuk, V.I.; Mansilla, A.; Menzies, F.M.; Rubinsztein, D.C. Autophagy inhibition compromises degradation of ubiquitin-proteasome pathway substrates. *Mol. Cell* **2009**, *33*, 517–527. [CrossRef] [PubMed]

cells MDPI

Article

p62 is Negatively Implicated in the TRAF6-BECN1 Signaling Axis for Autophagy Activation and Cancer Progression by Toll-Like Receptor 4 (TLR4)

Mi-Jeong Kim [1,†], Yoon Min [1,†], Ji Seon Im [1,†], Juhee Son [1], Joo Sang Lee [2,3] and Ki-Young Lee [1,3,4,*]

1 Department of Immunology, Sungkyunkwan University School of Medicine, 2066 Seobu-ro, Jangan-gu, Suwon, Gyeonggi-do 16419, Korea; kmjj0107@skku.edu (M.-J.K.); unizzzang@skku.edu (Y.M.); jsunny1319@skku.edu (J.S.I.); kimthsh@skku.edu (J.S.)

2 Department of Precision Medicine, Samsung Biomedical Research Institute, Sungkyunkwan University School of Medicine, 2066 Seobu-ro, Jangan-gu, Suwon, Gyeonggi-do 16419, Korea; lee.joosang@gmail.com

3 Samsung Medical Center, Seoul 06351, Korea

4 Department of Health Sciences and Technology, Samsung Advanced Institute for Health Sciences & Technology, Samsung Medical Center, Sungkyunkwan University, 81 Irwon-ro, Gangnam-gu, Seoul 06351, Korea

* Correspondence: thylee@skku.edu; Tel.: +82-31-299-6225

† Authors contributed equally to this work.

Received: 26 March 2020; Accepted: 2 May 2020; Published: 6 May 2020

Abstract: Toll-like receptors (TLRs) induce the activation of nuclear factor kappa-light-chain-enhancer of activated B cells (NF-κB) and autophagy through the TNF (Tumor necrosis factor) receptor-associated factor 6 (TRAF6)-evolutionarily conserved signaling intermediate in Toll pathways (ECSIT) and TRAF6-BECN1 signaling axes, respectively. Having shown that p62 negatively regulates Toll-like receptor 4 (TLR4)-mediated signaling via TRAF6-ECSIT signaling axis, we herein investigated whether p62 is functionally implicated in the TRAF6-BECN1 signaling axis, thereby regulating cancer cell migration and invasion. p62 interacted with TRAF6 and BECN1, to interrupt the functional associations required for TRAF6-BECN1 complex formation, leading to inhibitions of BECN1 ubiquitination and autophagy activation. Importantly, p62-deficient cancer cells, such as p62-knockdown ($p62^{KD}$) SK-HEP-1, $p62^{KD}$ MDA-MB-231, and p62-knockout ($p62^{KO}$) A549 cells, showed increased activation of autophagy induced by TLR4 stimulation, suggesting that p62 negatively regulates autophagy activation. Moreover, these p62-deficient cancer cells exhibited marked increases in cell migration and invasion in response to TLR4 stimulation. Collectively, these results suggest that p62 is negatively implicated in the TRAF6-BECN1 signaling axis, thereby inhibiting cancer cell migration and invasion regulated by autophagy activation in response to TLR4 stimulation.

Keywords: toll-like receptor 4; p62; TRAF6; BECN1; Autophagy

---

1. Introduction

p62 (SQSTM1 gene and sequestosome-1) is a ubiquitin-binding protein and a versatile adaptor protein with multiple functions for regulating cellular events [1,2]. These include autophagic flux through the interaction with autophagic substrates, apoptosis, cellular redox regulation through the Kelch-like ECH-associated protein 1-nuclear factor (erythroid-derived 2)-like 2 (KEAP1-NRF2) pathway, adipogenesis by the interaction with extracellular-signal-regulated kinase 1 (ERK1), and nuclear factor kappa-light-chain-enhancer of activated B cells (NF-κB) signaling through the interaction with protein kinase C ζ (PKCζ) [1–8]. Although the regulatory role of p62 in inflammatory responses is controversial [9,10], it is thought to be involved in the induction of inflammatory cytokine production via TNF Tumor necrosis factor) receptor-associated factor 6 (TRAF6) polyubiquitination and, thereby, NF-κB activation [11,12]. A recent report has shown that p62 is negatively implicated in Toll-like receptor 4 (TLR4)-mediated signaling through inhibition of TRAF6-evolutionarily conserved signaling intermediate in Toll pathways (ECSIT) association and the ubiquitination of ECSIT by TRAF6 [13]. Importantly, $p62^{-/-}$ KO mice exhibited a higher mortality rate following LPS challenge [13], suggesting that p62 might negatively regulate TLR4-mediated signaling for the activation of NF-κB.

TLRs act as the first line of host defense against microbial infections and play pivotal roles in the initiation of innate immunity and the induction of adaptive immune responses by recognizing distinct pathogen-associated molecular patterns (PAMPs) [14–16]. The downstream signaling cascades of TLRs are essentially mediated by several key molecules and these include myeloid differentiation primary response 88 (MyD88), TRAF6, TAK1, mitogen-activated protein kinase kinase kinase 7 (MAP3K7), and IκB kinase (IKK) complex [14,15]. Among them, TRAF6 as an E3 ubiquitin ligase and a scaffold protein plays a key role in the TLR4-mediated activation of NF-κB [17,18]. Interestingly, recent reports have demonstrated that TLR4 and TLR3 signals induce autophagy activation, and promote migration and invasion of lung cancer cells through TRAF6 ubiquitination and MAP3K7 activation [19]. Upon TLR4 stimulation, TRAF6 promotes the K63-linked ubiquitination of BECN1 to induce TLR4-mediated autophagy [19–23]. In addition, TLR4 signaling promotes proliferation of A549 lung cancer cells through PTGS2/COX-2 (prostaglandin-endoperoxide synthase 2 (prostaglandin G/H synthase and cyclooxygenase)) and epidermal growth factor receptor (EGFR) activation [19]. These results strongly indicate that TLR4-induced autophagy activation and cancer cell progression might be critically linked to the TRAF6-BECN1 signaling axis.

TLR4-mediated signaling induces the activation of NF-κB and autophagy induction via TRAF6-ECSIT signaling axis and TRAF6-BECN1 signaling axis, respectively [20–26]. Since p62 negatively regulates TLR4-mediated signals through the inhibition of TRAF6-ECSIT signaling [13], we therefore investigated whether p62 is functionally implicated in the TRAF6-BECN1 signaling axis, thereby regulating autophagy activation, and cancer cell migration and invasion induced by TLR4. We found that p62 interrupted the association of the TRAF6-BECN1 complex, and that inhibited BECN1 ubiquitination, leading to inhibition of autophagy activation induced by TLR4. Interestingly, p62-deficient cancer cells, p62-knockdown ($p62^{KD}$) SK-HEP-1, $p62^{KD}$ MDA-MB-231, and p62-knockout ($p62^{KO}$) A549 cells, exhibited increases in autophagic activation, and cancer cell migration/invasion induced by TLR4 stimulation. Taken together, these results suggest that p62 negatively regulates autophagy activation, thereby functionally affecting cancer cell migration and invasion induced by TLR4.

2. Materials and Methods

2.1. Cells

Human embryonic kidney (HEK) 293T cells were obtained from the American Type Culture Collection (Ca# CRL-11268, ATCC, Manassas, VA, USA). The cells were cultured in Dulbecco's modified Eagle's medium (DMEM; Ca#11965092, Thermo Fisher Scientific, Waltham, MA, USA). THP-1 cells, human monocytic cells, were purchased from ATCC (Ca# TIB-202), and cultured in RPMI (Roswell Park Memorial Institute) 1640 medium (Ca#11875093, Thermo Fisher Scientific) containing 10% fetal bovine serum (FBS; Fisher Scientific HyClone, Ca#11306060), 2 mM L-glutamine (GIBCO, Ca#A2916801), 100 units/mL penicillin (GIBCO, Ca#15140122), 100 μg/mL streptomycin (GIBCO, Ca#15140122), and 5×10^{-5} M β-mercaptoethanol (GIBCO, Ca#21985023). Human hepatic adenocarcinoma cell line SK-HEP-1 (ATCC, HTB-52), human breast adenocarcinoma cell line MDA-MB-231 (ATCC, HTB-26), and human lung cancer cell line A549 (ATCC, CCL-185) were purchased from ATCC, and cultured in DMEM or RPMI contained with 10% FBS.

2.2. Generation of p62-Knockdown Cell Line

Lentivirus containing small hairpin RNA (shRNA) targeting human SQSTM1 (p62, Ca# sc-29679-V) and control shRNA lentivirus (Ca# sc-108080) were purchased from Santa Cruz Biotechnology (Santa Cruz, CA, USA). Cells were cultured in wells of a 24-well plate (2×10^4 cells per well), and infected with lentivirus, according to the manufacturer's protocol. Control (Ctrl) THP-1, Ctrl SK-HEP-1, Ctrl MDA-MB-231 cells, p62 knockdown ($p62^{KD}$) THP-1, $p62^{KD}$ SK-HEP-1, and $p62^{KD}$ MDA-MB-231 cells were cultured in puromycin-containing (4–8 μg/mL) medium, and selected as previously described [21].

2.3. Generation of p62-Knockout Cell Line with CRISPR/Cas9

Guide RNA sequences for CRISPR/Cas9 were designed at the CRISPR design web site (http://crispr.mit.edu/), provided by the Feng Zhang Lab. Insert oligonucleotides for human SQSTM1 (p62) gRNA were 5′-CACCGTGGCTCCGGAAGGTGAAACACGG-3′/3′-CACCGAGGCC TTCCACTTTGTCAAA-5′. The p62 guide RNA targets exon 2 of p62 gene. The complementary oligonucleotides for guide RNAs (gRNAs) were annealed, and cloned into lenti CRISPR v2 vector (Addgene plasmid, Ca#52961). Lenti CRISPR v2/gRNA was transfected into A549 cells by using Lipofectamine 2000, according to the manufacturer's instructions. After two days, cells were treated with puromycin (2 μg/mL) and cultured for three days. Colonies were isolated after two weeks, and the p62 expression in the cells was confirmed by using western blot.

2.4. Antibodies and Reagents

Anti-p62 antibody (Ca# ab91526) was purchased from Abcam (Cambridge, MA, USA), anti-Myc (Ca# 2276) antibody was purchased from Cell Signaling Technology (Danvers, MA, USA), and anti-Flag (Ca# F3165) and anti-HA (Ca# H3663) antibodies were purchased from Sigma-Aldrich (St Louis, MO, USA). Lipopolysaccharide (LPS, Ca#L2887), 3-methyladenine (3-MA, Ca# M9281), chloroquine (CQ, Ca# C6628), dimethyl sulfoxide (DMSO, Ca# 472301), puromycin (Ca# P9620), paraformaldehyde (Ca# P6148), Triton X-100 (Ca# T9284), gentamicin (Ca# G1272), deoxycholate (Ca# D6750), and Dulbecco's phosphate-buffered saline (DPBS, Ca# D8537) were purchased from Sigma-Aldrich (St Louis, MO, USA). Lipofectamine 2000 (Ca# 11668-019) was purchased from Thermo Scientific (Rockford, IL, USA).

2.5. Plasmid Constructs

Flag-tagged TRAF6 (Ca# 21624), Flag-tagged BECN1 (Ca# 24388), and HA-tagged p62 (Ca# 28027) were purchased from Addgene (Cambridge, MA, USA). Myc-tagged ECSIT, HA-tagged Ub, and Flag-tagged ECSIT were obtained from Dr. Jae-Hyuck Shim (University of Massachusetts Medical School, USA). The constructs coding for full-length p62 with the Flag tag or Myc tag were cloned into the pCMV-3Tag-6 vector (Agilent Technologies) or pCMV-3Tag-7 vector, respectively, using HA-tagged p62 plasmid as a template. The constructs coding for full-length BECN1 with the Myc tag were cloned into the pCMV-3Tag-7 vector, using Flag-tagged BECN1 plasmid as a template. Flag-tagged TRAF6 truncated mutants and Myc-tagged BECN1 truncated mutants were generated as previously described [21,22].

2.6. Western Blotting Analysis and Immunoprecipitation (IP) Assays

Western blotting analysis and IP assays were carried out as previously described [13,20–23]. HEK293T cells were transfected with mock vector as control vector, Myc-tagged p62, or Flag-tagged TRAF6, and mock vector, Flag-tagged p62, and Myc-tagged BECN1 using Lipofectamine 2000. At 38 h after transfection, transfected cells were harvested, and cell lysates were immunoprecipitated with anti-Flag antibody. HEK293T cells were transfected with mock vector, Myc-tagged p62, or Flag-tagged TRAF6 wild type (WT) and Flag-tagged TRAF6 truncated mutants using Lipofectamine 2000. At 38 h after transfection, transfected cells were harvested, and cell lysates were immunoprecipitated with anti-Myc antibody. HEK293T cells were transfected with mock vector, Flag-tagged p62, or Myc-tagged BECN1 WT and Myc-tagged BECN1 truncated mutants using Lipofectamine 2000. At 38 h after transfection, transfected cells were harvested, and IP assay was performed with anti-Myc antibody. IP complexes were separated by SDS-PAGE (6–10%), and immune-probed with antibodies specific for anti-Myc or anti-Flag. HEK293T cells were transfected with mock vector, Flag-tagged TRAF6, and Myc-tagged BECN1, along with different concentrations of Myc-tagged p62, using Lipofectamine 2000. At 38 h after transfection, transfected cells were harvested, and IP assay was performed with with anti-Flag antibody. Immunoprecipitated complexes were separated by 6–10% SDS-PAGE, and probed with anti-Flag, anti-p62, or anti-BECN1 antibody. For ubiquitination assay, HEK293T cells were transfected with mock vector, Myc-tagged BECN1, Flag-tagged TRAF6, and HA-tagged Ub, along with different concentrations of Flag-tagged p62, using Lipofectamine 2000. At 38 h after transfection, transfected cells were harvested, and cell lysates were immunoprecipitated with anti-Myc antibody. Immunoprecipitated complexes were separated by 6–10% SDS-PAGE, and probed with anti-Myc, anti-HA, anti-p62, or anti-TRAF6 antibody. Control (Ctrl) THP-1 and $p62^{KD}$ THP-1 cells, Ctrl SK-HEP-1 and $p62^{KD}$ SK-HEP-1 cells, Ctrl MDA-MB-231 and $p62^{KD}$ MDA-MB-231 cells, or Ctrl A549 and $p62^{KO}$ A549 cells were treated with or without vehicle, 3MA (5mM), or CQ (10 μM), in the presence or absence of LPS (10 μg/mL), for 6 h. Whole cell lysates were immunoblotted with anti-LC3A/B antibody and anti-GAPDH as a loading control.

2.7. Wound-Healing and Transwell Migration Assay

A wound-healing assay was carried out as previously described [21,22]. Ctrl SK-HEP-1 and $p62^{KD}$ SK-HEP-1 cells, Ctrl MDA-MB-231 and $p62^{KD}$ MDA-MB-231 cells, or Ctrl A549 and $p62^{KO}$ A549 cells were cultured in 12-well plates, and inoculated to confluence. Cell monolayers were gently scratched by using a sterile yellow Gilson-pipette tip to make a wide gap (approximately 400 μm). Cells were washed with culture medium, and floating cells and debris were removed from plates. Cells were treated with vehicle (DMSO, <0.2% in DMEM culture medium), 3-MA (5 mM), or CQ (10 μM) in the presence or absence of LPS (10 μg/mL), and images were captured after different times as indicated in each experiment. Transwell inserts (8μm pore; Corning, 3422) were sited into wells for cell migration assay. 5×10^4 cells per well were suspended in culture medium (DMEM) including vehicle, 3-MA (5 mM), or CQ (10 μM) in the presence or absence of LPS (10 μg/mL), and placed into the top chambers of the 24-transwell plates. Culture medium, DMEM contained 10% FBS, was added to the bottom chambers. After an overnight incubation, the non-migrated cells to be remained in the top chamber were removed. The migrated cells to be existed in the bottom chamber were fixed. To visualize the nuclei, cells were stained by using crystal violet. All experiments were performed in triplicate. The experiments were repeated twice times.

2.8. Reverse Transcription-Quantitative Polymerase Chain Reaction (RT-qPCR)

Control (Ctrl) and $p62^{KO}$ A549 cells were treated with or without 10 μg/mL LPS for 6 h. Total RNA was extracted from cells using an RNA isolation kit (A&A Biotechnology, Gdynia, Poland) according to the manufacturer's protocol. cDNA was obtained by RT using a amfiRivert II cDNA Synthesis Master Mix (genDEPOT, R550), according to the manufacturer's protocol. Primers for hIL-6 (PPH 00560C), hMMP2 (PPH 00151B), and hCCL2 (PPH 00192F) were purchased from Qiagen, Inc. (Chatsworth, CA, USA). Fluorescence detection was performed using the ABI PRISM 7700 Sequence Detector (PerkinElmer; Applied Biosystems; Thermo Fisher Scientific, Inc.). The mRNA expressions were calculated and normalized to the level of GAPDH.

2.9. Statistical Analysis

In vitro data are expressed as mean ±SEM of triplicate samples. Statistical significance of experiments was analyzed by using ANOVA or Student's *t*-test using GraphPad Prism 5.0 (GraphPad Software, San Diego, CA, USA).

3. Results

3.1. P62 Interacts with TRAF6 and BECN1

It has been reported that p62 plays an inhibitory role in TLR4 signaling through interrupting the association of TRAF6 with ECSIT, eventually leading to inhibition of NF-κB activation [13]. TRAF6 is an essential regulator for the induction of NF-κB and autophagy in TLR4 signaling [14–18]. Based on these earlier reports, we asked whether p62 is involved in TLR4-induced activation of autophagy. To investigate this, we first examined the molecular association between p62 and autophagy regulator proteins, such as TRAF6 and BECN1, in TLR4 signaling. Myc-tagged p62 or Flag-tagged p62 expressing vector were transfected into HEK293T cells along with Flag-tagged TRAF6 or Myc-tagged BECN1 expressing vector, and then immunoprecipitation (IP) was performed with anti-Flag antibody. Consistent with a previous report [13], Flag-tagged TRAF6 precipitated with Myc-tagged p62 (Supplementary Materials Figure S1A, lane 4). Additionally, Flag-tagged p62 precipitated with Myc-tagged BECN1 (Figure 1A, lane 4). To determine the specific binding site of p62 to TRAF6 or BECN1, truncated mutants of TRAF6 or BECN1 were generated, and IP assay was performed between p62 and these truncated proteins (Supplementary Materials Figure S2A,B). As shown in Supplementary Materials Figure S3, Flag-tagged TRAF6 wild type (WT) and Flag-tagged TRAF6 truncated mutants were significantly precipitated with Myc-tagged p62 (Supplementary Materials Figure S3A, lane 6-8), suggesting that p62 interacts with the TRAF-C domain of TRAF6 (Supplementary Materials Figure S3B). The results were consistent with a previous report [13]. As well, Myc-tagged BECN1 WT and Myc-tagged BECN1 1-269 truncated mutant were precipitated with Flag-tagged p62 (Figure 1B, lane 6 and 7), whereas no significant interaction could be observed with Myc-tagged BECN1 1-127 truncations (Figure 1B, lane 8), indicating that p62 interacts with the coiled-coil domain of BECN1 (Figure 1C).

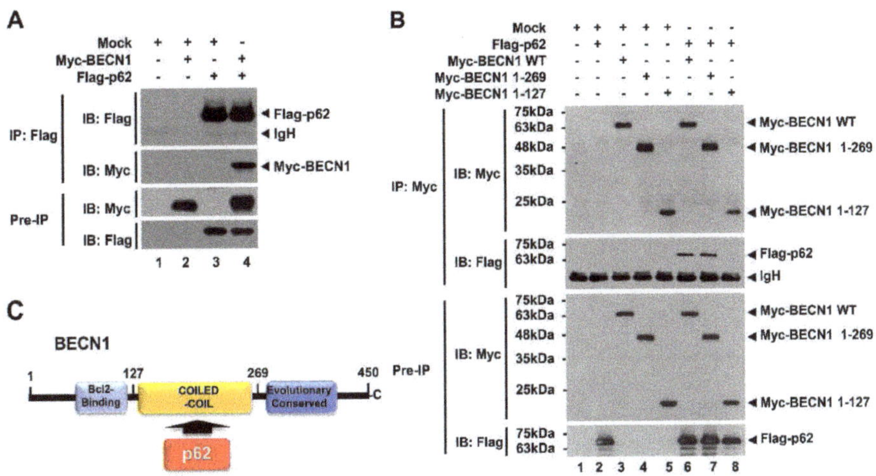

Figure 1. p62 interacts with BECN1 proteins. (**A**) HEK293T cells were transfected with mock control vector, Flag-tagged p62, or Myc-tagged BECN1, as indicated. Transfected cells were harvested, and cell lysates were immunoprecipitated with anti-Flag antibody and probed with anti-Myc or anti-Flag antibody. (**B**) HEK293T cells were transfected with mock vector, Flag-tagged p62, or Myc-tagged BECN1 WT and Myc-tagged BECN1 truncated mutants. Transfected cells were harvested, and cell lysates were immunoprecipitated with anti-Myc antibody. Immunoprecipitated complexes were separated by SDS-PAGE, and probed with anti-Myc or anti-Flag antibody. (**C**) A schematic view of the molecular interaction between BECN1 and p62.

3.2. p62 Interrupts the Association of BECN1-TRAF6 and Inhibits the Ubiquitination of BECN1 Induced by TRAF6

A previous study showed that TRAF6 interacts with the BECN1 and induces the ubiquitination of BECN1, leading to induction of autophagy [21]. Consistently, Flag-tagged TRAF6 was significantly precipitated with Myc-tagged BECN1 (Supplementary Materials Figure S4A). Importantly, TRAF6 interacted with the coiled-coil domain of BECN1 [21], as depicted in Supplementary Materials Figure S4B. As shown in Figure 1B,C, p62 interacted with the coiled-coil domain of BECN1, which was the same binding domain as TRAF6 [21]. Therefore, we raised the possibility that p62 affects the molecular association of TRAF6-BECN1, thereby inhibiting BECN1 ubiquitination, as depicted in Figure 2A. To examine this possibility, we performed a competitive binding assay with different concentrations of p62. Flag-tagged TRAF6 and Myc-tagged BECN1 vectors were transfected into HEK293T cells along with different concentrations of Myc-tagged p62 vectors, as indicated Figure 2B. IP was then performed with anti-Flag antibodies in cell lysates. Correlating to increases of Myc-tagged p62 vector, the interactions between Flag-tagged TRAF6 and Myc-tagged BECN1 were significantly attenuated (Figure 2B, lane2-4; IB: BECN1 in IP with Flag-tagged TRAF6), suggesting that p62 interrupts the molecular association between TRAF6 and BECN1.

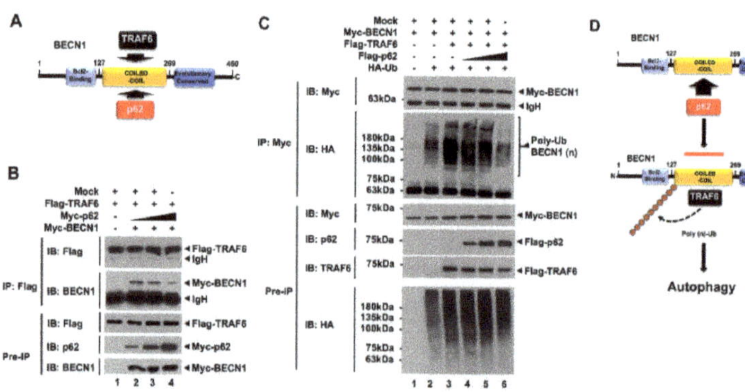

Figure 2. p62 interrupts the association of TNF (Tumor necrosis factor) receptor-associated factor 6 (TRAF6)-BECN1 complex and inhibits the ubiquitination of BECN1. (**A**) TRAF6 and p62 interact with the coiled-coil domain of BECN1. (**B**) HEK293T cells were transfected with mock vector, Flag-tagged TRAF6, and Myc-tagged BECN1, along with different concentrations of Myc-tagged p62, as indicated. Transfected cells were harvested, and cell lysates were immunoprecipitated with anti-Flag antibody and probed with anti-Flag, anti-p62, or anti-BECN1 antibody. (**C**) HEK293T cells were transfected with mock vector, Myc-tagged BECN1, Flag-tagged TRAF6, and HA-tagged Ub, along with different concentrations of Flag-tagged p62, as indicated. Transfected cells were harvested, and cell lysates were immunoprecipitated with anti-Myc antibody and probed with anti-Myc, anti-HA, anti-p62, or anti-TRAF6 antibody. (**D**) A schematic model for how p62 interrupts the association of TRAF6-BECN1 complex and inhibits the ubiquitination of BECN1. TRAF6 interacts with the coiled-coil domain of BECN1 and induces the ubiquitination of BECN1, leading to autophagy activation. Simultaneously, p62 can interact with the coiled-coil domain of BECN1, and that inhibits the interaction of TRAF6 to BECN1 and the ubiquitination of BECN1.

Since TRAF6 interacted with BECN1 and induced its ubiquitination [19–22], we examined whether inhibiting the interaction between TRAF6 and BECN1 by p62 affects BECN1 ubiquitination. To do that, Myc-tagged BECN1, Flag-tagged TRAF6, and HA-tagged Ub vectors were transfected into HEK293T cells along with different concentrations of Flag-tagged p62, and then IP was performed with anti-Myc antibodies. The marginal ubiquitination of BECN1 could be seen in the absence of Flag-tagged TRAF6 (Figure 2C, lane 2), whereas a marked increase could be seen in the presence of Flag-tagged TRAF6

(Figure 2C, lane 3). Based on increasing Flag-tagged p62 expression, the ubiquitination of BECN1 was gradually attenuated (Figure 2C, lane 4-6). These results suggest that p62 interrupts the molecular interaction between TRAF6 and BECN1, and that in turn inhibits the ubiquitination of BECN1 induced by TRAF6, as depicted in Figure 2D.

Studies have shown that the TRAF6-induced ubiquitination of BECN1 plays a key role in TLR-induced autophagy activation, thereby functionally implicating cancer progression and migration [19,21]. We found that p62 interrupted the association between BECN1 and TRAF6, and inhibited the ubiquitination of BECN1 (Figure 2B,C). Therefore, we explored the functional role of p62 in autophagy induction, cancer progression, and migration, all of which are regulated by TLR4 signaling [19,21], as depicted in Figure 2D.

3.3. p62-Knockdown THP-1 ($p62^{KD}$ THP-1), SK-HEP-1 ($p62^{KD}$ SK-HEP-1), and MDA-MB 231 ($p62^{KD}$ MDA-MB-231) Cells Exhibit Elevated Autophagy Activation Induced by TLR4 Stimulation

To investigate the functional role of p62 in autophagy activation induced by TLR4 stimulation, we generated p62-knockdown THP-1 ($p62^{KD}$ THP-1 cells) by using the lentivirus containing shRNA targeted to p62, as described in Materials and Methods. The efficiency of p62 knockdown was significant compared to control cells (Figure 3A, lane 1 versus lane 2). Control (Ctrl) THP-1 and $p62^{KD}$ THP-1 cells were treated with or without LPS and an autophagy inhibitor, chloroquine (CQ), as indicated in Figure 3B, and autophagy activation was assessed by western blotting with anti-LC3 antibody. Upon LPS stimulation, the level of LC3-II increased in both cell lines, (Figure 3B, lane 1 versus lane 2 in Ctrl THP-1 and lane 4 versus lane 5 in $p62^{KD}$ THP-1), but the increase in LC3-II was significantly higher in $p62^{KD}$ THP-1 cells than in Ctrl THP-1 cells (Figure 3B, lane 2 versus lane 5; Figure 3C, open bars versus closed bars in LPS). As expected, CQ treatment induced the marked accumulation of LC3-II in both cell lines (Figure 3B, lane 3 and lane 6; Figure 3C, open bars versus closed bars in LPS plus CQ), suggesting that p62 negatively regulates autophagy activation induced by TLR4 stimulation.

To investigate further the role of p62 in activating autophagy, we generated p62-knockdowns in two cancer cell lines, $p62^{KD}$ SK-HEP-1 and $p62^{KD}$ MDA-MB 231 cells, as described in Materials and Methods. The efficacy of p62 knockdown in SK-HEP-1 and MDA-MB-231 cells was significant as compared to control cells (Figure 3D in SK-HEP-1 and 3G in MDA-MB-231 cells, lane 1 versus lane 2). In similar fashion to $p62^{KD}$ THP-1 cells (Figure 3B,C), the levels of LC3-II were significantly enhanced in $p62^{KD}$ SK-HEP-1 and $p62^{KD}$ MDA-MB-231 cells in the presence of LPS, as compared to their controls (Figure 3E,F in SK-HEP-1 lane 2 versus lane 5 and open bars versus closed bars in LPS treated; Figure 3H,I in MDA-MB-231, lane 2 versus lane 5 and open bars versus closed bars in LPS treated). These results suggest that p62 negatively regulates autophagy activation induced by TLR4 stimulation, presumably by the inhibition of the ubiquitination of BECN1 as demonstrated in Figure 2D.

3.4. p62-Deficient Cancer Cells Exhibit Increased Cancer Cell Migration and Invasion, Induced by TLR4 Stimulation

Having shown that p62 negatively regulated autophagy activation, we asked whether the inhibitory effect was functionally associated with cancer cell migration and invasion. To do that, migration and invasion assays were performed in $p62^{KD}$ SK-HEP-1 and $p62^{KD}$ MDA-MB-231 cells. Ctrl SK-HEP-1 and $p62^{KD}$ SK-HEP-1 were treated with vehicle, LPS, LPS plus a 3-methyladenine (3-MA) autophagy inhibitor, and LPS plus a CQ autophagy inhibitor, and then wound healing assay was performed. Based on LPS treatment, cancer cell migratory behavior was significantly higher in $p62^{KD}$ SK-HEP-1 than the Ctrl SK-HEP-1 cells in a time dependent manner (Figure 4A,B, Ctrl versus $p62^{KD}$ SK-HEP-1 in LPS treatment). These results were consistently observed in Ctrl and $p62^{KD}$ MDA-MB-231 cells (Figure 4C,D, Ctrl versus $p62^{KD}$ MDA-MB 231 in LPS treatment). As expected, marked attenuations could be seen in co-treatments with 3-MA or CQ (Figure 4A,B, Ctrl versus $p62^{KD}$ SK-HEP-1 in LPS plus 3-MA or CQ: Figure 4C,D, Ctrl versus $p62^{KD}$ MDA-MB-231 in LPS plus 3-MA or

CQ). We next assessed invasiveness following TLR4 stimulation. Ctrl SK-HEP-1 and p62^{KD} SK-HEP-1 were treated with vehicle, LPS, LPS plus 3-MA, and LPS plus CQ, and then transwell migration assay was performed. Similar to the wound healing assay, progressive invasiveness was significantly higher in p62^{KD} SK-HEP-1 than in Ctrl SK-HEP-1 in the presence of LPS (Figure 5A,B, Ctrl versus p62^{KD} SK-HEP-1 in LPS treatment). Consistent results were observed in Ctrl and p62^{KD} MDA-MB-231 cells (Figure 5C,D, Ctrl versus p62^{KD} MDA-MB-231 in LPS treatment).

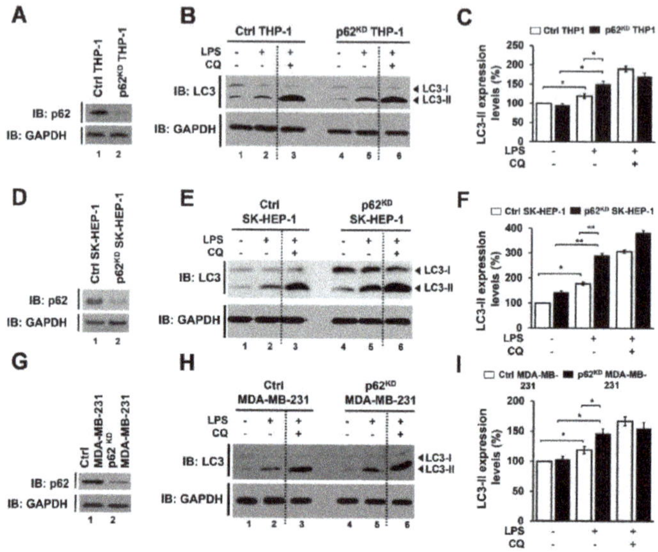

Figure 3. p62-deficient cells, p62^{KD} THP-1, p62^{KD} SK-HEP-1, and p62^{KD} MDA-MB-231 cells, exhibit enhanced autophagy activation in response to TLR4 stimulation. (**A**) p62^{KD} THP-1 cells were generated, and the knockdown efficacy of p62 was confirmed with anti-p62 antibody. (**B**,**C**) Ctrl and p62^{KD} THP-1 cells were treated with or without vehicle or CQ (10 μM), in the presence or absence of LPS (10 μg/mL), for 6 h. Whole cell lysates were immunoblotted with anti-LC3A/B antibody and anti-GAPDH antibody as a loading control (**B**). The LC3II levels were analyzed with Image J program (**C**). Data shown are averages from a minimum of 3 independent experiments (± SEM). *, $p < 0.05$. (**D**) p62^{KD} SK-HEP-1 cells were generated, and the knockdown efficacy of p62 was confirmed with anti-p62 antibody. (**E**,**F**) Ctrl and p62^{KD} SK-HEP-1 were treated with or without vehicle or CQ, in the presence or absence of LPS. Whole cell lysates were immunoblotted with anti-LC3A/B and anti-GAPDH antibodies (**E**). The LC3II levels were analyzed with Image J program (**F**). Data shown are averages from a minimum of 3 independent experiments (± SEM). *, $p < 0.05$ and **, $p < 0.01$. (**G**) p62^{KD} MDA-MB-231 cells were generated, and the knockdown efficacy of p62 was confirmed with anti-p62 antibody. (**H**,**I**) Ctrl and p62^{KD} MDA-MB-231 were treated with or without vehicle or CQ, in the presence or absence of LPS. Whole cell lysates were immunoblotted with anti-LC3A/B and anti-GAPDH antibodies (**H**). The LC3II levels were analyzed with Image J program (**I**). Data shown are averages from a minimum of 3 independent experiments (± SEM). * $p < 0.05$.

To verify the above results, we generated p62-knockout (p62^{KO}) A549 cells using Crispr cas9 (Figure 6A). Upon TLR4 stimulation, the levels of LC3-II were significantly increased in both Ctrl A459 and p62^{KO} A549 cells (Figure 6B, lane 1 versus lane 2 in Ctrl A549 and lane 4 versus lane 5 in p62^{KO} A549 cells). Moreover, the LC3-II ratio relative to control was significantly higher in p62^{KO} A549 cells than in Ctrl A549 (Figure 6C, closed bar versus open bar in LPS). As expected, treatments of autophagy inhibitor CQ or 3-MA induced increased or decreased LC3-II levels, respectively (Figure 6B, lane 3 and lane 4 in Ctrl; lane 6 and 7 in p62^{KO} A549). Consistent with results of p62^{KD} SK-HEP-1 and p62^{KD}

MDA-MB 231, the cell migration and invasion activities induced by LPS were markedly enhanced in $p62^{KO}$ A549 cells compared to controls, whereas significant attenuations were observed in co-treatments with 3-MA or CQ (Figure 6D,E, Ctrl versus $p62^{KO}$ A549 in LPS treatment; Figure 6F,G, Ctrl versus $p62^{KO}$ A549 in LPS treatment). It has been previously reported that TLR4-induced autophagy activation promoted migration and invasion of lung cancer by induction of chemokines and immunosuppressive factors including CCL2, CCL20, IL-6, VEGFA, and MMP2 [19,27,28]. Therefore, we evaluated the production of IL-6, MMP2, and CCL2 in Ctrl and $p62^{KO}$ A549 cells in the presence or absence of LPS stimulation. Consistent with the earlier report, the levels of IL-6 mRNA, MMP2 mRNA, CCL2 mRNA were significantly elevated in Ctrl A549 cells treated with LPS (Figure 6H, without LPS versus with LPS in Ctrl A549). Interestingly, these levels were markedly higher in $p62^{KO}$ A549 cells than in Ctrl A549 cells under LPS stimulation (Figure 6H, Ctrl versus $p62^{KO}$ A549 with LPS). Collectively these results suggest that p62 negatively regulates autophagy activation, cancer cell migration and invasion induced by TLR4 stimulation.

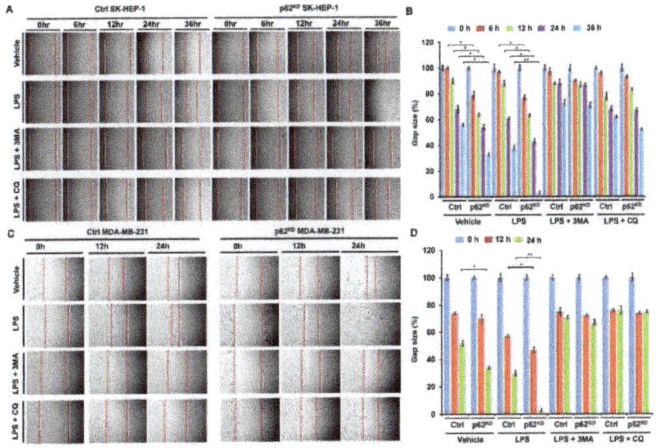

Figure 4. $p62^{KD}$ SK-HEP-1 and $p62^{KD}$ MDA-MB-231 cells exhibit increased cell migration in response to TLR4 stimulation. (**A**,**B**) Ctrl and $p62^{KD}$ SK-HEP-1 cells were seeded into 12-well cell culture plates. Confluent monolayers were scraped with a sterile yellow Gilson-pipette tip, and the wound was then treated with vehicle (DMSO, <0.2% in culture medium), LPS (10 μg/mL), 3-MA (5 mM) plus LPS (10 μg/mL), and CQ (10 μM) plus LPS (10 μg/mL) for different time periods, as indicated. A representative experiment is shown (**A**). The residual gap between the migrating cells from the opposing wound edge was expressed as a percentage of the initial scraped area (± SEM, n = 3) (**B**). *, $p < 0.05$ and **, $p < 0.01$. (**C**,**D**) Ctrl and $p62^{KD}$ MDA-MB-231 were seeded into 12-well cell culture plates. Confluent monolayers were scraped with a sterile yellow Gilson-pipette tip, and the wound was then treated with vehicle (DMSO, <0.2% in culture medium), LPS (10 μg/mL), 3-MA (5 mM) plus LPS (10 μg/mL), and CQ (10 μM) plus LPS (10 μg/mL) for different time periods, as indicated. A representative experiment is shown (**C**). The residual gap between the migrating cells from the opposing wound edge was expressed as a percentage of the initial scraped area (± SEM, n = 3) (**D**). * $p < 0.05$ and ** $p < 0.01$.

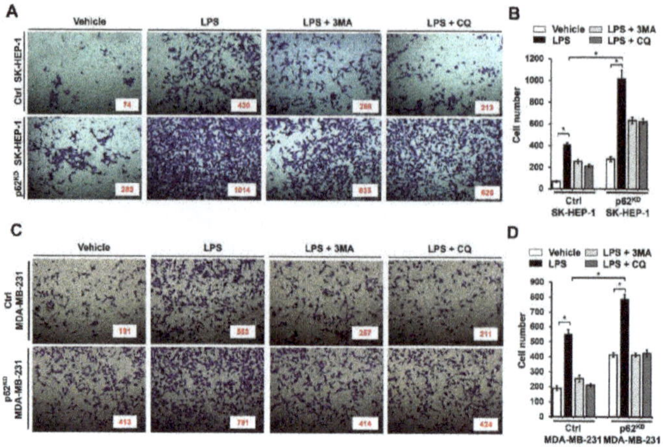

Figure 5. p62^{KD} SK-HEP-1 and p62^{KD} MDA-MB-231 cells exhibit increased invasiveness in response to TLR4 stimulation. (**A**,**B**) Ctrl and p62^{KD} SK-HEP-1 cells were suspended in DMEM culture medium including vehicle, LPS (10 μg/mL), 3-MA (5 mM) plus LPS (10 μg/mL), and CQ (10 μM) plus LPS (10 μg/mL). Cells were placed into the top chambers of 24-transwell plates and incubated for overnight. Fixed cells were stained by using crystal violet (**A**). Numbers of migrated cells were counted, and results are represented as mean ± SEM (**B**). * $p < 0.05$. (**C**,**D**) Ctrl and p62^{KD} MDA-MB-231 cells were suspended in culture medium of RPMI including vehicle, LPS (10 μg/mL), 3-MA (5 mM) plus LPS (10 μg/mL), and CQ (10 μM) plus LPS (10 μg/mL). Cells were placed into top chambers of 24-transwell plates and further incubated for overnight. Fixed cells were stained with crystal violet (**C**). Numbers of migrated cells were counted, and results are represented as mean ± SEM (**D**). * $p < 0.05$.

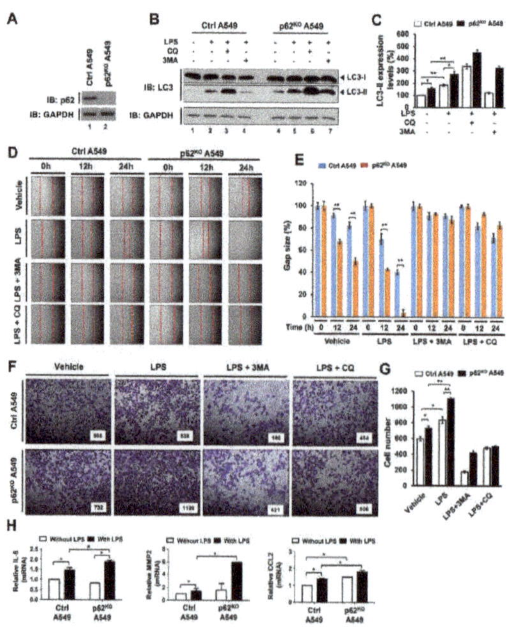

Figure 6. p62^{KO} A549 cells exhibited increased autophagy activation, migration, and invasion in response to TLR4 stimulation. (**A**) p62^{KO} A549 cells were generated, and the knockout efficacy of p62

was confirmed with anti-p62 antibody. (**B,C**) Ctrl and p62^{KO} A549 cells were treated with or without vehicle, CQ (10 μM), or 3-MA (5 mM), or, in the presence or absence of LPS (10 μg/mL), for 6 h. Whole cell lysates were immunoblotted with anti-LC3A/B antibody and anti-GAPDH antibody as a loading control (**B**). The LC3II levels were analyzed with Image J program (right, histogram) (**C**). Data shown are averages from a minimum of 3 independent experiments (± SEM). * $p < 0.05$ and ** $p < 0.01$. (**D,E**) Ctrl and p62^{KO} A549 cells were seeded into 12-well cell culture plates. Confluent monolayers were scraped with a sterile yellow Gilson-pipette tip to make wounds, and then incubated with vehicle (DMSO, <0.2% in culture medium), LPS (10 μg/mL), 3-MA (5 mM) plus LPS (10 μg/mL), and CQ (10 μM) plus LPS (10 μg/mL) for different time periods. A representative experiment is represented (**D**). The residual gap between the migrating cells from the opposing wound edge was represented as a percentage of the initial scraped area (± SEM, $n = 3$) (**E**). ** $p < 0.01$. (**F,G**) Ctrl and p62^{KO} A549 cells were suspended in RPMI medium including vehicle, LPS (10 μg/mL), 3-MA (5 mM) plus LPS (10 μg/mL), and CQ (10 μM) plus LPS (10 μg/mL), and placed to the top chambers of 24-transwell plates. After an overnight incubation, cells were fixed and stained with crystal violet (**F**). Number of migrating cell were counted, and results are presented as mean ± SEM of 3 independent experiments (**G**). * $p < 0.05$ and ** $p < 0.01$. (**H**) Control (Ctrl) and p62^{KO} A549 cells were treated with or without 10 μg/mL LPS for 6 h. Total RNA was extracted, cDNA was obtained, as described in materials and methods, and RT-qPCR analysis performed with specific primers, such as hIL-6, hMMP2, and hCCL2. * $p < 0.05$.

4. Discussion

In this study, we demonstrate that p62, which is a known negative regulator for the activation of autophagy by TLR4 signaling, inhibits the ubiquitination of BECN1 mediated by TRAF6. We demonstrate this occurs through an interruption to the molecular association between TRAF6 and BECN1 and, thereby, involves cancer cell migration and invasion, activities that are facilitated by autophagy, as depicted in Figure 7. Through the biochemical studies, we found that p62 competitively interacted with the coiled-coil domain of BECN1 affecting binding to TRAF6, and that induced an attenuation of BECN1 ubiquitination. Levels of LC3-II revealed increased autophagy induction in the presence of TLR4 stimulation by p62^{KD} THP-1 and p62^{KD} cancer cells. More importantly, these p62-deficient cancer cells, including p62^{KO} A549 cells, showed enhanced migration and invasion in response to TLR4 stimulation. Taken together, these results suggest that p62 is negatively implicated in the activation of autophagy by TLR4 signaling, thereby demonstrating an involvement in cancer cell invasiveness facilitated by autophagy induction.

Accumulating evidence suggests that TLR signaling might be critical for induction and activation of autophagy, thereby facilitating cancer cell migration and invasion [19,21]. Among signal cascades, TRAF6 plays a pivotal role in NF-κB activation and autophagy induction through the activation of TLR downstream molecules and the ubiquitination of BECN1, respectively [19–26]. A previous study demonstrated that p62 attenuated the ubiquitination of ECSIT, which is known as one of the regulators in TLR4-mediated signaling for NF-κB activation, induced by TRAF6 [13]. Interestingly, we found in the current study that p62 interacted with BECN1, as well as TRAF6. The molecular associations between p62 and BECN1 or TRAF6 revealed that p62 interacted with the coiled-coil domain of BECN1, which also surprisingly served as the interaction site for TRAF6 [21], possibly explaining the inhibitory mechanism of p62 in the association of TRAF6-BECN1. p62 interference in TRAF6-BECN1 complex formation, eventually induced the attenuation of BECN1 ubiquitination by TRAF6.

Another important finding in the current study was that p62 regulated cancer cell invasiveness was facilitated by TLR4 stimulation through autophagy activation. Although there is some controversy over whether autophagy activation is either positively or negatively involved in cancer metastasis [29–33], recent studies have demonstrated that TLRs induced cancer cell migration and invasion through facilitating autophagy induction, which was mechanistically associated with the TRAF6-BECN1 signaling axis [19,21]. Consistent with these reports, we found that p62-deficient cancer cells, exhibited elevated autophagy activation, cancer cell migration, and invasiveness in response to TLR4 stimulation. Moreover, TCGA (The Cancer Genome Atlas) data revealed that the expression of p62 was significantly lower in primary tumors, such as colorectal cancer, breast cancer, prostate adenocarcinoma/prostate cancer, and bladder cancer/bladder urothelial carcinoma, than those of normal primary cells (Supplementary Materials Figure S5A–D). In addition, metastatic tumors, such as clear cell renal cell carcinoma, kidney cancer, and colorectal cancer, showed greater downregulation of p62, compared to primary tumors (Supplementary Materials Figure S6A,B). These results strongly suggest that p62 may be negatively associated with autophagy activation in cancer cells induced by TLR4 stimulation, and thereby be a key regulator of cancer cell metastasis.

In summary, we propose a molecular mechanism for, and functional effects of, p62 in autophagy activation and cancer progressions induced by TLR4 stimulation, as depicted in Figure 7. There are some controversies related to p62 and its involvement in inflammatory versus anti-inflammatory responses [9–12]. Considering p62 is a versatile adaptor protein with multiple cellular functions [1–8], the outstanding issues need to be clarified in the near future. Our results contribute to a growing understanding of the multi-functional role of p62 in autophagy and cancer progression. This new understanding may prove useful for the development of promising therapeutic approaches for treatment of inflammatory diseases and cancers in certain pathological conditions.

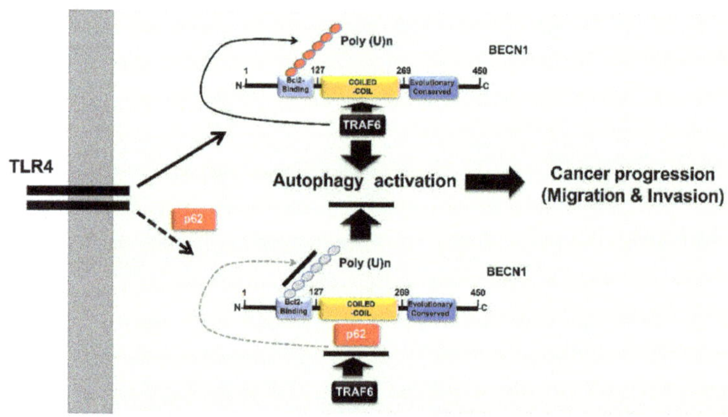

Figure 7. p62 is negatively involved in autophagy activation, and cancer cell migration and invasion in response to TLR4 stimulation. Engagement of TLR4 ligand leads to the association of TRAF6-BECN1 complex, and that induces the ubiquitination of BECN1. BECN1 ubiquitination induces the activation of autophagy, thereby potentially regulating cancer progression, via migration and invasion, as depicted in the upper panel. However, the interaction between p62 and BECN1 inhibits the association of TRAF6 to BECN1, and that inhibits the ubiquitination of BECN1, leading to inhibitions of autophagy activation, and cancer cell migration and invasion, as depicted in the lower panel.

Supplementary Materials: The following are available online at http://www.mdpi.com/2073-4409/9/5/1142/s1, Figure S1: TRAF6 interacts with p62.; Figure S2: Generations of TRAF6 or BECN1 truncated mutants.; Figure S3: p62 interacts with the TRAF-C domain of TRAF6.; Figure S4: TRAF6 interacts with the coiled-coil domain of BECN1.; Figure S5: The association between SQSTM1/p62 expression and primary tumors.; Figure S6: The association between SQSTM1/p62 expression and tumor metastasis.

Author Contributions: Conceptualization, K.-Y.L.; validation, K.-Y.L., M.-J.K., and Y.M.; formal analysis, M.-J.K., and Y.M.; investigation, M.-J.K., Y.M., J.S.L., and J.S.I.; data curation, K.-Y.L., M.-J.K., and Y.M.; writing-original draft preparation, K.-Y.L., and J.S.L.; writing- review and editing, K.-Y.L., and J.S.L.; visualization, K.-Y.L., M.-J.K., Y.M., and J.S.I.; supervision, K.-Y.L.; project administration, K.-Y.L., and M.-J.K.; funding acquisition, K.-Y.L., and J.S.L. All authors have read and agreed to the published version of the manuscript.

Funding: This work was supported by a grant of the National Research Foundation of Korea (NRF), funded by the Korean Government (NRF-2018R1D1A1B07042470) and the Korean Government (MSIP) (2016R1A5A2945889).

Acknowledgments: We would like to thank Hyehwa Forum members for their helpful discussions.

Conflicts of Interest: The authors declare no conflict of interest.

References

1. Geetha, T.; Wooten, M.W. Structure and functional properties of the ubiquitin binding protein p62. *FEBS Lett.* **2002**, *512*, 19–24. [CrossRef]
2. Seibenhener, M.L.; Babu, J.R.; Geetha, T.; Wong, H.C.; Krishna, N.R.; Wooten, M.W. Sequestosome 1/p62 is a polyubiquitin chain binding protein involved in ubiquitin proteasome degradation. *Mol. Cell. Biol.* **2004**, *24*, 8055–8068. [CrossRef]
3. Pugsley, H.R. Assessing Autophagic Flux by Measuring LC3, p62, and LAMP1 Co-localization Using Multispectral Imaging Flow Cytometry. *J. Vis. Exp.* **2017**, *125*, e55637. [CrossRef]
4. Jiang, P.; Mizushima, N. LC3- and p62-based biochemical methods for the analysis of autophagy progression in mammalian cells. *Methods* **2015**, *75*, 13–18. [CrossRef]
5. Ichimura, Y.; Waguri, S.; Sou, Y.S.; Kageyama, S.; Hasegawa, J.; Ishimura, R.; Saito, T.; Yang, Y.; Kouno, T.; Fukutomi, T.; et al. Phosphorylation of p62 activates the Keap1-Nrf2 pathway during selective autophagy. *Mol. Cell* **2013**, *51*, 618–631. [CrossRef]
6. Durán, A.; Serrano, M.; Leitges, M.; Flores, J.M.; Picard, S.; Brown, J.P.; Moscat, J.; Diaz-Meco, M.T. The atypical PKC-interacting protein p62 is an important mediator of RANK-activated osteoclastogenesis. *Dev. Cell* **2004**, *6*, 303–309. [CrossRef]
7. Puls, A.; Schmidt, S.; Grawe, F.; Stabel, S. Interaction of protein kinase C zeta with ZIP, a novel protein kinase C-binding protein. *Proc. Natl. Acad. Sci. USA* **1997**, *94*, 6191–6196. [CrossRef]
8. Sanz, L.; Diaz-Meco, M.T.; Nakano, H.; Moscat, J. The atypical PKC-interacting protein p62 channels NF-kappaB activation by the IL-1-TRAF6 pathway. *EMBO J.* **2000**, *19*, 1576–1586. [CrossRef]
9. Into, T.; Inomata, M.; Niida, S.; Murakami, Y.; Shibata, K. Regulation of MyD88 aggregation and the MyD88-dependent signaling pathway by sequestosome 1 and histone deacetylase 6. *J. Biol. Chem.* **2010**, *285*, 35759–35769. [CrossRef]
10. Kim, J.Y.; Ozato, K. The sequestosome 1/p62 attenuates cytokine gene expression in activated macrophages by inhibiting IFN regulatory factor 8 and TNF receptor-associated factor 6/NF-kappaB activity. *J. Immunol.* **2009**, *182*, 2131–2140. [CrossRef]
11. Duran, A.; Linares, J.F.; Galvez, A.S.; Wikenheiser, K.; Flores, J.M.; Diaz-Meco, M.T.; Moscat, J. The signaling adaptor p62 is an important NF-kappaB mediator in tumorigenesis. *Cancer Cell* **2008**, *13*, 343–354. [CrossRef] [PubMed]
12. Wooten, M.W.; Seibenhener, M.L.; Mamidipudi, V.; Diaz-Meco, M.T.; Barker, P.A.; Moscat, J. The atypical protein kinase C-interacting protein p62 is a scaffold for NF-kappaB activation by nerve growth factor. *J. Biol. Chem.* **2001**, *276*, 7709–7712. [CrossRef] [PubMed]
13. Kim, M.J.; Min, Y.; Kwon, J.; Son, J.; Im, J.S.; Shin, J.; Lee, K.Y. p62 Negatively Regulates TLR4 Signaling via Functional Regulation of the TRAF6-ECSIT Complex. *Immune Netw.* **2019**, *19*, e16. [CrossRef] [PubMed]
14. Akira, S.; Takeda, K.; Kaisho, T. Toll-like receptors: Critical proteins linking innate and acquired immunity. *Nat. Immunol.* **2001**, *2*, 675–680. [CrossRef]
15. Kopp, E.B.; Medzhitov, R. The Toll-receptor family and control of innate immunity. *Curr. Opin. Immunol.* **1999**, *11*, 13–18. [CrossRef]
16. Ninomiya-Tsuji, J.; Kishimoto, K.; Hiyama, A.; Inoue, J.; Cao, Z.; Matsumoto, K. The kinase TAK1 can activate the NIK-I kappaB as well as the MAP kinase cascade in the IL-1 signalling pathway. *Nature* **1999**, *398*, 252–256. [CrossRef]
17. Kawai, T.; Akira, S. TLR signaling. *Cell Death Differ.* **2006**, *13*, 816–825. [CrossRef]
18. Kawasaki, T.; Kawai, T. Toll-like receptor signaling pathways. *Front. Immunol.* **2014**, *5*, 461. [CrossRef]
19. Zhan, Z.; Xie, X.; Cao, H.; Zhou, X.; Zhang, X.D.; Fan, H.; Liu, Z. Autophagy facilitates TLR4- and TLR3-triggered migration and invasion of lung cancer cells through the promotion of TRAF6 ubiquitination. *Autophagy* **2014**, *10*, 257–268. [CrossRef]
20. Kim, M.J.; Min, Y.; Shim, J.H.; Chun, E.; Lee, K.Y. CRBN Is a Negative Regulator of Bactericidal Activity and Autophagy Activation Through Inhibiting the Ubiquitination of ECSIT and BECN1. *Front Immunol.* **2019**, *10*, 2203. [CrossRef]
21. Min, Y.; Kim, M.J.; Lee, S.; Chun, E.; Lee, K.Y. Inhibition of TRAF6 ubiquitin-ligase activity by PRDX1 leads to inhibition of NFKB activation and autophagy activation. *Autophagy* **2018**, *14*, 1347–1358. [CrossRef]

22. Min, Y.; Lee, S.; Kim, M.J.; Chun, E.; Lee, K.Y. Ubiquitin-Specific Protease 14 Negatively Regulates Toll-Like Receptor 4-Mediated Signaling and Autophagy Induction by Inhibiting Ubiquitination of TAK1-Binding Protein 2 and Beclin 1. *Front Immunol.* **2017**, *8*, 1827. [CrossRef]
23. Min, Y.; Wi, S.M.; Kang, J.A.; Yang, T.; Park, C.S.; Park, S.G.; Chung, S.; Shim, J.H.; Chun, E.; Lee, K.Y. Cereblon negatively regulates TLR4 signaling through the attenuation of ubiquitination of TRAF6. *Cell Death Dis.* **2016**, *7*, e2313. [CrossRef]
24. Xiao, C.; Shim, J.H.; Klüppel, M.; Zhang, S.S.; Dong, C.; Flavell, R.A.; Fu, X.Y.; Wrana, J.L.; Hogan, B.L.; Ghosh, S. Ecsit is required for Bmp signaling and mesoderm formation during mouse embryogenesis. *Genes Dev.* **2003**, *17*, 2933–2949. [CrossRef]
25. Wi, S.M.; Moon, G.; Kim, J.; Kim, S.T.; Shim, J.H.; Chun, E.; Lee, K.Y. TAK1-ECSIT-TRAF6 complex plays a key role in the TLR4 signal to activate NF-κB. *J. Biol. Chem.* **2014**, *289*, 35205–35214. [CrossRef]
26. Mi, W.S.; Park, J.; Shim, J.H.; Chun, E.; Lee, K.Y. Ubiquitination of ECSIT is crucial for the activation of p65/p50 NF-κBs in Toll-like receptor 4 signaling. *Mol. Biol. Cell* **2015**, *26*, 151–160. [CrossRef]
27. Kelly, M.G.; Alvero, A.B.; Chen, R.; Silasi, D.A.; Abrahams, V.M.; Chan, S.; Visintin, I.; Rutherford, T.; Mor, G. TLR-4 signaling promotes tumor growth and paclitaxel chemoresistance in ovarian cancer. *Cancer Res.* **2006**, *66*, 3859–3868. [CrossRef]
28. Hattar, K.; Savai, R.; Subtil, F.S.; Wilhelm, J.; Schmall, A.; Lang, D.S.; Goldmann, T.; Eul, B.; Dahlem, G.; Fink, L.; et al. Endotoxin induces proliferation of NSCLC in vitro and in vivo: Role of COX-2 and EGFR activation. *Cancer Immunol. Immunother.* **2013**, *62*, 309–320. [CrossRef]
29. Peng, Y.F.; Shi, Y.H.; Ding, Z.B.; Ke, A.W.; Gu, C.Y.; Hui, B.; Zhou, J.; Qiu, S.J.; Dai, Z.; Fan, J. Autophagy inhibition suppresses pulmonary metastasis of HCC in mice via impairing anoikis resistance and colonization of HCC cells. *Autophagy* **2013**, *9*, 2056–2068. [CrossRef]
30. Peng, Y.F.; Shi, Y.H.; Shen, Y.H.; Ding, Z.B.; Ke, A.W.; Zhou, J.; Qiu, S.J.; Fan, J. Promoting colonization in metastatic HCC cells by modulation of autophagy. *PLoS ONE* **2013**, *8*, e74407. [CrossRef]
31. Galavotti, S.; Bartesaghi, S.; Faccenda, D.; Shaked-Rabi, M.; Sanzone, S.; McEvoy, A.; Dinsdale, D.; Condorelli, F.; Brandner, S.; Campanella, M.; et al. The autophagy-associated factors DRAM1 and p62 regulate cell migration and invasion in glioblastoma stem cells. *Oncogene* **2013**, *32*, 699–712. [CrossRef]
32. Kiyono, K.; Suzuki, H.I.; Matsuyama, H.; Morishita, Y.; Komuro, A.; Kano, M.R.; Sugimoto, K.; Miyazono, K. Autophagy is activated by TGF-beta and potentiates TGF-beta-mediated growth inhibition in human hepatocellular carcinoma cells. *Cancer Res.* **2009**, *69*, 8844–8852. [CrossRef]
33. Li, J.; Yang, B.; Zhou, Q.; Wu, Y.; Shang, D.; Guo, Y.; Song, Z.; Zheng, Q.; Xiong, J. Autophagy promotes hepatocellular carcinoma cell invasion through activation of epithelial-mesenchymal transition. *Carcinogenesis* **2013**, *34*, 1343–1351. [CrossRef]

cells

MDPI

Article

A Conserved LIR Motif in Connexins Mediates Ubiquitin-Independent Binding to LC3/GABARAP Proteins

Steve Catarino [1,2,3,†], Teresa M Ribeiro-Rodrigues [1,2,3,†], Rita Sá Ferreira [1,2], José Ramalho [4], Christine Abert [5], Sascha Martens [5] and Henrique Girão [1,2,3,*]

1 Coimbra Institute for Clinical and Biomedical Research (iCBR), Faculty of Medicine, University of Coimbra, 3000-548 Coimbra, Portugal; scatarino@fmed.uc.pt (S.C.); teresamrrodrigues@gmail.com (T.M.R.-R.); rita_saferreira@hotmail.com (R.S.F.)

2 Center for Innovative Biomedicine and Biotechnology (CIBB), University of Coimbra, 3000-548 Coimbra, Portugal

3 Clinical Academic Centre of Coimbra (CACC), University of Coimbra, 3000-548 Coimbra, Portugal

4 CEDOC, Chronic Diseases Research Centre, NOVA Medical School|Faculdade de Ciências Médicas, Universidade NOVA de Lisboa, 1150-082 Lisboa, Portugal; jose.ramalho@nms.unl.pt

5 Department of Biochemistry and Cell Biology, Max Perutz Labs, University of Vienna, 1030 Vienna, Austria; chr.abert@gmail.com (C.A.); sascha.martens@univie.ac.at (S.M.)

* Correspondence: hmgirao@fmed.uc.pt; Tel.: +35-12-3948-0221

† These authors contributed equally to this work.

Received: 27 February 2020; Accepted: 3 April 2020; Published: 7 April 2020

check for updates

Abstract: Gap junctions (GJ) are specialized cell-cell contacts formed by connexins (Cxs), which provide direct communication between adjacent cells. Cx43 ubiquitination has been suggested to induce the internalization of GJs, as well as the recruitment of the autophagy receptor p62 to mediate binding to LC3B and degradation by macroautophagy. In this report, we describe a functional LC3 interacting region (LIR), present in the amino terminal of most Cx protein family members, which can mediate the autophagy degradation of Cx43 without the need of ubiquitin. Mutation of the LIR motif on Cx37, Cx43, Cx46 and Cx50 impairs interaction with LC3B and GABARAP without compromising protein ubiquitination. Through in vitro protein-protein interaction assays, we demonstrate that this LIR motif is required for the binding of Cx43 to LC3B and GABARAP. Overall, our findings describe an alternative mechanism whereby Cxs interact with LC3/GABARAP proteins, envisioning a new model for the autophagy degradation of connexins.

Keywords: autophagy; Cx43; GABARAP; gap junction; MAPLC3

1. Introduction

Connexins (Cx) are a family of proteins that directly connect the cytoplasm of adjacent cells through the formation of gap junction (GJ) channels. In humans, twenty members form the connexin family, while at least seventeen members have been reported in rats. Based on sequence homology, connexin genes can be grouped into four classes, α, β, γ and δ [1]. Each connexin consists of four transmembrane regions, two extracellular loops and one intracellular loop, with both the amino and carboxyl terminals facing the cytosol. After synthesis, connexin proteins oligomerize into hexameric structures termed hemichannels and are transported to the plasma membrane, where they may dock to hemichannels from adjacent cells to form a functional GJ channel. The carboxyl terminal is where most differences between connexins are found, and a large part of the protein interactions and post-translational modifications of connexins are thought to occur in this region [2]. Although traditionally associated with establishing direct communication between adjacent cells, the localization of connexins at the plasma membrane as undocked connexin hemichannels as well as its presence in mitochondria, nucleus and extracellular vesicles, suggest that connexins play other non-canonical biological roles [3]. Given all of the above, fine-tuned regulatory mechanisms are required to modulate connexin levels, function and localization.

Although GJ intercellular communication can be modulated through the gating of the channel pore, mechanisms that regulate the number of connexin-containing channels at the plasma membrane have also been implicated in the regulation of intercellular communication. Given the short half-life of Cx43, mechanisms of protein degradation play an important role in the regulation of Cx43 levels and intercellular communication. We and others have shown that ubiquitination of Cx43, modulated by E3 ligases, including neural precursor cell expressed developmentally down-regulated protein 4 (Nedd4), and deubiquitinating enzymes, such as associated molecule with the SH3 domain of STAM (AMSH), dictates the final fate of the protein [4–7]. At the plasma membrane, Cx43 is modified by Lysine 63-linked polyubiquitin chains, which are recognized by the endocytic adaptor epidermal growth factor receptor substrate 15 (Eps15) to trigger the internalization of the protein [7]. Furthermore, Cx43 sorting from early endosomes to the lysosome is also reliant on ubiquitination to mediate the interaction with the endosomal sorting complex required for transport (ESCRT) components hepatocyte growth factor-regulated tyrosine kinase substrate (Hrs) and tumour susceptibility gene 101 (Tsg101) [8,9]. The role of Cx43 ubiquitination in signalling lysosomal GJ degradation is important not only in the endocytic pathway but also in the autophagy process. Indeed, GJ degradation by macroautophagy is also dependent, at least partially, on prior ubiquitination of Cx43 [10,11], a process that requires not only the ubiquitin-binding domain containing autophagy receptor p62, but also, and surprisingly, the endocytic adaptor Eps15 [11]. However, preventing the ubiquitination of Cx43 did not fully abrogate its autophagic degradation, suggesting that alternative mechanisms for targeting Cx43 to autophagosomes exist.

Autophagy is one of the major catabolic pathways in the cell, in which substrates are delivered to the lysosome for degradation. In macroautophagy, a double membrane structure termed the phagophore, grows around the target substrate, eventually closing to form an autophagosome, which subsequently fuses with the lysosome. During phagophore formation, the ubiquitin-like protein microtubule-associated protein 1 light chain 3 (MAPLC3) is conjugated to phosphatydilethanolamine (PE) present on the nascent phagophore, through a mechanism closely resembling ubiquitin conjugation. The E1-like protein autophagy-related protein 7 (ATG7) activates LC3, to allow its transfer to the E2-like protein ATG3. In the final step of the conjugation process, the E3-like complex ATG12-ATG5-ATG16L1 facilitates the transfer of LC3 to PE [12]. The LC3/GABARAP (γ-aminobutyric acid receptor–associated protein) family is comprised of 6 members in humans: LC3A, LC3B, LC3C, GABARAP, GABARAPL1 and GABARAPL2. Although macroautophagy was initially thought of as an unselective process for the bulk degradation of cytosolic content, it is now understood that this process can be highly selective. Substrates to be degraded are labelled with ubiquitin chains, which, in turn, are recognized by ubiquitin-binding domains on autophagy receptors such as p62, neighbour of BRCA1 gene 1 (NBR1), NDP52 and Optineurin. These receptors bridge the substrate to the nascent phagophore by binding to LC3/GABARAP family proteins present in the inner membrane through LC3-interacting region (LIR) motifs. The core consensus sequence of the LIR motif consists of [W/F/Y]-X1-X2-[L/I/V] [12,13]. More recently, a GABARAP interacting motif (GIM) has been described, consisting of the consensus sequence [W/F]-[V/I]-X2-V [14]. Although the ubiquitin-dependent binding of Cx43 to LC3 was previously shown to be mediated by p62 [11], the existence of a putative LIR motif in the amino terminal as well as the fact that hampering Cx43 ubiquitination and p62 silencing does not completely abolish the interaction of Cx43 with LC3, prompted us to investigate whether Cx43 can directly bind to LC3 through a LIR motif, without the need of ubiquitin.

In this study we identify a functional LIR domain in several members of the connexin family, which mediates binding to both LC3B and GABARAP proteins. We also show that connexin ubiquitination alone is insufficient to promote binding to LC3 proteins. Mutation of the LIR motif of Cx43 also rendered the protein resistant to nutrient deprivation-induced degradation. Altogether, the results presented herein support a model in which connexin proteins are targeted for autophagosomal degradation through direct binding to LC3/GABARAP family proteins present on the nascent phagophore membrane.

2. Materials and Methods

2.1. Antibodies and Reagents

Goat polyclonal antibodies against Cx43 (AB1600), GST (AB9919-200), V5 (AB0096-500) and calnexin (AB0041-500) were obtained from SICGEN (Cantanhede, Portugal). Rabbit polyclonal antibodies against Cx43 (H-150) and mouse monoclonal antibodies against p62 (sc-28359) were obtained from Santa Cruz Biotechnology (Heidelberg, Germany). Rabbit polyclonal antibodies against GABARAP (ab109364) and Nedd4 (ab14592) were obtained from Abcam (Cambridge, UK). Rabbit polyclonal antibodies against LC3B (PA1-16930) and mouse monoclonal antibodies against V5 (46-0705) were obtained from Invitrogen (Paisley, UK). Mouse monoclonal antibodies against ubiquitin (P4D1) were obtained from Covance (San Diego, CA, USA). Mouse monoclonal antibodies against EEA1 (610457) were obtained from BD Transduction Laboratories (San Jose, CA, USA). Rabbit polyclonal antibodies against ATG7 (A2856), Phorbol 12-myristate 13-acetate (PMA), 4′,6-Diamidino-2-phenylindole dihydrochloride (DAPI) and cycloheximide (CHX) were obtained from Sigma (Saint Louis, MO, USA). Bafilomycin A1 was obtained from Bioaustralis (Smithfield, Australia).

2.2. Cell Culture and Transfections

HEK293A cells were cultured in Dulbecco's modified Eagle's medium (DMEM) supplemented with 10% fetal bovine serum and antibiotics (100 U/mL penicillin, 100 g/mL streptomycin and 8 μg/mL blasticidin) and maintained at 37 °C under 5% CO_2. Transient transfection of cells was performed with Lipofectamine 2000 (Grand Island, NY, Invitrogen), according to the manufacturer's recommendations.

2.3. Plasmid Constructions

Rat Cx43 cDNA was cloned into a modified pENTR GFP C2 vector [15]. Site-directed mutagenesis was performed to generate the $Cx43^{W4A+L7A}$, $Cx43^{Y265A}$, $Cx43^{Y286A}$ and $Cx43^{Y265,286A}$ mutants from Cx43 cDNA. Plasmids expressing V5-$Cx43^{WT}$ and V5-$Cx43^{W4A+L7A}$ were generated by cloning the appropriate cDNA into a pENTR V5 vector. Rat Cx37 and Cx40 cDNA were amplified from rat aorta tissue using RT-PCR and cloned into a pENTR V5 vector. Rat Cx46 and Cx50 cDNA were amplified from rat lens using RT-PCR and cloned into a pENTR V5 vector. Site-directed mutagenesis was performed to generate the W4A+L7A mutants from Cx37, Cx40, Cx46 and Cx50. Plasmids expressing GST-$Cx43^{WT}$_NT and GST-$Cx43^{W4A+L7A}$_NT were generated by subcloning the first 22 amino acids of $Cx43^{WT}$ and $Cx43^{W4A+L7A}$ into a pGEX4T1 vector. All constructs were verified by DNA sequencing. GFP-LC3B and GFP-GABARAP were obtained by insertion of human LC3B and GABARAP cDNAs into pEGFP-C1. Fusion proteins were subsequently cloned into pETDuet-1 for bacterial expression. The last five amino acids of the LC3B coding sequence and the last amino acid of GABARAP were deleted to mimic ATG4 cleavage. A 6xHis-tag was added C-terminally to purify the proteins [16]. Plasmids expressing GFP-LC3B were kindly provided by Dr. Tamotsu Yoshimori (National Institute for Basic Biology, Okazaki, Japan).

2.4. Viral shRNA Infection

HEK293A cells were incubated with the lentiviral vectors and 8 μg/mL of polybrene for 20 min at room temperature. After 30 min centrifugation at 800× *g* and 32 °C, cells were plated and monitored for the expression of GFP. Lentiviral vectors containing shRNA targeting ATG7 and the control empty vector were kindly provided by Dr. A.M. Cuervo (Albert Einstein College of Medicine, Yeshiva University, New York, USA).

2.5. *siRNA-Mediated Knockdown*

siRNA targeting p62 (s16960 or s16961) and a non-targeting control sequence were obtained from Ambion. Cells were transfected with 20 nM siRNA using Lipofectamine 2000 (Grand Island, NY, Invitrogen) according to manufacturer's recommendations. p62 knockdown was achieved after 24 h of transfection.

2.6. *Immunoprecipitation and Western Blotting*

Cells were rinsed with phosphate buffered saline (PBS) at 4 °C, resuspended in lysis buffer (190 mM NaCl, 50 mM Tris-HCl, 6 mM EDTA, 1% Triton X-100, pH 8.3) supplemented with protease inhibitor cocktail (Roche), 2 mM PMSF, 10 mM iodoacetamide, and incubated on ice for 10 min. The samples were then centrifuged at 10,000× *g* for 10 min and the supernatants used for immunoprecipitation. Briefly, protein G was incubated with goat polyclonal antibodies directed against Cx43 or V5. Incubations proceeded for 1 h at 4 °C, followed by incubation with supernatants for 3 h at 4 °C. The samples were then centrifuged and the protein G-sepharose sediments washed 3 times in an appropriate washing buffer (500 mM NaCl, 50 mM Tris-HCl, 6 mM EDTA, 1% Triton X-100, pH 8.3), resuspended in Laemmli buffer and denatured at 70 °C for 10 min.

For Western blot analysis of the immunoprecipitated proteins, samples were separated using SDS-PAGE, transferred to a nitrocellulose membrane and probed with appropriate antibodies. Inputs represent about 10% of the total amount of protein in the lysates before immunoprecipitation. Immunoprecipitation controls were performed by pooling the lysates of two samples transfected and/or treated in the same conditions, separating them in two fractions and then proceeding with the immunoprecipitation without adding antibody to one of the samples (No Ab). The corresponding pooled lysate appears in the Western blot panels to the right of the No Ab samples.

2.7. *Bacterial Protein Expression and Purification*

GST-Cx43^{WT}_NT and GST-Cx43$^{W4A+L7A}$_NT proteins were expressed in Escherichia coli BL21-CodonPlus (DE3)-RILP Cells (Agilent Technologies, Santa Clara, CA, USA). Bacteria were grown in Luria broth (LB) medium until OD600 ≈ 0.8–1, induced with 0.1 mM isopropylthiogalactoside (IPTG) and grown at 37 °C for 4 h. GST constructs were isolated from harvested cells using Glutathione Sepharose 4B (GE Healthcare, Buckinghamshire, UK) according to manufacturer's recommendations. GFP-LC3B and GFP-GABARAP proteins were expressed in Escherichia coli Rosetta (DE3) pLysS cells. Cells were induced at an OD600 of 0.5 for 16 h at 18 °C with 0.1 mM IPTG. Harvested cells were resuspended in lysis buffer 50 mM 4-(2-hydroxyethyl)-1-piperazineethanesulfonic acid (HEPES) at pH 7.5, 500 mM NaCl, 10 mM imidazole, 2 mM $MgCl_2$, 2 mM β-mercaptoethanol, complete protease inhibitor (Roche, Basel, Switzerland) and DNase I and lysed using a freeze–thaw cycle followed by brief 30 s sonication. Lysates were cleared using ultracentrifugation at 140,000 g for 30 min at 4 °C (Beckman, Brea, CA, USA, Ti45 rotor). Supernatants were applied to Ni-NTA columns (GE Healthcare, Buckinghamshire, UK), and constructs were eluted via a stepwise imidazole gradient (50, 75, 100, 150, 200, and 300 mM) [16].

2.8. *GFP-Trap Pull-Down Assay*

Five microlitres of GFP-Trap beads slurry (ChromoTek, Munich, Germany) were mixed with a 5 μM solution of GFP-fused bait proteins (GFP-LC3B or GFP-GABARAP) and incubated on a rotating wheel at 4 °C for 1 h. Subsequently the beads were washed twice with 150 mM NaCl, 50 mM Tris at pH 7.4, and incubated with 20 μg of prey solution (GST-Cx43^{WT}_NT or GST-Cx43$^{W4A+L7A}$_NT). Precipitates were analysed using Western blot using goat polyclonal antibodies against GST.

2.9. Microscopy-Based Protein-Protein Interaction Assay

Five microlitres of glutathione Sepharose 4B beads slurry (GE Healthcare, Buckinghamshire, UK) were mixed with 20 µg of GST-fused bait proteins (GST-Cx43^{WT}_NT or GST-Cx43$^{W4A+L7A}$_NT) and incubated on a rotating wheel at 4 °C for at least 30 min. Subsequently the beads were washed twice with 150 mM NaCl, 50 mM Tris at pH 7.4 and resuspended in 6 µL of the same buffer. Of a 5 µM prey solution (GFP-LC3B or GFP-GABARAP), 10 µL was plated into the well of a 384-well glass-bottom microplate (Greiner Bio-One, Frickenhausen, Germany), after which 10% of the previous beads solution was added. Samples were then imaged on a confocal microscope. To quantify the protein recruitment to beads the maximum brightness along a straight line drawn through a single bead was taken (maximal fluorescence). Next, the average brightness of an empty portion of each picture was measured (background fluorescence) and subtracted from the maximal fluorescence for each bead [16].

2.10. Triton X-100 Fractionation Assay

The detergent solubility assay with 1% Triton X-100 was performed essentially as described previously by others [17]. Cells were resuspended in lysis buffer (190 mM NaCl, 50 mM Tris-HCl, 6 mM EDTA, 1% Triton X-100, pH 8.3) supplemented with protease inhibitor cocktail (Roche, Basel, Switzerland), 2 mM PMSF and 10 mM iodoacetamide. Samples were then ultracentrifuged at 100,000 g for 50 min and the supernatant recovered (Triton X-100 soluble fraction). The detergent-insoluble pellets were resuspended in lysis buffer supplemented with 0.1% SDS (Triton X-100 insoluble fraction) and sonicated. Laemmli buffer was added to 10% of each Triton X-100 soluble and insoluble fractions and denatured at 70 °C for 10 min before SDS-PAGE analysis. The remainder of each sample was used for immunoprecipitation.

2.11. Immunofluorescence

HEK293A cells grown on glass coverslips were fixed with 4% paraformaldehyde in PBS. The samples were then washed with PBS, permeabilised with 0.2% v/v Triton X-100 in PBS and blocked with 2% w/v BSA in PBS for 20 min prior to incubation with primary antibodies overnight at 4 °C. The samples were then washed three times with PBS before incubation with the secondary antibody for 1 h at room temperature. The specimens were rinsed in PBS and mounted with MOWIOL 4-88 Reagent (Calbiochem, San Diego, CA, USA). Nuclei were stained with DAPI. For controls primary antibodies were omitted. Imaging was performed on a laser-scanning confocal (Zeiss LSM 710, Carl Zeiss, Oberkochen, Germany) with Plan-Apochromat 63X/1.4 oil objective (Carl Zeiss, Oberkochen, Germany). Colocalization of Cx43 and GFP-LC3B or EEA1 was quantitated using Pearson's correlation coefficient (PCC) using the Coloc 2 plugin of ImageJ software (http://imagej.net/Coloc_2). Cells were selected as regions of interest (ROIs) using the freehand selection tool before running the plugin. Pearson's R values (no threshold) reported were plotted in a graph.

2.12. Statistical Analysis

Data were displayed with means on a scatter dot plot. For data with non-Gaussian distribution, statistical significance was determined using a non-parametric Kruskall-Wallis test followed by a Dunn's multiple comparison test, non-parametric Mann-Whitney test or multiple Student's t-test with Holm-Sidak's correction. Data was analysed and graphs were assembled with GraphPad Prism 6 for Windows, version 6.01 (GraphPad Software, Inc., San Diego, CA, USA). Results were considered significantly different for $p < 0.05$. The tests used in each experiment are indicated in the figure legends. Graphical data depict individual experiments except when noted in the figure legend.

3. Results

3.1. The Connexin Protein Family Contains a Conserved LIR Motif in Its Amino Termini

In silico analysis of the amino acid sequence of rat Cx43 through the iLIR webtool [18] revealed a potential LIR motif in the amino terminal of the protein with core residues corresponding to tryptophan 4 (W4) and leucine 7 (L7). Further analysis of the amino acid sequence of the human [1] and rat connexin families (Figure S1) evidenced that these residues were strongly conserved within the family, being present in 17 of the 20 human connexins and 15 of the 17 rat connexins. Given these observations we proceeded to explore the interaction between several connexins of the alpha subfamily with LC3B. For this purpose we transiently transfected human embryonic kidney 293A (HEK293A) cells (which express low levels of endogenous Cx43) with a vector encoding a green fluorescent protein (GFP) tagged LC3B chimera (to overcome the low level expression of LC3B in this cell line), in addition to several connexins tagged with V5. Lysates were then subjected to immunoprecipitation against the V5 tag, after which we assessed the levels of co-immunoprecipitated GFP-LC3B. As depicted in Figure 1A, all connexins tested (Cx37, Cx40, Cx43, Cx46 and Cx50) interact with LC3B. It has been shown that the nature of the amino acids surrounding the core LIR motif, namely where hydrophobicity and charge is concerned, affects binding to LC3 [13]. Therefore, it is conceivable that variations in the levels of interaction with LC3 among the different connexins tested reflect differences in the amino acids surrounding the LIR motif on each connexin. Importantly, we also show that all of these connexins are modified with ubiquitin, which, to our knowledge, has never been reported for Cx37, Cx46 and Cx50 [4,19]. Having established that connexin binds to LC3B, we then investigated whether the putative LIR motif found in the amino terminal of these proteins was important to mediate this interaction. For this purpose we used site-directed mutagenesis to mutate the core LIR motif residues tryptophan 4 and leucine 7 to alanine, to generate the corresponding $Cx^{W4A+L7A}$ mutants. As depicted in Figure 1B–F, mutation of the LIR motif in all connexins tested, with the exception of Cx40, led to a substantial decrease in the amount of LC3B that is co-immunoprecipitated. Given our previous observation that Cx43 ubiquitination induces its interaction with LC3B [11], there was the possibility that mutation of the LIR motif was reducing the interaction with LC3B due to a defect in ubiquitination. However, the W4A+L7A mutation had no significant effect on the ubiquitination levels of all connexins tested (Figure 1B–F).

3.2. Cx43 Amino Terminal Peptides Interact with LC3B and GABARAP In Vitro

To further ascertain that the LIR motif present in connexins can interact directly with LC3 proteins and is thus a bona fide functional motif we proceeded to conduct in vitro experiments using recombinant glutathione S-transferase (GST) proteins fused to the amino terminal of mutant or wild type Cx43. Cx43 was chosen as the model for these assays due to being the most widely expressed connexin in the body and the main focus in connexin research. Given that Cx43 is an integral transmembrane protein, it is difficult to produce and isolate the full-length protein from bacterial or insect protein production systems, and thus only the amino terminal was used in these pull-down experiments. As a first approach, we carried out pull-down experiments in which recombinant GFP-LC3B and GFP-GABARAP were bound to GFP-TRAP beads to serve as bait. GST-$Cx43^{WT}$_NT and GST-$Cx43^{W4A+L7A}$_NT were used as prey in the experiment. As depicted in Figure S2, there is a very clear interaction between GFP-GABARAP and GST-$Cx43^{WT}$_NT. Importantly, when the LIR motif is mutated, this interaction is substantially decreased. Unexpectedly, we were not able to detect any interaction between GST-$Cx43^{WT}$_NT and GFP-LC3B in this assay, even though the full-length form of Cx43 readily interacted with GFP-LC3B when co-expressed in HEK293A cells (Figure 1A,D). To further confirm this in vitro data, we performed a microscopy based protein-protein interaction assay [16], in which the GST-tagged $Cx43^{WT}$_NT or $Cx43^{W4A+L7A}$_NT constructs are first bound to glutathione beads and subsequently incubated with recombinant GFP-LC3 or GFP-GABARAP. The intensity of fluorescence formed around the glutathione beads reflects the interaction between Cx43 and either

LC3B or GABARAP. Representative images displayed in Figure 2 show a strong interaction between GST-Cx43[WT]_NT and GFP-GABARAP, which is almost entirely abolished in the LIR mutant (middle row panels). We were also able to detect a weak interaction between GST-Cx43[WT]_NT and GFP-LC3B, which is also abolished when the LIR mutant is used (top row panels). No interaction was observed when Cx43 coated beads were incubated with GFP alone.

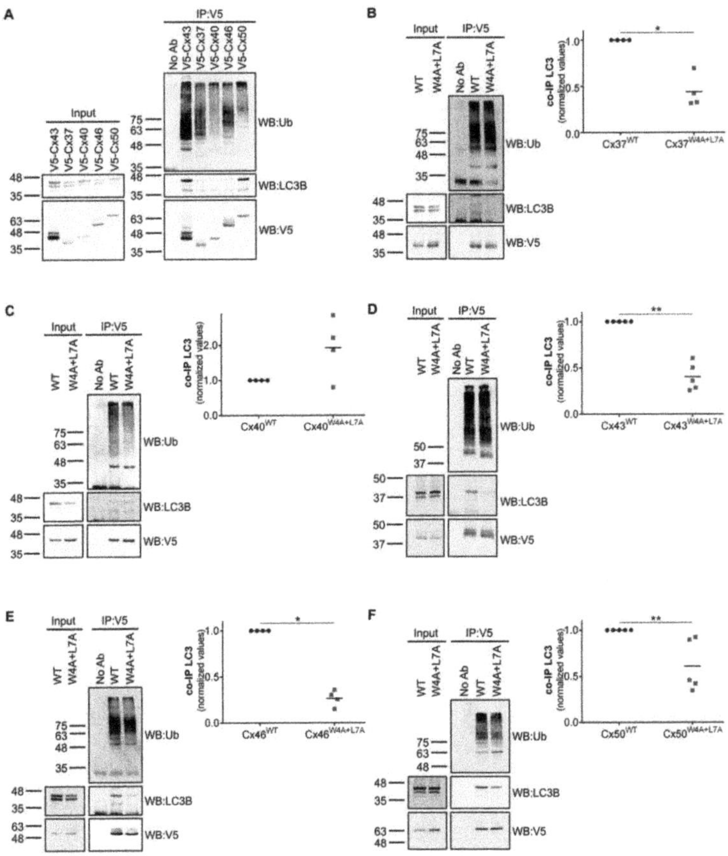

Figure 1. Connexin family members interact with LC3B through a LC3-interacting region (LIR) motif on their amino terminal. HEK293A cells were co-transfected with several different V5-connexin constructs and GFP-LC3B. Cell lysates were then subjected to immunoprecipitation with goat polyclonal antibodies against V5. The precipitates were analysed using Western blot using mouse monoclonal antibodies against V5 and ubiquitin, or rabbit polyclonal antibodies against LC3B (**A**). HEK293A cells were co-transfected with GFP-LC3B and V5 constructs of either wild type or W4A+L7A mutants of Cx37 (**B**), Cx40 (**C**), Cx43 (**D**), Cx46 (**E**) and Cx50 (**F**). Cell lysates were then subjected to immunoprecipitation with goat polyclonal antibodies against V5. The precipitates were analysed using Western blot using mouse monoclonal antibodies against V5 and ubiquitin, or rabbit polyclonal antibodies against LC3B. The level of co-immunoprecipitated LC3B with each connexin was quantified, normalized to the immunoprecipitated connexin and plotted. *$p < 0.05$, **$p < 0.01$, ns: non-significant using the Mann-Whitney test.

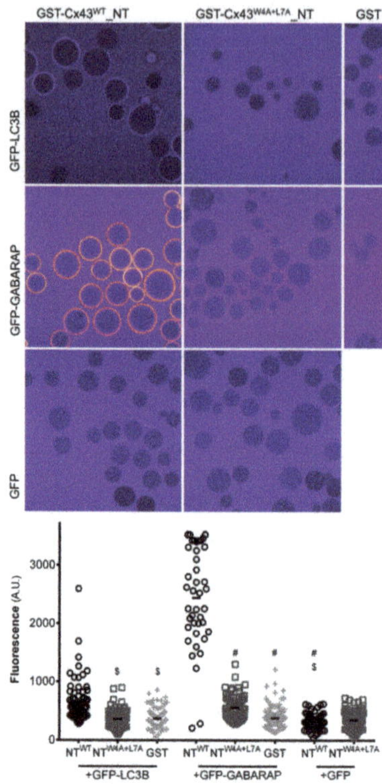

Figure 2. The LIR motif in the amino terminal of Cx43 interacts with GABARAP in vitro. Peptides comprising either the amino terminal of wild type (NT^{WT}) or W4A+L7A mutant ($NT^{W4A+L7A}$) Cx43 fused to GST were bound to glutathione beads and subsequently incubated with recombinant GFP-LC3B or GFP-GABARAP, after which the beads were transferred to a glass-bottom microplate and imaged using a confocal microscope. The GFP signal is shown in false colour (ImageJ: Fire). The intensity of the GFP signal surrounding the glutathione beads was measured and plotted in a graph. \$ $p < 0.0001$ vs. NT^{WT}+GFP-LC3B, # $p < 0.0001$ vs. NT^{WT}+GFP-GABARAP using a Kruskal-Wallis test, followed by Dunn's multiple comparison test. Data represents four individual experiments.

3.3. Cx43 and LC3B Interaction Occurs in Endocytic Compartments

Having established that the LIR motif present in Cx43 is functional, we sought to investigate where in the cell this interaction preferentially occurs. Triton X-100 fractionation assays are a crude method to separate free connexin hemichannels, which are predominantly recovered in the Triton X-100 soluble fraction, from GJ channels localized at plaques and internalized vesicles, which partition into the insoluble fraction [17]. Results depicted in Figure 3A show a clear preference for LC3B to interact with Cx43 present in the insoluble fraction. Serum starvation also appeared to induce an increase in the interaction in the insoluble fraction. Given that GJ plaques are internalized as a double membrane structure, these results suggest that LC3B interaction with Cx43 could occur at the plasma membrane or during the endocytosis/lysosomal pathway. To clarify this we evaluated the subcellular localization of Cx43 and GFP-LC3B through confocal immunofluorescence imaging. While the presence of Cx43WT and GFP-LC3B in the same intracellular compartments is easily apparent (Figure 3B, top panels), colocalization between $Cx43^{W4A+L7A}$ and GFP-LC3B is much less discernible (Figure 3B, bottom panels), which was confirmed by quantification of the colocalized fluorescence signals. Taken together, these results suggest that Cx43 interaction with LC3B occurs mainly in internalized vesicles.

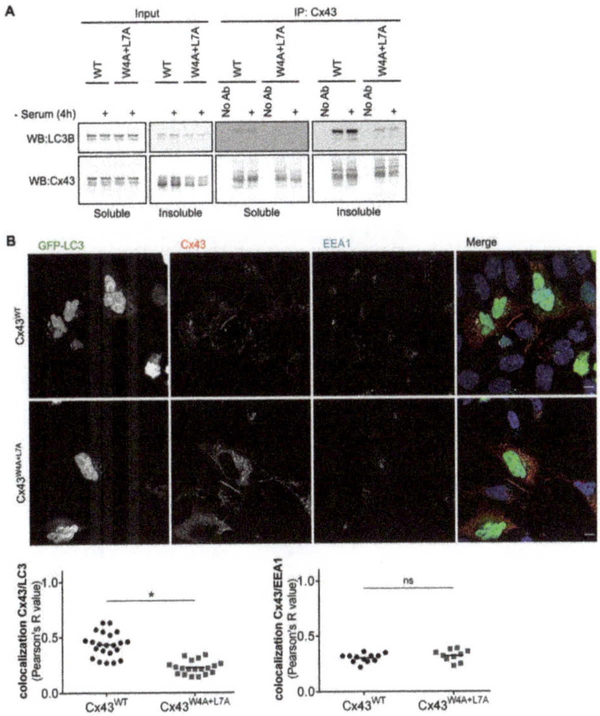

Figure 3. LC3B preferentially interacts with Cx43 present in post-endocytic vesicles. (**A**) HEK293A cells co-transfected with GFP-LC3B and wild type (WT) or mutant (W4A+L7A) Cx43 were incubated in medium without serum for 4 h (-Serum). Cell lysates were then separated into Triton X-100 soluble and insoluble fractions and immunoprecipitated with goat polyclonal antibodies directed against Cx43. Precipitates were then analysed using Western blot using rabbit polyclonal antibodies against Cx43 or LC3B. (**B**) HEK293A cells grown on coverslips were co-transfected with plasmids expressing GFP-LC3B and either $Cx43^{WT}$ or $Cx43^{W4A+L7A}$. Subcellular distribution of the proteins was evaluated using confocal microscopy, and co-localization of Cx43 (red) with GFP-LC3B (green) or EEA1 (cyan) was quantified and plotted in a graph. Scale bars, 10 µm. Dots represent values obtained for individual images (n = 4). *$p < 0.05$ using the Mann-Whitney test.

Mutation of the LIR motif of Cx43 appeared to not affect the stability of Cx43 at the plasma membrane, as demonstrated by its accumulation in GJ plaques at the cell surface. However, given the lower colocalization of $Cx43^{W4A+L7A}$ with LC3-positive vesicles, when compared to $Cx43^{WT}$, we sought to elucidate whether this mutation impairs Cx43 endocytosis. To address this question we performed confocal microscopy experiments using antibodies directed against early endosome antigen 1 (EEA1), a marker of early endosomes. Data presented in Figure 3B shows no significant difference between $Cx43^{WT}$ and $Cx43^{W4A+L7A}$ colocalization with EEA1, suggesting that mutation of the LIR motif of Cx43 does not affect its sorting into early endosomes and presumably its internalization.

3.4. Tyrosine Residue 265 of Cx43 is Important for the Interaction with LC3B

In an attempt to narrow the possible cellular locations where the binding between Cx43 and LC3B occurs, we mutated tyrosine residues Y265 and Y286 on Cx43, which are part of two independent endocytic tyrosine-sorting signals, leading to the accumulation of Cx43 at the plasma membrane [20]. Tyrosine residue Y286 is also part of a proline-rich motif, which is required for Nedd4 binding and Cx43 ubiquitination (Table S1) [5,6,21]. Results presented in Figure 4A show that the ubiquitination defective mutant Cx43^{Y286A}, with an intact LIR motif, displayed an intermediate behaviour between Cx43^{WT} and Cx43$^{W4A+L7A}$ concerning the interaction with LC3B, reflecting the importance of ubiquitin in mediating the interaction between LC3 and Cx43. On the other hand, mutants Cx43^{Y265,286A} and Cx43^{Y265A} present a strong decrease in the interaction with LC3B. Considering that mutation of the Y265 residue does not significantly affect ubiquitination of Cx43^{Y265A} and the amino acids surrounding Y265 on Cx43 do not resemble the core consensus sequence for a LIR motif, structural changes likely underlie this effect.

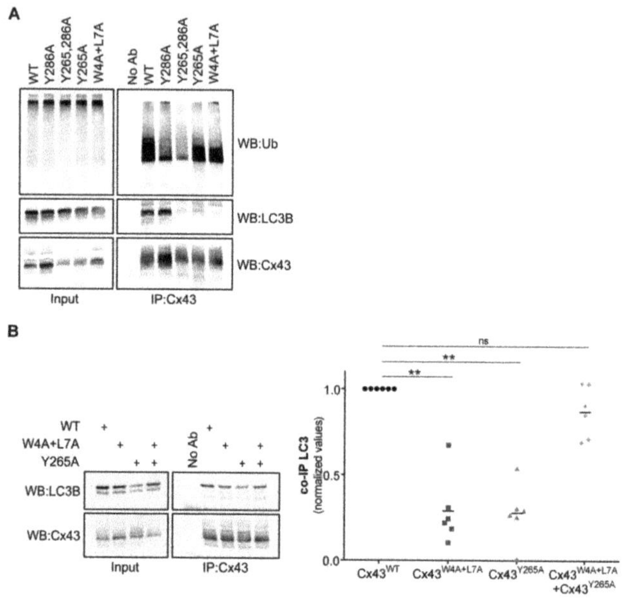

Figure 4. Tyrosine 265 of Cx43 is important for the interaction with LC3B. (**A**) HEK293A cells were co-transfected with GFP-LC3B and either wild type (WT) or mutant Cx43 (Y286A, Y265,286A, Y265A, W4A+L7A). Cell lysates were then subjected to immunoprecipitation with goat polyclonal antibodies against Cx43, and the precipitates were analysed using Western blot using mouse monoclonal antibodies against ubiquitin, or rabbit polyclonal antibodies against Cx43 and LC3B. (**B**) HEK293A cells co-transfected with GFP-LC3B and several combinations of wild type (WT) or mutant Cx43 (Y265A, W4A+L7A). Cell lysates were then subjected to immunoprecipitation with goat polyclonal antibodies against Cx43, and the precipitates were analysed using Western blot using rabbit polyclonal antibodies against Cx43 and LC3B. The level of co-immunoprecipitated LC3B with each connexin was quantified, normalized to the immunoprecipitated connexin and plotted in a graph. ** $p < 0.01$, ns: non-significant using a Kruskal-Wallis test, followed by Dunn's multiple comparison test.

Before being delivered to the plasma membrane, connexin proteins may oligomerize into hexameric structures containing different connexins called heteromeric hemichannels. Taking this into account, we evaluated the effect of co-expressing connexin mutants Cx43$^{W4A+L7A}$ and Cx43^{Y265A} in the binding with LC3B. After establishing that these connexin mutants co-oligomerized into the same hemichannels (data not shown), we then co-expressed GFP-LC3B with both Cx43$^{W4A+L7A}$ and Cx43^{Y286A} in HEK293A cells, and immunoprecipitated Cx43 from the lysates. Results depicted in Figure 4B show that when the two connexin mutants are co-expressed the binding to LC3B is restored, presumably due to the formation of heteromeric channels in which the non-mutated region of one protein can somehow compensate for the mutated region of the other. Thus, both the amino and carboxyl terminals of Cx43 play a role in LC3 binding.

3.5. LC3 Lipidation Is Not A Pre-Requisite for Binding to Cx43

LC3 family proteins exist in the cell as a non-lipidated form (LC3-I) or as a lipidated form when they are conjugated to PE (LC3-II). LC3-II is present on the membrane of phagophores where it can act as a binding location for LIR-motif containing autophagic receptors such as p62 [12]. Although the GFP-LC3B construct is indeed capable of being conjugated to PE, distinguishing its LC3-I and LC3-II forms on a Western blot is not always possible. Therefore, we sought to investigate if Cx43 binds preferentially to one of these forms through the use of HEK293A cells knocked down for ATG7, to impede the conjugation of LC3-I to PE. As observed in Figure S3, cells with lower levels of ATG7 present a defect in the lipidation of both GFP-LC3B and endogenous LC3B, as represented by the decrease in the lower molecular weight band corresponding to LC3B-II, even when autophagic degradation is interrupted by treatment with Bafilomycin A1 (Baf A1), which inhibits the fusion of autophagosomes with the lysosome. However, ATG7 depletion has no significant impact on the binding of GFP-LC3B to Cx43, suggesting that Cx43 interacts preferentially with LC3B-I.

3.6. Role of Phosphorylation in the Binding of Cx43 to LC3B

It has been established that Cx43 phosphorylation induces the ubiquitination of the protein and promotes its interaction with p62 [4,11]. Given that LC3B can interact directly with the LIR motif present on Cx43, we assessed the effect of Cx43 phosphorylation on this interaction. HEK293A cells co-transfected with GFP-LC3B and either wild type or Cx43 mutated in the LIR motif were treated with phorbol 12-myristate 13-acetate (PMA) to promote Cx43 phosphorylation and ubiquitination. As demonstrated in Figure 5A, PMA treatment leads to the accumulation of higher molecular weight forms of Cx43, consistent with Cx43 phosphorylation [22]. Furthermore, although PMA treatment induced an increase in the interaction of $Cx43^{WT}$ with LC3B, this effect was milder for $Cx43^{W4A+L7A}$, and insufficient to reach the interaction levels of the wild type protein. Notably, interaction of $Cx43^{W4A+L7A}$ with p62 was only slightly decreased, indicating that binding to p62 does not rely on this LIR motif. Thus, although Cx43 phosphorylation does indeed promote the interaction between Cx43 and LC3 (in accordance with previous reports), it is insufficient to compensate for the absence of the LIR motif on Cx43.

3.7. Role of p62 in the Binding of Cx43 to LC3B

It has been previously shown that ubiquitination of Cx43 leads to the recruitment of the macroautophagy adaptor p62 and subsequent degradation of the protein [11]. As p62 can recruit cargo to nascent phagophores through its LIR-mediated binding to LC3, we evaluated the impact of depleting p62 in Cx43 binding to LC3B. Results presented in Figure 5B show that p62 depletion leads to a decrease in the interaction between LC3B and $Cx43^{WT}$, whereas the already low interaction with the Cx43 mutants was not significantly altered. This suggests that although Cx43 interaction with LC3B is partially mediated by ubiquitination and p62 binding, this alone cannot overcome the loss of the LIR motif on Cx43.

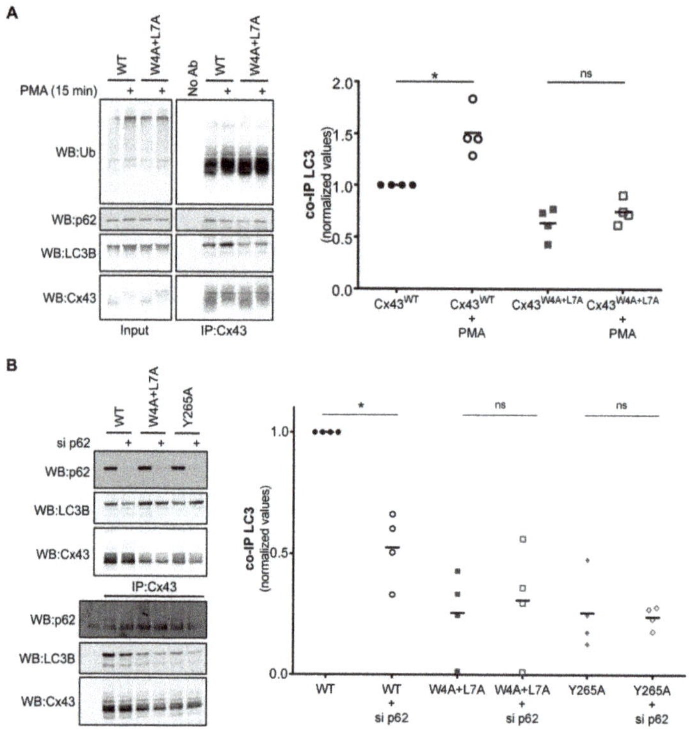

Figure 5. Role of phosphorylation and p62 in the binding of Cx43 to LC3B. (**A**) HEK293A cells co-transfected with GFP-LC3B and either wild type (WT) or mutant Cx43 (W4A+L7A) were incubated with 100 ng/mL PMA for 15 min. Cell lysates were then subjected to immunoprecipitation with goat polyclonal antibodies against Cx43, and the precipitates were analysed using Western blot using mouse monoclonal antibodies against ubiquitin, or rabbit polyclonal antibodies against Cx43, Nedd4 and LC3B. The level of co-immunoprecipitated LC3B with connexin in each condition was quantified, normalized to the immunoprecipitated connexin and plotted in a graph. *$p < 0.05$, ns: non-significant using the Mann-Whitney test. (**B**) HEK293A cells were co-transfected with GFP-LC3B, either wild type (WT) or mutant Cx43 (W4A+L7A), and with siRNA directed against p62. Cells were harvested after 24 h, the cell lysates were then subjected to immunoprecipitation with goat polyclonal antibodies against Cx43 and the precipitates were analysed using Western blot using mouse monoclonal antibodies against p62, or rabbit polyclonal antibodies against Cx43 and LC3B. The level of co-immunoprecipitated LC3B with connexin in each condition was quantified, normalized to the immunoprecipitated connexin and plotted in a graph. *$p < 0.05$, ns: non-significant using the Mann-Whitney test.

3.8. GABARAP Interacts with the LIR Motif of Connexins

In vitro assays showed that GABARAP had a much stronger interaction with the LIR motif of Cx43 than LC3B (Figure 2 and Figure S2). To investigate this interaction in vivo, HEK293A cells were co-transfected with GFP-GABARAP and either wild type or mutant connexin. In accordance with the results obtained with LC3B, Cx43 binds to GABARAP and this interaction is reduced upon mutation of the LIR motif, and also when residue Tyrosine 265 is mutated to alanine (Figure 6A). Given that we observed a much stronger interaction between Cx43 and GABARAP in the in vitro assays, we investigated if co-expression of both LC3B and GABARAP could exert a dominant negative effect over the other for binding to Cx43. As observed in Figure 6B, when GFP-LC3B and GFP-GABARAP are co-expressed in HEK293A cells, neither protein has a special predominance for binding with Cx43.

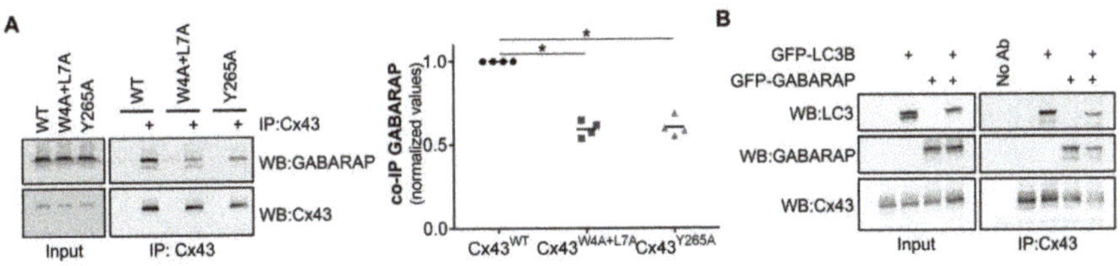

Figure 6. Mutation of the LIR motif of Cx43 impairs its binding to GABARAP and confers resistance to nutrient deprivation induced autophagy. (**A**) HEK293A cells were co-transfected with GFP-GABARAP and either wild type (WT) or mutant Cx43 (W4A+L7A, Y286A). Cell lysates were then subjected to immunoprecipitation with goat polyclonal antibodies against Cx43, and the precipitates were analysed using Western blot using rabbit polyclonal antibodies against Cx43 and GABARAP. The level of co-immunoprecipitated GABARAP with each connexin was quantified, normalized to the immunoprecipitated connexin and plotted in a graph. $*p < 0.05$, ns: non-significant using the Mann-Whitney test. (**B**) HEK293A cells were co-transfected with Cx43 and either GFP-LC3B, GFP-GABARAP or both. Cell lysates were then subjected to immunoprecipitation with goat polyclonal antibodies against Cx43, and the precipitates were analysed using Western blot using rabbit polyclonal antibodies against Cx43, LC3B and GABARAP.

We also investigated the binding of GABARAP to other connexins by co-transfecting HEK293A cells with GFP-GABARAP and either wild type or mutant Cx37, Cx40, Cx46 and Cx50 (Figure S4). In accordance with what was described above for LC3B, all connexins tested interacted with GABARAP. Likewise, mutation of the LIR motif inhibited this interaction for all connexins, with the exception of Cx40. Thus, the conserved LIR motif in the amino terminal of the connexin family appears to mediate the direct interaction with members of the LC3/GABARAP family.

3.9. Mutation of the LIR Motif of Cx43 Renders the Protein Resistant to Nutrient Deprivation Induced Autophagic Degradation

Previous work from our laboratory demonstrated that nutrient deprivation induced Cx43 ubiquitination, p62 and Eps15 recruitment and, ultimately, Cx43 autophagic degradation [11]. We next investigated the role of the LIR motif of Cx43 during nutrient deprivation induced autophagy degradation of Cx43. HEK293A cells transfected with plasmids encoding either $Cx43^{WT}$ or $Cx43^{W4A+L7A}$ were incubated with media without serum for 4 h or with the protein synthesis inhibitor cycloheximide (CHX) for 2 or 4 h. As displayed in Figure 7A, CHX assays show that mutation of the LIR motif of Cx43 has a moderate effect in prolonging the half-life of the protein in basal conditions. In contrast, mutation of the LIR motif of Cx43 rendered the protein strongly resistant to nutrient deprivation induced autophagic degradation, suggesting a role for the motif in directing Cx43 to degradation in stress conditions.

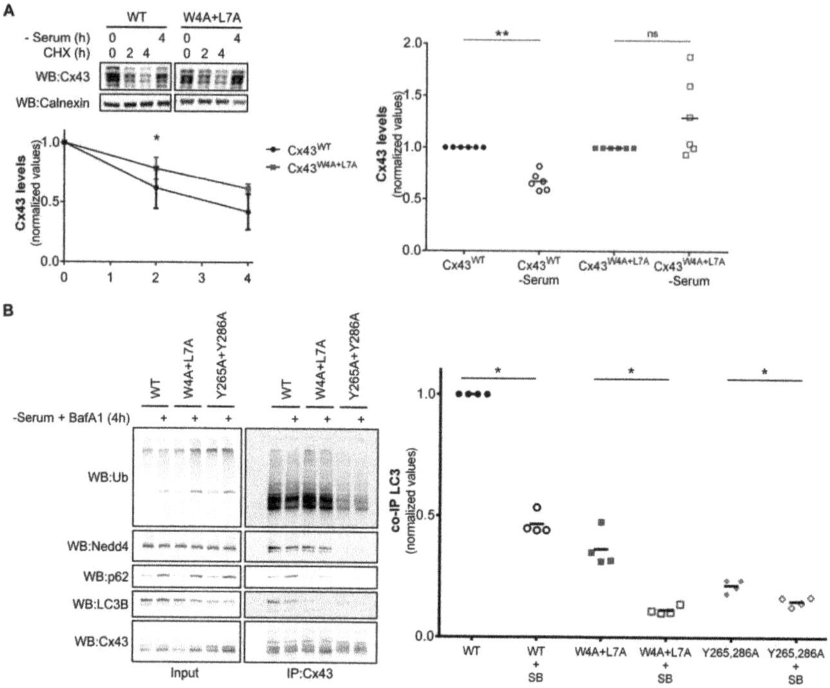

Figure 7. Mutation of the LIR motif of Cx43 renders the protein resistant to nutrient deprivation induced autophagic degradation. (**A**) HEK293A cells transfected with either wild type (WT) or mutant Cx43 (W4A+L7A) were either incubated with 50 μg/mL cycloheximide (CHX) for 2 or 4 h or subjected to starvation with media without serum for 4 h. Lysates were analysed using Western blot using goat polyclonal antibodies against Cx43 and calnexin. Cx43 levels were quantified, normalized to calnexin and plotted in a graph. $^{*}p < 0.05$ using the multiple Student's t-test with Holm-Sidak's correction. (**B**) HEK293A cells co-transfected with GFP-LC3B and either wild type (WT) or mutant Cx43 (W4A+L7A, Y265,286A) were incubated in media without serum supplemented with 50 μM BafA1 for 4 h (SB). Cell lysates were then subjected to immunoprecipitation with goat polyclonal antibodies against Cx43, and the precipitates were analysed using Western blot using mouse monoclonal antibodies against ubiquitin and p62, or rabbit polyclonal antibodies against Cx43, Nedd4 and LC3B. The level of co-immunoprecipitated LC3B with connexin in each condition was quantified, normalized to the immunoprecipitated connexin and plotted in a graph. $^{*}p < 0.05$ using the Mann-Whitney test.

We further assessed the importance of the LIR motif in the binding to LC3B during nutrient deprivation induced autophagy activation. In an attempt to preserve the interaction between LC3B and Cx43, HEK293A cells were incubated with the autophagosome/lysosome fusion inhibitor BafA1. In accordance, interaction with p62 is increased when the autophagy pathway is blocked, even in the case of Cx43$^{W4A+L7A}$, as expected, since ubiquitination is not affected. Likewise, the Cx43^{Y265,286A} mutant presents a reduced interaction with Nedd4 and impaired ubiquitination, leading to less interaction with p62 and LC3B (Figure 7B). Surprisingly, we observed a decrease in the interaction between LC3B and Cx43, both in the wild type and the mutants, in the presence of BafA1, suggesting that Cx43 binding to LC3B, either directly or through ubiquitin/p62, is a transient process that occurs and is likely disrupted before the final autophagosome/lysosome fusion step. Alternatively, given that p62 interaction with Cx43 in the presence of BafA1 increases with the concomitant decrease in LC3B interaction, it is possible that binding of p62 to Cx43 is displacing LC3B from Cx43.

4. Discussion

Previous studies demonstrated that connexin proteins are degraded by autophagy. We and others have reported that degradation of Cx43 by macroautophagy requires prior ubiquitination of Cx43 and its further recognition by p62 [10,11]. However, preventing Cx43 ubiquitination by siRNA depletion of the E3 ubiquitin ligase Nedd4, or mutation of Tyrosine 286, a core residue in the PPxY motif used for Nedd4 binding to Cx43, failed to completely abrogate degradation by autophagy, suggesting that a parallel pathway, independent of Cx43 ubiquitination, may exist to deliver the protein to the autophagosome. In this manuscript we describe a LIR motif, present in the amino terminal, that is conserved in almost all human and rat connexin family members. Mutation of the core consensus residues tryptophan 4 and leucine 7 to alanine in Cx37, Cx43, Cx46 and Cx50 lead to a substantial decrease in binding to both LC3B and GABARAP. Moreover, in the case of Cx43, interaction with LC3B occurs mostly in internalized vesicles, as evidenced by the Triton X-100 fractionation experiments and immunofluorescence imaging. Lastly, mutation of the LIR motif of Cx43 also renders the protein resistant to nutrient deprivation induced autophagy. Data from each mutant is summarized in Table S1.

The results of in vitro protein-protein interaction assays suggest that the LIR motif present on the amino terminal of Cx43 binds preferentially to GABARAP as opposed to LC3B. A subset of LIR motifs that bind preferentially to GABARAP (GIM motif), characterized by the consensus sequence [W/F]-[V/I]-X2-V, was recently described [14]. Curiously, none of the LIR motifs present in the human and rat connexin family members can be considered as GIMs. More intriguing is the observation that this preference for binding to GABARAP present in the in vitro assays performed using only the amino terminal of Cx43 is not carried over to the interaction experiments using full-length Cx43 expressed in mammalian cells. Indeed, when GFP-LC3B is co-expressed with GFP-GABARAP, Cx43 does not display any binding preference. It is plausible that post-translational modifications, which are not available during the production of the GST-Cx43^{WT}_NT chimera in bacteria, are necessary to facilitate the binding of Cx43 to LC3B. For instance, although the LIR consensus sequence is commonly defined as [W/F/Y]-X1-X2-[L/I/V], the presence of negative residues surrounding the core residues of the LIR motif has been described to modulate the interaction with LC3B [13]. Accordingly, it is conceivable that in vivo phosphorylation of amino acid residues surrounding the LIR motif of Cx43 promote interaction with LC3. Furthermore, changes in the topology and conformation adopted by Cxs when expressed in cells may justify the discrepancy of in vitro and in vivo data. Along the synthetic pathway, Cx monomers oligomerize to form a hexameric structure, the connexon, before reaching the plasma membrane. Thus, a fully mature Cx channel exists as a structure that would contain 6 LIR motifs in close proximity. Binding to LC3B in vivo may be facilitated by the presence of several LIR motifs clustered in a limited region and in a specific orientation, which is not easily replicated on the surface of the beads used in the in vitro experiments. Other regions of the protein may also facilitate the binding of LC3 to Cx43, as will be discussed further below.

Although all the five alpha subfamily connexins (Cx37, Cx40, Cx43, Cx46 and Cx50) tested can bind to LC3 proteins, only mutation of the LIR motif of Cx40 did not impair its interaction with either LC3B or GABARAP. It is possible that in the case of Cx40, the interaction we observed with LC3B/GABARAP occurred through other regions of the protein or was mediated by ubiquitination. Moreover, differences in the amino terminal of Cx40 may also contribute to render the LIR motif nonfunctional. It will be interesting to investigate this issue in the future considering that Cx40 can co-oligomerize and is found co-expressed with Cx37 and Cx43 in several tissues.

Surprisingly, we demonstrated that mutation of Tyrosine 265 on the carboxyl terminal of Cx43 also prevented binding to LC3B and GABARAP, although this residue was not part of any discernible LIR motif. Tyrosine 265 and 286 are part of two different endocytic tyrosine-sorting signals that were previously shown to play a role in the endocytosis of Cx43 [6,20], whereas Tyrosine 286 is additionally part of a PPxY motif required for binding to Nedd4 and subsequent Cx43 ubiquitination [5,21]. At a first glance, the impaired interaction of Cx43^{Y265A} with LC3B may suggest that the interaction could only occur after Cx43 internalization. Nevertheless, Cx43^{Y286A}, which also accumulates at the plasma membrane, still retains a relatively high binding level to LC3B, although to a lesser extent when compared to $Cx43^{WT}$, likely due to ubiquitination defects that hamper p62-mediated binding. Thus, some other mechanism, beyond the mere subcellular localization of $Cx43^{Y265A}$ must be in play to explain its diminished binding to LC3B. Curiously, when $Cx43^{W4A+L7A}$ was co-expressed with $Cx43^{Y265A}$, binding to LC3B was restored to normal levels. Presumably, when the two mutants oligomerize into the same hemichannel, they are able to compensate for each other's mutation. This suggests a model in which direct LC3B binding to Cx43 depends not only on the LIR motif in the amino terminal, but also on the carboxyl terminal of another Cx present in the hemichannel. Furthermore, it is plausible that more than a strict dependence on Y265, mutation of this residue simply induces a conformational change in the carboxyl terminal that physically prevents access of LC3B to the LIR motif. It has been reported that Y265 phosphorylation by Src kinase impacts on different aspects of the Cx43 life cycle [23], raising the possibility that Y265 phosphorylation may also modulate binding to LC3B. Another intriguing possibility is that the interaction of LC3B with Cx43 requires two binding sites, the LIR motif on the amino terminal, which would bind to the LIR docking site on LC3B, and an additional region in the carboxyl terminal, either including or modulated by Y265. Notably, LC3B does not bind to other proteins exclusively through its LIR docking site [12]. Indeed, it has recently been reported that LC3B can interact with ubiquitin interacting motifs (UIM) through an UIM-docking site (UDS) [24], and also with cardiolipin [25] and Lamin B1 [26].

Triton X-100 fractionation experiments showed that the interaction occurred mainly in the Triton X-100 resistant fraction, which contains Cx43 present in GJ plaques and also internalized GJs. In addition, immunofluorescence imaging showed that the two proteins colocalized mainly in intracellular vesicles, thus supporting a model in which Cx43/LC3B interaction occurs mainly in internalized compartments. Treatment with PMA, which induces Cx43 phosphorylation, ubiquitination and internalization, lead to an increase in LC3B binding to $Cx43^{WT}$, while having no effect on binding with $Cx43^{W4A+L7A}$, suggesting that inducing the phosphorylation and ubiquitination of Cx43 cannot overcome the lack of a functional LIR motif. However, and in accordance with previous reports [11], Cx43 ubiquitination still plays an important role in LC3B binding, as shown by the decrease in interaction when p62 is depleted. Our ATG7 silencing experiments also suggest that Cx43 binds preferentially to the non-conjugated form of LC3B, as while ATG7 depletion decreased the levels of LC3B-II, binding of LC3B to Cx43 was not impaired. This data raises an important question as to the function of this direct interaction of Cx43 with LC3B, within the context of the existence of a parallel pathway involving ubiquitination and p62 binding. An important fact to consider is that an autophagy substrate containing multiple points of LC3 interaction is not without precedent. For example, during mitophagy, ubiquitinated proteins on the surface of the mitochondria recruit autophagy receptors to assist in targeting the mitochondria to autophagosomes [12]. However, in addition to this mechanism, LIR motif containing proteins on the surface of the mitochondria, such as Nix and FUNDC1, can also bind directly to LC3 to assist in mitophagy [27]. In addition, the phospholipid cardiolipin, which is transported to the outer mitochondrial membrane following damage, can also bind to LC3 in a non-LIR motif dependent manner [25]. Although plasma membrane proteins committed for degradation usually reach the lysosome through the endolysosomal pathway, GJs can be delivered to lysosomes through autophagy. This particularity of Cx proteins may stem from their unique internalization mechanism. Plasma membrane proteins are normally internalized through single membrane structures, whereas GJ plaques are internalized into one of the adjacent cells through a double membrane structure commonly referred to as an annular gap junction (AGJ). Perhaps the structure of AGJs is unsuited for easy processing through the endolysosomal pathway, requiring their degradation through autophagy. In such a case, as with mitophagy, the presence of multiple binding spots for LC3 may allow for more efficient engulfment of the AGJ by the phagophore. Another intriguing possibility is that this interaction is used by Cx43 to promote the recruitment of non-conjugated LC3B to AGJs to assist with its degradation by macroautophagy. Notably, Cx43 is also known to interact with several autophagy-related proteins, including ATG16, Vps35, Beclin-1 and Vps15 [28], raising the possibility that Cx43 may function as a scaffold for autophagy machinery to further facilitate phagophore formation around AGJs. Although LC3/GABARAP family members offer some redundancy in function, specific functions for particular family members have been described, with the LC3 subfamily participating in the elongation of the phagophore membrane, while the GABARAP subfamily is involved in later stages of autophagosome formation [12]. The ability of Cxs to interact with both subfamilies envisions a model in which the sequential binding of Cx43 to different LC3/GABARAP subfamily members is important to ensure AGJ engulfment and processing during autophagosome maturation.

Altogether, the results presented in this work lead us to propose a model in which Cxs bind directly to LC3/GABARAP proteins to facilitate their degradation by autophagy. Given the importance of GJIC in multiple tissues and organs, and its involvement in specific pathologies such as heart failure, understanding the mechanisms that regulate connexin degradation may shed new light upon the role of the connexin family in these diseases.

Supplementary Materials: The following are available online at http://www.mdpi.com/2073-4409/9/4/902/s1, Figure S1: Comparison of the amino acid sequences of the amino terminal domains of human and rat connexins, Figure S2: The LIR motif in the amino terminal of Cx43 interacts with GABARAP in vitro, Figure S3: LC3B lipidation is not required for the interaction with Cx43, Figure S4: Mutation of the LIR motif of Cx37, 46 and 50 impairs their interaction with GABARAP, Table S1: Description of connexin mutants and their effect on binding to LC3/GABARAP proteins, Graphical Abstract.

Author Contributions: Conceptualization, S.C., T.M.R.-R. and H.G.; validation, S.C., T.M.R.-R. and R.S.F.; formal analysis, S.C. and T.M.R.-R.; investigation, S.C., T.M.R.-R., R.S.F. and C.A.; resources, J.R. and S.M.; writing—original draft preparation, S.C.; writing—review and editing, S.C., T.M.R.-R., S.M. and H.G.; visualization, S.C. and T.M.R.-R.; supervision, S.M. and H.G.; project administration, H.G.; funding acquisition, H.G. All authors have read and agreed to the published version of the manuscript.

Funding: This research was supported by the European Regional Development Fund (ERDF) through the Operational Program for Competitiveness Factors (COMPETE) [under the projects PAC "NETDIAMOND" POCI-01-0145-FEDER-016385; HealthyAging2020 CENTRO-01-0145-FEDER-000012-N2323; POCI-01-0145-FEDER-007440, CENTRO-01-0145-FEDER-032179, CENTRO-01-0145-FEDER-032414, FCTUID/NEU/04539/2013 and UID/NEU/04539/2019]. This research is based upon work from COST Action (PROTEOSTASIS BM1307), supported by COST (European Cooperation in Science and Technology). S.C. was supported by Action BM1307 – 39607 from COST Action PROTEOSTASIS BM1307, T.M.R.-R. was supported by PD/BD/52294/2013 from Fundação para a Ciência e a Tecnologia (FCT). C.A. was supported by a DOC Fellowship of the Austrian Academy of Sciences.

Conflicts of Interest: S.M. is member of the scientific advisory board of Casma Therapeutics. The funders had no role in the design of the study; in the collection, analyses, or interpretation of data; in the writing of the manuscript, or in the decision to publish the results. The authors declare no conflicts of interest

References

1. Beyer, E.C.; Lipkind, G.M.; Kyle, J.W.; Berthoud, V.M. Structural organization of intercellular channels II. Amino terminal domain of the connexins: Sequence, functional roles, and structure. *Biochim. Biophys. Acta Biomembr.* **2012**, *1818*, 1823–1830. [CrossRef] [PubMed]
2. Hervé, J.C.; Bourmeyster, N.; Sarrouilhe, D.; Duffy, H.S. Gap junctional complexes: From partners to functions. *Prog. Biophys. Mol. Biol.* **2007**, *94*, 29–65. [CrossRef] [PubMed]
3. Martins-Marques, T.; Ribeiro-Rodrigues, T.; Batista-Almeida, D.; Aasen, T.; Kwak, B.R.; Girao, H. Biological Functions of Connexin43 Beyond Intercellular Communication. *Trends Cell Biol.* **2019**, *29*, 835–847. [CrossRef] [PubMed]
4. Ribeiro-Rodrigues, T.M.; Catarino, S.; Pinho, M.J.; Pereira, P.; Girao, H. Connexin 43 ubiquitination determines the fate of gap junctions: Restrict to survive. *Biochem. Soc. Trans.* **2015**, *43*, 471–475. [CrossRef] [PubMed]
5. Girão, H.; Catarino, S.; Pereira, P. Eps15 interacts with ubiquitinated Cx43 and mediates its internalization. *Exp. Cell Res.* **2009**, *315*, 3587–3597. [CrossRef] [PubMed]
6. Catarino, S.; Ramalho, J.S.; Marques, C.; Pereira, P.; Girão, H. Ubiquitin-mediated internalization of connexin43 is independent of the canonical endocytic tyrosine-sorting signal. *Biochem. J.* **2011**, *437*, 255–267. [CrossRef]

7. Ribeiro-Rodrigues, T.M.; Catarino, S.; Marques, C.; Ferreira, J.V.; Martins-Marques, T.; Pereira, P.; Girão, H. AMSH-mediated deubiquitination of Cx43 regulates internalization and degradation of gap junctions. *FASEB J.* **2014**, *28*, 4629–4641. [CrossRef]
8. Auth, T.; Schlüter, S.; Urschel, S.; Kussmann, P.; Sonntag, S.; Höher, T.; Kreuzberg, M.M.; Dobrowolski, R.; Willecke, K. The TSG101 protein binds to connexins and is involved in connexin degradation. *Exp. Cell Res.* **2009**, *315*, 1053–1062. [CrossRef]
9. Leithe, E.; Kjenseth, A.; Sirnes, S.; Stenmark, H.; Brech, A.; Rivedal, E. Ubiquitylation of the gap junction protein connexin-43 signals its trafficking from early endosomes to lysosomes in a process mediated by Hrs and Tsg101. *J. Cell Sci.* **2009**, *122*, 3883–3893. [CrossRef]
10. Fong, J.T.; Kells, R.M.; Gumpert, A.M.; Marzillier, J.Y.; Davidson, M.W.; Falk, M.M. Internalized gap junctions are degraded by autophagy. *Autophagy* **2012**, *8*, 794–811. [CrossRef]
11. Bejarano, E.; Girao, H.; Yuste, A.; Patel, B.; Marques, C.; Spray, D.C.; Pereira, P.; Cuervo, A.M. Autophagy modulates dynamics of connexins at the plasma membrane in a ubiquitin-dependent manner. *Mol. Biol. Cell* **2012**, *23*, 2156–2169. [CrossRef]
12. Johansen, T.; Lamark, T. Selective Autophagy: ATG8 Family Proteins, LIR Motifs and Cargo Receptors. *J. Mol. Biol.* 2019. [CrossRef] [PubMed]
13. Birgisdottir, Å.B.; Lamark, T.; Johansen, T. The LIR motif - crucial for selective autophagy. *J. Cell Sci.* **2013**, *126*, 3237–3247. [PubMed]
14. Rogov, V.V.; Stolz, A.; Ravichandran, A.C.; Rios-Szwed, D.O.; Suzuki, H.; Kniss, A.; Löhr, F.; Wakatsuki, S.; Dötsch, V.; Dikic, I.; et al. Structural and functional analysis of the GABARAP interaction motif (GIM). *EMBO Rep.* **2017**, *18*, 1382–1396. [CrossRef]
15. Lopes, V.S.; Ramalho, J.S.; Owen, D.M.; Karl, M.O.; Strauss, O.; Futter, C.E.; Seabra, M.C. The ternary Rab27a-Myrip-Myosin VIIa complex regulates melanosome motility in the retinal pigment epithelium. *Traffic* **2007**, *8*, 486–499. [CrossRef] [PubMed]
16. Wurzer, B.; Zaffagnini, G.; Fracchiolla, D.; Turco, E.; Abert, C.; Romanov, J.; Martens, S. Oligomerization of p62 allows for selection of ubiquitinated cargo and isolation membrane during selective autophagy. *Elife* **2015**, *4*, 1–28. [CrossRef] [PubMed]
17. Vanslyke, J.K.; Musil, L.S. Analysis of connexin intracellular transport and assembly. *Methods* **2000**, *20*, 156–164. [CrossRef]
18. Jacomin, A.C.; Samavedam, S.; Promponas, V.; Nezis, I.P. iLIR database: A web resource for LIR motif-containing proteins in eukaryotes. *Autophagy* **2016**, *12*, 1945–1953. [CrossRef]
19. Totland, M.Z.; Rasmussen, N.L.; Knudsen, L.M.; Leithe, E. Regulation of gap junction intercellular communication by connexin ubiquitination: Physiological and pathophysiological implications. *Cell. Mol. Life Sci.* **2019**, *77*, 573–591. [CrossRef]
20. Fong, J.T.; Kells, R.M.; Falk, M.M. Two tyrosine-based sorting signals in the Cx43 C-terminus cooperate to mediate gap junction endocytosis. *Mol. Biol. Cell* **2013**, *24*, 2834–2848. [CrossRef]
21. Leykauf, K.; Salek, M.; Bomke, J.; Frech, M.; Lehmann, W.D.; Dürst, M.; Alonso, A. Ubiquitin protein ligase Nedd4 binds to connexin43 by a phosphorylation-modulated process. *J. Cell Sci.* **2006**, *119*, 3634–3642. [CrossRef] [PubMed]
22. Girão, H.; Pereira, P. Phosphorylation of connexin 43 acts as a stimuli for proteasome-dependent degradation of the protein in lens epithelial cells. *Mol. Vis.* **2003**, *9*, 24–30. [PubMed]
23. Solan, J.L.; Lampe, P.D. Specific Cx43 phosphorylation events regulate gap junction turnover in vivo. *FEBS Lett.* **2014**, *588*, 1423–1429. [CrossRef] [PubMed]
24. Marshall, R.S.; Hua, Z.; Mali, S.; McLoughlin, F.; Vierstra, R.D. ATG8-Binding UIM Proteins Define a New Class of Autophagy Adaptors and Receptors. *Cell* **2019**, *177*, 766–781.e24. [CrossRef]
25. Chu, C.T.; Ji, J.; Dagda, R.K.; Jiang, J.F.; Tyurina, Y.Y.; Kapralov, A.A.; Tyurin, V.A.; Yanamala, N.; Shrivastava, I.H.; Mohammadyani, D.; et al. Cardiolipin externalization to the outer mitochondrial membrane acts as an elimination signal for mitophagy in neuronal cells. *Nat. Cell Biol.* **2013**, *15*, 1197–1205. [CrossRef]
26. Dou, Z.; Xu, C.; Donahue, G.; Shimi, T.; Pan, J.A.; Zhu, J.; Ivanov, A.; Capell, B.C.; Drake, A.M.; Shah, P.P.; et al. Autophagy mediates degradation of nuclear lamina. *Nature* **2015**, *527*, 105–109. [CrossRef]

27. Liu, L.; Feng, D.; Chen, G.; Chen, M.; Zheng, Q.; Song, P.; Ma, Q.; Zhu, C.; Wang, R.; Qi, W.; et al. Mitochondrial outer-membrane protein FUNDC1 mediates hypoxia-induced mitophagy in mammalian cells. *Nat. Cell Biol.* **2012**, *14*, 177–185. [CrossRef]
28. Bejarano, E.; Yuste, A.; Patel, B.; Randy, R.F.; Spray, D.C.; Cuervo, A.M. Connexins modulate autophagosome biogenesis. *Nat. Cell Biol.* **2014**, *16*, 401–414. [CrossRef]

MDPI
St. Alban-Anlage 66
4052 Basel
Switzerland
Tel. +41 61 683 77 34
Fax +41 61 302 89 18
www.mdpi.com

Cells Editorial Office
E-mail: cells@mdpi.com
www.mdpi.com/journal/cells

www.ingramcontent.com/pod-product-compliance
Lightning Source LLC
Chambersburg PA
CBHW061817210926
51850CB00006B/732
* 9 7 8 3 0 3 6 5 1 3 4 2 3 *